Tissue Culture Techniques and Medicinal Plants

Nature is the most potent source of cure and care. With the advent of the 'herbal renaissance,' the dependence on medicinal plants has increased in all spheres ranging from their use as raw material for medicinal, nutraceutical, cosmetic preparations, and food. The tissue culture technique has proven to be extremely advantageous not only for large-scale propagation of medicinal plants in a limited space and time, but also for conservation, storage for prolonged use, genetic improvement, and metabolite production.

This book describes in detail the different methods of plant tissue culture that have been employed for the mass production of medicinal plants. The chapters on methods of organogenesis, embryogenesis, haploid plant production, virus-free medicinal plant production, and so forth will be of interest to a wide range of readers. Dedicated chapters on the tissue culture of high-altitude medicinal plants, cannabis, and so on are some of the book's highlights. The book also discusses the implementation of advanced biotechnological interventions like the production of phytochemicals, cryopreservation, nanotechnology, and bioinformatics in medicinal plant studies.

With contributions from world-leading specialists in plant tissue culture, the knowledge presented in this book will aid in the translation of plant tissue culture techniques for the production and improvement of medicinal plants. The information presented has the potential to be utilized for commercial production and formulations.

Key features:

- Comprehensive guide for tissue culture systems as potential methods for the production of clones, new varieties, improved germplasms, haploids, and *in vitro* preservation of medicinal plants
- Elaborates on the standardization of *in vitro* systems for propagation and transplantation of important medicinal plants and enhanced secondary metabolite production
- Enriches the understanding of advanced technologies like cryopreservation, bioinformatics, and nanotechnology in medicinal plants tissue culture

Exploring Medicinal Plants

Series Editor: Azamal Husen, Wolaita Sodo University, Ethiopia

Medicinal plants render a rich source of bioactive compounds used in drug formulation and development; they play a key role in traditional or indigenous health systems. As the demand for herbal medicines increases worldwide, supply is declining as most of the harvest is derived from naturally growing vegetation. Considering global interests and covering several important aspects associated with medicinal plants, the Exploring Medicinal Plants series comprises volumes valuable to academia, practitioners, and researchers interested in medicinal plants. Topics provide information on a range of subjects including diversity, conservation, propagation, cultivation, physiology, molecular biology, growth response under extreme environments, handling, storage, bioactive compounds, secondary metabolites, extraction, therapeutics, mode of action, and healthcare practices.

Led by Azamal Husen, PhD, this series is directed to a broad range of researchers and professionals consisting of topical books exploring information related to medicinal plants. It includes edited volumes, references, and textbooks available for individual print and electronic purchases.

Medicinal Roots and Tubers for Pharmaceutical and Commercial Applications
Edited by Rakesh Kumar Bachheti and Archana Bachheti

Pharmacological Aspects of Essential Oils: Current and Future Trends
Edited by Sunita Singh, Pankaj Kumar Chaurasia, and Shashi Lata Bharati

Promising Antiviral Herbal and Medicinal Plants
Edited by Nadeem Akhtar, Azamal Husen, Vagish Dwibedi, and Santosh Kumar Rath

Antimalarial Medicinal Plants
Edited by Azamal Husen

Medicinal Spice and Condiment Crops
Edited by Azamal Husen

Therapeutic Medicinal Plants in Traditional Persian Medicine
Edited by Roja Rahimi and Roodabeh Bahramsoltani

Medicinal Plants for the Management of Neurodegenerative Diseases
Edited by Jamal Akhtar, Maryam Sarwat, and Fouzia Bashir

Nutraceuticals Inspiring the Contemporary Therapy for Lifestyle Diseases
Edited by Mala Trivedi, Sachidanand Singh, Parul Johri, and Pedro López-Sánchez

Plants as Medicine and Aromatics: Uses of Botanicals
Edited by Mohd Kafeel Ahmad Ansari, Mushtaq Ahmad, and Gary Owens

Exploring Traditional Wild Edible Plants
Edited by Vibhor Agarwal, Sachidanand Singh, and Rhaul Datta

Propagation to Pharmacopeia: Modern Approaches in Medicinal Plants
Edited by Abha Manohar K, Gopal Shukla, Saikat Gantait, A P Das, and Sumit Chakravarty

Exploring Medicinal Orchids
Edited by Azamal Husen and Manu Pant

Symbiotic Association of Microorganisms with Medicinal and Herbal Plants
Edited by Allah Ditta, Sajid Mehmood, and Azamal Husen

Tissue Culture Techniques and Medicinal Plants: Enhancing Propagation and Production
Edited by Azamal Husen and Manu Pant

For more information about this series, please visit:
www.routledge.com/Exploring-Medicinal-Plants/book-series/CRCEMP

Tissue Culture Techniques and Medicinal Plants

Enhancing Propagation and Production

Edited by
Azamal Husen and Manu Pant

CRC Press
Taylor & Francis Group
Boca Raton London New York

CRC Press is an imprint of the
Taylor & Francis Group, an **informa** business

Designed cover image: ©Azamal Husen

First edition published 2025
by CRC Press
2385 NW Executive Center Drive, Suite 320, Boca Raton FL 33431

and by CRC Press
4 Park Square, Milton Park, Abingdon, Oxon, OX14 4RN

CRC Press is an imprint of Taylor & Francis Group, LLC

ISBN: 978-1-032-55558-4 (hbk)
ISBN: 978-1-032-56651-1 (pbk)
ISBN: 978-1-003-43662-1 (ebk)

DOI: 10.1201/b23374

Typeset in Times
by KnowledgeWorks Global Ltd.

Dedicated to Reena Pant (mother of Dr. Manu Pant)

Contents

Preface

Medicinally active plant species have been cultivated and used as therapeutic agents by humans since ancient times. Medicinal plants have gained significance as raw materials for the pharmaceutical sector, cosmetics, and the nutraceutical industry, besides being utilized as herbal medicines. As a result, the international trade in medicinal plants is at an all-time high. Consequently, these ecological resources have become critical not only for social growth but also for economic development. It has become imperative to make concerted efforts in the conservation and judicious use of medicinal plants, as their loss impoverishes us all.

Plant tissue culture is an important part of *ex situ* conservation strategy, circumventing many problems associated with conventional conservation strategies of medicinal plants, and has the merit of rapid propagation and replenishment of plants. *In vitro* tools in plant tissue culture also form the basis for germplasm conservation, genetic manipulation, and improvement of the plants under study. Clonal propagation ensures a huge number of 'true-to-type' plants that can be raised through tissue culture technology in a relatively short time, hardened, and made available for (re)introduction in the wild. Manipulation of culture conditions, use of explants, choice of plant growth regulators, and hardening procedures play a significant role in determining regeneration efficiency and multiplication rates, along with better *ex vitro* survival in medicinal plant cultivation. With increasing dependency on medicinal plant supplies from tissue culture labs, the *in vitro* method of cryopreservation has also been successfully employed for prolonged storage of medicinal plant germplasm. Somatic embryo formation and full plantlet regeneration in medicinal plants have also been achieved. Encapsulation systems to develop artificial seeds are also being utilized for prolonged storage, with their small size providing additional benefits of easy storage and transport. These can be considered 'gene repositories' for posterity.

Tissue culture technology has also played an important role in providing superior quality material for medicinal plant cultivation. Viruses pose a serious threat to plant propagation and quality control of medicinal plant products. Different means of virus control like chemotherapy, thermotherapy, and recombinant DNA technology have not been effective in controlling virus infection. In contrast, plant tissue culture techniques like meristem tip culture have been used to successfully develop 'virus-free' plants from infected germplasms. Techniques of anther ovary culture have been shown to develop haploid plants that can be further used for the production of agriculturally important homozygous diploids in medicinal plants. Further, genetic variability generated through somaclonal variations has also been used for the development of new varieties. Besides, the role of *in vitro* cultures has been extensively studied for enhanced metabolite production.

This book presents a compendium of aforementioned tissue culture studies on medicinal plants (herbs, shrubs, and trees), with individual chapters dedicated to different organogenic pathways like callus culture, somatic embryogenesis, haploid culture, etc., for medicinal plant cultivation and advancements like secondary metabolite production and implementation of *in silico* studies and nanotechnological interventions in medicinal plants research. In addition, chapters on tissue culture of high-altitude medicinal plants, cannabis, and so on would be of interest to the readers.

We deeply admire the efforts of all the authors from different parts of the world who have contributed significantly to producing a valuable resource on medicinal plant tissue culture by providing detailed insights into their respective subjects in the form of chapters. Our heartfelt gratitude goes to the publishers for their continuous support in bringing out this important book that addresses the captivating and concurrent subject of tissue culture of medicinally important plant species.

We are hopeful that our sincere efforts in bringing out a combination of conventional plant tissue culture methods and implementation of advanced techniques in medicinal plant tissue culture will be appreciated by a large audience for the clarity of concepts, applications, and recent advancements in the field of tissue culture of medicinal plants.

Wishing for a better, more sustainable future with medicinal plants!

Azamal Husen
Manu Pant

About the Editors

Azamal Husen is a professor at Sankalchand Patel University, Visnagar, India, and an adjunct professor at Graphic Era (Deemed to be University), Dehradun, Uttarakhand, India. He is also a visiting professor at University Putra Malaysia, Selangor, Malaysia, and is a foreign delegate at Wolaita Sodo University, Wolaita, Ethiopia. Previously, he served as a professor and head of the Department of Biology, University of Gondar, Ethiopia. He also worked as a visiting faculty of the Forest Research Institute and the Doon College of Agriculture and Forest at Dehra Dun, India. His research and teaching experience of 25 years encompasses biogenic nanomaterial fabrication and application; plant responses to nanomaterials; plant adaptation to harsh environments at the physiological, biochemical, and molecular levels; herbal medicine; and clonal propagation for improvement of tree species. He has conducted research sponsored by the World Bank, the National Agricultural Technology Project, the Indian Council of Agriculture Research, the Indian Council of Forest Research Education, and the Japan Bank for International Cooperation. Husen has published extensively (~300 works) and served on the editorial board and as a reviewer of reputed journals published by Elsevier, Frontiers Media, Taylor & Francis, Springer Nature, RSC, Oxford University Press, Sciendo, the Royal Society, CSIRO, PLOS, MDPI, John Wiley & Sons, and UPM Journals. He is on the advisory board of Cambridge Scholars Publishing, UK. He is a fellow of the Plantae group of the American Society of Plant Biologists and a member of the International Society of Root Research, Asian Council of Science Editors, and International Natural Product Sciences. He is Editor-in-Chief of the *American Journal of Plant Physiology* and a series editor of *Exploring Medicinal Plants* (Taylor & Francis Group, USA); *Plant Biology, Sustainability, and Climate Change* (Elsevier, USA); and *Smart Nanomaterials Technology* (Springer Nature, Singapore). He has achieved the distinguished honor of being recognized as one of the 'World's Top 2% Scientists' for the year 2022 and again for 2023 by Stanford University, USA. This recognition has also been prominently featured in the Elsevier Data Repository.

Manu Pant is an associate professor and head of the Department of Biotechnology at Graphic Era (Deemed to be University), Dehradun, Uttarakhand, India. She received her master's degree in applied microbiology and biotechnology from Banasthali Vidyapith in Rajasthan, India, and her doctorate in biotechnology from HNB Garhwal University (a Central University) in Srinagar, India. She has extensive experience in the field of plant tissue culture, having worked as a research fellow at the Forest Research Institute and the Indian Council of Forestry Research and Education on nationally funded projects. Her research work on orchids has been funded by the state governments, earning her recognition and awards at the national level. She has several peer-reviewed international papers to her credit, published patents, and numerous oral and poster presentations, keynote addresses, and invited talks at national and international scientific events. Pant has conducted many national and international conferences, seminars, and trainings related to the significant areas in plant sciences. She has supervised PhD, MSc, and BSc students in their research work on different aspects of orchids, medicinal plants, and indigenous crop species. For her work in the field of plant sciences, Pant has received appreciation from state- and national-level bodies.

Contributors

Rambod Abiri
Department of Forestry Science and Biodiversity
Faculty of Forestry and Environment
Universiti Putra Malaysia
Selangor, Malaysia

and

Department of Plant Agriculture
University of Guelph
Guelph, Canada

Urja Agarwal
Department of Biotechnology
Graphic Era (Deemed to Be University)
Dehradun, India

Hashitha N. Aluthgamage
Department of Crop Science
Faculty of Agriculture
University of Ruhuna
Kamburupitiya, Matara, Sri Lanka

H. I. G Kokila Anuruddi
Department of Crop Science
Faculty of Agriculture
University of Ruhuna
Kamburupitiya, Matara, Sri Lanka

Mohamed Sabri Bensaad
Department of Biology of Organism
Faculty of Natural and Life Sciences
University of Batna 2
Batna, Algeria

Alok Bharadwaj
Department of Biotechnology
GLA University
Mathura, India

Indra Dutt Bhatt
Center for Biodiversity Conservation and Management
G. B. Pant National Institute of Himalayan Environment
Almora, India

Subhash Chander
ICAR – National Bureau of Plant Genetic Resources
New Delhi, India

Suman Chandra
National Center for Natural Products Research
School of Pharmacy
University of Mississippi
Oxford, Mississippi, USA

Nital Nilesh Chankeshwara
Department of Biotechnology
Graphic Era (Deemed to Be University)
Dehradun, India

Shweta Chaudhary
School of Biotechnology
Gautam Buddha University
Greater Noida, India

Akansha Chauhan
Department of Biotechnology
Graphic Era (Deemed to Be University)
Dehradun, India

D. A. Deepak
Division of Plant Genetic Resources
ICAR – Indian Agricultural Research Institute
New Delhi, India

Mahmoud A. ElSohly
National Center for Natural Products Research
University of Mississippi
Oxford, Mississippi, USA

and

Department of Pharmaceutics and Drug Delivery
School of Pharmacy
University of Mississippi
Oxford, Mississippi, USA

Alison M. R. Ferrie
National Research Council of Canada
Saskatoon, Canada

D. L. C Kumari Fonseka
Department of Crop Science
Faculty of Agriculture
University of Ruhuna
Kamburupitiya, Matara, Sri Lanka

Gunjan Garg
School of Biotechnology
Gautam Buddha University
Greater Noida, India

Aarushi Gautam
Department of Biotechnology
Graphic Era (Deemed to Be University)
Dehradun, India

Ravi Gowthami
Division of Germplasm Conservation
ICAR – National Bureau of Plant Genetic Resources
New Delhi, India

Sandhya Gupta
ICAR – National Bureau of Plant Genetic Resources
New Delhi, India

Nirmal Joshee
Agricultural Research Station
Fort Valley State University
Fort Valley, Georgia, USA

Mohamed Amine Kahoul
Laboratory of Food Sciences (LSA)
Institute of Agriculture and Veterinary Sciences
University Batna 1
Batna, Algeria

Sanjay Kataria
Department of Botany and Microbiology
BSA College
Mathura, India

Ikhlas A. Khan
National Center for Natural Products Research

and

Department of BioMolecular Sciences
School of Pharmacy
University of Mississippi
Oxford, Mississippi, USA

Sonam Khan
Department of Biotechnology
Graphic Era (Deemed to Be University)
Dehradun, India

Priti Krishna
Department of Microbiology
Graphic Era (Deemed to Be University)
Dehradun, India

Hemant Lata
National Center for Natural Products Research
School of Pharmacy
University of Mississippi
Oxford, Mississippi, USA

Abhilasha Mishra
Department of Chemistry
Graphic Era (Deemed to Be University)
Dehradun, India

Debasis Mitra
Department of Microbiology
Graphic Era (Deemed to Be University)
Dehradun, India

Pradeep Kumar Das Mohapatra
Department of Microbiology
Raiganj University
Raiganj, India

Shivansh Negi
Department of Biotechnology
Graphic Era (Deemed to Be University)
Dehradun, India

Vedanshi Pal
Centre for Crop and Disease Management
Curtin University
Bentley, Australia

and

Department of Biotechnology
Graphic Era (Deemed to Be University)
Dehradun, India

Sushma Pandey
Central Department of Botany
Tribhuvan University
Kathmandu, Nepal

and

Butwal Multiple Campus
Tribhuvan University
Kathmandu, Nepal

Periyasamy Panneerselvam
Crop Production Division
Indian Council of Agricultural Research – National Rice Research Institute
Cuttack, India

Bijaya Pant
Central Department of Botany
Tribhuvan University
Kathmandu, Nepal

and

Butwal Multiple Campus
Tribhuvan University
Kathmandu, Nepal

Gaurav Pant
Department of Microbiology
Graphic Era (Deemed to Be University)
Dehradun, India

Kumud Pant
Department of Biotechnology
Graphic Era (Deemed to Be University)
Dehradun, India

G. M. Puneeth
Division of Plant Genetic Resources
ICAR – Indian Agricultural Research Institute
New Delhi, India

Abhijeet Rathod
Department of Biotechnology
Graphic Era (Deemed to Be University)
Dehradun, India

Pramod Rawat
Department of Biotechnology
Graphic Era (Deemed to Be University)
Dehradun, India

Sonalika Roy
Infrastructure and Planning
Indian Institute of Technology
Kanpur, India

Muthusamy Shankar
Division of Plant Genetic Resources
ICAR – Indian Agricultural Research Institute
New Delhi, India

Aditi Sharma
Department of Biotechnology
Graphic Era (Deemed to Be University)
Dehradun, India

Mahipal S. Shekhawat
Plant Biotechnology Unit
Kanchi Mamunivar Government Institute for Postgraduate Studies and Research
Puducherry, India

Arundhati Singh
Centre for Crop and Disease Management
Curtin University
Bentley, Australia

and

Department of Biotechnology
Graphic Era (Deemed to Be University)
Dehradun, India

Kartar Singh
Indian Council of Agricultural Research – National Bureau of Plant Genetic Resources
New Delhi, India

Laxman Singh
Center for Biodiversity Conservation and Management
G. B. Pant National Institute of Himalayan Environment
Almora, India

Yashaswi Singh
Department of Biotechnology
Graphic Era (Deemed to Be University)
Dehradun, India

Vartika Srivastava
ICAR – National Bureau of Plant Genetic Resources
New Delhi, India

Chandra Bahadur Thapa
Central Department of Botany
Tribhuvan University
Kathmandu, Nepal

and

Butwal Multiple Campus
Tribhuvan University
Kathmandu, Nepal

Deep Chandra Tiwari
Center for Biodiversity Conservation and Management
G. B. Pant National Institute of Himalayan Environment
Almora, India

1 Tissue Culture Techniques for Medicinal Plant Propagation

Manu Pant, Nital Nilesh Chankeshwara, Abhijeet Rathod, and Azamal Husen

1.1 INTRODUCTION

The discovery of a cell and the establishment of cell theory paved the way for the future science and art of plant tissue culture. After the declaration that a cell can function independently, it was perceived that it should be able to regenerate into a whole plant in the right conditions (Ribatti, 2018). Based on this idea, German scientist Gottlieb Haberlandt cultivated leaf palisade cells on a nutrient medium comprising Knop's salt solution supplemented with sucrose. Interestingly, the cells grew larger, collected starch, and continued to function for up to a month, but were unable to divide (Argaye, 2021). The work established Haberlandt as the 'Father of Plant Tissue Culture' for establishing the groundwork for tissue culture technique (Hussain et al., 2012).

Consequently, plant tissue culture technology progressed on the phenomenon of the 'totipotency' of plants and encompasses all the methods used to grow and develop selected plants from small plant tissue, excised organs, cells, and protoplasts under a regulated environment. Selection of the starting material, called the explant, in most cases, defines the organogenic pathway that would be adopted during plantlet development *in vitro*. However, the success of plant propagation depends on the establishment of an efficient regeneration protocol that ensures 'lab-to-land' transfer of the *in vitro* regenerated plants (Figure 1.1). The protocol refers to the standardization of nutrient media (to fulfill the mineral requirements for plant growth), requisition of natural and synthetic phytohormones, intensity and duration of light and temperature (this includes artificial light and temperature conditions under laboratory conditions that replicate the natural environment, reducing the fluctuations in growth), and the hardening and acclimatization procedure for shifting *in vitro* raised plants under field conditions.

Over the years, the technique has been applied for large-scale production of various plant species. In this context, medicinal plant tissue culture has its significance. Being an effective technique of mass multiplication from limited tissue amounts, tissue culture for medicinal plants has proven to be extremely useful in the conservation of rare, endangered, and threatened (RET) plant species. There are a large number of instances where 'true-to-type' medicinal plants have been regenerated through tissue culture (Palni et al., 2018). Besides, the production of disease-free and healthy plants throughout the year ensures a sustained supply of superior quality planting material for pharmaceutical and therapeutic purposes. The tissue culture system is also shown to be instrumental in playing with the secondary metabolite production in medicinal plants, optimizing the production (Cardoso et al., 2019). The tissue culture material also serves as the starting material for any further genetic transformation for medicinal plant improvement strategies. Consequently, several studies have focused on the mass production of important medicinal plants through different explants and organogenic differentiation pathways (Table 1.1).

This chapter provides an overview of the different strategies of plant tissue culture that are used for medicinal plant propagation and further improvement.

DOI: 10.1201/b23374-1

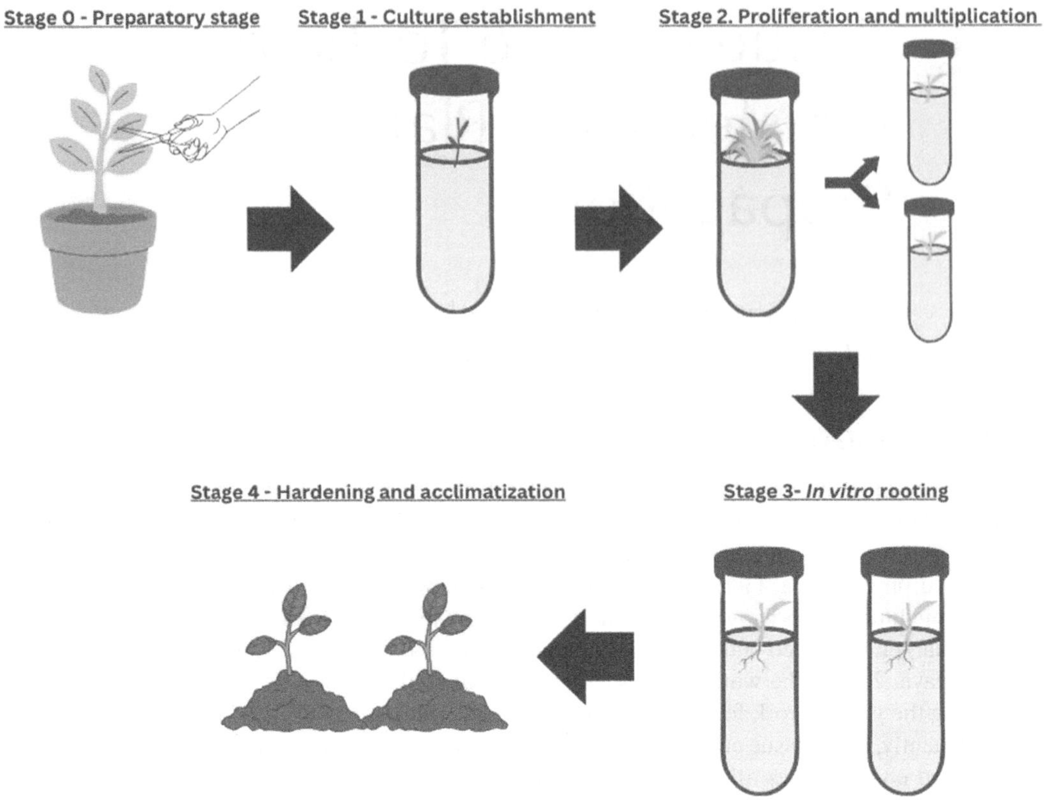

FIGURE 1.1 General stages of plant tissue culture.

1.2 AXILLARY BUD CULTURE

Axillary buds, also known as the lateral buds, are the buds present in the axil of the leaf on the nodal segments of the shoot. These are the precursors of lateral shoots and proliferate once the apical dominance is inhibited. Axillary buds are minuscule preformed buds that are difficult to excise from the mother plant with high chances of mortality on subjection to surface sterilization treatment. It is therefore preferred to use nodal segments carrying the axillary buds for inoculation purposes, thereby making the term 'nodal explant culture' synonymous with axillary bud culture (Figure 1.2a). Axillary bud culture is the most commonly utilized and reliable technique for micropropagation due to its simplicity and efficacy. The method concentrates on the growth and development of new shoots from existing meristems as opposed to methods that involve the dedifferentiation of differentiated cells. This approach to rapid clonal propagation with minimal risk of genetic instability makes it one of the most favored methods of plant propagation (Ngezahayo and Liu, 2014).

Over the years, different media and phytohormone combinations have been tried for effective propagation of medicinal plants. In most cases, Murashige and Skoog (MS) medium (Murashige and Skoog, 1962) has been reported to be efficient, while in cases of woody species, woody plant medium (WPM) has been used. In general, cytokinins alone or with lower concentrations of auxins are mostly used for axillary bud proliferation and shoot multiplication, while auxins alone are sufficient for *in vitro* rooting. In a study on axillary bud culture of *Swertia chirata*, an endangered medicinal plant used as a therapeutic agent against chronic liver diseases, bronchial asthma, chronic fever, anemia, and other stomach-related diseases, it was shown that MS medium supplemented with the

TABLE 1.1
Different Culture Techniques are Used for Various Medicinal Plants along with Their Medicinal Properties

S. No.	Plant Name	Culture Technique	Medicinal Properties	Reference(s)
1	*Swertia chirata*	Axillary bud culture	Liver disorders, diabetes	Pant et al., 2010
2	*Ceropegia candelabrum* L.		Antioxidant, anti-bacterial	Binish and Nayagi, 2019
3	*Prunus africana*		Anti-inflammatory and analgesic properties	Komakech et al., 2020
4	*Bacopa monnieri* L. Pennell		Insomnia, epilepsy	Sape et al., 2020
5	*Pluchea lanceolata*		Anti-inflammatory diseases	Singh et al., 2021
6	*Smilax zeylanica* L.		Antimicrobial, analgesic, antidepressant, antioxidant	Azad et al., 2022
7	*Saussurea costus*	Direct organogenesis	Anticancer, antiulcer, anti-inflammatory, hepatoprotective activities	Sharma et al., 2019
8	*Artemisia arborescens* L.		Anti-inflammatory	Riahi et al., 2022
9	*Celastrus paniculatus* Willd.	Indirect organogenesis	Improves memory and cognitive function	Moola and Kumari, 2020
10	*Thymus persicus*	Callus culture	Anti-rheumatic, anti-spasmodic, diuretic	Bakhtiar et al., 2016
11	*Ficus krishnae*		Aphrodisiac, antidiabetic, anticancer	Khan et al., 2019
12	*Zanthoxylum armatum*		Stomach pain, inflammation	Purohit et al., 2020
13	*Atropa acuminata*		Anti-inflammatory, anticholinergic, antispasmodic	Rajput and Agrawal, 2020
14	*Picrorhiza kurroa*		Liver and respiratory disorders, dyspepsia, chronic diarrhea	Partap et al., 2020
15	*Kelussia odoratissima* Mozaff,	Somatic embryogenesis	Hypertension, anti-inflammatory	Ebrahimi et al., 2018
16	*Asparagus racemosus* Willd.		Antispasmodic, aphrodisiac, galactogogue	Chaudhary and Dantu, 2019
17	*Ferula jaeschkeana*		Antiseptic, abortifacient, antihistaminic	Sharma and Khajuria, 2020
18	*Rauvolfia serpentina*	Embryonic cell suspension culture	Antipsychotic, hypertension	Uikey et al., 2016
19	*Mondia whitei*		Galactogogue, antidepressant	Baskaran et al., 2017
20	*Celastrus paniculatus*		Improves memory and cognitive function	Fonseka et al., 2019
21	*Eryngium alpium* L.		Diuretic agent, stimulant	Kikowska et al., 2020
22	*Glycyrrhiza glabra* L.	Hairy root culture	Antimicrobial, antitussive, antimutagenic	Srivastava et al., 2019
23	*Swertia chirata*		Liver disorders, diabetes	Chouhan et al., 2020
24	*Decalepis salicifolia*	Adventitious root culture	Skin diseases, intestinal disorders	Rodrigues et al., 2021
25	*Plumbago zeylanica* L.		Anti-atherogenic, hepatoprotective, cardiotonic	Katoch et al., 2022

(Continued)

TABLE 1.1 (*Continued*)
Different Culture Techniques are Used for Various Medicinal Plants along with Their Medicinal Properties

S. No.	Plant Name	Culture Technique	Medicinal Properties	Reference(s)
26	*Dendrobium lasianthera*	Seed culture	Anticancer	Utami et al., 2017
27	*Aerides multiflora* Roxb. and *Rhynchostylis retusa* L. Blume		Heal cuts and wounds, anti-rheumatic	Lal et al., 2020
28	*Araucaria araucana*		Ulcers, wounds	Gallia et al., 2021
29	*Sinopodophyllum hexandrum*		Anti-cancer, fever, jaundice	Sharma et al., 2022
30	*Tylophora indica*	Protoplast culture	Asthma, bronchitis, whooping cough	Thomas, 2009
31	*Phellodendron amurense* Rupr		Anti-stomachic, anti-inflammatory, and anti-pyretic agent	Ikuta et al., 1998; Azad et al., 2006

cytokinin benylaminopurine (BAP), auxin indole-3 acetic acid (IAA), and additive adenine sulfate (Ads) was most suitable for shoot multiplication, while half-strength MS medium containing auxin indole-3 butyric acid (IBA) alone was effective for root induction from regenerated shoots (Pant et al., 2010). In contrast, BAP alone proved to be effective for *in vitro* shoot multiplication in *Bacopa monnierri*, a medicinal plant well recognized as a nerve tonic (Figure 1.2b). *Gymnema sylvestre*, known for its medicinal properties like anticancer, anti-inflammatory, antimicrobial, antioxidant, gastro- and hepatoprotective, etc. (Khan et al., 2019), is widely used in herbal medicinal preparations. Murab and Chandurkar (2018) reported that the plant can be successfully micropropagated from nodal explants on media containing a combination of cytokinin and auxin (BAP:kinetin:IAA).

Centella asiatica, another renowned medicinal plant that addresses neurological, endocrine, skin, cardiovascular, and gastrointestinal diseases when propagated through axillary buds, was shown to respond well to the treatments of MS medium enriched with BAP and auxin naphthalene acetic acid (NAA), and MS medium containing BAP and cytokinin kinetin (Kn) (Siddiqui et al., 2019).

In *Prunus africana*, an endangered medicinal shrub endemic to sub-Saharan Africa (Jimu, 2011; Komakech et al., 2017) 100% axillary bud induction in cultured nodal segments could be achieved on WPM medium containing BAP alone (1 mg/L), while rooting was best on a medium containing IAA (Komakech et al., 2020). *Pluchea lanceolata*, a medicinal plant showing activity against cough, neurological diseases, edema, sciatica, and psoriasis, showed the best results of axillary bud proliferation on MS medium fortified with BAP and NAA. However, optimal shoot multiplication

FIGURE 1.2 (a) Nodal explant culture of lemon and (b) *in vitro* shoot multiplication of *Bacopa monnieri*.

was achieved with MS medium containing BAP and Kn (Singh et al., 2021). Successful micropropagation from nodal segments of *Dalbergia congestiflora*, a woody medicinal plant, was achieved on MS medium with very low concentrations of BAP (0.05 mg/L), while shoot multiplication was best on MS medium containing BAP and NAA, and *in vitro* rooting could be achieved with low-strength MS medium containing IBA (Hernández-García et al., 2021). More recently, Riahi et al. (2022) showed that nodal segments of *Artemisia arborescens* proliferated when the shoots were inoculated onto MS medium with BAP and IAA, while half-strength MS medium containing IBA proved to be most efficient for long and healthy root development.

1.3 SHOOT TIP CULTURE

The shoot tip, also referred to as the shoot apical meristem tip, is the shoot primordium meristem tissue protected by a few leaf primordia. The meristem helps in vertical growth of the plant and also exhibits dominance over lateral bud growth. The use of meristem tip for micropropagation was first shown by Morel and Martin (1952) in dahlia, followed by cymbidium orchid propagation through meristem tip culture (Morel, 1966). The method is of special interest when micropropagation is done for virus elimination (as the shoot tip is considered free of any viral infection, despite the whole plant being infested) and also for rapid clonal propagation (Rout et al., 2006). The system outlines how meristems can develop into organs like shoots, leaves, roots, etc.

Babaei et al. (2014) reported that shoot tip culture was an effective method to mass-propagate *Curculigo latifolia*, a medicinal plant with anticancer and antidiabetic properties and valued as a natural sweetener. They standardized MS medium with thidiazuron (TDZ) and IBA to be most suitable for shoot proliferation, while 0.25 mg/L of IBA alone was sufficient to induce long roots. Grzegorczyk-Karolak et al. (2017) showed that shoot tip culture was best suited not only for micropropagation but also for secondary metabolite production in the medicinal plant *Scutellaria alpina*. In their study, a liquid nanocellulose support system proved to be very beneficial for the highest shoot proliferation within 5 weeks. Sharma et al. (2019) showed that shoot tip explants of *Saussurea costus*, used in the treatment of stomach and respiratory ailments, rheumatoid arthritis, etc., can be successfully used for shoot regeneration on MS medium containing TDZ and NAA. The regenerated shoots of *S. costus* multiply well on MS medium with BAP, Kn, and gibberellic acid, while the micro shoots rooted well on IBA-supplemented medium. In *Andrographis producta*, a medicinal plant known to harbor several therapeutic compounds, shoot tip explants resulted in shoot proliferation on MS medium containing 2-isopentenyl adenine (2-ip), while quarter-strength medium with IBA was most optimal for rooting (Kadapatti and Murthy, 2022).

1.4 SEED CULTURE

Employing seeds as explants is often considered a more convenient and efficient method of plant propagation. The ease of collection, storage, and prolonged use of seeds as explants make them popular as the starting material for the micropropagation of medicinal plants. Some studies on seed culture–based medicinal plant propagation are mentioned next.

Utami et al. (2017) employed mature seeds of *Dendrobium lasianthera* as an explant for large-scale propagation. They used Vacin and Went medium (VW medium) supplemented with 2 g/L peptone for seed germination and shoot induction from protocorms. VW medium with 15% coconut water was found to be the best for root and leaf development followed by acclimatization of tissue culture–raised plantlets. In a study on *Araucaria araucana* micropropagation, Gallia et al. (2021) showed that the seeds of this exceptional source of antioxidants could be used for callus induction in MS medium under dark conditions, followed by plantlet regeneration in medium with 1.5 mg/L BAP and 1 mg/L NAA. The mass-regenerated plantlets were found to be non-cytotoxic to mammalian cells and were successfully acclimatized for field growth. *Podophyllum hexandrum* is an important medicinal plant well known for the treatment of pneumonia and cervical and testicular cancers (Liu et al., 2022). In a recent study, Sharma et al. (2022) established the micropropagation protocol of *S. hexandrum* using seeds. Delayed germination was observed after the 77th day of culture; however, the highest shoot multiplication was obtained after the 4th subculture on MS medium containing BAP (2.0 mg/L), NAA (0.2 mg/L), and calcium chloride (0.7 mg/L). The *in vitro* regenerated leaves were further used for callus induction for podophyllotoxin production on Gambrog's (B5 medium) fortified with BAP, NAA, and 2,4-D.

1.5 PROTOPLAST CULTURE

The use of protoplasts to initiate *in vitro* cultures has been realized as an effective method of plant regeneration in cases where other methods have been difficult or non-responsive due to recalcitrance of the cell wall. More importantly, protoplast culture allows for protoplast fusion to create hybrids or cybrids with beneficial traits at an exceptionally faster pace as compared to conventional methods (Mukami et al., 2022). Protoplasts also prove to be a preferable explant for genetic transfer to create novel traits, as they overcome the challenge posed by the presence of rigid cell walls in plant cells. Protoplast culture has also been tried in some medicinal plant species. In their study on *Phellodendron amurense,* a medicinal tree, Azad et al. (2006) used the leaves of micropropagated shoots as well as plantlets as plant materials for cellulase-based protoplast isolation. Gellan gum solidified or liquid MS medium fortified with mannitol, BAP, NAA, IBA, and 2,4-D could be used for sustained protoplast division for further regeneration. In a study on *Tylophora indica* (Thomas, 2009), mesophyll cells were used to isolate protoplasts and cultured on MS medium containing 4 μM 2,4-D, 0.4 M mannitol, and 3% (w/v) sucrose with pH 5.8 for organogenic calli development. The calli differentiated into shoots on medium supplemented with TDZ and NAA, with an average of 12 shoots per callus. The regenerated shoots could be rooted *in vitro* on half-strength medium containing IBA to facilitate complete plantlet development from cultured protoplasts.

1.6 CALLUS CULTURE

A callus is the undifferentiated mass of cells originating from a cultured tissue segment and has the potential to redifferentiate into an organ (Figure 1.3a). This scheme of organ development under *in vitro* conditions is referred to as indirect organogenesis. Optimization of media components

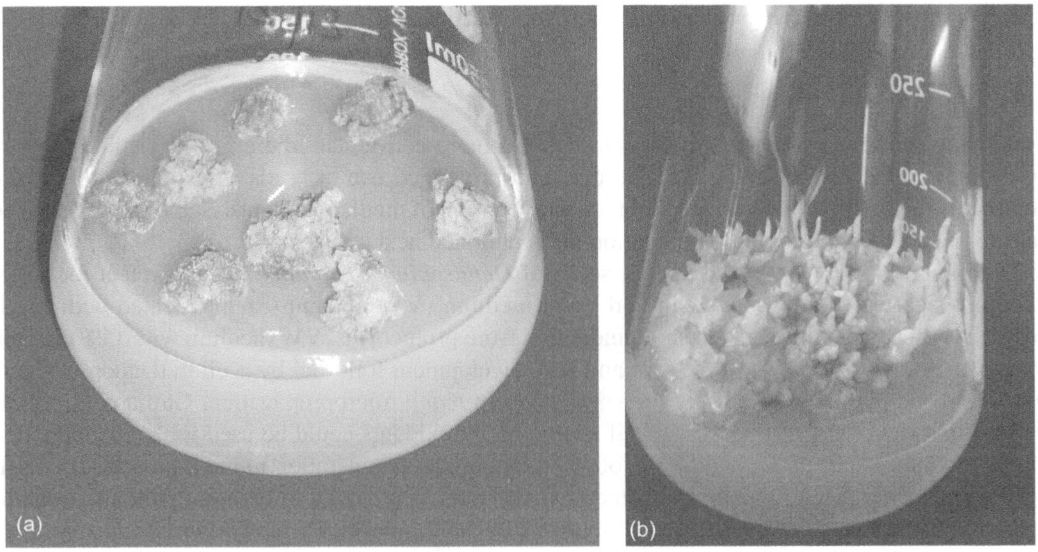

FIGURE 1.3 (a) *In vitro* callus development from leaf explants and (b) indirect organogenesis from callus.

remains a crucial factor in evaluating the effect of different media on the growth rate of the callus. Commonly used parameters for determining callus growth include growth index and fresh and dry weight (Bhatia et al., 2015). Callus culture is often linked to possible somaclonal variations that might occur in the progeny, and therefore is not a method of choice for germplasm conservation. However, caulogenesis is desirable for *in vitro* production of plants' secondary metabolites, giving room for media manipulations that may enhance metabolite production. Since callus culture is also reported to be suitable for enhanced and prolonged shoot regeneration under *in vitro* conditions, many studies on medicinal plant tissue culture focused on callus-based shoot regeneration (Figure 1.3b). While the leaf remains the most preferred explant for callus induction owing to its ease of collection and uniformity of cells, other explants can also be used.

In vitro, callus induction was used in a study by Bakhtiar et al. (2016) to design an effective regeneration protocol for the threatened medicinal plant *Thymus persicus,* well known for its anticancer properties. They used MS medium enriched with Kn (0.5 mg/L) and NAA (2.0 mg/L) for 100% callus induction from internodal explants, while a medium containing BAP (2.0 mg/L) and NAA (1.0 mg/L) was standardized to be the best for shoot induction and multiplication from callus, and half-strength MS medium with IBA was the best for rooting. Thakur et al. (2020) showed that MS medium with BAP (7.5 μM) and NAA (1.5 μM) was the best to induce callus from hypocotyl segments of *Hippophae salicifolia*, while BAP (7.5 μM) and NAA (3 μM) was most efficient for shoot induction and multiplication. For *in vitro* rooting of regenerated shoots, half-strength MS medium with IAA (3 μM) was most efficient.

Koul and Mallubhotla (2020) stated that *Bacopa monnieri* leaves could be used for primary callus induction on Gamborg's B5 medium fortified with 2,4-dichlorophenoxyacetic acid (2,4-D) (1.0 mg/L). The callus could be used to generate a cell suspension culture, which was utilized for efficient bacoside production using the elicitor salicylic acid in the suspension medium. The superiority of B5 medium in callus induction from leaf explants was also shown in *Dysosma pleiantha*, acknowledged for its anticancer properties (Karuppaiya and Tsay, 2020). They used B5 medium containing 1 mg/L 2,4-D and 0.2 mg/L Kn for callus induction under dark conditions. Further, the inclusion of 2 g/L peptone in the medium significantly elevated the synthesis of the therapeutic compounds podophyllotoxin, kaempferol, and quercetin into the medium. Leaf segments have also been used to induce callus in *Thalictrum minus* L., recognized for its medicinal efficacy in treating bacterial and fungal infections, tuberculosis, and lung inflammation (Badamjav et al., 2020). In a study conducted by Sargazi Moghadam et al. (2022), a callus could be induced from leaf bits of *T. minus* in liquid MS medium and was further harnessed for berberine production from cultured calli.

In *Oroxylum indicum*, a medicinal tree renowned for its efficacy in treating various ailments such as asthma, bronchitis, diarrhea, dysentery, and rheumatoid arthritis (Jagetia, 2021), cotyledonary leaf explants from raised seedlings were reported to be efficient for callus induction on MS medium supplemented with 0.5–2.0 mg/L BAP alone or in combination with auxins, specifically 0.5 mg/L 2,4-D and 0.5 mg/L IAA. (Smitha et al., 2021). They reported that best bud formation (61.33 buds per explant) could be obtained on MS medium supplemented with the combination of 2 mg/L BAP and 0.5 mg/L 2,4-D. Gibberellic acid (GA_3) was found to enhance shoot elongation suitable for initiating *in vitro* rooting on half-strength MS containing 1.0 mg/L NAA.

In a recent study, leaf explant–mediated callus culture was established in *Ceropegia bulbosa*, a threatened medicinal herb esteemed for its medicinal efficacy against kidney stones, dysentery, and deafness (Kashyap et al., 2023). Optimal callus induction could be obtained on a medium containing BAP and NAA. The callus culture was then used to initiate cell suspension cultures, which were used for secondary metabolite production using zinc oxide nanoparticles as elicitors.

1.7 EMBRYOGENIC CELL SUSPENSION CULTURE

The embryonic cell suspension is a plant tissue culture method that involves the cultivation of embryonic cells in a liquid medium. This technique is beneficial for both plant regeneration and genetic modification. In the process, embryogenic cells are propagated in a liquid medium under precise conditions, such as temperature, light, and nutrient supply. In most cases, these embryogenic cells are produced from zygotic immature embryos that have been cultivated on a semi-solid media to facilitate callus induction. Subsequently, the callus is placed in a liquid media to create a suspension culture of embryonic cells. To preserve and encourage cell proliferation, the culture can be subcultured regularly. This technique has several benefits, such as the capacity to generate a high number of cells in a constrained amount of space, the simplicity of genetic alteration and metabolite production, and the possibility of producing recombinant advantageous proteins (Woo et al., 2021). The method is of special interest in medicinal plants, as it offers the advantage of enhanced secondary metabolite production.

The endangered medicinal shrub *Rauvolfia serpentina*, which is used to treat a variety of ailments, including insomnia, hypertension, and hypochondriasis, could be propagated through embryonic cell suspension culture (Uikey et al., 2016). A study showed that MS medium with 2,4-D (2.0 mg/L) and BAP (0.5 mg/L) was most effective in establishing suspension cultures. The use of a lower concentration of 2,4-D with BAP was shown to accelerate somatic embryo development, while medium nourished with BAP, TDZ, and NAA helped in frequent plantlet regeneration.

Using somatic embryogenesis in cell suspension culture, Baskaran et al. (2017) created an effective *in vitro* mass propagation protocol for the medicinal herb *Mondia whitei*, an endangered medicinal plant. In their study, leaf explants were utilized to induce embryogenic callus on MS medium containing picloram and 2,4-D, and when shifted to a liquid medium containing NAA and meta-topolin riboside (mTR), a high incidence of somatic embryo development could be observed. The maximum embryo germination rate between 97% and 98% could be obtained in a liquid half-strength MS medium containing NAA.

Cell suspension cultures have also been increasingly utilized for metabolite production. In a study on *Angelica sinensis,* a plant that offers health advantages like angiogenesis, neurogenesis, antioxidative stress, anti-inflammation, and antiplatelet aggregation (Han et al., 2021), cell suspension cultures established on half-strength MS medium with 30 g/L sucrose was most efficient for biomass growth and ferulic acid (FA) production (Zhang et al., 2019).

1.8 SOMATIC EMBRYOGENESIS

Somatic embryogenesis is a process by which plants differentiate somatic cells into bipolar structures called embryos, which parallel the developmental path of zygotic embryos. The cultured tissues are categorized as pre-embryonic determined cells (PEDCs) and induced embryonic determined cells (IEDCs), depending on the spontaneity of the cells to *in vitro* treatments for somatic embryo induction. While PEDCs exhibit a more rapid response to the treatments, IEDCs call for a more concerted effort in terms of the use of plant growth regulators (PGRs), physicochemical conditions, stress treatment, etc., for the inoculated tissue to differentiate as somatic embryos (Nic-Can et al., 2016). Somatic embryogenesis has important applications in plant biotechnology, including the production of genetically identical plants, the propagation of rare or endangered species, providing embryos for encapsulation and subsequent use as artificial seeds, ease of gene transfer in somatic embryos, and rapid regeneration of genetically engineered plantlets (Loyola-Vargas, 2016).

Somatic embryogenesis has been tried in some medicinal plants. Ebrahimi et al. (2018) reported that the mass production of *Kelussia odoratissima*, a medicinal herb, could be achieved through somatic embryogenesis. Green, spherical, embryogenic callus could be obtained on cotyledonary leaves when cultured on MS medium having 1 mg/L of 2,4-D, 0.25 mg/L of kinetin, and MS media. The study showed that alterations in light and carbon source treatment were crucial for somatic embryo proliferation and development.

In *Asparagus racemosus*, somatic embryos could be obtained from zygotic embryos grown in the dark on MS media (Murashige and Skoog, 1962; Chaudhary and Dantu, 2019) Compact embryogenic callus which had been differentiated on MS containing 0.1 mg L^{-1} NAA and 0.5 mg L^{-1} kinetin resulted in a significant rise in somatic embryo development without any abnormalities when supplemented with 0.75 mg/L ancymidol. In contrast, Balamurugan et al. (2019) reported that the use of cytokinin BAP in a liquid medium was more advantageous to induce somatic embryos of *Gloriosa superba*, a plant well known to address ailments like gonorrhea, cholera, arthritis, kidney issues, and even cancer.

Viola canescens Wall. ex. Roxb., valued for its medicinal properties, including its effectiveness in treating diarrhea (Ahmad et al., 2023), was mass-propagated through somatic embryogenesis using leaf-derived calli by Khajuria et al. (2021). In this study, a basal medium containing L-glutamine (2%) and casein hydrolysate (50 mg/L) proved to be most efficient in inducing the highest average number of somatic embryos. These results superseded the otherwise superior results obtained with a combination of 2,4-D and Kn in the medium, indicating that the use of PGRs may not be imperative for somatic embryo induction, depending on the genotype and explant. However, the use of abscisic acid (ABA) was established to be important for the maturation of somatic embryos, while a combination of ABA and $AgNO_3$ was efficient for somatic embryo conversion into plantlets, followed by good results of plantlet development on medium containing low concentrations of BAP.

In a recent study on the propagation of *Pluchea lanceolata*, a medicinal herb valued for treating rheumatism, arthritis, fever, edema, muscular pains, etc., callus-mediated somatic embryo development was reported on medium containing NAA (1.0 mg/L), which subsequently resulted in plantlet development on medium supplemented with 0.5 mg/L GA_3 (Mamgain et al., 2023).

1.9 ARTIFICIAL SEED TECHNOLOGY

Synthetic seed (also called artificial seed) technology presents a new method for the long-term preservation and conservation of beneficial and important plant material. This method can restore plant species with valuable elite germplasm with substantial economic and therapeutic values for future generations. Besides long-term storage, the ability of synthetic seeds to directly multiply at the field level makes them an essential plant propagation tool in modern times. Additionally, plant species that are unable to generate seeds can be propagated using this technique (Garg and Maheshwari, 2023).

Artificial seed formation involves the utilization of *in vitro* culture techniques for encapsulation of plant propagules. A variety of methods, such as somatic embryogenesis; organogenesis; and the proliferation of apical bud, protocorm, and protocorm-like bodies (PLBs), are employed to generate plant micro propagules for synthetic seed production (Sharma et al., 1992; Corriealri and Tandonalri, 1993; Nagananda et al., 2011; Mani et al., 2023). These propagules are then enclosed in a protective coating developed from hydrogel, sodium alginate, or other materials, which safeguards them from mechanical injury, microbial invasion, and desiccation. The resulting artificial seeds can be preserved for extended periods and used for either direct sowing or transplantation. This technology has also been successfully used for medicinal plant conservation and propagation. In *Scutellaria montana*, classified as a threatened medicinal plant with exceptional antioxidant properties that are used to treat cancer, cardiovascular disease, and osteoarthritis (Vaidya et al., 2014; Irvin et al., 2019), alginate bead encapsulation has shown promise for synthetic seed development (Joshee et al., 2013). In this study, shoot tips and nodal explants were encapsulated in 3% sodium alginate and subjected to a 60-day cold treatment at 4–6 °C, demonstrating robust regeneration potential upon transfer to the regeneration medium.

In *Celestrus paniculatus*, a plant widely used for therapeutic purposes, *in vitro* regenerated shoot tips were encapsulated in 4% sodium alginate enriched with BAP and NAA to develop synthetic seeds (Fonseka et al., 2019). A germination rate of 80% could be achieved when the seeds were sown on a medium containing BAP (2 mg/L) and NAA (0.2 mg/L). In *Eryngium alpinum*, a highly valuable endangered medicinal plant, conventional propagation is challenging due to prolonged dormancy. Kikowska et al. (2020) showed that *in vitro* grown axillary buds of *E. alpinum*, when encapsulated with sodium–alginate (4%) and 300 mM calcium chloride (300 mM), could be stored for as long as 2 months followed by a germination rate of 74% when inoculated on MS medium.

Crinum malabaricum is a valuable plant used in the treatment of Parkinson's and Alzheimer's diseases and could be conserved using the artificial seed method. According to Priyadharshini et al. (2020), synthetic seeds developed from somatic embryos encapsulated with 2.0% sodium alginate and 100 mM calcium chloride could be preserved for up to a year at −4 °C and showed 92.5% post-thawing germination potential on MS medium containing BAP and NAA (Priyadharshini et al., 2020). Behera et al. (2020) showed that artificial seeds could be developed from axenic branch segments of *Hedychium coronarium*. This plant has proven antimycobacterial activity and works well in treating respiratory infections (Sharma et al., 2023). The study showed that the standardized encapsulation matrix for axenic shoots was an MS medium containing 100 mM $CaCl_2$ and 3% sodium alginate. The artificial seeds demonstrated 90% unique conversion in artificial seeds at 4 °C and 25° C, with conversion rates of 73% and 62%, respectively.

1.10 ROOT CULTURE

Adventitious and hairy root cultures hold great promise in addressing the critical issue of declining populations of medicinal plants due to overexploitation. Overcoming time and environmental limitations, these cultures can produce higher quantities of bioactive compounds. Bioreactor-grown root cultures have produced more secondary metabolites as well as greater biomass accumulation (Sui et al., 2018). Researchers are investigating the use of elicitors like temperature and light stresses to increase the production of secondary metabolites in adventitious root culture (Ali et al., 2019). With advancements in bioreactor systems and optimization strategies, it is now possible to mass-produce valuable compounds from endangered medicinal plants. The media composition, pH, temperature, oxygen concentration, carbon source, and optimization strategies have been studied to optimize secondary metabolite production in adventitious root cultures (Hussain et al., 2022).

Tissue culture methods can be used to increase the root biomass of medicinal plants for increased extraction of bioactives. For example, *Gentiana scabra* has immense potential for treating inflammation, tuberculosis, rheumatism, cholecystitis, and hepatitis due to its metabolite contents. A study by Hao et al. (2021) reported that leaf explants of *G. scabra* could be cultivated on MS media with NAA (3 mg/L) and TDZ (0.25 mg/L) and induced to yield 53.20 ± 3.15 roots per explant. High-performance liquid chromatography (HPLC) analysis of these induced adventitious roots showed that they accumulated high concentrations of secoiridoids within 4 weeks viz. sweroside (4.42 ± 0.11 mg g^{-1} DW), swertiamarin (1.61 ± 0.04 mg g^{-1} DW), and gentiopicroside (25.59 ± 0.65 mg g^{-1} DW). The valepotriates content of *Valeriana jatamansi* is known to impart the plant its insect-repelling and antihelminthic properties, along with its sedative, neurotoxic, cytotoxic, antidepressant, antioxidant, and antimicrobial activities (Jugran et al., 2019). Pandey et al. (2022) established the root cultures of *V. jatamansi* for enhanced accumulation of these bioactives. Among the tested PGRs and elicitors, 2.0 mg/L of NAA and methyl jasmonate (MeJA) were found to induce a higher biomass of the root cultures. Root extracts grown on liquid medium supplemented with 2.0 mg/L of NAA were shown to have the highest total phenolic content (TPC), total flavonoid content (TFC), and antioxidant potential, as confirmed by gas chromatography–mass spectrometry (GC-MS) analysis. Similarly, *Plumbago zeylanica* is valued for its plumbagin content, which is mostly accumulated in the roots. A study reported that leaf explants of *P. zeylanica* could be used for rapid production of a maximum number of roots when inoculated on half-strength MS medium containing 6.0 μM BAP and 2.5 μM NAA (Katoch et al., 2022).

Alternatively, *Agrobacterium rhizogenes*–mediated transformation of plant tissues has been well established to be a potential method of secondary metabolite production in medicinal plants. In a study by Srivastava et al. (2019), hairy root cultures of *Glycyrrhiza glabra* were established using the *A. rhizogenes* A4 strain, and it was shown that the use of polyethylene glycol (PEG) at 1% concentration increased the yield of glycyrrhizin (triterpenoid saponin) up to 5.4 times after 24 hours of exposure, while 200 μg mL^{-1} cellulase increased the yield of glycyrrhizin to 8.6 times after 7 days of treatment.

The therapeutic properties of *Mentha spicata*, like treatment of cough, cold, fever, digestive problems, jaundice, etc. (Mahendran et al., 2021), are mainly linked to its phenolic content. Yousefian et al. (2020) standardized the protocol for enhanced phenolic content production in *M. spicata* through hairy root culture. They used different strains of *A. rhizogenes* for hairy root induction in cultured leaf and stem explants. The study confirmed that while the highest infection efficiency (almost 75% of the explants) was exhibited by A13 among the tested strains, the highest biomass production was shown by A13 and R318 (nearly 60 mg/flask), and the highest content of phenolic acids was observed with GMI 9534. Simultaneously, treatments with IBA and the elicitor MeJA were found to be effective for a substantial increase in root growth and phenolic acid production. In another study on *Centella asiatica*, a prized neuro-tonic, hairy roots were induced both from leaves and petiole explants after being suspended in agro-stock for 30 minutes and then co-cultured with *A. rhizogenes* for 3 days (Baek et al., 2022). The study showed that petiole-derived lines showed the highest triterpenoid concentration (46.57 $mg \cdot g^{-1}$) and the highest growth index (37.8). It was also reported that the explant's endogenous triterpenoid concentration did not influence the bioactive accumulation in hairy roots, hence confirming *A. rhizogenes*–mediated root induction to be an extremely powerful method for *in vitro* bioactive production.

1.11 HARDENING AND ACCLIMATIZATION OF TISSUE CULTURE–RAISED PLANTLETS

The process of acclimatization can be difficult, and certain circumstances and attention may be needed to guarantee the plants' survival and effective establishment. *In vitro*–grown medicinal plants need to be suitably hardened and acclimatized before being shifted under field conditions. In general, the method of *in vitro* hardening includes bringing down the concentration of nutrients in the medium, carbon source, and gelling agent. Pant et al. (2010) used liquid MS medium at one-fourth strength, devoid of any PGR and containing only 2% sucrose, for *in vitro* hardening of *Swertia chirata* plantlets, followed by *ex vitro* transfer on a combination of soil:sand:manure (1:1:1). The combination of equal portions of cow dung, autoclaved river sand, and garden soil has been demonstrated to be the most successful substrate for the acclimatization of *in vitro* regenerated *Ceropegia candelabrum*, achieving an 85% survival rate (Binish and Nayagi, 2019).

The plantlets of *Zanthoxylum armatum* were transferred to PGR-free half-strength MS medium and then treated with 0.5% bavistin solution (to prevent fungal contamination) and then transferred into small plastic cups with a 3:1 ratio of soil and farmyard manure and covered with a polythene bag to maintain humidity and watered every 3 days in a polyhouse for more than a month. The plants were then shifted to shade and a mist chamber in earthen pots containing garden soil and finally transferred to natural sunlight after around 28 days (Purohit et al., 2017).

Lijalem and Feyissa (2020) showed that plantlets of *Securidaca longipedunculata* (Fresen) grew well on medium composed of red soil, sand, and compost (2:1:1) when kept in a polyhouse after initial maintenance under laboratory conditions. It was shown that overwatering may prove to be lethal for the plants due to rotting and fungal contamination.

The studies indicate that careful selection of growing substratum, levels of acclimatization, and maintenance of physical parameters is a must for successful transplantation of tissue culture–raised plantlets.

1.12 CONCLUSION

Plant tissue culture has enormous potential for the conservation and sustainable utilization of medicinal plants. To ensure the genetic fidelity and long-term survival of medicinal plants that are rare, threatened, or endangered, plant tissue culture can help with mass propagation and *ex situ* conservation. Different plant tissues can be used for *in vitro* plant propagation and serve specific purposes.

Plant tissue culture technology also enables the production of valuable secondary metabolites in a completely controlled environment, reducing the need for wild harvesting. The development of plant-based pharmaceuticals and bioproducts is made achievable by the advancement in plant tissue culture techniques, genetic engineering, and metabolic engineering. However, the application of this technique is not without challenges. These include challenges with the collection of explants from rare, endangered, and threatened species; product removal; maintenance of aseptic conditions; and continuous operation. The costs of supplies, electricity, etc., are some other obstacles, which, however, have been tackled by the use of low-cost alternatives (Pant, 2016). Nevertheless, the advantages of plant tissue culture far exceed the minuscule challenges and present invaluable techniques for medicinal plant propagation and utilization.

REFERENCES

Ahmad, I., Alotaibi, B. S., Malak, N., Asad, F., Ullah, B., Nasreen, N., … Chen, C. C. (2023). Antidiarrheal potential of *Viola canescens*: *In vivo* and in silico approaches. *Pharmaceuticals*, *16*(4), 489.

Ali, H., Khan, M. A., Kayani, W. K., Dilshad, E., Rani, R., & Khan, R. S. (2019). Production of biomass and medicinal metabolites through adventitious roots in *Ajuga bracteosa* under different spectral lights. *Journal of Photochemistry and Photobiology B: Biology*, *193*, 109–117.

Argaye, S. (2021). Development and the role of tissue culture in plant breeding: A review. *International Journal of Research Studies in Agricultural Sciences*, *7*(5), 27–36.

Azad, M. A. K., Adhikari, M., Sarker, M. N. A., Hasan, A. M., & Amin, M. N. (2022). *In vitro* shoot regeneration from nodal and shoot tip explants of a threatened medicinal plant-*Smilax Zeylanica* L. *Journal of Bio-Science*, *30*(1), 69–78.

Azad, M. A. K., Yokota, S., Ishiguri, F., & Yoshizawa, N. (2006). Plant regeneration from mesophyll protoplasts of a medicinal plant, *Phellodendron amurense* Rupr. *In Vitro Cellular & Developmental Biology-Plant*, *42*, 502–507.

Babaei, N., Psyquay Abdullah, N. A., Saleh, G., & Lee Abdullah, T. (2014). An efficient *in vitro* plantlet regeneration from shoot tip cultures of *Curculigo latifolia*, a medicinal plant. *The Scientific World Journal*, *2014*, 275028. https://doi.org/10.1155/2014/275028

Badamjav, R., Sonom, D., Wu, Y., Zhang, Y., Kou, J., Yu, B., & Li, F. (2020). The protective effects of *Thalictrum minus* L. On lipopolysaccharide-induced acute lung injury. *Journal of Ethnopharmacology*, *248*, 112355.

Baek, S., Han, J. E., Ho, T. T., & Park, S. Y. (2022). Development of hairy root cultures for biomass and triterpenoid production in *Centella asiatica*. *Plants*, *11*(2), 148.

Bakhtiar, Z., Mirjalili, M. H., & Sonboli, A. (2016). *In vitro* callus induction and micropropagation of *Thymus persicus* (Lamiaceae), an endangered medicinal plant. *Crop Breeding and Applied Biotechnology*, *16*, 48–54.

Balamurugan, V., Amal, T. C., Karthika, P., Selvakumar, S., & Vasanth, K. (2019). Somatic embryogenesis and plant regeneration in *Gloriosa superba* L.: An endangered medicinal plant. In: Kumar, M., Muthusamy, A., Kumar, V., Bhalla-Sarin, N. (eds), *In Vitro Plant Breeding towards Novel Agronomic Traits*, Springer, 27–42.

Baskaran, P., Kumari, A., & Van Staden, J. (2017). Rapid propagation of *Mondia whitei* by embryonic cell suspension culture *in vitro*. *South African Journal of Botany*, *108*, 281–286.

Behera, S., Rout, K. K., Panda, P. C., & Naik, S. K. (2020). Production of non-embryogenic synthetic seeds for propagation and germplasm transfer of *Hedychium coronarium* J. Koenig. *Journal of Applied Research on Medicinal and Aromatic Plants*, *19*, 100271.

Bhatia, S., Sharma, K., Dahiya, R., & Bera, T. (2015). Plant tissue culture. In: Bhatia, S., Sharma, S., Dahiya, R., Bera, T. (eds.), *Modern Applications of Plant Biotechnology in Pharmaceutical Sciences*, Academic Press, 31, 107.

Binish, T., & Nayagi, N. J. (2019). Micropropagation of endangered medicinal plant *Ceropegia candelabrum* L. *Pharma Innovation*, *8*(3), 273–277.

Cardoso, J. C., Oliveira, M. E., & Cardoso, F. D. C. (2019). Advances and challenges on the *in vitro* production of secondary metabolites from medicinal plants. *Horticultura Brasileira*, *37*, 124–132.

Chaudhary, J., & Dantu, P. K. (2019). Induction of somatic embryos in cultures of *Asparagus racemosus* Willd: An endangered medicinally important plant. *Bulletin of the National Research Centre*, *43*(1), 5.

Chouhan, R., Kumar, N., Kumar, A., Ahmed, S., Sen, Y., & Gandhi, S. G. (2020). Establishment of hairy roots of endangered Himalayan plant Swertia chirata: A sustainable alternative to extraction from nature. In: Srivastava, V., Mehrotra, S., Mishra, S. (eds.), *Hairy Root Cultures Based Applications: Methods and Protocols*, Springer, 55–63.

Corriealri, S., & Tandonalri, P. (1993). Propagation of *Cymbidium giganteum* Wall. through high frequency conversion of encapsulated protocorms under *in vivo* and *in vitro* conditions. *Indian Journal of Experimental Biology*, *31*(1), 61–64.

Ebrahimi, M., Mokhtari, A., & Amirian, R. (2018). A highly efficient method for somatic embryogenesis of *Kelussia odorotissima* Mozaff., an endangered medicinal plant. *Plant Cell, Tissue and Organ Culture (PCTOC)*, *132*, 99–110.

Fonseka, D. L. C. K., Wickramaarachchi, W. W. U. I., & Madushani, R. P. S. (2019). Synthetic seed production as a tool for the conservation and domestication of *Celastrus paniculatus*: A rare medicinal plant. *Annual Research & Review in Biology*, *32*, 1–8.

Gallia, M. C., Echeverri Del Sarto, J., & Bongiovanni, G. A. (2021). Sustainable and efficient protocols for *in vitro* germination and antioxidants production from seeds of the endangered species *Araucaria araucana. Journal of Genetic Engineering and Biotechnology*, *19*(1), 1–10.

Garg, R., & Maheshwari, S. (2023). Synthetic seed technology, application and future trends. *EPH-International Journal of Agriculture and Environmental Research*, *9*(1), 1–10.

Grzegorczyk-Karolak, I., Rytczak, P., Bielecki, S., & Wysokińska, H. (2017). The influence of liquid systems for shoot multiplication, secondary metabolite production and plant regeneration of *Scutellaria alpina. Plant Cell Tissue and Organ Culture*, *128*, 479–486.

Hao, N., Piao, Z., Zang, J., Li, H., & Zhou, R. (2021). Establishment of adventitious root cultures and assessment of secoiridoid production in the Chinese medicinal plant *Gentiana scabra. In Vitro Cellular & Developmental Biology-Plant*, *57*(6), 864–873.

Han, Y., Chen, Y., Zhang, Q., Liu, B.-W., Yang, L., Xu, Y.-H., & Zhao, Y. (2021). Overview of therapeutic potentiality of *Angelica sinensis* for ischemic stroke. *Phytomedicine*. 90, 153652.

Hernández-García, A., Ambriz-Parra, E., López-Albarrán, P., Cruz-de León, J., & Salgado-Garciglia, R. (2021). *In vitro* propagation from axillary buds of the endangered tree *Dalbergia congestiflora* Pittier (Fabaceae). *Plant Biotechnology*, *38*(4), 409–414.

Hussain, A., Qarshi, I. A., Nazir, H., & Ullah, I. (2012). Plant tissue culture: Current status and opportunities. *Recent Advances in Plant in Vitro Culture*, *6*(10), 1–28.

Hussain, M. J., Abbas, Y., Nazli, N., Fatima, S., Drouet, S., Hano, C., & Abbasi, B. H. (2022). Root cultures, a boon for the production of valuable compounds: A comparative review. *Plants*, *11*(3), 439.

Ikuta, A., Nakamura, T., & Urabe, H. (1998). Indolopyridoquinazoline, furoquinoline and canthinone type alkaloids from *Phellodendron amurense* callus tissues. *Phytochemistry*, *48*(2), 285–291.

Irvin, L., Jackson, C., Hill, A. L., Bajaj, R., Mahmoudi, C., Vaidya, B. N., & Joshee, N. (2019). Skullcaps (*Scutellaria* spp.): Ethnobotany and current research. In: *Medicinal Plants: From Farm to Pharmacy*, Springer, 141–168.

Jagetia, G. C. (2021). A review on the medicinal and pharmacological properties of traditional ethnomedicinal plant sonapatha, *Oroxylum indicum. Sinusitis*, *5*(1), 71–89.

Jimu, L. (2011). Threats and conservation strategies for the African cherry (*Prunus africana*) in its natural range-a review. *Journal of Ecology and the Natural Environment*, *3*(4), 118–130.

Joshee, N., Tascan, A., Medina-Bolivar, F., Parajuli, P., Rimando, A. M., Shannon, D. A., & Adelberg, J. W. (2013). *Scutellaria*: Biotechnology, phytochemistry and its potential as a commercial medicinal crop. In: *Biotechnology for Medicinal Plants: Micropropagation and Improvement*, Springer, 69–99.

Jugran, A. K., Rawat, S., Bhatt, I. D., & Rawal, R. S. (2019). *Valeriana jatamansi*: An herbaceous plant with multiple medicinal uses. *Phytotherapy Research*, *33*(3), 482–503.

Kadapatti, S. S., & Murthy, H. N. (2022). Micropropagation of *Andrographis producta* through axillary and adventitious shoot regeneration. *Journal of Genetic Engineering and Biotechnology*, *20*(1), 152.

Karuppaiya, P., & Tsay, H. S. (2020). Enhanced production of podophyllotoxin, kaempferol, and quercetin from callus culture of *Dysosma pleiantha* (Hance) Woodson: An endangered medicinal plant. *Biotechnology and Applied Biochemistry*, *67*(1), 95–104.

Kashyap, K., Parihar, S., & Shekhawat, G. S. (2023). *In vitro* establishment of cell suspension culture of *Ceropegia bulbosa* for improved production of cerpegin content through elicitation of engineered carbon and ZnO nanoparticles. *Environmental Science and Pollution Research*, *30*, 118263–118279.

Katoch, K., Gupta, S., Nazir, R., Kumar, V., Sanyal, R., Dey, A., & Pandey, D. K. (2022). Establishment of adventitious root culture from leaf explants of *Plumbago zeylanica*: An endangered medicinal plant. *Vegetos*, *35*, 276–280.

Khajuria, A. K., Hano, C., & Bisht, N. S. (2021). Somatic embryogenesis and plant regeneration in *Viola canescens* Wall. Ex. Roxb.: An endangered Himalayan herb. *Plants*, *10*(4), 761.

Khan, F., Sarker, M. M. R., Ming, L. C., Mohamed, I. N., Zhao, C., Sheikh, B. Y., ... Rashid, M. A. (2019). Comprehensive review on phytochemicals, pharmacological and clinical potentials of *Gymnema sylvestre*. *Frontiers in Pharmacology*, *10*, 1223.

Kikowska, M., Sliwinska, E., & Thiem, B. (2020). Micropropagation and production of somatic seeds for short-term storage of the endangered species *Eryngium alpinum* L. *Plants*, *9*(4), 498.

Komakech, R., Kang, Y., Lee, J. H., & Omujal, F. (2017). A review of the potential of phytochemicals from Prunus africana (Hook f.) Kalkman stem bark for chemoprevention and chemotherapy of prostate cancer. *Evidence-Based Complementary and Alternative Medicine*, 2017, 3014019.

Komakech, R., Kim, Y. G., Kim, W. J., Omujal, F., Yang, S., Moon, B. C., ... Kang, Y. (2020). A micropropagation protocol for the endangered medicinal tree Prunus africana (Hook f.) Kalkman: Genetic Fidelity and Physiological Parameter Assessment. *Frontiers in Plant Science*, *11*, 548003.

Koul, A., & Mallubhotla, S. (2020). Elicitation and enhancement of bacoside production using suspension cultures of *Bacopa monnieri* (L.) Wettst. *3 Biotech*, *10*(6), 256.

Lal, A., Pant, M., Palni, L. M. S., & Kumar, A. (2020). Development of rapid micropropagation protocol for germplasm conservation of two orchid species – *Aerides multiflora* Roxb. and *Rhynchostylis retusa* (L.) Blume. *Asian Journal of Conservation Biology*, *9*(2), 341–347.

Lijalem, T., & Feyissa, T. (2020). *In vitro* propagation of *Securidaca longipedunculata* (Fresen) from shoot tip: An endangered medicinal plant. *Journal of Genetic Engineering and Biotechnology*, *18*, 1–10.

Liu, W., Zhang, Z., Zhang, T., Qiao, Q., & Hou, X. (2022). Phenolic profiles and antioxidant activity in different organs of *Sinopodophyllum hexandrum*. *Frontiers in Plant Science*, *13*, 1037582.

Loyola-Vargas, V. M. (2016). The history of somatic embryogenesis. In: Loyola-Vargas, V., Ochoa-Alejo, N. (eds.), *Somatic Embryogenesis: Fundamental Aspects and Applications*, Springer, 11–22.

Mahendran, G., Verma, S. K., & Rahman, L. U. (2021). The traditional uses, phytochemistry and pharmacology of spearmint (*Mentha spicata* L.): A review. *Journal of Ethnopharmacology*, *278*, 114266.

Mamgain, J., Mujib, A., Syeed, R., Ejaz, B., Malik, M. Q., & Bansal, Y. (2023). Genome size and gas chromatography-mass spectrometry (GC-MS) analysis of field-grown and in vitro regenerated *Pluchea lanceolata* plants. *Journal of Applied Genetics*, *64*(1), 1–21.

Mani, M., Mathiyazhagan, C., Dey, A., Faisal, M., Alatar, A. A., Alok, A., & Shekhawat, M. S. (2023). Micro-morpho-anatomical transitions at various stages of *in vitro* development of *Crinum malabaricum* Lekhak and Yadav: A critically endangered medicinal plant. *Plant Biology*, *25*(1), 142–151.

Moola, A. K., & Kumari, B. R. (2020). Rapid propagation of *Celastrus paniculatus* Willd.: An endangered medicinal plant through indirect organogenesis. *Vegetos*, *33*(2), 277–285.

Morel, G. (1966). Meristem culture clonal propagation of orchids. *Orchid Dig*, *30*, 45–49.

Morel, G., & Martin, C. (1952). Guérison de dahlias atteints d'unemaladie à virus. *Comptes rendus de l'Académie des Sciences*, *235*, 1324–1325.

Mukami, A., Juma, B. S., Mweu, C., Ngugi, M., Oduor, R., & Mbinda, W. M. (2022). Plant regeneration from leaf mesophyll-derived protoplasts of cassava (*Manihot esculenta* Crantz). *PLoS One*, *17*(12), e0278717.

Murab, T., & Chandurkar, P. (2018). Invitro micropropogation of nodal explant of *Gymnema sylvestre* R. Br. a multidimensional plant. *International Journal for Research in Applied Science & Engineering Technology*, *6*(4), 4195–4197

Murashige, T., & Skoog, F. (1962). A revised medium for rapid growth and bio assays with tobacco tissue cultures. *Physiologia Plantarum*, *15*, 473–497.

Nagananda, G. S., Nalini, S., & Rajath, S. (2011). Regeneration of encapsulated protocorm like bodies of medicinally important vulnerable orchid *Flickingeria nodosa* (Dalz.) Seidenf. *International Journal of Botany*, *7*(4), 310–313.

Ngezahayo, F., & Liu, B. (2014). Axillary bud proliferation approach for plant biodiversity conservation and restoration. *International Journal of Biodiversity*, *2014*, 1–9.

Nic-Can, G. I., Avilez-Montalvo, J. R., Aviles-Montalvo, R. N., Márquez-López, R. E., Mellado-Mojica, E., Galaz-Ávalos, R. M., & Loyola-Vargas, V. M. (2016). The relationship between stress and somatic embryogenesis. In: Loyola-Vargas, V., Ochoa-Alejo, N. (eds.), Somatic Embryogenesis: Fundamental Aspects *and* Applications, Springer, 151–170.

Palni, L. M. S., Lal, A., & Pant, M. (2018). Ex situ approaches to conservation of Himalayan biodiversity with particular reference to medicinal plants. In: Das, A. P., Bera, S. (eds.), *Plant Diversity in the Himalaya Hotspot Region*, Vol II, 903–932

Pandey, S., Sundararajan, S., Ramalingam, S., & Pant, B. (2022). Elicitation and plant growth hormone-mediated adventitious root cultures for enhanced valepotriates accumulation in commercially important medicinal plant *Valeriana jatamansi* Jones. *Acta Physiologiae Plantarum*, *44*(1), 13.

Pant, M. (2016). A minimal cost micropropagation protocol for Dianthus caryophyllus L. — A commercially significant venture. *Indian Journal of Experimental Biology*, *54*, 203–211.

Pant, M., Bisht, P., & Gusain, M. P. (2010). *In vitro* propagation through axillary bud culture *of Swertia chirata* Buch.-Ham ex Wall: An endangered medicinal herb. *International Journal of Integrative Biology*, *10*(1), 48–53.

Partap, M., Kumar, P., Ashrita, Kumar, P., Kumar, D., & Warghat, A. R. (2020). Growth kinetics, metabolites production and expression profiling of picrosides biosynthetic pathway genes in friable callus culture of *Picrorhiza kurroa* Royle ex Benth. *Applied Biochemistry and Biotechnology*, *192*(4), 1298–1317.

Priyadharshini, S., Manokari, M., & Shekhawat, M. S. (2020). *In vitro* conservation strategies for the critically endangered Malabar river lily (*Crinum malabaricum* Lekhak & Yadav) using somatic embryogenesis and synthetic seed production. *South African Journal of Botany*, *135*, 172–180.

Purohit, S., Jugran, A.K., Bhatt, I.D. , Palni, L. M. S., Bhatt, A., & Nandi, S. K. (2017). In vitro approaches for conservation and reducing juvenility of *Zanthoxylum armatum* DC: An endangered medicinal plant of Himalayan region. *Trees*, *31*, 1101–1108.

Purohit, S., Joshi, K., Rawat, V., Bhatt, I. D., & Nandi, S. K. (2020). Efficient plant regeneration through callus in *Zanthoxylum armatum* DC: An endangered medicinal plant of the Indian Himalayan region. *Plant Biosystems-An International Journal Dealing with all Aspects of Plant Biology*, *154*(3), 288–294.

Rajput, S., & Agrawal, V. (2020). Micropropagation of *Atropa acuminata* Royle ex Lindl. (a critically endangered medicinal herb) through root callus and evaluation of genetic fidelity, enzymatic and non-enzymatic antioxidant activity of regenerants. *Acta Physiologiae Plantarum*, *42*(11), 160.

Riahi, L., Chograni, H., Ben Rejeb, F., Ben Romdhane, M., Masmoudi, A. S., & Cherif, A. (2022). Efficient *in vitro* regeneration of the endangered species *Artemisia arborescens* L. through direct organogenesis and impact on secondary metabolites production. *Horticulture, Environment, and Biotechnology*, *63*(3), 439–450.

Ribatti, D. (2018). An historical note on the cell theory. *Experimental Cell Research*, *364*(1), 1–4.

Rodrigues, V., Kumar, A., Prabhu, K. N., Pragadheesh, V. S., Shukla, A. K., & Sundaresan, V. (2021). Adventitious root cultures of *Decalepis salicifolia* for the production of 2-hydroxy-4-methoxybenzaldehyde, a vanillin isomer flavor metabolite. *Applied Microbiology and Biotechnology*, *105*, 3087–3099.

Rout, G. R., Mohapatra, A., & Jain, S. M. (2006). Tissue culture of ornamental pot plant: A critical review on present scenario and future prospects. *Biotechnology Advances*, *24*(6), 531–560.

Sape, S. T., Kandukuri, A. V., & Owk, A. K. (2020). Direct axillary shoot regeneration with nodal explants of *Bacopa monnieri* (L.) Pennell – A multi medicinal herb. *Journal of Applied Biological Sciences*, *14*(2), 190–197.

Sargazi Moghadam, Z., Moshtaghi, N., Sharifi, A., Zahedi, A., & Khadem, A. (2022). *In vitro* propagation and callus induction of medicinal endangered plant meadow rue (*Thalictrum minus* L.) for producing berberine. *Journal of Medicinal Plants*, *21*(81), 92–101.

Sharma, A., Tandon, P., & Kumar, A. (1992). Regeneration of dendrobium *Wardianum warne* (*orchidaceae*) from synthetic seeds. *Indian Journal of Experimental Biology*, *30*(8), 747–748.

Sharma, N., Thakur, M., Sharma, P., Dutt, B., & Sharma, Y. P. (2022). *In vitro* propagation from seeds and enhanced synthesis of podophyllotoxin from root callus of *SinoPodophyllum hexandrum* Royle TS Ying (Himalayan Mayapple) – An endangered medicinal plant. *Industrial Crops and Products*, *186*, 115300.

Sharma, R. K., & Khajuria, A. K. (2020). Somatic embryogenesis and plant regeneration in *Ferula jaeschkeana* Vatke: A threatened medicinal herb. *Vegetos*, *33*, 658–664.

Sharma, S., Khandelwal, V., Ahirwar, R. K., & Gupta, V. K. (2023). *Hedychium coronarium* J. Koenig produced extracellular and intracellular anti-tuberculosis activity with morphological alteration of the tubercle bacillus. *South African Journal of Botany*, *162*, 701–706.

Sharma, S., Sharma, R., Sharma, P., Thakur, K., & Dutt, B. (2019). Direct shoot organogenesis from seedling derived shoot tip explants of endangered medicinal plant *Saussurea costus* (Falc.) Lipsch. *Proceedings of the National Academy of Sciences, India Section B: Biological Sciences*, *89*, 755–764.

Siddiqui, S., Thomas, T., & Khan, S. (2019). Standardization of an efficient protocol for *in vitro* micropropagation of *Centella asiatica* L. – An important medicinal plant. *Pharmaceutical and Biosciences Journal*, *7*(4), 22–26.

Singh, S., Mishra, R., & Agnihotri, R. K. (2021). In-vitro propagation of *Pluchea lanceolata* (DC) CB Clarke a potent antiarthritic medicinal herb through axillary bud. *Trends in Phytochemical Research*, *5*(1), 24–30.

Smitha, C. K., Udayan, P. S., Bindu, T. K., & Maju, T. T. (2021). Callus induction and plant regeneration from cotyledonary leaf explants of *Oroxylum indicum* (L.) Vent.-An Endangered Medicinal Tree. *Tropical Plant Research*, *8*(1), 36–41.

Srivastava, M., Singh, G., Sharma, S., Shukla, S., & Misra, P. (2019). Elicitation enhanced the yield of glycyrrhizin and antioxidant activities in hairy root cultures of *Glycyrrhiza glabra* L. *Journal of Plant Growth Regulation*, *38*, 373–384.

Sui, Q., Jiang, C., Yu, D., Chen, M., Zhang, J., Wang, Y., & Wei, Y. (2018). Performance of a sequencing-batch membrane bioreactor (SMBR) with an automatic control strategy treating high-strength swine wastewater. *Journal of Hazardous Materials*, *342*, 210–219.

Thakur, A., Pant, M., & Lal, A. (2020). Direct and indirect organogenesis in *Hippophae salicifolia* D. Don. – A nutraceutically important plant. *Research Journal of Biotechnology*, *15*(5): 33–41.

Thomas, T. D. (2009). Isolation, callus formation and plantlet regeneration from mesophyll protoplasts of *Tylophora indica* (Burm. f.) Merrill: An important medicinal plant. *In Vitro Cellular & Developmental Biology-Plant*, *45*, 591–598.

Uikey, D. S., Tripathi, M. K., Tiwari, G., Patel, R. P., & Ahuja, A. (2016). Embryogenic cell suspension culture induction and plantlet regeneration of *Rauvolfia serpentina* (L.) Benth.: Influence of different plant growth regulator concentrations and combinations. *Medicinal Plants-International Journal of Phytomedicines and Related Industries*, *8*(2), 153–162.

Utami, E. S. W., Hariyanto, S., & Manuhara, Y. S. W. (2017). *In vitro* propagation of the endangered medicinal orchid, *Dendrobium lasianthera* JJ Sm through mature seed culture. *Asian Pacific Journal of Tropical Biomedicine*, *7*(5), 406–410.

Vaidya, B. N., Brearley, T. A., & Joshee, N. (2014). Antioxidant capacity of fresh and dry leaf extracts of sixteen *Scutellaria* species. *Journal of Medicinally Active Plants*, *2*(3), 42–49.

Woo, H. A., Ku, S. S., Jie, E. Y., Kim, H., Kim, H. S., Cho, H. S., … Kim, S. W. (2021). Efficient plant regeneration from embryogenic cell suspension cultures of *Euonymus alatus*. *Scientific Reports*, *11*(1), 15120.

Yousefian, S., Lohrasebi, T., Farhadpour, M., & Haghbeen, K. (2020). Production of phenolic acids in hairy root cultures of medicinal plant *Mentha spicata* L. in response to elicitors. *Molecular Biology Research Communications*, *9*(1), 23.

Zhang, Y., Lu, Y., He, C. & Gao, S. (2019). A method for cell suspension culture and plant regeneration of *Angelica sinensis* (Oliv.) Diels. *Plant Cell, Tissue and Organ Culture*, *136*, 313–322.

2 Tissue Culture–Mediated Interventions in High-Altitude Himalayan Medicinal Plants

Indra Dutt Bhatt, Laxman Singh, and Deep Chandra Tiwari

2.1 INTRODUCTION

As of today, the two most interlinked environmental challenges faced by humanity are climate change and biodiversity loss. Increasing lines of evidence now show that there is no deceleration in the pace of either (Singh et al., 2022). Amid this, numerous national and international programs have been launched to combat biodiversity loss. One significant step is to understand the threats and their levels, whether new or conventional, levels of intensity, mitigation possible at local levels, and risk assessment on species extinction (Singh et al., 2023). Identifying these threat levels facilitates the determination of conservation priorities and the formulation of corresponding action plans (Rodrigues et al., 2006). Intriguingly, *in situ* conservation is perceived as the principal and the most effective conservation effort; however, this is not absolute because conserving imperiled species coexists with *ex situ* germplasm preservation and *in situ* conservation measures. The *ex situ* conservation method has now become a prominent part of several threatened species survival plan (SSP) programs (where the wild populations have dwindled to critical levels). It also acts as an important backup to some diversity plans for the conservation of species that are at risk of loss in an ecosystem predominated by humans (Cohen et al., 1991). The role of *ex situ* conservation has become even more significant considering the challenges in the natural habitat of many species, for example, shifts in seed set, seed dormancy, biotic and abiotic stresses, ineffective propagation systems, and loss of habitats (Arora et al., 2022). Thus, it is a complementary conservation strategy, especially desirable to facilitate conservation in controlled conditions for contributing to species reintroduction programs. Recent developments in biotechnological techniques, such as plant tissue culture, have made it possible to conserve and propagate several economically significant species, including highly valued aromatic and medicinal plants. The major applications of plant tissue culture techniques include micropropagation, producing genetically pure plants, *in vitro* bioactive compound production, cryopreservation, and the development of DNA banks for genomic resource conservation (Rohini, 2020). Besides these, pathogen-free plant production for ease of germplasm exchanges across international boundaries is another advantage (Bhojwani and Dantu, 2013). Several tissue culture–based propagation techniques like axillary bud culture, meristem tip culture, somatic embryogenesis, adventitious shoot multiplication, anther culture, ovary culture, etc., are now applied for scale-up multiplication of priority plants (Vengadesan and Pijut, 2009). The developed propagation methods hold utmost importance. For instance, the demand for raw materials of medicinal plants can only be achieved when there is a continuous supply of quality planting materials for cultivation and to achieve conservation objectives. For centuries medicinal plants have been an integral part of indigenous communities, particularly in developing nations (about 80% as per the World Health Organization [WHO]), to support their health problems and also have contributed significantly in alleviating the socioeconomic status of the different communities (Singh et al., 2017, 2021). In recent decades, there has been a significant increase in the use of nature therapy, leading to a growing prevalence of herbal or plant-derived remedies. With the rise in demand for medicinal plant resources, the supply is diminished by habitat loss, loss of natural populations, illicit harvesting from the wild and overexploitation,

DOI: 10.1201/b23374-2

FIGURE 2.1 Pictorial depiction of Himalayan medicinal plants used in the study (a) *Nardostachys jatamansi*, (b) *Paris polyphylla*, (c) *Hedychium spicatum*, (d) *Podophyllum hexandrum*, and (e) *Swertia chirayita*.

and competition with alien species, to name a few (Singh et al., 2021, 2022). In India alone, a meager amount 10% of medicinally active plants are under planned cultivation. At the same time, the remaining is directly collected from the wild, thus endangering many species with some risk of total local extinction (Singh et al., 2022). This necessitates the conservation of medicinal plant species (Figure 2.1a–e), leading to the application of biotechnological tools for the propagation and conservation of these species. The Himalayan ecosystem is enriched with a plethora of medicinally active plant species. Several plant tissue culture methods have been employed to conserve threatened plant germplasms (Figures 2.2 and 2.3; https://www.vosviewer.com/).

2.2 TISSUE CULTURE TECHNIQUES FOR PROPAGATION AND MULTIPLICATION OF HIMALAYAN MEDICINAL PLANTS

The biotechnology of medicinal plants employs three primary techniques. The first approach involves micropropagation techniques, which are utilized to achieve large-scale production of the elite germplasm. The second approach entails genetic transformation aimed at improving the

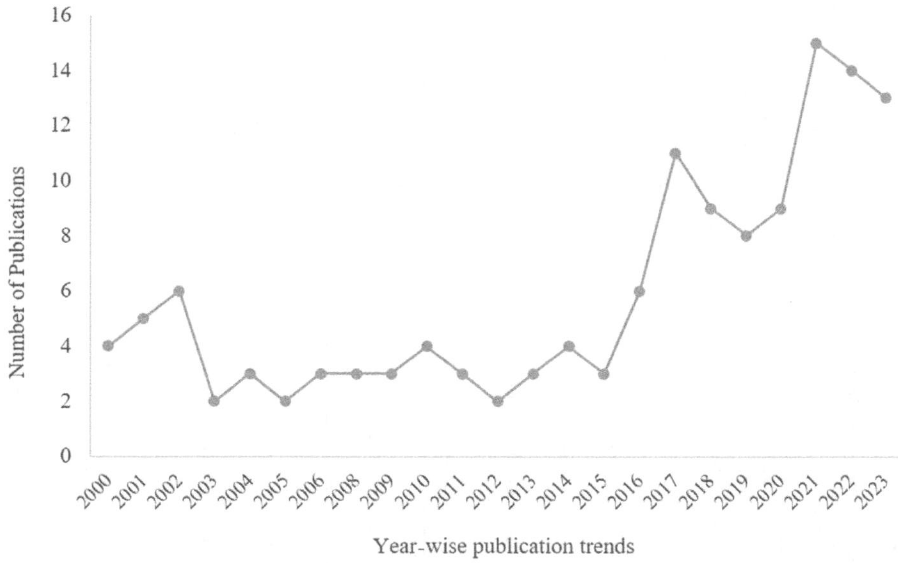

FIGURE 2.2 Plant tissue culture publication trends in high-altitude Himalayan medicinal plants for the period January 2000 to October 2023.

quality as well as quantity of the product. Another approach is to artificially synthesize secondary metabolites that possess therapeutic properties. Among these, the utilization of plant tissue culture technology emerges as a promising approach for ensuring the sustainable utilization of plant resources. Using this technique, several important threatened Himalayan medicinal plants have been propagated on large scales and subsequently conserved (Table 2.1). A wide range of explants,

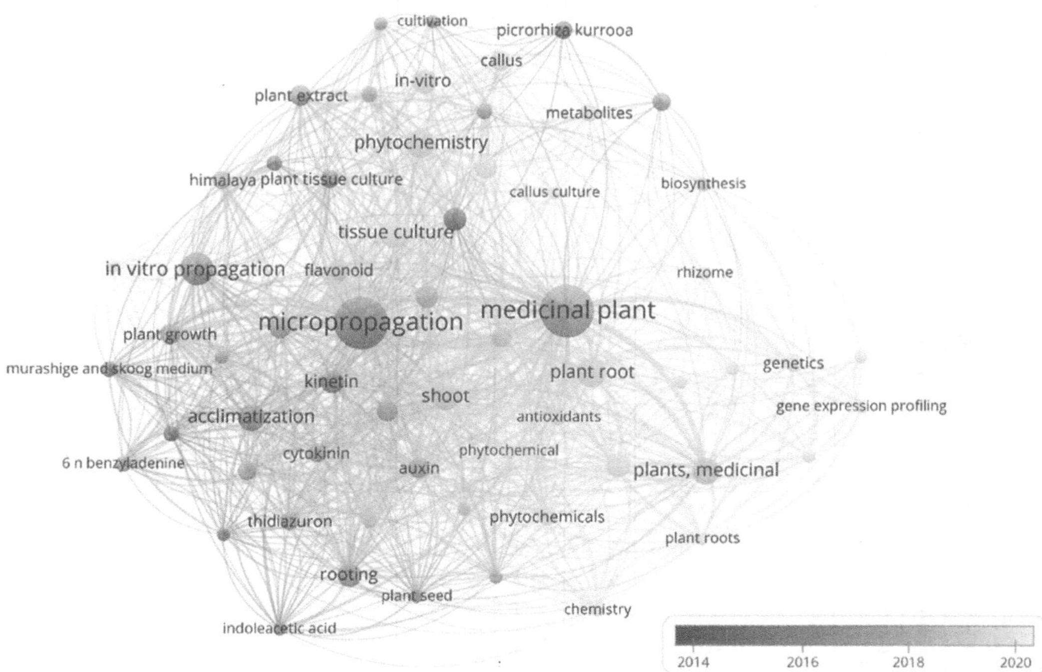

FIGURE 2.3 Prominent terminological map of publications in the field of plant tissue culture of high-altitude Himalayan medicinal plants for the period January 2000 to October 2023.

TABLE 2.1
Tissue Culture–Mediated Interventions in Threatened Himalayan Medicinal and Aromatic Plants

S. No	Name of Species	*In Vitro* Regeneration Condition	*In Vitro* Rooting and Hardening Strategies	References
1	*Gentiana kurroo*	MS medium + 0.10 mg/L NAA and 1.0 mg/L TDZ	• Half-strength MS medium + 1.0 mg/L IBA • 100 % survival rate was recorded on autoclaved cocopeat	Sharma et al., 2014a
2	*Saussurea costus*	MS medium + 2.0 mg/L BAP + 1.0 mg/L NAA + 0.25 mg/L GA_3	• MS medium + BAP (0.5 mg/L) + IAA (0.5 mg/L) + IBA (0.5 mg/L) • 87% survival rate was recorded in autoclaved garden soil, farmyard soil, and sand (2:1:1) mixture	Khan et al., 2021
3	*Swertia chirayita*	MS medium + 5.0 µM BAP	• MS medium + 5.0 µM IBA • 58.70% survival rate was reported in mixture of soil, sand, and FYM (2:1:1) under 75% agro shade net	Kumar et al., 2018
4	*Aconitum balfourii*	MS medium + 4.5 µM BAP and 1.4 µM NAA	• MS medium + 12.3 µM IBA • 50% survival rate was recorded in a mixture of soil and FYM (1:1)	Pandey et al., 2004
5	*Aconitum heterophyllum*	MS medium + 0.25 mg/L + 0.50 mg/L BAP	• MS medium + 1.0 mg/L IAA • Hardening was carried out in mixture of sand:soil:vermiculite (1:1:1)	Jabeen et al., 2006
6	*Picrorhiza kurrooa*	MS medium + 1.0 µM BAP or 2.5 µM Kn	• PGR-free medium • Soil and sand (2:1) were used as hardening medium in a greenhouse (25 °C; RH, 90 %)	Chandra et al., 2004
7	*Podophyllum hexandrum*	MS medium + 1.0–4.0 µM IAA and 1.0 µM BAP	• MS medium + 0.5 µM IAA • Polybags containing vermiculite were used	Nadeem et al. 2000
		MS medium + 5 mg/L GA_3	• MS medium + 5 mg/L GA_3 • Mixture of perlite, peat moss, and vermiculate mixture in the ratio 1:1:1 showed 38 % survivability	Rajesh et al., 2014
		MS medium + 0.2 mg/L BA and 0.1 mg/L GA_3	• MS + 2.0 mg/L BA + 0.2 mg/L NAA • Mixture of soil:sand:FYM (1:1:1) showed 100% survival after 1 month	Sharma et al., 2022
8	*Saussurea obvallata*	MS medium + 1.0 µM Kn + 0.25 µM NAA	• Half-strength MS medium + 2.5 µM IBA • 66.7% survival was reported during *ex vitro* acclimatization process	Joshi and Dhar, 2003
		MS medium + 2.5 µMBA and 1.0 µMNAA	• Half-strength MS medium + 2.5 µM IBA	Dhar and Joshi, 2005

9	*Saussurea lappa*	MS medium + 0.45 µM TDZ	• MS medium + 1.07 µM NAA • 66.7% survival was reported upon hardening in soil:sand (1:1) mixture	Johnson et al., 1997
10	*Habenaria edgeworthii*	MS medium + 0.1 1 µM BA + 0.01 µM NAA	• Half-strength MS medium + NAA (0.01–0.50 µM) or IBA (0.01–0.50 µM) • 68% survival was reported upon hardening in soil:sand:perlite (1:1:1 ratio)	Giri et al., 2012b
11	*Crocus sativus*	MS medium + 26.64 µM BAP	• MS medium + paclobutrazol (PBZ) was required for induction and development of *in vitro* cormlets	Gautam and Bhattacharya, 2021
12	*Nardostachys jatamansi*	MS medium + 0.5 mg/L Kn and 0.75 mg/L TDZ	• Half-strength MS medium + 0.5 mg/L IAA • 83% survival was recorded in sand or sand plus soil (1:1) mixture	Dhiman et al., 2021
		MS medium + 1.5 mg/L NAA and 1.0 mg/L meta-topolin	• MS medium + 1.0 mg/L meta-topolin +1.0 mg/L IAA • 83% survival was recorded in mixture of soil, sand, and leaf litter mixture (2:1:1) covered with moss to retain humidity	Bose et al., 2016
13	*Polygonatum verticillatum*	MS medium + 1.0 mg/L BAP with 0.5 mg/L NAA	• Half-strength MS medium + 1.0 mg/L NAA • Forest soil with organic compost (1:2 v/v) was used achieve high survival percentage	Bisht et al., 2012
14	*Paris polyphylla*	MS medium + 0.5 mg/L GA_3	• MS medium + 0.05 mg/L BAP and 0.1 mg/L NAA • 95% survival rate was reported in sterilized mixture of garden soil, compost, and peat moss (2:1:1)	Raomai et al., 2014
		Half-strength MS medium + 0.5 mg/L BAP	• Plant growth regulator–free medium • 95% survival rate was reported in potting mixtures of soil, compost, and peat moss (2:1:1)	Raomai et al., 2015
15	*Valeriana jatamansi*	MS medium + 1.5 µM BAP + 0.5 µM NAA + and 0.1 µM GA3	• MS medium + 1.5 µM BAP + 0.5 µM NAA and 0.1 µM GA_3 • 90% survival was reported in mixture of sterile soil:vermiculite:perlite in 1:1:1 ratio	Purohit et al. 2015
16	*Hedychium spicatum*	MS medium + 1.0 µM TDZ	• MS medium + 2.5 µM IBA • 90.0% survival was reported when transferred to the field condition	Giri and Tamta, 2011
17	*Mahonia jaunsarensis*	MS medium + 0.75 µM TDZ	• MS medium + 0.01 µM IBA • 55% survival percentage was reported in mixture of vermiculite + garden soil + farmyard manure (1:1:1) under greenhouse condition	Bisht et al:, 2023b

(*Continued*)

TABLE 2.1 (*Continued*)
Tissue Culture–Mediated Interventions in Threatened Himalayan Medicinal and Aromatic Plants

S. No	Name of Species	*In Vitro* Regeneration Condition	*In Vitro* Rooting and Hardening Strategies	References
18	*Angelica glauca*	MS medium + 8.0 µM BAP and 0.1 µM IAA	• MS medium + 0.5 µM IAA and 0.1 µM NAA • Survival rate of 72% was reported in mixture of soil, sand, and compost (1:1:1)	Rawat et al., 2018a
		MS medium + 2.0 µM BA and 2.0 µM NAA	• Half-strength MS medium • Under field conditions, mixture of soil and sand (1:1) reported 85% survival	Pandey et al., 2011
		MS medium + 2.0 µM BAP and 2.0 µM NAA	• MS medium +1.0 µM BAP and 2.0 µM NAA • Plantlets transferred to autoclaved mixture of sand, soil, and peat moss (1:1:1 ratio) reported 75% survival	Bisht et al., 2015
19	*Arnebia euchroma*	MS medium + 1 µM IBA + 2.5 µM BA	• Half-strength MS medium + 2.0 µM IBA • A sterile mixture of soil, sand, and peat (1:1:1) was reported to record 72% survival	Manjkhola et al., 2005
20	*Malaxis acuminata*	MS medium + 4 µM NAA	• MS medium + 3 µM NAA + 3 µM BA • 75% survival rate was reported in homogenous mixture of charcoal, chopped forest litter and coconut husk, and sand (1:1)	Arenmongla and Deb, 2012
21	*Aconitum ferox*	MS medium + 6 µM BAP	• Liquid MS medium + 3 µM IAA + 6 µM BAP • Peat moss and autoclaved soil (1:3) used in plantlet acclimatization reported 70% survival rate	Singh et al., 2020
22	*Rhodiola imbricata*	MS medium + 5 µM BAP and 1 µM TDZ	• MS medium + 0.5 µM NAA • Proportionate mixture of soil:perlite:vermiculite (1:1:1) reported 60% survival rate	Bhardwaj et al., 2018
23	*Lilium oxypetalum*	MS medium + 2.0 µM BA and 0.1 µM NAA	• MS medium with 4.5 (w/v) sucrose + 2.0 µM BA + 0.1 µM NAA • 66.7% survival rate was reported in soil:sand:vermiculite:perlite (1:1:2:2) mixture	Joshi and Dhar, 2009
24	*Heracleum candicans*	MS medium + 10 µM BA	• MS medium + 8 µM IBA • The survival of plantlets *ex vitro* was 70% soil:sand:peat (1:1:1)	Joshi et al., 2004

25	*Aconitum chasmanthum*	MS medium + 0.5 mg/L BAP	• MS basal media • Sterilized garden soil, vermicompost, and sand (1:1:1) resulted in 60% survival and multiplication	Rafiq et al., 2021
26	*Rheum emodi*	MS medium + 12 µM BAP	• MS medium + 12 µM BAP + 10 µM TDZ + 12 µM IBA • 35% survival was reported after 4 months of acclimatization in autoclaved sand:soil:peat (1:2:1) mixture	Tabin et al., 2018
27	*Rheum spiciforme*	MS medium + 12.5 µM BA and 0.5 µM NAA + 25 µM glutamine + 25 µM adenine sulfate	• Half-strength liquid MS medium + 2.5 µM NAA	Khan et al., 2022
28	*Aconitum violaceum*	MS medium + 0.5 µM BAP + 0.1 µM NAA	• MS medium + 0.5 µM NAA + 0.1 µM IAA • Soil, sand, and compost (1:1:1) demonstrated 85% survival rate	Rawat et al., 2013
29	*Atropa acuminata*	MS medium + 2.0 mg/L BAP	• MS medium + 0.5 mg/L IBA • 80% survival was achieved in compost only	Maqbool et al., 2016

phytohormones, and different media types have been used to initiate *in vitro* cultures. The details of some prominent species are summarized in Table 2.1.

2.2.1 Case Studies of Medicinal Plant Propagation by Tissue Culture Techniques

2.2.1.1 *Nardostachys jatamansi* (D. Don) DC

Nardostachys jatamansi (D. Don) DC (family Caprifoliaceae)—a critically endangered species (IUCN, 2024)—is highly prized, with an annual consumption of 528.11 MT in India (Goraya and Ved, 2017). The species finds applications in different traditional medicines and recently has received high impetus in modern therapeutics for treatments of disorders of the central nervous system (e.g., antispasmodic hysteria, epilepsy, stress, insomnia), cardiovascular functioning (e.g., heart palpitations and blood problems), gastrointestinal problems (e.g., spasms, stomachache), respiratory problems, urinary infections, and other integumentary systems (Wang et al., 2021). In light of this information and industrial prospects, the species is in the 20 most traded species list (Kaur et al., 2020), with the National Medicinal Plant Board, India, providing a 75% subsidy to encourage cultivators. Conversely, the plant is collected from the wild, leading to a serious impact on its natural populations. Here, tissue culture strategy has proven to be very helpful for scaled-up plantlet propagation desired for cultivation (Moyo et al., 2011). Different explants have been used to propagate *N. jatamansi* in an *in vitro* environment. Using petiole explants, Mathur (1992) achieved callusing and rhizogenesis on MS medium with 3.0 mg/L NAA and 0.25 mg/L Kn and medium containing 0.25–1.0 mg/L IAA or IBA. In the later experiment, Mathur (1993) reported somatic embryogenesis and plantlet regeneration upon a sequential subculture of the callus over 7 months. The callus was inoculated on varied phytohormone concentrations, i.e., by consistently decreasing NAA (16.1–1.34 μM) concentration and, in parallel, increasing the cytokinin kinetin (1.16–9.30) concentrations. Among these, 1.34 μM NAA and increased Kn (9.301 μM) concentrations proved effective in somatic embryo development and its conversion to plantlet formation (Mathur, 1993). In an attempt to develop an efficient propagation and hardening system, Bose et al. (2016) developed approaches such as direct and indirect shoot organogenesis for the plantlet establishment. The plantlets thus developed and subcultured and were tested for genetic fidelity. Explants (leaf and petiole) inoculated in MS medium complemented with 1.5 mg/L NAA and 1.0 mg/L meta-topolin started callusing and achieved 88% callusing at low temperature (13 ± 1° C) and a 14-hour photoperiod. The ability to regenerate plantlets by proliferative calli was tested in different concentrations of plant hormones. Among others, meta-topolin (1.0 mg/L) and NAA (0.5 mg/L) were reported to obtain the highest (92%) calli regeneration frequencies, and the maximum number of 25 shoots and 6.1 leaves per shoot were achieved when cultured at 13 ± 1 °C and a 14-hour photoperiod (Bose et al., 2016). Likewise, the use of meta-topolin was more prevalent in direct shoot organogenesis than cytokinins. MS medium with 1.0 mg/L meta-topolin produced the highest number of multiple shoots per explant (14.50 ± 0.8) after 6 weeks of culture and further developed adventitious shoots directly within 2 weeks of culture. Rooting frequency was enhanced (82.40%) with MS medium fortified with 1.0 mg/L of IAA. The rooted plantlet was acclimatized to the mixture (2:1:1) of autoclaved soil, sand, and leaf litter, and an 80% survival rate was reported (Bose et al., 2016). Recently, Dhiman et al. (2021) developed a cost-efficient, industry-ready micropropagation protocol for commercial exploitation. Rhizome segments with shoot buds (explants) were multiplied on a liquid MS medium fortified with Kn or thidiazuron (TDZ) wherein 0.5 mg/L Kn and 0.75 mg/L TDZ were reported to develop the maximum number (6.7 shoots per explant) of shoots after 30 days, while ½ MS with 1.0 mg/L IAA achieved 100% (3.07 roots/shoot) rooting (Dhiman et al., 2021). The study further achieved a 3% higher survival rate (i.e., 83%) when compared with the study (80%) of Bose et al. (2016), who used sand alone or in combination with soil (1:1). Also, the study rationalized the use of medium volume from 100 mL to 30 mL for micropropagation to reduce the total cost per plantlet.

2.2.1.2 *Swertia chirayita* (Roxb.) H. Karst.

Swertia chirayita (family Gentianaceae) is one of the 32 high-priority medicinal plants in trade in India (Joshi and Dhawan, 2007; Kala and Sajwan, 2007; Shukla et al., 2017). The species is native to sub-temperate regions of the Himalayas and grows at high altitudes between 1200 and 3000 m asl (Gaur, 1999). By distribution, it is spread all across the Indian Himalayan region, Nepal, and Bhutan, with the preferred habitat being moist shady slopes (Edwin and Chungath, 1988). The species is mentioned in several traditional systems of medicine, such as the AYUSH system and Tibetan system, as effective against epilepsy, high fever, hepatitis and liver disorders, ulcers, colds, skin diseases, etc. (Kirtikar and Basu, 1984; Pradhan and Badola, 2008; Kumar et al., 2018). Likewise, the pharmacological importance listed in the literature is hepato-protection and properties such as anthelmintic, antidiabetic, antifungal, antibacterial, anti-inflammatory, and antiviral (Banerjee et al., 2000; Verma et al., 2008; Das et al., 2012; Verma et al., 2013; Zhou et al., 2015). The medicinal importance of the species is accredited to the presence of several active constituents, i.e., glycosides, xanthones, triterpenoids, iridoids, and flavonoids (Negi and Singh, 2011; Li et al., 2017; Kaur et al., 2019a, b). Among these, some of the prominent bioactive compounds thoroughly studied for therapeutic importance include swertiamarin, swerchirin, amarogentin, and mangiferin; a xanthone C-glucoside; ursolic acid; and oleanolic acid (Joshi and Dhawan, 2005; Li et al., 2017). These medicinal values result in the demand for the species across the globe compounding every year. Goraya and Ved (2017) estimated the annual demand for *S. chirayita* in India alone to be over 404.7 MT. As with other medicinal plants of the region, and to meet the demand, wild-resource *S. chirayita* are being directly or indirectly exploited indiscriminately; hence, the Government of India has prioritized its conservation management (Goraya and Ved, 2017). In contrast, poor seed germination, lack of natural vegetative propagation, and high mortality rate limit commercial cultivation (Raina et al., 1994; Chaudhuri et al., 2007). To solve the problem, biotechnological approaches are used in the micropropagation of *S. chirayita*. For example, Wawrosch et al. (1999) reported micropropagation of *S. chirayita* using root explants (6- to 8 week-old plants). Explant-cultured 3 mM BAP for 3 weeks followed by hormone-free basal medium for another 3 weeks was more effective for adventitious shoot regeneration than shoot multiplication. The regenerated shoots dipped in NAA (15 ppm) solution followed by a 3- to 4-week culture on basal ½ MS was an efficient method for rooting (Wawrosch et al., 1999). Further, acclimatization at low temperatures of 17 °C was reported best for enhanced survival percent (i.e., 90% to ~100%), whereas at 25 °C, a high mortality rate was reported. Nodal explants/shoot apices of *S. chirayita* cultured in 0.5 mg/L BA + 1.0 mg/L GA_3 (Ahuja et al., 2003); 2 mg/L BA (Koul et al., 2009); 4.0 μM BA + 1.5 μM 2iP (Joshi and Dhawan, 2007); 0.44 μM BA + 4.65 μM KN (Chaudhuri et al., 2007); 1.0 mg/L BAP; and 0.5 mg/L IAA (Sharma et al., 2016) were found to be effective for rapid propagation. Likewise, an improved shoot proliferation and propagation strategy was developed with *in vivo* grown plants. Explants such as leaves, root tips, and shoot apices cultured in 2.22 μM BA + 11.6 μM KN + 0.5 μM NAA (Chaudhuri et al., 2008), 13.32 μM BA + 0.54 μM NAA (Wang et al., 2009), 1.0 mg/L BA + 0.1 mg/L KN (Balaraju et al., 2009), 1.0 mg/L 2,4-D and 0.5 mg/L 2,4-D + 0.5 mg/L BA (Balaraju et al., 2011), 0.5 mg/L BA + 1.0 mg/L GA_3 (Kumar and Chandra, 2014), 0.5 mg/L 2,4-D + 0.5 mg/L KN (Kumar and Chandra, 2014), 1.0 mg/L BA + 100 mg/L-adenine sulfate + 0.1 mg/L IAA (Kumar et al., 2014), and 13.32 μM BA + 0.90 μM 2,4-D (Pant et al., 2012) were effective methods for plant regeneration either by direct shoot proliferation or through somatic embryogenesis and/or through indirect organogenesis. Further, *in vitro* plantlet acclimatization was influenced by temperature, humidity, and bed mixture. A consortia of these are reported to have increased the survival percent from 80% (Chaudhuri et al., 2007), to 90% (Koul et al., 2009), and to 90–95% (Joshi and Dhawan, 2007).

2.2.1.3 *Paris polyphylla* Sm

Paris polyphylla Sm. (Vern. Satwa) is a perennial herbaceous species (in total, 27 *Paris* species are reported worldwide) of the family Melanthiaceae, found growing in humid places with shade and in soil that is rich in humus (Madhav et al., 2010). The species is native to the Himalayan region, with reported distribution from Pakistan, India (from Indian Himalayan region provinces), Nepal, Bhutan, China, Myanmar, Laos, Thailand, and Vietnam between 1400 and 3500 m in altitude (Kunwar et al., 2020). *P. polyphylla* has a rich history of use in various traditional and modern therapeutics and is widely used as expectorant, anthelmintic, and digestive and in the treatment of spams, gastric lesions, internal bleeding, and tumors (Lee and Lin, 1988; Matsuda et al., 2003; Wu et al., 2004; Fu et al., 2008; Uprety et al., 2010). The curative pharmacological properties of *P. polyphylla* have been attributed to steroidal saponins, dioscin, balanitin 7, polyphyllin D, flavonoids, glycosides, and ecdysteroids (Deng et al., 1999; Li et al., 2001; Gao et al., 2011; Tariq et al., 2018). Due to its enormous therapeutic value, there has been wholesale trading and exploitation of *P. polyphylla* rhizomes (Singh et al., 2017). As a result, there has been a continuous decrease in the natural population of *P. polyphylla.* In the latest International Union for Conservation of Nature (IUCN) (https://www.iucnredlist.org/) threatened species report, *P. polyphylla* is listed as a vulnerable species (IUCN, 2024). In nature, *P. polyphylla* reproduces through seeds or rhizome buds; however, by seeds, it encounters prolonged and double seed dormancy (over 18 months), poor viability, and low germination rate (Tariq et al., 2023; Wang et al., 2023). Hence, large-scale multiplication via seeds is less attractive. The vegetative method (rhizome splitting) is an alternative, simple, and cost-effective option; however, it has a shortcoming in terms of mass cultivation. Propagation protocols using tissue culture are reported, though there are not many papers in the database. Using immature green capsules of *P. polyphylla*, Raomai et al. (2014) excised immature zygotic embryos and cultured them on different plant growth media. Of the different media, MS medium was reported best for somatic embryogenesis (32.6%) and somatic embryos (SEs; 28.7) per explant. Further, on changing the solidifying agent from agar-agar to gelrite, the study reported a significant increase in primary SEs, i.e., from 32.6% to 40.7%. Culturing cotyledonary SEs on medium supplemented with ½ MS + 0.5 mg/L GA_3 was reported to germinate into a complete seedling; however, the seedlings were weak and could not transform into healthy plantlets. The seedlings had to be shifted to a medium (MS) supplemented with different concentrations of BAP and NAA, wherein 0.05 mg/L BAP and 0.1 mg/L NAA proved to be most effective in producing healthy plantlets with a 95% survival (Raomai et al., 2014). Further, Raomai et al. (2015) used the sliced basal and middle stem portion as an explant for plant regeneration. When transverse thin cell layer (tTCL) explants were cultured on different plant growth hormones (i.e., BAP, TDZ, and Kn), the frequency of mini-rhizome (MR) development was found to be highest (86.6%) on medium containing 0.5 mg/L BAP. Further, the same concentration is best suited for shoot bud development (5.6 ± 0.4) followed by TDZ (0.5 mg/L). The sprouted plantlets transferred to the soil were reported to show a 95% survival rate (Raomai et al., 2015). Recently, Tariq et al. (2023) used young rhizome buds cultured on ½ MS medium fortified with different plant growth regulator (i.e., BAP, NAA, and GA_3). Among the various concentrations, bud initiation and leaf emergence were reported within 2–6 weeks of inoculation on ½ MS medium containing BAP (0.5 μM), NAA (0.1 μM), and GA_3 (0.1 μM). On repeated culture, multiple shoots and root emergence were reported after 8–10 weeks (Tariq et al., 2023). The 12-week-old *in vitro*–developed plants were maintained under a greenhouse and reintroduced to their natural habitat/field demonstration sites.

2.2.1.4 *Podophyllum hexandrum* Royle

Podophyllum hexandrum (Himalayan mayapple or bankakri) belongs to the family Berberidaceae. The species is a 'critically endangered' perennial rhizomatous herb (IUCN, 2024), having a restricted distribution in alpine and sub-alpine meadows (2400–4500 m) (Singh et al., 2017). In the Indian Himalaya region (IHR), the reported locations and affinities of the species are Uttarakhand, Himachal Pradesh, Jammu and Kashmir, Ladakh, Sikkim, and Arunachal Pradesh. While at the outskirts of IHR, the species has a sparse distribution in Nepal, Bhutan, and the Tibetan plateau (Airi, 1997). Chemically, the roots and rhizomes of the species contain aryltetralin lignin and podophyllotoxin (~4–5% on dry weight), used for treating tumors, lung cancer, and testicular cancer by inhibiting topoisomerase II enzyme and microtubule assembly formation during mitosis, resulting in cytotoxicity (Fay and Ziegler, 1985; Giri and Lakshmi Narasu, 2000; Chaurasia et al., 2012). Habitat specificity, long seed dormancy (10 months), poor regeneration rate, dearth of cultivation practices, and exploitation (unsystematic and illegal harvesting) have whittled the natural population (Singh et al., 2021). An ancillary approach to *in vitro* multiplication of the species is via shoot multiplication and SE formation. *In vitro* plantlet development of *P. hexandrum* is reported through seed (Guo et al., 2012; Deb et al., 2018; Ahmad et al., 2021), callus (Sultan et al., 2006), and SE (Arumugam and Bhojwani, 1990; Nadeem et al., 2000; Rajesh et al., 2014) and direct organogenesis from rhizome buds (Chakraborty et al., 2010; Sharma et al., 2022). Arumugam and Bhojwani (1990) excised embryos from freshly collected seeds and cultured them on basal medium (BM) containing 2 μM BA. The excised embryos were reported to show 90% germination within 7 days of inoculation, as compared to decorated seeds (embryos surrounded by endosperm) with no germination even after 8 weeks. Moreover, embryos cultured on BM medium with BA (2 μM) and IAA (0.5 μM) were reported to develop into a friable and creamy callus. On subsequent culture with BM + 2.5 μM NAA, the callus formed dicotyledonous embryos, and multiple shoots were reported with BM + BA (1 μM) + GA (2.8 μM) within 4 weeks of culture (Arumugam and Bhojwani, 1990). Similar results were obtained by Nadeem et al. (2000), where MS medium fortified with 5.0 mM NAA + 0.5 mM BA induced embryogenesis and embryo development. Arumugam and Bhojwani (1990) did not report rooting of the shoots, whereas Nadeem et al. (2000) did report rooting of the shoots when cultured on MS medium supplemented with 0.5 mM IAA. Guo et al. (2012) developed a specific protocol for root development using a WPM medium fortified with 1.5 mg/L of IAA and 0.5 mg/L of NAA. Taking this forward, Chakraborty et al. (2010) suggested 100 μM IBA in ½ MS medium to be the best for rooting in *in vitro*–developed shoots, while Sharma et al. (2022) suggested the use of MS medium with BA (2.0 mg/L) and NAA (0.2 mg/L) for both shoot and root multiplication.

2.2.1.5 *Valeriana jatamansi* Jones ex Roxb.

Valeriana jatamansi (Syn. *V. wallichii*; family Caprifoliaceae) is a creeping and perennial rhizomatous herb. Attitudinally, the species is distributed between 1000 and 3000 m in the subtropical and temperate region of the Himalayas. *V. jatamansi* harbors over 150–200 chemical constituents that elicit significant pharmacological properties, including valerenic acids (sesquiterpenoid derivatives), valepotriates (epoxy iridoid esters), essential oils, alkaloids, flavonoids, and lignans (Singh et al., 2006; Bhatt et al., 2012; Rawat et al., 2017). The ethno-pharmacological accreditation to the species is due to antidyspeptic, antidepressant, anticonvulsive, antispasmodic, and antimicrobial properties (Xu et al., 2011; Dong et al., 2015; Vishwakarma et al., 2016). In the AYSUH system, the species is used against hysteria, epilepsy, leprosy, etc. (Dhiman et al., 2020). Due to these inherent characteristics, in India, the species is one of 178 highly traded medicinal plants (100 and 200 MT) per year (Goraya and Ved, 2017). Over time, different multiplication methodologies have been adopted for mass multiplication of *V. jatamansi* (Figure 2.4). For instance, conventionally, seed germination and vegetative propagation are used, although they are less appealing considering the high gestation time and difficulty in maintaining uniformity in the population's germplasm. To maintain uniformity among propagules, the species has been clonally propagated and tested for genetic intactness (Table 2.2). Chen et al. (2014) reported *in vitro* propagation and multiplication of *V. jatamansi* through indirect organogenesis. MS medium supplemented with 2,4-D (1.0–7.5 μM) was used for callus induction, and shoot development was obtained with 5.0 μM BA and 0.25 μM NAA. While using plant nodal segments, Purohit et al. (2015) reported an implicit and updated propagation protocol with genetic fidelity analysis. Of the various concentrations used, MS medium fortified with BAP (1.5 μM), NAA (0.5 μM), and GA_3 (0.1 μM) obtained the highest shooting and 100% rooting. In the earlier concentration, maximum shoot length (3.20 cm), shoot number (2.33), leaf number (15.33), highest average root number (27.5), and length of longest root (50.00 cm) were reported. Further genetic fidelity analysis revealed the clonally propagated plantlets were identical to the mother plant. Recently, Pandey et al. (2020a) used coconut water to propagate the species

FIGURE 2.4 Different stages during *in vitro* propagation of *Valeriana jatamansi*. Initiation of shoot multiplication from cotyledonary node explants to multiplication with profuse shoot formation (a–c), hardening and acclimatization under polyhouse (d), and cultivation of tissue culture–raised plants in hydroponic conditions (e).

with sodium nitroprusside. Medium supplemented with 10% coconut water and 15 μM sodium nitroprusside was reported to be effective (89%) in shoot and root multiplication. In another experiment, Pandey et al. (2020b) reported that a medium alone with 10% coconut water was much more effective in shoot multiplication and root development than when used with sodium nitroprusside. Micronutrients present in coconut water were accredited for maximum shoot length (6 cm), shoot number (13.0), root length (7.5 cm), and root numbers (19.6). Similarly, various other propagation strategies of *V. jatamansi* are reported, along with hardening and acclimatization strategies adopted (Table 2.2).

2.2.1.6 *Hedychium spicatum* Sm.

Hedychium spicatum (family Zingiberaceae), is another perennial rhizomatous herb, commonly known as 'kapoor kachari/vanhaldi' or 'spiked ginger lily'. The species is native to southeast Asia and found between 1000 m and 2800 m. The species has wide applications in traditional and modern medicine and is extensively used in the cosmetic and perfumery industries (Chopra et al., 1986). The ethnopharmacological importance of the species is used in treating fever, diarrhea, asthma, and bronchitis and as a hepato-protective agent (Bisht et al., 2006; Reddy et al., 2009; Rawat et al., 2018b). The essential oil extracted from the species is reported to have anti-inflammatory, antimicrobial, and antioxidant activities (Bisht et al., 2006). Goraya and Ved (2017) estimated the annual

TABLE 2.2
Strategies Adopted for *In Vitro* Propagation of *Valeriana jatamansi* under Different Time Scales

S. No.	Explants	Hormonal Concentration Used	Results	Hardening and Acclimatization	References
1	Nodal segments of *in vitro* grown plant	MS medium + Kn or BAP (5.0 mg/L) and IAA (1.0 mg/L)	• Kn (5.0 mg/L) or without IAA (1.0 mg/L) resulted in best w.r.t. optimal responses and overall plantlet development (i.e., shoot length and leaf and root number)	• The influence of environmental parameters was studied during the hardening process • Optimum temperature (20 ± 2 °C), relative humidity (60–70%), and 14-hour photoperiod in a growth chamber resulted in 100% establishment of plantlets	Mathur et al., 1988
2	Shoot buds	i. MS + 1.0 mg/L BA ii. MS + 0.75 mg/L NAA	• *In vitro* bud proliferation w.r.t. mean number of leaves per explant (7.9), mean number of shoots per explant (5.1), and mean shoot length (6.4 mm) were highlighted with MS + 1.0 mg/L BA • Maximum (98.6%) percent rooting was reported with MS + 0.75 mg/L NAA. Also, the tested concentration produced the highest root length (28.3 mm) and number of roots per explant (13.5).	Not reported	Kaur et al., 1999
3	Leaf, petiole, and rhizomes	i. MS + 0.5 mg/L 2,4-D ii. MS + 0.75 mg/L TDZ + 0.5 mg/L NAA iii. MS medium + NAA (0.1 mg/L)	• Maximum (86 %) callus induction frequency was reported [concentration (i)] • MS medium + 0.75 mg/L TDZ and 0.5 mg/L NAA reported the highest (88.6%) shoot regeneration frequency and produced the highest number of shoot buds (15.20 ± 0.20) which were capable of growing into single plants • MS medium fortified with 0.1 mg/L NAA produced the highest number of roots (17.30 ± 2.01) and root length (9.30 cm)	• Under greenhouse conditions in garden soil • Survival frequency of 100%	Das et al., 2013

(Continued)

TABLE 2.2 *(Continued)*
Strategies Adopted for *In Vitro* Propagation of *Valeriana jatamansi* under Different Time Scales

S. No.	Explants	Hormonal Concentration Used	Results	Hardening and Acclimatization	References
4	*In vitro* leaves	i. MS + 5.0 µM NAA ii. MS + 5.0 µM BA + 0.25 µM NAA iii. MS + NAA 0.5–5.0 µM or IAA (0.5 µM) and BA (0.5 µM)	• 100% callus induction (treatment i) and after 60 days adventitious shoots and roots were reported • Shoot proliferation coefficient (SPC) in MS medium with 5.0 µM BA and 0.25 µM NAA, the highest shoot proliferation coefficient—measured as the number of adventitious shoots after inoculation divided by the number of adventitious shoots before inoculation—of 7.8 was attained after 45 days of inoculation • 100% rooting was observed with basal MS media as well as MS media with NAA (0.5–5.0 µM), IAA (0.5 µM), and BA (0.5 µM)	• A ratio mixture of 1:1:1 of humus:vermiculite:peat moss in 90% of light (<500 mol $m^{-2}s^{-1}$) and temperatures (18–25 °C) were reported to have 94.5–99.7% survival rate	Chen et al., 2014
5	Nodal segment	i. MS + 1.5 µM BAP + 0.5 µM NAA + and 0.1 µM GA_3	• After 10 weeks of culture, the highest shooting w.r.t average shoot length (3.20 cm), shoot number (2.33), and leaf number (15.33) • 100% rooting was reported with all the tested concentrations; however, MS medium + 1.5 µM BAP + 0.5 µM NAA and 0.1 µM GA_3 resulted in the highest average root number (27.5) and length of longest root (50 .00 cm)	• A mixture of sterile soil:vermiculite:perlite in 1:1:1 (v/v/v), maintained under high humid conditions, i.e., 95–100% RH, 25 ± 1 °C, 18 hours/6 hours day/night cycle, and under 50% ambient light • After 2 months, 90% survival was reported	Purohit et al., 2015
6	Leaf explants from *in vitro* plants	i. MS medium + 1.5 mg/L NAA + 15 µM SNP ii. 10% coconut water + 15 µM sodium nitroprusside	• In treatment (i) 91.18% callusing was reported • Treatment (ii) was highlighted as the best (89.32%) for *in vitro* shoot multiplication, and the same medium was reported to induce root development	Not reported	Pandey et al., 2020a

7	Nodal segments excised from *in vitro*–grown plants	MS medium + 10% coconut water	• Among the tested growth regulators, supplementation of 10% coconut water resulted in maximum shoot length (6 cm), shoot number (13.0), root length (7.5 cm), and root numbers (19.6). Further, supplementation of coconut water also enhanced the formation of roots.	• Plants in coconut water (10%) acclimatized well and grew proficiently • Rhizome development was reported in these plants within 45 days of acclimatization	Pandey et al., 2020b
8	Shoot buds	MS medium + 2 mg/L BAP	MS medium supplemented with BAP (2 mg/L) produced maximum shoot bud regeneration with shoot numbers (6), shoot lengths (3 cm), root numbers (6.3), and root length (1.4 cm)	A mixture of coco peat:soil (3:1), used for hardening	Pandey and Pant, 2020
9	Rhizome	i. MS medium + 2 mg/L BAP + 1 mg/L NAA ii. MS medium +1.5 mg/L BAP + 1.5 mg/L NAA	• The fortified combination (i) resulted in the highest mean shoot length (3.71 cm) • Combination (ii) was reported best for the mean root number (2.57) • The concentration of higher BAP (1–2 mg/L) when used along with NAA (0–1 mg/L) was reported to promote shoot length, leaf number, and leaf growth	A mixture of soil, sand, and farmyard manure used was not reported; however, acclimatization under the greenhouse was reported	Nazir et al., 2022

FIGURE 2.5 *In vitro* propagation strategy for *Hedychium spicatum* (a–c) stages of multiplication, hardening and acclimatization in the plant growth chamber (d), and well-acclimatized plants transferred to the field for cultivation (e).

trade volume of *H. spicatum* to be between 200 and 500 MT, with the wild population being a key component of this demand. In nature, the plant propagates either by seed germination or by vegetative propagation. Under laboratory conditions, Bisht et al. (2015) reported that the seed germination percent could be enhanced to 95.83% by treating seeds with IAA (100 µg/mL) for 24 hours, with a mean germination time of 84.00 days, while only 37.50–60.42% germination was reported in non-treated seeds (Bisht et al., 2015). Alternatively, to meet the large-scale availability of quality material for mass cultivation, some tissue culture-based protocols are developed for the species (Figure 2.5). Koul et al. (2005) reported a high-frequency axillary shoot regeneration protocol (bud culture) of *H. spicatum* when augmented in MS medium containing BAP (0.5 mg/L) and IAA (1.0 mg/L). Prolonged cultivation on the same medium or transfer to a hormone-free medium was reported to induce root/rhizome formation. The plantlets thus developed and, when transferred to the field, showed 80–85% field survival percent (Koul et al., 2005). Using different combinations (BA, Kn, and TDZ) and concentrations of phytohormones, Giri and Tamta (2011) developed a complete micropropagation protocol. Shoot tips obtained from seedlings germinated *in vitro* showed reduced concentration of TDZ (1.0 µM) and exhibited the most favorable outcomes in terms of high shoot multiplication (83.33%), shoots per explant (3.86), and shoots per flask (19.33). Similarly, a lower concentration of IBA (2.5 µM) proved effective for rooting. Upon transfer to the field at an elevation of 1990 meters, well-rooted and robust plantlets displayed a remarkable survival rate of 90.0% (Giri and Tamta, 2011).

2.3 APPLICATION OF PLANT TISSUE CULTURE INTERVENTIONS

2.3.1 *In vitro* Secondary Metabolite Production

Besides producing quality plant materials, plant tissue culture offers an alternative strategy for the production of various secondary metabolites, given its precise control over synthesis across different culture conditions (Giri et al., 2022a). Additionally, it facilitates rapid production cycles and is not limited by seasonal factors or cultivation location. As a result, it has gained widespread recognition because the chemical synthesis of many of the compounds is either time consuming, economically infeasible, or the reaction mechanism of many is unknown (Bisht et al., 2023a). Over the last decade, studies have now focused on elucidating metabolites' biosynthesis machinery, biotransformation, and genetic engineering studies (Rastegari et al., 2019). Significant research and development work has been carried out to increase the secondary product scale from laboratory to scale-up level, e.g., taxol production in cell cultures (Lee et al., 2010). However, in numerous instances, the production of secondary metabolites has significantly undershot anticipated levels, failing to attain commercial viability in the process. This is so because in *in vitro* cultures, metabolite production depends on parameters such as cell lines, culture strategies, media and elicitator type, etc. (Giri et al., 2022b). Among the different techniques, the elicitation of cell suspension culture has emerged as a straightforward and economically efficient approach for generating secondary metabolites (Rawat et al., 2020; Bisht et al., 2023a). Elicitation is a crucial factor in the transcript expression of the many genes involved in synthesizing secondary metabolites (Singh and Kumaria, 2021). Several commonly utilized precursors include jasmonic acid (JA), methyl jasmonate (MeJA), 2-hydroxyethyl jasmonate, salicylic acid (SA), acetylsalicylic acid (ASA), yeast extract (YE), chitosan, and sodium nitroprusside (SNP) (Giri et al. 2022b). Using these precursors, metabolite production in some of the Himalayan species is reported (Table 2.3). Bisht et al. (2023a) employed varying concentrations of SA and MeJA ranging from 50 μM to 200 μM to elicit *Berberis lycium* callus. The study further revealed that lower concentrations of SA and MeJA (specifically 50 μM and 100 μM, respectively) were significantly effective ($p < 0.05$) for biomass accumulation, alkaloid, and polyphenol production. Similar results are reported for *Habenaria edgeworthii*. In a study by Giri et al. (2012a), the effect of MeJA (0–10.0 mg L^{-1}) on *H. edgeworthii* callus culture was investigated. Higher MeJA concentrations were correlated with reduced fresh and dry biomass, while the phenolic compound increased more significantly than the control (Giri et al., 2012b). Likewise, Kumar et al. (2020) conducted expression profiling of steroidal alkaloids in bulbs of *Fritillaria roylei*. This study marked the first exploration of the species for such purposes, revealing the potential of callus culture as a significant source of steroidal alkaloids, notably sipeimine and peimine. Sharma et al. (2014b) delved into unraveling the intricate biosynthesis mechanism of aconitine within *in vitro* root cultures of *Aconitum balfourii*. Their findings reveal a significant rise in aconitine percentage when exposed to acetyl CoA, a precursor of the mevalonate pathway, compared to sodium pyruvate, a precursor of the non-mevalonate pathway. Similarly, several Himalayan species have been investigated for cell culture studies including *N. jatamansi* (D.Don) DC. (Rawat et al., 2020), *Rhodiola imbricate* (Rattan and Warghat, 2023), *Mahonia jaunsarensis* (Bisht et al., 2023b), *Picrorhiza kurroa* (Partap et al., 2022), and *A. balfourii* (Pandey et al., 2004), etc. (Table 2.3). In general, a major limitation of cell culture studies is not being able to breach the laboratory boundary and culminate into a scale-up production. Here, only a limited number of studies are reported, e.g., Chattopadhyay et al. (2002) investigated the synthesis of podophyllotoxin in a 3-liter stirred tank bioreactor, employing a low-shear steric impeller. The study reported an increase in volumetric productivity (27%), i.e., biomass (6.5 g/L) and podophyllotoxin (4.26 mg/L) accumulation, compared to that of shake flask cultivation.

TABLE 2.3
***In Vitro* Secondary Metabolite Production from Selected Himalayan Medicinal Plants**

S. No.	Plant Species	Bioactive Compound	Explant Used/Culture Medium/Condition	References
1	*Berberis lycium*	Gallic acid (4.19 mg/g), catechin (33.66 mg/g), *m*-coumaric acid (5.14 mg/g), vanillin (5.8 mg/g), phloridzin (9.19 mg/g), *p*-coumaric acid (3.6 mg/g), salicylic acid (16.59 mg/g), naringin (7.08 mg/g), *o*-coumaric acid (11.67 mg/g), berberine (10.41 mg/g), palmatine (0.68 mg/g)	*In vitro*–grown leaves/WPM medium + 1 µM 2,4-D + 0.5 µM BAP + 50 µM salicylic acid/photoperiod 16/8 hours light/dark	Bisht et al., 2023a
2	*Arnebia benthamii*	Shikonin derivatives (669.17 ± 26.71 mg DW/L)	Leaf/MS medium + 10 µM BAP and 5 µM IBA/temperature 25 ± 2 °C and relative humidity (60–65%)	Kumar et al., 2023
3	*Paris polyphylla*	Dioscin (5.43%) and diosgenin (21.09%)	Rhizome/MS medium + 50 µM MeJA/25 ± 2 °C for 16/8 hours light/dark photoperiod	Rawat et al., 2023
4	*Mahonia jaunsarensis*	Berberine 3.54 ± 0.00 mg/g FW and palmatine 0.58 ± 0.00 mg/g FW, total phenol (36.65 ± 0.45 mg GAE/g FW), tannin (10.35 ± 0.15 mg TAE/g FW), antioxidant activity such as ABTS (19.45 ± 0.75 mM AAE/100 g FW), FRAP (3.63 ± 0.11 mM AAE/100 g FW)	Leaf explants/MS medium + 0.75 µM TDZ and 0.01 µM IBA/ photoperiod 16/8 hours light/dark and at 25 ± 2 °C	Bisht et al., 2023b
5	*Rhodiola imbricata*	Salidroside (7.71 mg/g DW) and rosavin (0.99 mg/g DW)	50 µM salicylic acid (SA) elicitation	Rattan and Warghat, 2023
6	*Picrorhiza kurroa*	Picrosides (PI, PII, PIII; 1.887 mg/g DW), precursors (2.118 mg/g DW), *p*-hydroxyacetophenone (0.316 mg/g DW), and 4-hydroxy acetophenone (0.032 mg/g DW)	Rhizome explant	Partap et al., 2022
7	*Rhodiola imbricata*	Salidroside content (26.62 mg/g DW)	Root cell/MS medium + 0.5 mg/L TDZ + 1 mg/L NAA +3 mM tyrospol/photoperiod (16/8 hours light/dark, 70 µmol/m²/s)	Rattan et al., 2022
8	*Fritillaria roylei*	Sipeimine 1.98 µg/g	Bulb scales/MS medium + 0.5 mg/L TDZ +2.0 mg/L picloram	Kumar et al., 2020
9	*Rhodiola imbricata*	Salidroside (3.96 mg/g), tyrosol (0.99 mg/g), rosavin (1.62 mg/g), rosarin (0.92 mg/g), *p*-coumaric acid (0.88 mg/g)	MS + 0.5 mg/L TDZ and 1 mg/L NAA/ Methanol (70%) and sonicated for 30 minutes	Rattan et al., 2020
10	*Hypericum perforatum*	Flavonoid	Shoot, root/liquid MS + 1.2 mg/L BA + 0.02 mg/L NAA/25 °C, photoperiod 16/8 hours light/dark	Mohammed et al., 2019

11	*Nardostachys jatamansi*	Phenolics (9.7 mg GAE/g DW), flavonoids (12.8 mg QE/g DW), tannin (1.71 mg TAE/g DW)	Leaf derived callus/$^1/_2$ MS + 23.1 µM NAA + 4 µM BA + 4.16 µM Kn + 6.0 µM MeJA/methanol (80%) + rotatory shaker (24 hours)	Rawat et al., 2020
12	*Origanum vulgare*	3-hydroxy benzoic acid (0.96 mg/g), 3-hydroxycinnamic acid (0.05 mg/g), catechin (2.48 mg/g), chlorogenic acid (1.93 mg/g), ellagic acid (0.10 mg/g), ferulic acid (0.22 mg/g), gallic acid (0.16 mg/g), *p*-coumaric acid (0.08 mg/g), phloridzin (1.03 mg/g), vanillic acid (0.17 mg/g)	½ MS + 50 µM IBA/methanol (80%) and homogenized under ultrasonicator for 5 minutes	Pandey et al., 2019
13	*Rhodiola imbricata*	Phenol and flavonoid	Leaves/MS + 5 mg/L NAA and 5 mg/L BAP/blue light	Kapoor et al., 2018
14	*Aconitum balfourii*	Aconitine 0.012 % DW	*In vitro*–grown roots/MS + 1.43 µM IAA and1.23 µM IBA + 10 µM acetyl CoA treatment/25 °C, 65.0 % RH	Sharma et al., 2014
15	*Picrorhiza kurroa*	Picrotin (2.47 µg/g DW), picrotoxinin (45.2 µg/g DW)	Runners/MS liquid medium + yeast extract (0.5 mg/mL)/25 °C and 16/8 hours light/dark regimen with agitation (80 rpm)	Rawat et al., 2014
16	*Habenaria edgeworthii*	Gallic acid (143.63/100 g), catechin 2.98 mg/100 g), hydroxy benzoic acid (5.50 mg/100 g)	1/2 MS + 3.0 µM BA/methanol (80%); rotatory shaker (12 hours) + sonication (10 minutes)	Giri et al., 2012a
17	*Aconitum balfourii*	Aconitine 0.01 % DW, pseudaconitine 0.40 % DW	Tissue culture–raised plant/MS + 4.5 µM BA and NAA 1.4 µM/25 °C in 16.hours light	Pandey et al., 2004
18	*Podophyllum hexandrum*	Podophyllotoxin 4.9 mg/L	*In vitro*–grown seedling/MS + 2 mg/LIAA and 5 g/L activated charcoal + 10 g/L polyvinylpyrolidone (PVP)/20 °C and 16/8 hours light/dark regimen	Chattopadhyay et al., 2001
19	*Podophyllum hexandrum*	Podophyllotoxin of 4.9 mg/L	Seed/MS+ glucose (30 g/L) + PVP (5 g/L)/dark	Chattopadhyay et al., 2002
20	*Podophyllum hexandrum*	Podophyllotoxin 056% DW	Root/B5 medium + 4% sucrose and 4 mg/L NAA + 3 mM coniferyl alcohol/26 °C in the dark	Woerdenbag et al., 1990
21	*Podophyllum hexandrum*	Podophyllotoxin 1.88% DW	Rhizome/B5 medium + 4% sucrose and 4 mg/L NAA + 2.1 mM coniferin/26 °C dark-grown cultures	Van Uden et al., 1990

2.3.2 Hairy Root Culture or Genetic Transformation

Investigating the molecular causes of hairy root diseases and crown gall in plants, studies on horizontal gene transfer facilitated by the genus Agrobacterium (family Rhizobiaceae) are an area of active research (Mehrotra et al., 2015). An overview of the members of the genus Agrobacterium are gram-negative, rod-shaped, and pathogenic soil bacteria (Gelvin, 2003). Naturally, this bacterium causes hairy roots to emerge from the wounded site in higher plants through the process mediated by the transfer of a small (~25 base pairs) segment of transfer DNA (T-DNA) from the bacterium cell and translocated into the plant genome (Chandra, 2012). The advent of the hairy root development technique has revolutionized the approach to *in vitro* secondary metabolite production, ensuring enhanced and consistent metabolite yields (Mishra et al., 2011). Currently, researchers have explored hairy root induction and metabolite production in over 50 families spanning more than 150 genera of angiosperms (Mehrotra et al., 2015). However, there are not many published studies on the Himalayan medicinal plants. Some of the prominent studies are in *Picrorhiza kurroa*, an important and threatened Himalayan medicinal herb that was briefly studied by Mishra et al. (2011) for producing picrotin and picrotoxinin. *In vitro*–grown shoot tips (wounded) of *P. kurroa* co-cultivated with *Agrobacterium rhizogenes* strains (A 4 and PAT 405) on a hormone-free semi-solid (½) MS medium with 3% sucrose was reported to develop hairy roots after 14–21 days of inoculation. After 8 weeks of culture, the study reported the A4 strain as promising for inducing hairy roots, and picrotoxin (8.8 μg/g DW) and picrotoxinin (47.1 μg/g DW) yield. In another study, Verma et al. (2007) used the LBA9402 strain and reported 66.7% hairy root transformation frequency. Of the different hairy root clones studied, the 14P clone biosynthetic potential in terms of picroside-I and kutkoside content was reported to be three times higher as compared to that of the control (Verma et al., 2007). Likewise, swerchirin and 1,2,5,6-tetrahydroxyxanthone constitute important biomarker compounds in *S. chirayita*. In a study by Mahendran et al. (2022), when infected with *Rhizobium rhizogenes* LBA 9402 strain and stimulated with SA, MeJA, and SNP, swerchirin and 1,2,5,6-tetrahydroxyxanthone content exhibited a 1.80- to 6.0-fold increase in yield. Similarly, *Valeriana wallichii* transformation with A4 and LBA 9402 strains was reported to enhance valepotriate content by 2–3.3 times compared to the control (Banerjee et al., 1998). In *Aconitum heterophyllum* transformed culture, Giri et al. (1997) reported aconite yield enhancement by 3.75 times (2.96%), while in *Podophyllum hexandrum*, a three-fold increase (from 0.2% [control] to 0.7% [transformed cultures]) in podophyllotoxin content was reported (Giri et al., 2001).

2.3.3 Synthetic Seed Production

Synthetic seeds consist of encapsulated plant tissues such as somatic embryos, shoot tips, and protocorm-like bodies. These encapsulated tissues can be cultured or grown into a fully developed plant (Reddy et al., 2012; Qahtan et al., 2019). The advantage of synthetic seeds is their ability to serve as genetically identical materials (Gantait et al., 2015). Additionally, the produced seeds serve as inexpensive planting material, are simple to manage, offer both short- and long-term storage capabilities, and remain immune to seed-borne diseases and insect damage, simplifying the transportation and exchange of germplasm (Rai et al., 2009). Producing synthetic seeds holds utmost importance for species, particularly rare, endangered, and threatened (RET) species that face numerous obstacles in their seed development, reproduction, prolonged seed dormancy, seed dispersal, and successful establishment within their natural habitats (Saxena et al., 2019). The extension of synthetic seed technology to RET species has achieved significant success by encapsulating both embryogenic and non-embryogenic explants. As shown in Table 2.4, Kumar and Chandra (2014) devised a protocol for synthetic seed production of *S. chirayita* using somatic embryos and sodium alginate (4%) encapsulation. Following a brief incubation period, the regeneration percentage was assessed. The developed synthetic seeds, when inoculated in MS medium fortified with 0.5 mg/L NAA and 1 mg/L BA, exhibited an 84% germination rate and an impressive 80% survival rate when grown

TABLE 2.4
Synthetic Seed Production and Their Regeneration Percentage from Selected Himalayan Medicinal Plants

S. No.	Scientific Names	Family	Explants	Synthetic Medium	Regeneration (%)	References
1	*Arnebia euchroma*	Boraginaceae	Cotyledonary-stage embryos	3% sodium alginate + 100 mMCa(NO_3)$2H_2O$	60.6	Manjkhola et al., 2005
2	*Swertia chirayita*	Gentianaceae	Torpedo-stage somatic embryos	4% sodium alginate + 100 mM $CaCl_2$. $2H_2O$ solution	84	Kumar and Chandra, 2014
3	*Dendrobium nobile*	Orchidaceae	Protocorm-like body (PLBs)	3% sodium alginate + 7.5% mannitol + 7.5% sucrose + 100 mM $CaCl_2$. $2H_2O$ solution	78.2	Mohanty et al., 2013
4	*Picrorhiza kurrooa*	Plantaginaceae	Shoot tip, nodal segment	3% sodium alginate + 272.2 mM$CaCl_2$. $2H_2O$ solution	21.43	Mishra et al., 2011
5	*Picrorhiza kurrooa*	Plantaginaceae	Hairy root	Sodium alginate + $CaCl_2$. solution	73	Rawat et al. 2013
6	*Hedychium spicatum*	Zingiberaceae	Somatic embryos	2% sodium alginate and 100 mM calcium chloride	30	Giri and Tamta, 2013
7	*Angelica glauca*	Apiaceae	Somatic embryos	3% sodium alginate and 100 µM calcium nitrate	–	Bisht et al., 2015

in a mixture of soilrite and sand at a ratio of 1:2 (Table 2.4). Furthermore, Manjkhola et al. (2005) reported synthetic seed production of *Arnebia euchroma*, where early-stage cotyledonary embryos were encapsulated. The resultant synthetic seeds displayed a 60.6% germination rate in MS medium and a 72% survival rate when transferred to *ex vitro* conditions (Manjkhola et al., 2005). Likewise, protocorm-like bodies (PLBs) are encapsulated in orchid species. Mohanty et al. (2013) demonstrated the encapsulation of PLBs in 3% sodium alginate. Upon storage, the encapsulated artificial seeds showed 78.2% regeneration on ½ MS medium supplemented with 0.1 mg/L NAA and 1 mg/L BAP. Other than these, hairy root segments (3–4 mm) of synthetic seeds are reported by Rawat et al. (2013). Hairy root encapsulation resulted in adventitious shoot emergence (73%) on MS medium + 0.1 µM BAP. Further, the regenerated plantlets displayed a very high (85%) survival percentage and exhibited a morphotype comparable to non-transformed plants.

2.4 CONCLUSION AND FUTURE PERSPECTIVES

Plant tissue culture is a reliable and successful biotechnological approach for rapid multiplication of plant propagules (particularly for threatened and endangered plants), cryopreservation, breeding, and secondary metabolite production. During the past several decades, considerable progress has been made in the tools and techniques, media compositions, and biotransformation studies. Although to date, many optimized protocols have been developed, unfortunately, very few of them have been used commercially. The reason for this may be due to the higher cost of micropropagated plantlets. Also, many of the protocols developed are restricted to laboratory conditions or to greenhouse conditions, with a handful of successful case studies that span from laboratory to land (either cultivation or to facilitate rehabilitation). Here, the shortfall is in optimizing the hardening process. Among others, the use of plant growth–promoting endophytes is one of the important areas that need special attention. It becomes imminent for Himalayan species because a majority of the high-altitude species are habitat specific and are confined to narrow patches. Therefore, the study holds utmost importance and could facilitate a high establishment rate. Another key issue in *in vitro* propagation is selecting low-frequency explants for regeneration; however, this can be addressed by using explants with strong morphogenetic potential. Furthermore, because *in vitro*–produced plantlets are known to cause variability, regular screening and monitoring of genetic stability or somaclonal variation could be used to achieve elite (true-to-type) explants. Recent advancements in plant biotechnology include the use of multi-omics approaches, i.e., genomics, proteomics, and, more recently, metabolomics. These versatile tools have wide-ranging applications and can accelerate the discovery, isolation, and characterization of genes, resulting in the conferral of novel agronomic traits.

ACKNOWLEDGEMENT

The authors are thankful to the Director, G. B. Pant National Institute of Himalayan Environment (NIHE) for providing the necessary facilities and encouragement.

DECLARATION

Financial support from the National Mission for Sustaining the Himalayan Ecosystem, Task Force 3 (Phase-II) 'Forest Resources and Plant Biodiversity' (Department of Science and Technology, Govt. of India) and In-house-04 project of GBP-NIHE is gratefully acknowledged.

REFERENCES

Ahmad, R., Sharma, S., Kumar, V., Verma, C., & Bari, D. G. (2021). *In vitro* propagation of *Podophyllum hexandrum*. *International Journal of Pharmaceutical Research*, *6*, 169–172.

Ahuja, A., Koul, S., Kaul, B. L., Verma, N. K., Kaul, M. K., Raina, R. K., et al. (2003). Media Composition for Faster Propagation of *Swertia chirayita*. US Patent No. W0 03/045132 A1.

Airi, S. (1997). Population studies on *Podophyllum hexandrum* Royle-a dwindling, medicinal plant of the Himalaya. *Plant Genetic Resource Newsletter*, *110*, 29–34.

Arenmongla, T., & Deb, C. R. (2012). Germination of immature embryos and multiplication of *Malaxis acuminata* D. Don, an endangered therapeutically important orchid, by asymbiotic culture *in vitro*. *Indian Journal of Biotechnology*, *11*, 464–469.

Arora, K., Rai, M. K., & Sharma, A. K. (2022). Tissue culture mediated biotechnological interventions in medicinal trees: Recent progress. *Plant Cell, Tissue and Organ Culture (PCTOC)*, *150*(2), 267–287.

Arumugam, N., & Bhojwani, S. S. (1990). Somatic embryogenesis in tissue cultures of *Podophyllum hexandrum*. *Canadian Journal of Botany*, *68*(3), 487–491.

Balaraju, K., Agastian, P., & Ignacimuthu, S. (2009). Micropropagation of *Swertia chirata* Buch.-Hams. ex Wall.: A critically endangered medicinal herb. *ActaPhysiologiaePlantarum*, *31*, 487–494. https://doi.org/10.1007/s11738-008-0257-0

Balaraju, K., Saravanan, S., Agastian, P., & Ignacimuthu, S. (2011). A rapid system for micropropagation of *Swertia chirata* Buch-Ham. ex Wall.: An endangered medicinal herb via direct somatic embryogenesis. *ActaPhysiologiaePlantarum*, *33*, 1123–1133. https://doi.org/10.1007/s11738-010-0640-5

Banerjee, S., Rahman, L., Uniyal, G. C., & Ahuja, P. S. (1998). Enhanced production of valepotriates by *Agrobacterium rhizogenes* induced hairy root cultures of *Valeriana wallichii* DC. *Plant Science*, *131*(2), 203–208.

Banerjee, S., Sur, T. K., Mandal, S., Das, P. C., & Sikdar, S. (2000). Assessment of the anti-inflammatory effects of *Swertia chirata* in acute and chronic experimental models in male albino rats. *Indian Journal of Pharmacology*, *32*(1), 21–24.

Bhardwaj, A. K., Singh, B., Kaur, K., Roshan, P., Sharma, A., Dolker, D., & Chaurasia, O. P. (2018). *In vitro* propagation, clonal fidelity and phytochemical analysis of *Rhodiola imbricata* Edgew: A rare trans-Himalayan medicinal plant. *Plant Cell, Tissue and Organ Culture (PCTOC)*, *135*, 499–513.

Bhatt, I. D., Dauthal, P., Rawat, S., Gaira, K. S., Jugran, A., Rawal, R. S., & Dhar, U. (2012). Characterization of essential oil composition, phenolic content, and antioxidant properties in wild and planted individuals of *Valeriana jatamansi* Jones. *Scientia Horticulturae*, *136*, 61–68.

Bhojwani, S. S., & Dantu, P. K. (2013). *Plant Tissue Culture: An Introductory Text* (Vol. *318*). India: Springer.

Bisht, A., Singh, L., Pandey, A., Pandey, V., Dasila, K., Bhatt, I. D., & Pande, V. (2023a). Elicitor-induced phytochemicals production in *Berberis lycium* Royle. *Industrial Crops and Products, 199*, 116735.

Bisht, A., Singh, L., Singh, B., Bisht, M., Pandey, A., Bhatt, I. D., & Pande, V. (2023b). *In vitro* propagation of endemic species *Mahonia Jaunsarensis* Ahrendt through callus culture. *Applied Biochemistry and Biotechnology, 196*(1), 113–128.

Bisht, A. K., Bhatt, A., Dhar, U., Bhatt, A., & Dhar, U. (2015). Note on somatic embryogenesis and synthetic seed production in *Angelica glauca*: A valuable medicinal plant of Himalaya. *Journal of Medicinal Plants Research, 9*(12), 419–425.

Bisht, G. S., Awasthi, A. K., & Dhole, T. N. (2006). Antimicrobial activity of *Hedychium spicatum*. *Fitoterapia, 77*(3), 240–242.

Bisht, S., Bisht, N. S., & Bhandari, S. (2012). *In vitro* micropropagation in *Polygonatum verticillatum* (L.) All. an important threatened medicinal herb of Northern India. *Physiology and Molecular Biology of Plants, 18*, 89–93.

Bose, B., Kumaria, S., Choudhury, H., & Tandon, P. (2016). Assessment of genetic homogeneity and analysis of phytomedicinal potential in micropropagated plants of *Nardostachys jatamansi*, a critically endangered, medicinal plant of alpine Himalayas. *Plant Cell, Tissue and Organ Culture (PCTOC), 124*, 331–349.

Chakraborty, A., Bhattacharya, D., Ghanta, S., & Chattopadhyay, S. (2010). An efficient protocol for *in vitro* regeneration of *Podophyllum hexandrum*, a critically endangered medicinal plant. *Indian Journal of Biotechnology, 9*, 217–220.

Chandra, B., Palni, L. M. S., & Nandi, S. K. (2004). Micropropagation of *Picrorhiza kurrooa* Royle ex Benth., an endangered alpine herb, using cotyledonary node and shoot tip explants. *Phytomorphology, 54*(3&4), 303–316.

Chandra, S. (2012). Natural plant genetic engineer *Agrobacterium rhizogenes*: Role of T-DNA in plant secondary metabolism. *Biotechnology Letters, 34*, 407–415.

Chattopadhyay, S., Srivastava, A. K., Bhojwani, S. S., & Bisaria, V. S. (2001). Development of suspension culture of *Podophyllum hexandrum* for production of podophyllotoxin. *Biotechnology Letters, 23*, 2063–2066.

Chattopadhyay, S., Srivastava, A. K., Bhojwani, S. S., & Bisaria, V. S. (2002). Production of podophyllotoxin by plant cell cultures of *Podophyllum hexandrum* in bioreactor. *Journal of Bioscience and Bioengineering, 93*(2), 215–220.

Chaudhuri, R. K., Pal, A., & Jha, T. B. (2007). Production of genetically uniform plants from nodal explants of *Swertia chirata* Buch.-Ham. ex Wall.—An endangered medicinal herb. *In Vitro Cellular & Developmental Biology-Plant, 43*, 467–472.

Chaudhuri, R. K., Pal, A., & Jha, T. B. (2008). Conservation of *Swertia chirata* through direct shoot multiplication from leaf explants. *Plant Biotechnology Reports, 2*, 213–218.

Chaurasia, O. P., Ballabh, B., Tayade, A., Kumar, R., Kumar, G. P., & Singh, S. B. (2012). *Podophyllum* L.: An endergered and anticancerous medicinal plant–An overview. *Indian Journal of Traditional Knowledge, 11*(2), 234–241.

Chen, R., Zhang, M., Lü, J., Zhang, X., da Silva, J. A. T., & Ma, G. (2014). Shoot organogenesis and somatic embryogenesis from leaf explants of *Valeriana jatamansi* Jones. *Scientia Horticulturae, 165*, 392–397.

Chopra, R. N., Nayar, S. L., & Chopra, I. C. (1986). *Glossary of Indian Medicinal Plants (including the supplement)*. New Delhi, India: Council of Scientific & Industrial Research.

Cohen, J. I., Williams, J. T., Plucknett, D. L., & Shands, H. (1991). Ex situ conservation of plant genetic resources: Global development and environmental concerns. *Science, 253*(5022), 866–872.

Das, J., Mao, A. A., & Handique, P. J. (2013). Callus-mediated organogenesis and effect of growth regulators on production of different valepotriates in Indian valerian (*Valeriana jatamansi* Jones.). *ActaPhysiologiaePlantarum, 35*, 55–63.

Das, S. C., Bhadra, S., Roy, S., Saha, S. K., Islam, M. S., & Bachar, S. C. (2012). Analgesic and anti-inflammatory activities of ethanolic root extract of *Swertia chirata* (Gentianaceae). *Jordan Journal* of *Biological Sciences, 5*(1), 31–36.

Deb, C. R., Rout, G. R., Mao, A. A., Nandi, S. K., Singha, R. N., Vijayan, D., & Swain, D. (2018). *In vitro* propagation of some threatened plant species of India. *Current Science, 114*(3), 567–575.

Deng, S., Yu, B., Hui, Y., Yu, H., & Han, X. (1999). Synthesis of three diosgenylsaponins: Dioscin, polyphyllin D, and balanitin 7. *Carbohydrate Research, 317*(1–4), 53–62.

Dhar, U., & Joshi, M. (2005). Efficient plant regeneration protocol through callus for *Saussureaobvallata* (DC.) Edgew. (Asteraceae): Effect of explant type, age and plant growth regulators. *Plant Cell Reports, 24*, 195–200.

Dhiman, B., Sharma, P., & Pal, P. K. (2020). Biology, chemical diversity, agronomy, conservation and industrial importance of *Valeriana jatamansi*: A natural sedative. *Journal of Applied Research on Medicinal and Aromatic Plants, 16*, 100243.

Dhiman, N., Devi, K., & Bhattacharya, A. (2021). Development of low cost micropropagation protocol for *Nardostachys jatamansi*: A critically endangered medicinal herb of Himalayas. *South African Journal of Botany, 140*, 468–477.

Dong, F. W., Yang, L., Wu, Z. K., Zi, C. T., Yang, D., Luo, H. R., & Hu, J. M. (2015). Iridoids and sesquiterpenoids from the roots of *Valeriana jatamansi* Jones. *Fitoterapia, 102*, 27–34.

Edwin, R., & Chungath, J. I. (1988). Studies in *Swertia chirata*. *Indian Drugs, 25*, 143–146.

Fay, D. A., & Ziegler, H. W. (1985). Botanical source differentiation of *Podophyllum* resin by HPLC. *Journal of Liquid Chromatography, 8*(8), 1501–1505.

Fu, Y. L., Yu, Z. Y., Tang, X. M., Zhao, Y., Yuan, X. L., Wang, S., & Cong, Y. W. (2008). Pennogenin glycosides with a spirostanol structure are strong platelet agonists: Structural requirement for activity and mode of platelet agonist synergism. *Journal of Thrombosis and Haemostasis, 6*(3), 524–533.

Gantait, S., Kundu, S., Ali, N., & Sahu, N. C. (2015). Synthetic seed production of medicinal plants: A review on influence of explants, encapsulation agent and matrix. *ActaPhysiologiaePlantarum, 37*, 1–12.

Gao, L. L., Li, F. R., Jiao, P., Yao, S. T., Sang, H., & Si, Y. H. (2011). Apoptosis of human ovarian cancer cells induced by *Paris chinensis*dioscin via a Ca (2+)-mediated mitochondrion pathway. *Asian Pacific Journal of Cancer Prevention, 12*(5), 1361–1366.

Gaur, R. D. (1999). *Flora of the District Garhwal, North West Himalaya*. Transmedia.

Gautam, N., & Bhattacharya, A. (2021). Molecular marker based assessment of genetic homogeneity within the *in vitro* regenerated plants of *Crocus sativus* L.–A globally important high value spice crop. *South African Journal of Botany, 140*, 461–467.

Gelvin, S. B. (2003). *Agrobacterium*-mediated plant transformation: The biology behind the "gene-jockeying" tool. *Microbiology and Molecular Biology Reviews, 67*(1), 16–37.

Giri, A., Banerjee, S., Ahuja, P. S., & Giri, C. C. (1997). Production of hairy roots in *Aconitum heterophyllum* Wall. Using *Agrobacterium rhizogenes*. *In Vitro Cellular & Developmental Biology-Plant, 33*, 280–284.

Giri, A., Giri, C. C., Dhingra, V., & Narasu, M. L. (2001). Enhanced podophyllotoxin production from *Agrobacterium rhizogenes* transformed cultures of *Podophyllum hexandrum*. *Natural Product Letters, 15*(4), 229–235.

Giri, A., & Lakshmi Narasu, M. (2000). Production of podophyllotoxin from *Podophyllum hexandrum*: A potential natural product for clinically useful anticancer drugs. *Cytotechnology, 34*, 17–26.

Giri, D., & Tamta, S. (2011). Effect of plant growth regulators (PGRs) on micropropagation of a vulnerable and high value medicinal plant *Hedychium spicatum*. *African Journal of Biotechnology, 10*(20), 4040–4045.

Giri, D., & Tamta, S. (2013). Induction of somatic embryogenesis in vulnerable medicinal plant *Hedychium spicatum* Buch-Ham ex Smith. *Plant Tissue Culture and Biotechnology, 23*(2), 147–155.

Giri, L., Dhyani, P., Rawat, S., Bhatt, I. D., Nandi, S. K., Rawal, R. S., & Pande, V. (2012a). *In vitro* production of phenolic compounds and antioxidant activity in callus suspension cultures of *Habenaria edgeworthii*: A rare Himalayan medicinal orchid. *Industrial Crops and Products, 39*, 1–6.

Giri, L., Jugran, A., Rawat, S., Dhyani, P., Andola, H., Bhatt, I. D., & Dhar, U. (2012b). *In vitro* propagation, genetic and phytochemical assessment of *Habenaria edgeworthii*: An important Astavarga plant. *ActaPhysiologiaePlantarum, 34*, 869–875.

Giri, L., Singh, L., & Bhatt, I. D. (2022a). *In Vitro* Production of Phenolic Compound. In *Nutraceuticals Production from Plant Cell Factory* (pp. 105–141). Singapore: Springer Nature Singapore.

Giri, L., Singh, L., Joshi, K., Bisht, A., & Bhatt, I. D. (2022b). Factors Affecting *In Vitro* Production of Nutraceuticals. In *Nutraceuticals Production from Plant Cell Factory* (pp. 83–101). Singapore: Springer Nature Singapore.

Goraya, G. S., & Ved, D. K. (2017). Medicinal plants in India: An assessment of their demand and supply. *National Medicinal Plants Board, Ministry of AYUSH, Government of India, New Delhi and Indian Council of Forestry Research & Education, Dehradun*, 1–430.

Guo, Q., Zhou, J., Wang, Z., & Yang, H. (2012). *In vitro* rooting of *Podophyllum hexandrum* and transplanting technique. *Engineering, 4*(10), 142–145.

IUCN. (2024). https://www.iucnredlist.org/ (accessed on February, 2024)

Jabeen, N., Shawl, A. S., Dar, G. H., Jan, A., & Sultan, P. (2006). Callus induction and organogenesis from explants of *Aconitum heterophyllum* medicinal plant. *Biotechnology, 5*(3), 287–291.

Johnson, T. S., Narayan, S. B., & Narayana, D. B. A. (1997). Rapid *in vitro* propagation of *Saussurealappa*, an endangered medicinal plant, through multiple shoot cultures. *In Vitro Cellular & Developmental Biology-Plant, 33*, 128–130.

Joshi, M., & Dhar, U. (2003). *In vitro* propagation of *Saussureaobvallata* (DC.) Edgew.—An endangered ethnoreligious medicinal herb of Himalaya *Plant Cell Reports, 21*, 933–939.

Joshi, M., Manjkhola, S., & Dhar, U. (2004). Developing propagation techniques for conservation of *Heracleumcandicans*–An endangered medicinal plant of the Himalayan region. *The Journal of Horticultural Science and Biotechnology, 79*(6), 953–959.

Joshi, P., & Dhawan, V. (2005). *Swertia chirayita*–An overview. *Current Science, 89*(4), 635–640.

Joshi, P., & Dhawan, V. (2007). Axillary multiplication of *Swertia chirayita* (Roxb. Ex Fleming) H. Karst., a critically endangered medicinal herb of temperate Himalayas. *In Vitro Cellular & Developmental Biology-Plant, 43*, 631–638.

Joshi, S. K., & Dhar, U. (2009). *In vitro* propagation from axenic explants of *Liliumoxypetalum* (D. Don) Baker, an endemic bulbous plant of high altitude Himalaya. *ActaPhysiologiaePlantarum, 31*, 833–838.

Kala, C. P., & Sajwan, B. S. (2007). Revitalizing Indian systems of herbal medicine by the National Medicinal Plants Board through institutional networking and capacity building. *Current Science, 93*(6), 797–806.

Kapoor, S., Raghuvanshi, R., Bhardwaj, P., Sood, H., Saxena, S., & Chaurasia, O. P. (2018). Influence of light quality on growth, secondary metabolites production and antioxidant activity in callus culture of *Rhodiolaimbricata* Edgew. *Journal of Photochemistry and Photobiology B: Biology, 183*, 258–265.

Kaur, H., Lekhak, M. M., Chahal, S., Goutam, U., Jha, P., Naidoo, D., Ochatt, S. J., & Kumar, V. (2020). *Nardostachys jatamansi* (D. Don) DC.: An invaluable and constantly dwindling resource of the Himalayas. *South African Journal of Botany, 135*, 252–267.

Kaur, P., Gupta, R. C., Dey, A., & Pandey, D. K. (2019a). Simultaneous quantification of oleanolic acid, ursolic acid, betulinic acid and lupeol in different populations of five *Swertia* species by using HPTLC-densitometry: Comparison of different extraction methods and solvent selection. *Industrial Crops and Products, 130*, 537–546.

Kaur, P., Pandey, D. K., Gupta, R. C., & Dey, A. (2019b). Simultaneous microwave assisted extraction and HPTLC quantification of mangiferin, amarogentin, and swertiamarin in *Swertia* species from Western Himalayas *Industrial Crops and Products, 132*, 449–459.

Kaur, R., Sood, M., Chander, S., Mahajan, R., Kumar, V., & Sharma, D. R. (1999). *In vitro* propagation of *Valeriana jatamansi. Plant Cell, Tissue and Organ Culture, 59*, 227–229.

Khan, A., Shah, A. H., & Ali, N. (2021). In-vitro propagation and phytochemical profiling of a highly medicinal and endemic plant species of the Himalayan region (*Saussureacostus*). *Scientific Reports, 11*(1), 23575.

Khan, M. I., Shahzad, A., Ganie, I. B., & Pandith, S. A. (2022). Organogenesis, direct somatic embryogenesis, and shoot proliferation of *Rheum spiciforme* Royle: An endemic and vulnerable medicinal herb from Indian Trans Himalayas. *In Vitro Cellular & Developmental Biology-Plant, 58*(1), 35–50.

Kirtikar K. R., & Basu B. D. (1984). *Indian Medicinal Plants*, Volume III. Bishen Singh Mahendra Pal Singh, Dehradun. Ecological work book.

Koul, S., Raina, V., & Sharma, S. K. (2005). Conservation and propagation of high altitude medicinal and aromatic plant: *Hedychium spicatum. Journal of Plant Biochemistry and Biotechnology, 14*, 57–59.

Koul, S., Suri, K. A., Dutt, P., Sambyal, M., Ahuja, A., & Kaul, M. K. (2009). Protocol for *in vitro* Regeneration and Marker Glycoside Assessment in *Swertia chirata* Buch-Ham. In: Jain, S. M., Saxena, P. K. (eds.), *Protocols for In Vitro Cultures and Secondary Metabolite Analysis of Aromatic and Medicinal Plants*, Vol. 547 (pp. 139–153). Humana Press, Totowa, NJ. https://doi.org/10.1007/978-1-60327-287-2_12

Kumar, P., Partap, M., Rana, D., Kumar, P., & Warghat, A. R. (2020). Metabolite and expression profiling of steroidal alkaloids in wild tissues compared to bulb derived *in vitro* cultures of *Fritillariaroylei*—high value critically endangered Himalayan medicinal herb. *Industrial Crops and Products, 145*, 111945.

Kumar, R., Devi, J., Kumar, D., & Bhushan, S. (2023). *Arnebia benthamii* cell suspension cultures as a source of natural red pigments: Optimization of shaking speed and inoculum density to maximize process productivity during sequential scaleup. *Plant Biotechnology Reports, 17*(3), 353–367.

Kumar, R. R., Purohit, V. K., Prasad, P., & Nautiyal, A. R. (2018). Efficient *in vitro* propagation protocol of *Swertia chirayita* (Roxb. ex Fleming) Karsten: A critically endangered medicinal plant. *National Academy Science Letters, 41*, 123–127.

Kumar, V., & Chandra, S. (2014). High frequency somatic embryogenesis and synthetic seed production of the endangered species *Swertia chirayita. Biologia, 69*, 186–192.

Kumar, V., Singh, S., Bandopadhyay, R., Sharma, M., & Chandra, S. (2014). *In vitro* organogenesis secondary metabolite production and heavy metal analysis in *Swertia chirayita*. *Open Life Sciences*, *9*(7), 686–698.

Kunwar, R. M., Adhikari, Y. P., Sharma, H. P., Rimal, B., Devkota, H. P., Charmakar, S., Acharya, R. P., Baral, K., Ansari, A. S., Bhattarai, R., & Jentsch, A. (2020). Distribution, use, trade and conservation of *Paris polyphylla* Sm. in Nepal. *Global Ecology and Conservation*, *23*, e01081.

Lee, E. K., Jin, Y. W., Park, J. H., Yoo, Y. M., Hong, S. M., Amir, R., & Loake, G. J. (2010). Cultured cambial meristematic cells as a source of plant natural products. *Nature Biotechnology*, *28*(11), 1213–1217.

Lee, H., & Lin, J. Y. (1988). Antimutagenic activity of extracts from anticancer drugs in Chinese medicine. *Mutation Research/Genetic Toxicology*, *204*(2), 229–234.

Li, B., Yu, B., Hui, Y., Li, M., Han, X., & Fung, K. P. (2001). An improved synthesis of the saponin, polyphyllin D. *Carbohydrate Research*, *331*(1), 1–7.

Li, J., Zhao, Y. L., Huang, H. Y., & Wang, Y. Z. (2017). Phytochemistry and pharmacological activities of the genus *Swertia* (Gentianaceae): A review. *The American Journal of Chinese Medicine*, *45*(04), 667–736.

Madhav, K. C., Phoboo, S., & Jha, P. K. (2010). Ecological study of *Paris polyphylla* Sm. *Ecoprint: An International Journal of Ecology*, *17*, 87–93.

Mahendran, G., Verma, N., Singh, M., Shanker, K., Banerjee, S., Kumar, B., & Rahman, L. (2022). Elicitation enhances swerchirin and 1, 2, 5, 6-tetrahydroxyxanthone production in hairy root cultures of *Swertia chirayita* (Roxb.) H. Karst. *Industrial Crops and Products*, *177*, 114488.

Manjkhola, S., Dhar, U., & Joshi, M. (2005). Organogenesis, embryogenesis, and synthetic seed production in *Arnebia euchroma*—A critically endangered medicinal plant of the Himalaya *In Vitro Cellular & Developmental Biology-Plant*, *41*, 244–248.

Maqbool, F., Singh, S., Kaloo, Z. A., & Jan, M. (2016). A rapid micropropagation protocol of *Atropa acuminata* Royle ex Lindl.—A threatened medicinal plant species of Kashmir Himalaya. *Indian Journal of Biotechnology*, *15*, 576–580.

Mathur, J. (1992). *In vitro* morphogenesis in *Nardostachys jatamansi* DC.: Shoot regeneration from callus derived roots. *Annals of Botany*, *70*(5), 419–422.

Mathur, J. (1993). Somatic embryogenesis from callus cultures of *Nardostachys jatamansi*. *Plant Cell, Tissue and Organ Culture*, *33*, 163–169.

Mathur, J., Ahuja, P. S., Mathur, A., Kukreja, A. K., & Shah, N. C. (1988). *In vitro* propagation of *Valeriana wallichii*. *Planta Medica*, *54*(01), 82–83.

Matsuda, H., Pongpiriyadacha, Y., Morikawa, T., Kishi, A., Kataoka, S., & Yoshikawa, M. (2003). Protective effects of steroid saponins from *Paris polyphylla* var. yunnanensis on ethanol-or indomethacin-induced gastric mucosal lesions in rats: Structural requirement for activity and mode of action. *Bioorganic & Medicinal Chemistry Letters*, *13*(6), 1101–1106.

Mehrotra, S., Srivastava, V., Rahman, L. U., & Kukreja, A. K. (2015). Hairy root biotechnology—Indicative timeline to understand missing links and future outlook. *Protoplasma*, *252*, 1189–1201.

Mishra, J., Bhandari, H., Singh, M., Rawat, S., Agnihotri, R. K., Mishra, S., & Purohit, S. (2011). Hairy root culture of *Picrorhiza kurroa* Royle ex Benth.: A promising approach for the production of picrotin and picrotoxinin. *ActaPhysiologiaePlantarum*, *33*, 1841–1846.

Mohammed, N. R., Dalar, A., Ozdemir, F. A., & Turker, M. (2019). *In vitro* propagation and secondary metabolites investigation of *Hypericum perforatum* L. *Fresenius Environ Bull*, *28*(7), 5569–5576.

Mohanty, P., Nongkling, P., Das, M. C., Kumaria, S., & Tandon, P. (2013). Short-term storage of alginate-encapsulated protocorm-like bodies of *Dendrobium nobile* Lindl.: An endangered medicinal orchid from North-east India. *3 Biotech*, *3*, 235–239.

Moyo, M., Bairu, M. W., Amoo, S. O., & Van Staden, J. (2011). Plant biotechnology in South Africa: Micropropagation research endeavours, prospects and challenges. *South African Journal of Botany*, *77*(4), 996–1011.

Nadeem, M., Palni, L. M. S., Purohit, A. N., Pandey, H., & Nandi, S. K. (2000). Propagation and conservation of *Podophyllum hexandrum* Royle: An important medicinal herb. *Biological Conservation*, *92*(1), 121–129.

Nazir, U., Gul, Z., Shah, G. M., & Khan, N. I. (2022). Interaction effect of auxin and cytokinin on *in vitro* shoot regeneration and rooting of endangered medicinal plant *Valeriana jatamansi* jones through tissue culture. *American Journal of Plant Sciences*, *13*(2), 223–240.

Negi, J. S., & Pramod Singh, P. S. (2011). HPLC fingerprinting of highly demanding medicinal plant Swertia: An overview. *International Journal of Medicinal and Aromatic Plants*, 1(3), 333–337.

Pandey, A., Belwal, T., Tamta, S., Bhatt, I. D., & Rawal, R. S. (2019). Phenolic compounds, antioxidant capacity and antimutagenic activity in different growth stages of *in vitro* raised plants of *Origanum vulgare* L. *Molecular Biology Reports*, *46*, 2231–2241.

Pandey, H., Nandi, S. K., Kumar, A., Palni, U. T., Chandra, B., & Palni, L. M. S. (2004). *In vitro* propagation of *Aconitum balfourii* Stapf.: An important aconite of the Himalayan alpines. *The Journal of Horticultural Science and Biotechnology*, *79*(1), 34–41.

Pandey, M., Dhar, U., Sher Singh, S., Mrudul Vijay, S., & ShubhadaRatnakar, T. (2011). Recurrent somatic embryogenesis and plant regeneration in *Angelica glauca* Edgew., a critically endangered medicinal plant of the Western Himalaya. *The Journal of Horticultural Science and Biotechnology*, *86*(5), 493–498.

Pandey, S., & Pant, B. (2020). Establishment of *in vitro* cultures of valuable medicinal plant *Valeriana jatamansi* Jones, its conservation and production of bioactive metabolites. *Ecology, Environment and Conservation*, *26*, S33–S38.

Pandey, S., Sundararajan, S., Ramalingam, S., Baniya, M. K., & Pant, B. (2020a). Rapid clonal propagation and valepotriates accumulation in cultures of *Valeriana jatamansi* Jones, a high-value medicinal plant. *Journal of Applied Botany & Food Quality*, *93*, 177–185.

Pandey, S., Sundararajan, S., Ramalingam, S., & Pant, B. (2020b). Effects of sodium nitroprusside and growth regulators on callus, multiple shoot induction and tissue browning in commercially important *Valeriana jatamansi* Jones. *Plant Cell, Tissue and Organ Culture (PCTOC)*, *142*, 653–660.

Pant, M., Bisht, P., & Gusain, M. P. (2012). *In vitro* propagation through root-derived callus culture of *Swertia chirata* Buch.-Ham. ex Wall. *African Journal of Biotechnology*, *11*(29), 7408–7416.

Partap, M., Chhimwal, J., Kumar, P., Kumar, D., Padwad, Y., & Warghat, A. R. (2022). Growth dynamics and differential accumulation of picrosides and its precursor metabolites in callus cell lines of *Picrorhiza kurroa* with distinct anti-steatotic potential. *Process Biochemistry*, *120*, 85–100.

Pradhan, B. K., & Badola, H. K. (2008). Ethnomedicinal plant use by Lepcha tribe of Dzongu valley, bordering Khangchendzonga Biosphere Reserve, in north Sikkim, India. *Journal of Ethnobiology and Ethnomedicine*, *4*(1), 1–18.

Purohit, S., Rawat, V., Jugran, A. K., Singh, R. V., Bhatt, I. D., & Nandi, S. K. (2015). Micropropagation and genetic fidelity analysis in *Valeriana jatamansi* Jones. *Journal of Applied Research on Medicinal and Aromatic Plants*, *2*(1), 15–20.

Qahtan, A. A., Abdel-Salam, E. M., Alatar, A. A., Wang, Q. C., & Faisal, M. (2019). An Introduction to Synthetic Seeds: Production, Techniques, and Applications. In: Faisal, M., Alatar, A. (eds.), *Synthetic Seeds: Germplasm Regeneration, Preservation and Prospects* (pp. 1–20). Springer, Cham. https://doi.org/10.1007/978-3-030-24631-0_1

Rafiq, S., Wagay, N. A., Bhat, I. A., Kaloo, Z. A., Rashid, S., Lin, F., & Elansary, H. O. (2021). *In vitro* propagation of *Aconitum chasmanthum* Stapf Ex Holmes: An endemic and critically endangered plant species of the western Himalaya *Horticulturae*, *7*(12), 586.

Rai, M. K., Asthana, P., Singh, S. K., Jaiswal, V. S., & Jaiswal, U. (2009). The encapsulation technology in fruit plants—A review. *Biotechnology Advances*, *27*(6), 671–679.

Raina, R., Johri, A. K., & Srivastava, L. J. (1994). Seed germination studies in *Swertia chirata* L. *Seed Research*, *22*, 62–63.

Rajesh, M., Sivanandhan, G., Subramanyam, K., Kapildev, G., Jaganath, B., Kasthurirengan, S., & Ganapathi, A. (2014). Establishment of somatic embryogenesis and podophyllotoxin production in liquid shake cultures of *Podophyllum hexandrum* Royle *Industrial Crops and Products*, *60*, 66–74.

Raomai, S., Kumaria, S., Kehie, M., & Tandon, P. (2015). Plantlet regeneration of *Paris polyphylla* Sm. via thin cell layer culture and enhancement of steroidal saponins in mini-rhizome cultures using elicitors. *Plant Growth Regulation*, *75*, 341–353.

Raomai, S., Kumaria, S., & Tandon, P. (2014). Plant regeneration through direct somatic embryogenesis from immature zygotic embryos of the medicinal plant, *Paris polyphylla* Sm. *Plant Cell, Tissue and Organ Culture (PCTOC)*, *118*, 445–455.

Rastegari, A. A., Yadav, A. N., & Yadav, N. (2019). Genetic Manipulation of Secondary Metabolites Producers. In *New and Future Developments in Microbial Biotechnology and Bioengineering* (pp. 13–29). Elsevier.

Rattan, S., Kumar, A., Kumar, D., & Warghat, A. R. (2022). Enhanced production of phenylethanoids mediated through synergistic approach of precursor feeding and light regime in cell suspension culture of *Rhodiola imbricata* (Edgew.). *Applied Biochemistry and Biotechnology*, *194*(7), 3242–3260.

Rattan, S., Sood, A., Kumar, P., Kumar, A., Kumar, D., & Warghat, A. R. (2020). Phenylethanoids, phenylpropanoids, and phenolic acids quantification vis-à-vis gene expression profiling in leaf and root derived callus lines of *Rhodiola imbricata* (Edgew.). *Industrial Crops and Products*, *154*, 112708.

Rattan, S., & Warghat, A. R. (2023). Comparative analysis of salidroside and rosavin accumulation and expression analysis of biosynthetic genes in salicylic acid and methyl jasmonate elicited cell suspension culture of *Rhodiola imbricata* (Edgew.). *Industrial Crops and Products*, *198*, 116667.

Rawat, J. M., Bhandari, A., Mishra, S., Rawat, B., Dhakad, A. K., Thakur, A., & Chandra, A. (2018a). Genetic stability and phytochemical profiling of the *in vitro* regenerated plants of *Angelica glauca* Edgew.: An endangered medicinal plant of Himalaya. *Plant Cell, Tissue and Organ Culture (PCTOC)*, *135*, 111–118.

Rawat, J. M., Pandey, S., Rawat, B., Purohit, S., Anand, J., Negi, A. S., Thakur, A., Mahmoud, M. H., El-Gazzar, A. M., & Batiha, G. E. S. (2023). *In vitro* production of steroidal saponin, total phenols and antioxidant activity in callus suspension culture of *Paris polyphylla* Smith: An important Himalayan medicinal plant. *Frontiers in Plant Science*, *14*, 1225612. https://doi.org/10.3389/fpls.2023.1225612

Rawat, J. M., Rawat, B., Agnihotri, R. K., Chandra, A., & Nautiyal, S. (2013). *In vitro* propagation, genetic and secondary metabolite analysis of *Aconitum violaceum* Jacq.: A threatened medicinal herb. *ActaPhysiologiaePlantarum*, *35*, 2589–2599.

Rawat, J. M., Rawat, B., & Mishra, S. (2014). Effect of elicitation on picrotin and picrotoxinin production from *in vitro* products of *Picrorhiza kurrooa*. *African Journal of Biotechnology*, *13*(51), 4612–4619.

Rawat, S., Jugran, A. K., Bhatt, I. D., & Rawal, R. S. (2018b). *Hedychium spicatum*: A systematic review on traditional uses, phytochemistry, pharmacology and future prospectus. *Journal of Pharmacy and Pharmacology*, *70*(6), 687–712.

Rawat, S., Jugran, A. K., Bhatt, I. D., Rawal, R. S., Andola, H. C., & Dhar, U. (2017). Essential oil composition and antioxidant activity in *Valeriana jatamansi* Jones: Influence of seasons and growing sources. *Journal of Essential Oil Research*, *29*(1), 101–107.

Rawat, V., Ghildiyal, A., Singh, L., Jugran, A. K., Bhatt, I. D., Nandi, S. K., & Pande, V. (2020). Methyl jasmonate induced polyphenols and antioxidant production in callus suspension culture of *Nardostachys jatamansi*. *Plant Biosystems-An International Journal Dealing With All Aspects of Plant Biology*, *154*(6), 851–859.

Reddy, M. C., Murthy, K. S. R., & Pullaiah, T. (2012). Synthetic seeds: A review in agriculture and forestry. *African Journal of Biotechnology*, *11*(78), 14254–14275.

Reddy, P. P., Rao, R. R., Shashidhar, J., Sastry, B. S., Rao, J. M., & Babu, K. S. (2009). Phytochemical investigation of labdanediterpenes from the rhizomes of *Hedychium spicatum* and their cytotoxic activity. *Bioorganic & Medicinal Chemistry Letters*, *19*(21), 6078–6081.

Rodrigues, A. S., Pilgrim, J. D., Lamoreux, J. F., Hoffmann, M., & Brooks, T. M. (2006). The value of the IUCN red list for conservation. *Trends in Ecology & Evolution*, *21*(2), 71–76.

Rohini, M. R. (2020). Biotechnological Interventions for Conservation and Multiplication of Threatened Medicinal Plants. In: Rajasekharan, P., Wani, S. (eds), *Conservation and Utilization of Threatened Medicinal Plants* (pp. 135–158). Springer, Cham. https://doi.org/10.1007/978-3-030-39793-7_6

Saxena, A., Shukla, M., & Saxena, P. (2019). Synthetic seeds: Relevance to Endangered Germplasm Conservation *in vitro*. In: Faisal, M., Alatar, A. (eds), *Synthetic Seeds: Germplasm Regeneration, Preservation and Prospects* (pp. 21–60). Springer, Cham. https://doi.org/10.1007/978-3-030-24631-0_2

Sharma, A., Kaur, R., & Sharma, N. (2014a). *In vitro* morphogenic response of different explants of *Gentiana kurroo* Royle from Western Himalayas—An endangered medicinal plant. *Physiology and Molecular Biology of Plants*, *20*, 249–256.

Sharma, E., Goel, A., & Gaur, A. K. (2014b). Effect of precursors and inhibitors on aconitine accumulation in *in vitro* root cultures of *Aconitum balfourii* Stapf. *Indian Journal of Plant Physiology*, *19*, 375–381.

Sharma, N., Thakur, M., Sharma, P., Sharma, Y. P., & Dutt, B. (2022). *In vitro* propagation from rhizomes, molecular evaluation and podophyllotoxin production in Himalayan May Apple (*Sinopodophyllum hexandrum* Royle TS Ying): An endangered medicinal plant. *Plant Cell, Tissue and Organ Culture (PCTOC)*, *149*(1–2), 159–173.

Sharma, V., Belwal, N., Kamal, B., Dobriyal, A. K., & Jadon, V. S. (2016). Assessment of genetic fidelity of *in vitro* raised plants in *Swertia chirayita* through ISSR, RAPD analysis and peroxidase profiling during organogenesis. *Brazilian Archives of Biology and Technology*, *59*, e16160389.

Shukla, J. K., Dhakal, P., Uniyal, R. C., Paul, N., & Sahoo, D. (2017). Ex-situ cultivation at lower altitude and evaluation of *Swertia chirayita*, a critically endangered medicinal plant of Sikkim Himalayan region, India. *South African Journal of Botany*, *109*, 138–145.

Singh, L., Bhatt, I. D., Negi, V. S., Nandi, S. K., Rawal, R. S., & Bisht, A. K. (2021). Population status, threats, and conservation options of the orchid *Dactylorhiza hatagirea* in Indian Western Himalaya *Regional Environmental Change*, *21*, 1–16.

Singh, L., Kanwar, N., Bhatt, I. D., Nandi, S. K., & Bisht, A. K. (2022). Predicting the potential distribution of *Dactylorhiza hatagirea* (D. Don) Soo-an important medicinal orchid in the West Himalaya, under multiple climate change scenarios. *PLos One, 17*(6), e0269673.

Singh, L., Nandi, S. K., Bhatt, I. D., & Bisht, A. K. (2023). Inventorization of Ecology, Ethnobotany, and Conservation Status of *Dactylorhiza hatagirea*: Problems, Progress, and Prospects. In: Jha, S., Halder, M. (eds), *Medicinal Plants: Biodiversity, Biotechnology and Conservation*, Vol. 33 (pp. 307–325). Springer, Singapore. https://doi.org/10.1007/978-981-19-9936-9_12

Singh, L., Tariq, M., Sekar, K. C., Bhatt, I. D., & Nandi, S. K. (2017). Ecological niche modelling: An important tool for predicting suitable habitat and conservation of the Himalayan medicinal herbs. *ENVIS Bulletin Himalayan Ecology, 25*, 154–155.

Singh, M., Chettri, A., Pandey, A., Sinha, S., Singh, K. K., & Badola, H. K. (2020). *In vitro* propagation and phytochemical assessment of *Aconitum ferox* Wall: A threatened medicinal plant of Sikkim Himalaya. *Proceedings of the National Academy of Sciences, India Section B: Biological Sciences, 90*, 313–321.

Singh, N., Gupta, A. P., Singh, B., & Kaul, V. K. (2006). Quantification of valerenic acid in *Valeriana jatamansi* and *Valeriana officinalis* by HPTLC. *Chromatographia, 63*, 209–213.

Singh, N., & Kumaria, S. (2021). Deciphering the role of stress elicitors on the differential modulation of chalcone synthase gene and subsequent production of secondary metabolites in micropropagated *Coelogyneovalis* Lindl., a therapeutically important medicinal orchid. *South African Journal of Botany, 140*, 336–348.

Sultan, P., Shawl, A. S., Ramteke, P. W., Jan, A., Chisti, N., Jabeen, N., & Shabir, S. (2006). *In vitro* propagation for mass multiplication of *Podophyllum hexandrum*: A high value medicinal herb. *Asian Journal of Plant Sciences, 5*(2), 179–184.

Tabin, S., Kamili, A. N., Gupta, R. C., Parray, J. A., & Bansal, A. (2018). *In vitro* culture of *Rheum emodi* Wall: An endangered medicinal plant of Northwestern Himalaya. *Proceedings of the National Academy of Sciences, India Section B: Biological Sciences, 88*, 995–1006.

Tariq, M., Belwal, T., Bhatt, I. D., Pande, V., & Nandi, S. K. (2018). Polyphenolics in leaves of *Paris polyphylla*: An important high value Himalayan medicinal herb. *Industrial Crops and Products, 117*, 66–74.

Tariq, M., Nandi, S. K., & Bhatt, I. D. (2023). Securing Conservation Status of *Paris polyphylla*, a Medicinally Important Plant of the Indian Himalayan Region. In *Medicinal Plants: Biodiversity, Biotechnology and Conservation* (pp. 133–154). Singapore: Springer Nature Singapore.

Uprety, Y., Asselin, H., Boon, E. K., Yadav, S., & Shrestha, K. K. (2010). Indigenous use and bio-efficacy of medicinal plants in the Rasuwa District, Central Nepal. *Journal of Ethnobiology and Ethnomedicine, 6*, 1–10.

Van Uden, W., Pras, N., & Malingré, T. M. (1990). On the improvement of the podophyllotoxin production by phenylpropanoid precursor feeding to cell cultures of *Podophyllum hexandrum* Royle *Plant Cell, Tissue and Organ Culture, 23*, 217–224.

Vengadesan, G., & Pijut, P. M. (2009). *In vitro* propagation of northern red oak (*Quercus rubra* L.). *In Vitro Cellular & Developmental Biology-Plant, 45*, 474–482.

Verma, H., Patil, P. R., Kolhapure, R. M., & Gopalkrishna, V. (2008). Antiviral activity of the Indian medicinal plant extract, *Swertia chirata* against herpes simplex viruses: A study by in-vitro and molecular approach. *Indian Journal of Medical Microbiology, 26*(4), 322–326.

Verma, P. C., Rahman, L., Negi, A. S., Jain, D. C., Khanuja, S. P. S., & Banerjee, S. (2007). *Agrobacterium rhizogenes*-mediated transformation of *Picrorhiza kurroa* Royle ex Benth.: Establishment and selection of superior hairy root clone. *Plant Biotechnology Reports, 1*, 169–174.

Verma, V. K., Sarwa, K. K., Kumar, A., & Zaman, M. K. (2013). Comparison of hepatoprotective activity of *Swertia chirayita* and *Andrographis paniculata* plant of North–East India against CCl4 induced hepatotoxic rats. *Journal of Pharmacy Research, 7*(7), 647–653.

Vishwakarma, S., Goyal, R., Gupta, V., & Dhar, K. L. (2016). GABAergic effect of valeric acid from *Valeriana wallichii* in amelioration of ICV STZ induced dementia in rats. *RevistaBrasileira de Farmacognosia, 26*, 484–489.

Wang, L., An, L., Hu, Y., Wei, L., & Li, Y. (2009). Influence of phytohormones and medium on the shoot regeneration from leaf of *Swertia chirata* Buch.-Ham. ex Wall. *in vitro*. *African Journal of Biotechnology, 8*(11), 2513–2517.

Wang, M., Li, W., Qiang, Q., Ma, J., Chen, J., Zhang, X., & Lin, L. (2023). Clonal propagation and assessment of biomass production and saponin content of elite accessions of wild *Paris polyphylla* var. *yunnanensis*. *Plants, 12*(16), 2983.

Wang, M., Yang, T. T., Rao, Y., Wang, Z. M., Dong, X., Zhang, L. H., & Wu, H. H. (2021). A review on traditional uses, phytochemistry, pharmacology, toxicology and the analytical methods of the genus *Nardostachys*. *Journal of Ethnopharmacology, 280*, 114446.

Wawrosch, C., Maskay, N., & Kopp, B. (1999). Micropropagation of the threatened Nepalese medicinal plant *Swertia chirata* Buch.–Ham. ex Wall. *Plant Cell Reports, 18*, 997–1001.

Woerdenbag, H. J., van Uden, W., Frijlink, H. W., Lerk, C. F., Pras, N., & Malingré, T. M. (1990). Increased podophyllotoxin production in *Podophyllum hexandrum* cell suspension cultures after feeding coniferyl alcohol as a β-cyclodextrin complex. *Plant Cell Reports, 9*, 97–100.

Wu, S. S., Gao, W. Y., & Duan, H. Q. (2004). Advances in studies on chemical constituents and pharmacological activities of rhizoma paridis. *Chinese Traditional and Herbal Drugs, 35*(3), 344–346.

Xu, J., Yang, B., Guo, Y., Jin, D. Q., Guo, P., Liu, C., & Sun, Z. (2011). Neuroprotective bakkenolides from the roots of *Valeriana jatamansi*. *Fitoterapia, 82*(6), 849–853.

Zhou, N. J., Geng, C. A., & Huang, X. Y. (2015). Anti-hepatitis B virus active constituents from *Swertia chirayita*. *Fitoterapia, 100*, 27–34.

3 Large-Scale Propagation of Medicinal Trees through Tissue Culture Techniques

Its Significance, Challenges and Applications

Urja Agarwal, Sonam Khan, Shivansh Negi, Pramod Rawat, and Yashaswi Singh

3.1 INTRODUCTION

Since time immemorial, trees, besides being a vital constituent of biodiversity, have been an indispensable part of human society and culture. Trees are a sustainable source of many tangible human needs such as food, timber, fuel wood, fodder, fibre and non-timber products at the same time providing various intangible benefits also, such as carbon sequestration, climate change mitigation, converting carbon dioxide (CO_2) into oxygen (O_2), soil conservation, habitat for wildlife and human recreation. Plants in general and medicinal trees in particular have been utilized for thousands of years by humans for treating diseases and maintaining a healthy lifestyle (Giri et al., 2004). The traditional medicine (TM) systems such as traditional Indian medicine (Ayurveda), traditional Chinese medicine (TCM) and traditional Arabic medicine (Unani) are some of the best-known TM systems in the world that utilize natural products for treating diseases and maintaining health (Majeed et al., 2023). According to the World Health Organization (WHO), more than 80% of the global population relies on such TM systems. A renewed interest in plant-based natural medicine has resulted in 40% or more pharmaceuticals being directly or indirectly derived from such natural sources (WHO, 2022).

Over the years, severe anthropogenic pressures such as rapid deforestation, change in land use patterns, over-exploitation, invasive species and climate change have resulted in reduction of tree cover, both in forests and outside forests. Many valuable medicinal plants are on the cusp of extinction due to a rapid decline in the wealth of plant genetic resources. There is an urgent need for conservation of such plants in general and forest genetic resources in particular for biodiversity and human survival. Conventional tree breeding approaches for tree propagation and improvement which are being used currently are limited to fast-growing and the most valuable tree species and suffer from inherent bottlenecks because tree species have slow growth rates, long maturation and life spans, high heterozygosity and sexual self-incompatibility. Tree species also suffer from inbreeding depression and high genetic load, which further restrict the use of traditional breeding methods for tree improvement and propagation. Ergo, traditional tree improvement methods find limited applications owing to their low productivity and high resource requirements (Giri et al., 2004).

Biotechnological tools such as plant tissue culture offer an alternative method for successful mass multiplication and clonal propagation of superior and elite tree genotypes within a stipulated time frame. Studies on biotechnological interventions for *in vitro* regeneration, multiplication and

DOI: 10.1201/b23374-3

clonal propagation have been promising and slowly gaining pace over the last decade. *In vitro* micropropagation of tress can be broadly classified into two approaches, direct organogenesis and indirect organogenesis. Direct organogenesis, or direct regeneration, is the preferred technique, since the plant propagation is done without an intermediate callus phase, unlike indirect organogenesis. Indirect organogenesis, or indirect regeneration, usually results in somaclonal variations, which makes the technique less desirable for mass clonal propagation. This makes direct regeneration the preferred practice for reliable true-to-type clonal propagation. Clonal propagation of trees provides a rapid means for producing planting stock for reforestation, afforestation and plantation programmes and conservation of rare, endangered and threatened plant genetic resources (Fenning and Gersheazon 2002; Giri et al., 2004). Genetic homogeneity of *in vitro*–raised cultures is another major consideration in micropropagation of trees for commercial purposes. Genetic variability within *in vitro*–grown plantlets arise due to the application of plant growth regulators (PGRs), stress and culture conditions. Molecular markers viz. start codon targeted (SCoT), random amplified polymorphic DNA (RAPD) and inter-simple sequence repeats (ISSR) have been successfully used for assessment of clonal/genetic fidelity. SCoT markers are preferred over the use of RAPD and ISSR, since they target the translational start site or the short-conserved regions surrounding the initiation codon (ATG), and hence the results obtained are reproducible and more reliable and, thus, eliminate the shortcomings of RAPD and ISSR markers (Collard and Mackill, 2009).

Much work has been carried out on *in vitro* propagation of important medicinal tree species such as *Ginkgo biloba* (Mantovani et al., 2013; Qaderi et al., 2021), *Taxus baccata* (Davarpanah et al., 2014; Baharak et al., 2014; Sahari Moghadam et al., 2022), *Juglans regia* (Al-Mizory and Mayi, 2012; Kepenek and Kolağasi, 2016), *Salix* sp. (Regueira et al., 2018; Capuana et al., 2022; Parzymies et al., 2023), *Rhododendron* sp. (Sandeepani et al., 2021; Wu et al., 2023), *Terminalia* sp. (Dangi et al., 2014; Gupta et al., 2014; Chaudhary et al., 2015), *Azadirachta indica* (Arora et al., 2011; Dhandapani et al., 2021; Bello et al., 2022), *Eucalyptus* sp. (Navroski et al., 2014; de Oliveira et al., 2022), *Acacia auriculiformis* (Yadav et al., 2016; Ismail et al., 2016), *Olea europea* (Mangal et al., 2014; Khalid et al., 2022) and *Ficus carica* (Dessoky et al., 2016; Shahcheraghi and Shekafandeh, 2016; Ling et al., 2022).

The present chapter is an attempt at making a comprehensive study of the traditional use of trees for therapeutic uses, important bioactive molecules isolated from them and development of *in vitro* regeneration protocols of the aforementioned medicinal tree species, highlighting their significance, challenges in tree micropropagation and future prospects.

3.2 MICROPROPAGATION OF MEDICINAL TREES

3.2.1 *Ginkgo biloba* L.

The endangered *Ginkgo biloba* L. tree is a member of the non-flowering seed plant group Gymnospermae. It is also known as silver apricot, maiden hair tree, and ginkgo. Being the sole member of the Ginkgoales family still alive that can live for a millennium, gingko is a living fossil and has existed for the past 200 million years. Originating in China, it has gradually extended to Korea, Japan and certain regions of Europe and the United States (Shahrajabian et al., 2022). There are only a few trees left in India, and they are mostly located in the country's hilly regions. It is included in the International Union for Conservation of Nature's (IUCN's; Talreja and Tiwari, 2023) Red List under the endangered category.

Ginkgo trees are deciduous and can grow to a height of 25–35 metres. The morphology of the male and female ginkgo trees differs. Compared to the male tree, the female tree is wider and broader. The tree is resistant to wind and snow damage because of its well-ingrained roots (Rahimi et al., 2023).

Its leaf extracts have evidence of being a significant aspect of Chinese medicine, primarily used for treating age-related problems. Its role has been seen in improving the blood circulation primarily to the cerebral parts of the body and in regulating the metabolic rate, which are the consequences of old age. Since it has effects in the cerebral parts of the body, it has been seen to be effective in curing cognitive issues and neurological disorders such as dementia, ischemic stroke, drug-induced Parkinson's disorder, etc. (Talreja and Tiwari, 2023).

G. biloba contains many chemical constituents such as flavonoids, bioflavonoids-ginkgolides (A, B, C, J and M), terpenoids, bilobalide and bilirubin and glycosides like quercetin, kaempferol, terpenes, trilactones, proanthocyanins, polyphenols and organic acids. The commercial production of the leaf extracts depends on the chemical constituent required for the job. Terpenoids (ginkgolides and bilobalides) and flavonoids are primarily extracted commercially. An extract of *G. biloba* (EGb) patent has also been granted with respect to the commercial production and extraction of the chemical constituents for medical purposes (Talreja and Tiwari, 2023; Rahimi et al., 2023; Table 3.2).

The extracts have antioxidant and platelet inhibiting factor (PIF properties (Kobus-Cisowska et al., 2014; Shiyong et al., 2015). They are capable of regulating corticosteroid production and maintaining the ionic gradient (Rodriguez-Enriquez et al., 2017). EGb has potential to decrease nitric oxide (NO) toxicity and fluoride toxicity, which helps in protecting cardiovascular tissues and haematological parameters, respectively (El-Boghdady, 2013; Raju et al., 2019). Since the 20th century, clinical trials including both humans and animals have been carried out to assess the effectiveness of EGb. Overall analysis of the trials has shown mixed results for the extract. However, positive results observed are the ones in which the dosage of 240 mg/day was given for a long period, i.e. over 5 months or so in the case of mild to moderate Alzheimer's dementia and cognitive-level problems (Maurer et al., 1997). The pharmacological benefits of EGb extract have been shown through various clinical trials.

Ginkgo tissue culture has been conducted since the 1900s in an effort to enhance the species' reputation on the IUCN Red List. The EGb extract that was previously mentioned is significant. However, the ginkgo leaf extract is synthesized in a 50:1 ratio (Rigney et al., 1999). Therefore, tissue culture procedures can aid in naturally increasing the extract's content in terms of ginkgo. The most effective medium for culturing the tissue of the micro-cutting segments, such as leaves and buds, was Murashige and Skoog (MS; Murashige and Skoog, 1962) media (3/4) supplemented with benzylaminopurine (BAP) at 5.0 mg/L, which demonstrated effective induction and proliferation of shoots (Sarhan et al., 2010). Using nodal segments in MS media with BAP (20 mg/L) and kinetin (KIN; 1 mg/L), another *in vitro* growth of a ginkgo species was conducted. The outcome was 80% growth efficiency of *in vitro* generated shoots (Ana-Maria and Radu, 2011). The location of the nodal segments from the part of the plant is also significant. The nodal explants used from the herbaceous shoot showed a more positive response than the woody nodal explants. Mantovani et al. (2013) used hydrolysed casein (500 mg/L), KIN (0.46 μM and 4.65 μM) and activated charcoal (1.5 g/L). The results showed better morphogenic response of the herbaceous nodal segments over woody ones and the preference of choosing explants during the winter season. Inhibitory effects of the high amount of cytokinin were also observed during the experiment. Lately, lateral buds have been employed as explants in woody plant media (WPM; Smith and McCown, 1983) media containing IBA (0.5 mg/L) and KIN (1 mg/L), which caused the lateral bud to induce shoots in addition to the leaves. The most effective media for the root induction was determined to be WPM with IBA at 1 mg/L and activated charcoal (AC; 2 g/L). When comparing the biochemical composition of the leaves of the *in vitro* raised plantlets to that of the mother plant, high-performance liquid chromatography (HPLC) analysis revealed no change, and the acclimation frequency was greater than 50% (Qaderi et al., 2021). By extending the sub-culturing period, Cheng et al. (2014) were able to successfully increase the gingko culture's flavonoid and terpene lactone content. They employed callus generated from induced embryos, which were subsequently grown in MS medium containing BAP (1 mg/L) and NAA (2 mg/L). In the quest of choosing the most appropriate explant, shoot meristem is found to be the best and petiole was found to be the least responsive to media (Table 3.1). This has paved way for the successful production of EGb in tissue culture regenerated plantlets.

TABLE 3.1
Micropropagation of Some Important Medicinal Tree Species

Sl. No.	Tree Species	Explant Source	Basal Medium	Medium Additives	Plant Growth Regulator(s)	Culture Response	References
1	*Ginkgo biloba*	Lateral buds and leaves	WPM	AC	1 mg/L KIN + 0.5 mg/mL IBA	Multiple shoots induced	Qaderi et al., 2021
		In vitro shoots	WPM	2 mg/L AC	1 mg/L IBA	*In vitro* induced roots	
		Nodal segments	MS	–	20 mg/L BAP	Shoot induction	Ana-Maria and Radu, 2011
		Leaf and buds	¾ MS	–	5 mg/L BAP	Multiple shoots induced	Sarhan et al., 2010
		Herbaceous nodal segments	MS	Hydrolysed casein 500 mg/L Activated Charcoal 1.5 g/L	0.46 µM KIN	Shoot induction	Mantovani et al., 2013
2	*Taxus baccata*	Branched twigs	MS	AC + Vitamins	2 mg/L BAP	Multiple shoots induced	Abbasin et al., 2010
		In vitro shoots	½ MS	Soil	8 mg/L IBA	Roots induced	
		Apical buds	MS	–	2 mg/L KIN + 1 mg/L IBA	Multiple shoots induced	Sahari Moghadam et al., 2022
		In vitro shoots	MS	–	1 mg/L IBA + 2 mg/L KIN	Roots induced	
		Seeds	½ MS	2 g/L AC	–	Embryo	Davarpanah et al., 2014
		Shoot tips	MS	2 g/L AC	3 mg/L KIN + 3 mg/L BAP	Multiple shoots induced	Baharak et al., 2014
		In vitro shoots	WPM	–	–	Roots induced	

3	*Juglans regia*	Axillary bud	NGE	–	0.5 mg/L BAP	Embryo	Toosi and Dilmagani 2010
		Shoot apices and leaf buds	NGE	–	2.5 mg/L TDZ + 2.5 mg/L IBA	Multiple shoots induced	Kepenek and Kolağasi, 2016
		Nodal segments	MS	–	2.5 mg/L BAP + 0.5 mg/L KIN	Shoots induced	Al-Mizory and Mayi, 2012
		In vitro shoots	MS	–	1 mg/L IBA	Roots induced	
		Nut	MS	–	BAP 4 μM + IBA 12 μM	Embryo	Amiri and Gharati, 2012
4	*Salix myritilloides*	Shoot	MS	–	KIN 0.5 mg/dm^3	Multiple shoots induced	Parzymies et al., 2023
	S. lapponnum	Shoot	MS	–	KIN 0.5 mg/dm^3 + IAA 0.5 mg/dm^3	Shoots induced	Parzymies et al., 2020
	S. acmophylla	Shoot tip	WPM	–	BAP 5 μM	Shoot induced	Capuana et al., 2022
		In vitro shoot	WPM	–	IBA 5 μM	Roots induced	
	S. vimanlis	Nodal segments	MS	½ Nitrates	0.22 μM BAP	Shoot induced	Regueira et al., 2018
		In vitro shoots	MS	½ Nitrates	0.22 μM BAP	Roots induced	
	S. tetrasperma	Nodal segments	WPM	–	1 μM BAP + 0.5 μM NAA + 2.5 μM TDZ	Shoots induced	Khan and Anis, 2012
		In vitro shoots	½ WPM	–	0.5 μM IBA	Roots induced	
	S. retusa and *S. alpina*	Nodal segments	MS	2 g/L AC	0.5 mg/L KIN	MSI	Chornobrov et al., 2019
5	*Rhododendron wattii* Cowan	Flower with ovary + pedicle	Anderson	–	Isopentyladenine 39.36 μM +4.90 μM IBA	Multiple shoots induced	Mao et al., 2011
		In vitro shoots	Anderson	–	IBA 2.45 μM	Roots induced	
	R. decorum	Flower buds	WPM	–	0.2 mg/L NAA + 0.5 mg/L TDZ	Multiple shoot induced	Wu et al., 2023
		In vitro shoots	WPM	–	0.1 mg/L NAA + 1 mg/L IBA	Roots induced	
	R. arboreum	Seeds	ADW+ agar	–	–	Shoots induced	Sandeepani et al., 2021
		In vitro shoots	ADW+ agar	–	–	Roots induced	–
	R. fourtunei	Nodal segments	ER or WPM	–	4 mg/L Zeatin + NAA 1 mg/L	Axillary shoots induced	Wei et al., 2018
		In vitro induced shoots	ER or WPM	–	IBA 1 mg/L	Roots induced	–

(Continued)

TABLE 3.1 (*Continued*)
Micropropagation of Some Important Medicinal Tree Species

Sl. No.	Tree Species	Explant Source	Basal Medium	Medium Additives	Plant Growth Regulator(s)	Culture Response	References
6	*Terminalia arjuna*	Apical bud, leaves	MS	TDZ 0.5 mg/L + GA_3	BAP 3.0 mg/L	Multiple shoots induced	Kumar et al., 2010
		In vitro shoots	MS	–	NAA +IBA + KIN 0.1–3 mg/L	Roots induced	–
		Nodal segments	Modified MS	-	0.5 mg/L BAP + 0.1 mg/L NAA	Multiple shoots induced	Gupta et al., 2014
		In vitro shoots	Modified MS	–	250 mg/L IBA + 250 mg/L NOA	Roots induced	–
		Axillary buds	MS	–	4.44 µM BAP + 0.54 µM NAA	Multiple shoots induced	Choudhary et al., 2015
		In vitro shoots	½ MS	100 mg/L AC	984 µM IBA	Roots induced	–
	T. bellerica	Nodal segments	MS	3.9 µM Phloroglucinol	4.4 µM BAP	Shoots induced	Dangi et al., 2014
		In vitro shoots	MS	–	2.5 µM IBA	Roots induced	–
7	*Azadirachta indica*	Shoot apices	MS	–	0.3 mg/L BAP + 0.005 mg/L NAA	Multiple shoots induced	Ashrafuzzaman et al., 2008
		In vitro shoots	MS	–	0.05 mg/l IBA + 0.05 mg/l IAA	Roots induced	–
		Axillary and terminal buds	MS	–	0.7 mg/L BAP	Multiple shoots induced	Bello et al., 2022
		In vitro shoots	LS	–	4 mg/L IBA	Roots induced	–
		Leaflets	MS	–	6.9 µ M KIN + 8.6 µ M IAA	Embryo	Shekhawat et al., 2009
		Root segments	MS	Adenine hemisulphate 8.43 µM + Putrescine 2.27 µM	8.88 µM BAP + 9.84 µM N^6-(2-isopentenyl) adenine + 5.71 µM IAA	Shoots induced	Arora et al., 2011
		Nodes and cotyledons	½ MS	1 g/L casein hydrolysate	3 mg/L BAP + 0.5 mg/L NAA	Shoot induced	Dhandapani et al., 2021
		In vitro induced shoots	½ MS	–	2 mg/L IBA	Roots induced	–

8	*Eucalyptus cloeziana*	Epicormic shoots	WPM	–	0.60 mg/L BAP + 0.10 mg/L NAA	Shoots induced	Silva de Oliveira et al., 2015
	E. benthamii	Nodal segments from sprouts	WPM	–	0.1 mg/L BAP + 0.05 mg/L NAA	Bud formation	Brondani et al. 2012
	E. dunnii	Nodal segments	MS	–	0.50 mg/L BAP	Bud formation	Navroski et al., 2014
	E. cloeziana	Adventitious buds	WPM	–	0.89 µM BAP + 0.05 µM NAA	Adventitious buds	de Oliveira et al., 2022
9	*Acacia auriculiformis*	Nodal segments	MS	–	2 mg/L BAP + 0.1 mg/L NAA	Multiple shoots induced	Girijashankar, 2011
		Shoot	MS	AC	2 mg/L KIN + 0.5 mg/L IAA	Multiple shoots induced	Yadav et al., 2016
		In vitro shoots	½ MS	–	0.1 mg/L IAA	Roots induced	–
		Seedlings	MS	–	0.5 mg/L BAP + 0.1 mg/L KIN	Multiple shoots induced	Ismail et al., 2016
		Nodal segments	MS	–	0.1 mg/L BAP + 0.1 mg/L KIN	Multiple shoots induced	Ismail et al., 2012
10	*Olea europea*	Nodal segments	MS	–	1 mg/L IBA + 1 mg/L KIN + 2 mg/L BAP	Shoots induced	Mangal et al., 2014
		In vitro shoots	MS	1.5 g/L AC	0.2 mg/L IBA or 1 mg/L NAA	Roots induced	
		Nodal segments	OM	–	1–4 mg/L zeatin	Shoots induced	Lambardi et al., 2012
		In vitro shoots	½ OM	–	2–3 mg/L IBA +1 mg/L NAA	Roots induced	–
		Nodal segments	MS	–	2.1 mg/L BAP + 0.6 mg/L GA_3	Shoots induced	Rostami and Shahsava, 2012
		In vitro shoots	½ MS	–	4 mg/L IBA	Roots induced	–
		Nodal segments	OM	–	Zeatin 9.12 µM + BAP 17.7 µM	Shoots induced	Khalid et al., 2022

(*Continued*)

TABLE 3.1 (*Continued*)
Micropropagation of Some Important Medicinal Tree Species

Sl. No.	Tree Species	Explant Source	Basal Medium	Medium Additives	Plant Growth Regulator(s)	Culture Response	References
11	*Ficus carica*	Leaf	MS	–	0.5 mg/L NAA +7 mg/L TDZ	Callus	Dhage et al., 2012
		In vitro callus	½ MS	2 g/L AC	1 mg/L IAA	Roots induced	–
		Shoot tip	MS	–	0.5 mg/l BAP	Shoots induced	Taha et al., 2013
		In vitro shoots	½ MS	–	0.1 mg/L NAA + 0.5 mg/L BAP	Roots induced	–
		Leaf disc	MS	–	0.5 mg/L BAP	Shoots induced	Danial et al., 2014
		In vitro shoots	MS	–	0.5 mg/L IBA + NAA	Roots induced	–
		Shoot tip	MS	–	0.5 mg/L BAP + 0.2 mg/L NAA	Shoots induced	Moniruzzaman et al., 2015
		Nodal segments	MS	0.2 mg/L 2iP	0.5 mg/L BAP	Shoots induced	Shahcheraghi and Shekafandeh 2016
		In vitro shoots	MS	–	1.5 mg/L IBA	Roots induced	–
		Nodal segments	MS	–	3 mg/L BAP + 0.1 mg/L KIN	Shoots induced	Dessoky et al., 2016
		In vitro shoots	½ MS	2 g/L AC	2 mg/L IAA + 0.1 mg/L IBA	Roots induced	–
		Shoot tip	MS	–	0.5 mg/L BAP	Multiple shoots induced	Ling et al., 2022
		In vitro shoots	WPM	–	3 mg/L IBA	Roots induced	
12.	*Cedrus deodara*	Stem segments	LP	–	5 µM BAP	Shoots induced	Tamta and Palni, 2004
		Seeds	MS	–	BAP 1 mg/L + GA_3 0.5 mg/L	Shoot induced	Ahmad et al., 2013
		Defoliated shoot tips	WPM	–	BAP 2.5 µM+ TDZ 0.8 µM	Shoots induced	Rafi and Salehi, 2018
		Nodal segments	MS	–	BAP 1 mg/L 2,4-D 0.2 mg/L	Shoots induced	Chauhan et al., 2022

Abbreviations: AC, Activated charcoal; ADW, Anderson's woody plant medium; BAP, 6-Benzylaminopurine; ER, Economou and Read medium; GA_3, Gibberellic acid; IAA, Indole-3-acetic acid; IBA, Indole-3-butyric acid; 2-iP, 2 Inorganic phosphate; KIN, Kinetin; LS, Linsmaier and Skoog medium; MS, Murashige and Skoog medium; NAA, Naphthalene acetic acid; NGE, New general epiphyte; OM, Olive medium; TDZ, Thidiazuron; WPM, Woody plant medium.

3.2.2 *Juglans regia* L.

An angiosperm species from the Old World, *Juglans regia* L. is a member of the Juglandaceae family. It is commonly called walnut tree. It originated in Iran and was then brought to Europe, the United States, India, China and other nations near China and India. This deciduous tree can grow to a height of 30–35 feet. There is only one species. On the same plant are the male and female flower blooms. The edible portion of the plant, known as the drupe, has significant health advantages. In addition to the fruit, phytochemicals have been found in the husks and leaves of the tree (Al-Snafi, 2018).

J. regia has been used traditionally to treat liver damage, arthritis, ulcers, and diarrhoea with an anti-helminthic effect. They have been recognized as possessing phenols, tannins, flavonoids, etc. Phytochemicals found in walnuts are flavonoids and its derivatives, glycerides and phenols. The main chemicals are juglone, quercetin, ferulic acid, gallic acid, pedunculagin, galactosidase and its derivatives. These are of medicinal importance due to their antioxidant activity, treatment of type 2 diabetes and has even shown anti-cancer properties (Panth et al., 2016; Table 3.2).

Due to their high polyunsaturated fatty acid (PUFA) content, walnuts are naturally an excellent agent for enhancing cardiovascular health. Mice consuming walnuts have lower levels of endothelin-1 (ET-1) according to an *in vivo* investigation. The vascular cells' generation of reactive oxygen species (ROS) is caused by this ET-1 (Davis et al., 2006; Hayes et al., 2016). The generation of ROS in the body causes a number of health issues. Neural toxicity may result, potentially impairing the patient's cognitive function and causing damage to the liver. Walnut leaf extracts enhance the body's capacity for cognitive function. By reducing the production of ROS, walnut leaf extract helps shield the body against oxidative stress, brain toxicity, heart issues and liver damage. This was shown in *in vivo* studies using mice that had CCl4-induced liver injury. They displayed lower ROS levels. It has been demonstrated that walnut protein peptides suppress angiotensin I-converting enzyme (ACE), preventing issues with hypertension (Gu et al., 2015; Wang et al., 2016). The lung cancer cell lines NCI-H322 and A549 demonstrate effective cytotoxicity at the early phases of the anti-cancer activity of the walnut extract juglone, according to Zhang et al. (2015). The terminal deoxynucleotidyl transferase dUTP nick end labelling (TUNEL) assay verified that the MDA-ME-231 breast cancer cell line had undergone apoptosis using a chloroform extract of the root bark of *J. regia* (RBJR). Bax and Bcl-2 proteins demonstrate apoptosis induction, illustrating how susceptible cells are to death signals (Alshatwi, 2011; Alshtawi et al., 2012). Additionally, *in vivo* studies using mice injected with HT-29 colorectal cancer cells have demonstrated that the mice administered with a combination of walnut extracts had slower tumour growth. *J. regia* leaf extracts have been used in an *in vivo* investigation and a 3-month clinical trial for the treatment of type 2 diabetes in mice and patients that were generated by toxins. Both demonstrated decreased levels of triglycerides, cholesterol, fasting response, and HbA1C. Furthermore, as previously observed in patients receiving drug dosages, the doses did not cause any significant adverse effects in the patients (Asgary et al., 2008; Hosseini et al., 2014). Because of these *in vivo* studies and clinical trials, the value of *Juglans* to humanity has become abundantly clear, necessitating the preservation and enhancement of the species' quality.

Toosi and Dilmagani (2010) conducted tissue culture of *J. regia* in New General Epiphyte (NGE) medium conditions, utilizing indole-3-butyric acid (IBA) for root induction and BAP (0.5 mg/L) for shoot induction. According to Kepenek and Kolağasi (2016), explants of shoot apices and leaves were utilized in NGE medium together with thidiazuron (TDZ; 2.5 mg/L) and IBA (2.5 mg/L) for the induction of shoots. Nuts have also been used for the *in vitro* propagation with the most popular MS media with BAP (4 μM) and IBA (12 μM). Embryo formation was also successfully demonstrated, indicating that more could be explored with walnuts (Amiri and Gharati, 2012). Additionally, root induction was carried out in the same medium for the propagules that had been shooted, using IBA (5 mg/L). Additionally, MS media has been employed for culture through nodal explants; the optimum response in terms of shooting and shoot length has been reported with BAP

TABLE 3.2
Chemical Structures and Uses of Some Important Bioactive Molecules Isolated from Medicinal Tree Species

S. No.	Species	Phytochemicals	Chemical Structure	Uses	References
1.	*Ginkgo biloba*	a. Quercetin	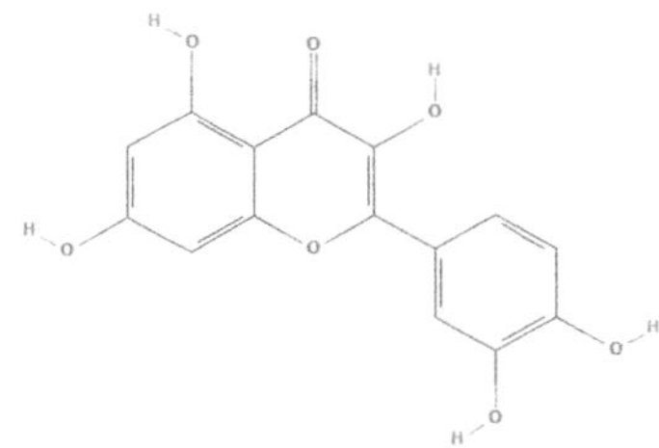	Oxidative stress and neurogenesis stimulation	Ude et al., 2013 Kang et al., 2010
		b. Luteolin	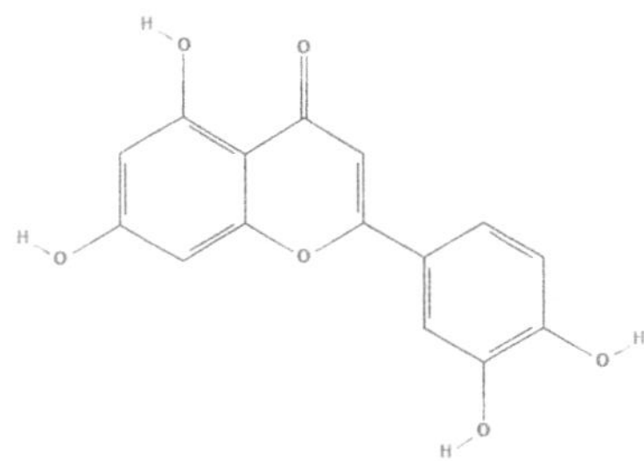	–	–
		c. Kaempferol	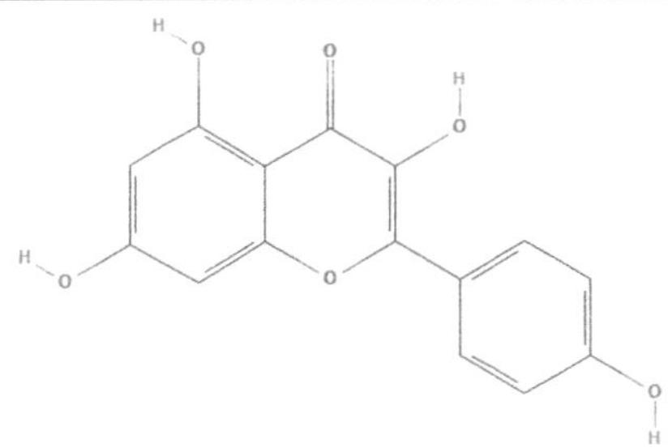	Prevention of pancreatic cancer and oral cancer	Zhang et al., 2008 Kang et al., 2010

		d. Ginkgolic acid	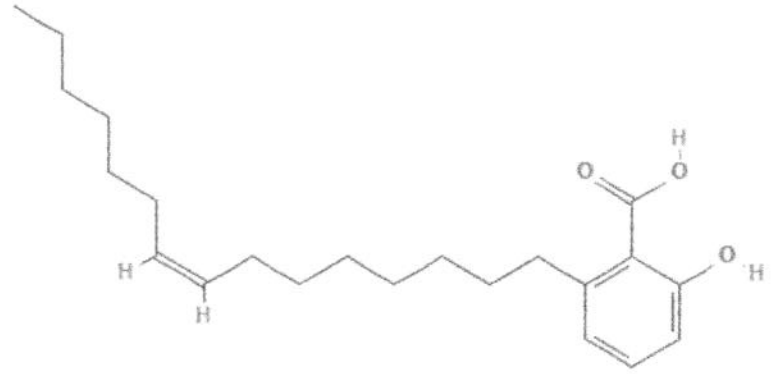	Prevention of hepatocellular carcinoma, cytotoxic effect against MCF-7, A549, HL-60	Yang et al., 2014 Oh et al., 2013
2.	*Juglans regia*	a. Juglone	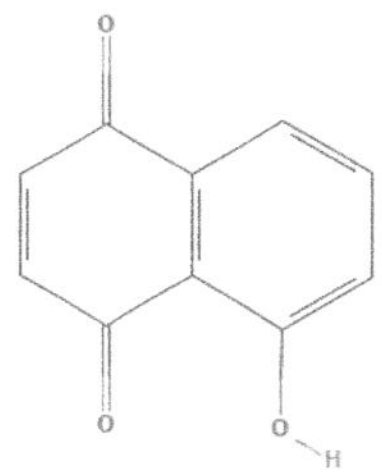	*Lung cancer cell lines*: A-549, NCI-H322, apoptosis of HL-60 cell line	Zhang et al., 2015 Soto-Maldonado et al., 2019
		b. Pedunculagin	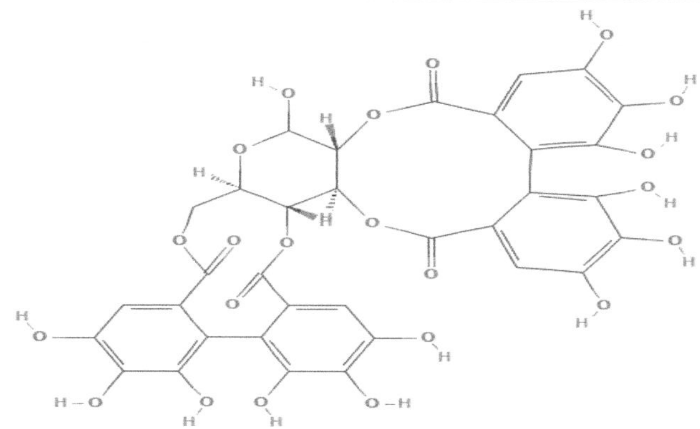	–	–

(*Continued*)

TABLE 3.2 (*Continued*)

Chemical Structures and Uses of Some Important Bioactive Molecules Isolated from Medicinal Tree Species

S. No.	Species	Phytochemicals	Chemical Structure	Uses	References
		c. Gallic acid	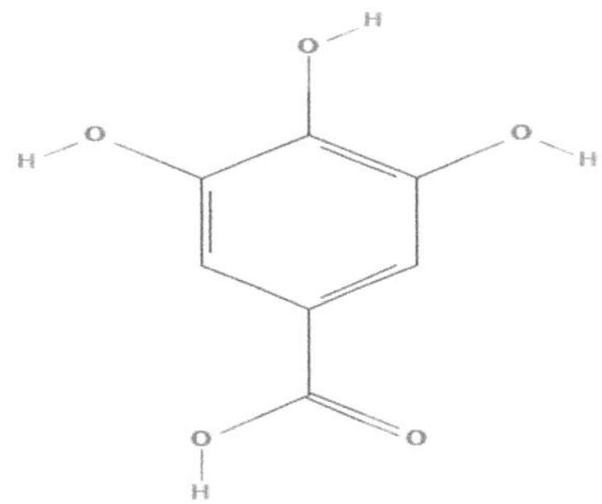	–	–
		d. Ferulic acid	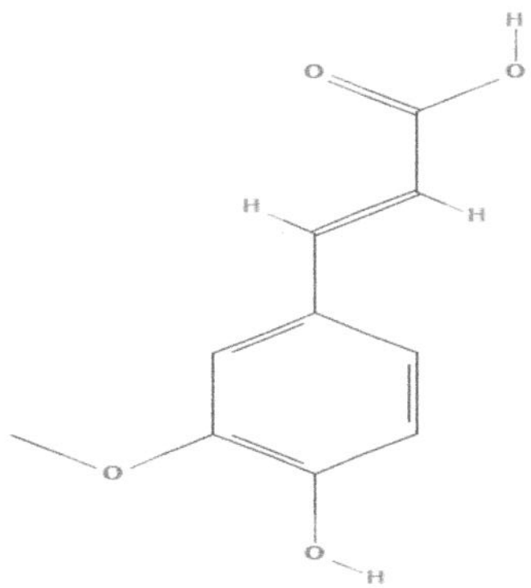	–	–

3.	*Azadirachta indica*	a. Nimbin	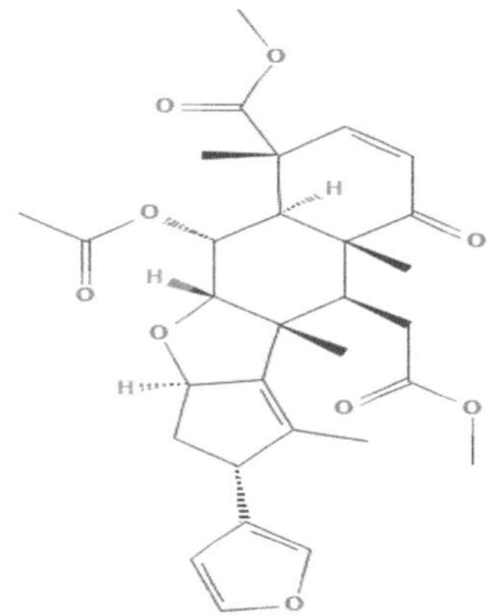	Anticancer activity observed *in vitro* in THP-1 cell lines	Sharma, 2020
		b. Nimbolide	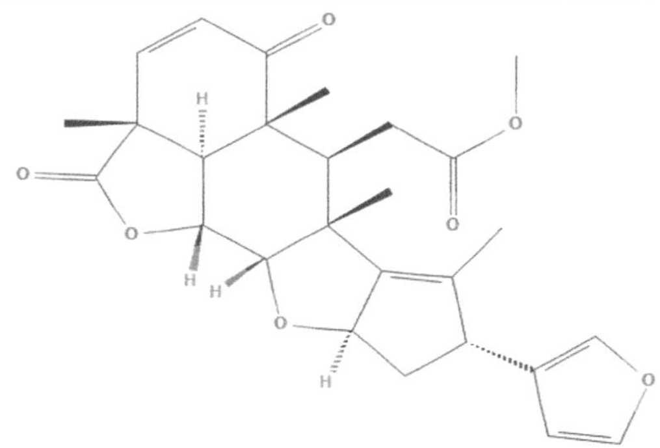	*In vitro* pulmonry fibrosis induction by TGF-B1 is controlled *in vivo* by bleomycin	Nagini et al., 2021

(*Continued*)

TABLE 3.2 (*Continued*)
Chemical Structures and Uses of Some Important Bioactive Molecules Isolated from Medicinal Tree Species

S. No.	Species	Phytochemicals	Chemical Structure	Uses	References
4.	*Terminalia arjuna*	a. Arjunetin	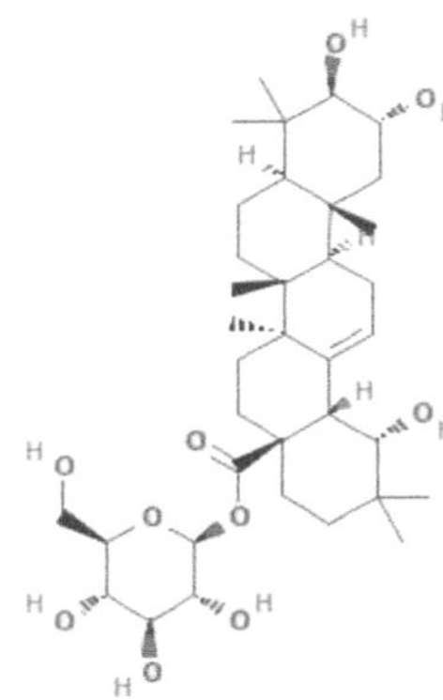	Inhibits alpha amylase enzyme activity (antidiabetic activity)	Shengule et al., 2018
		b. Arjunic acid Arjunolic acid	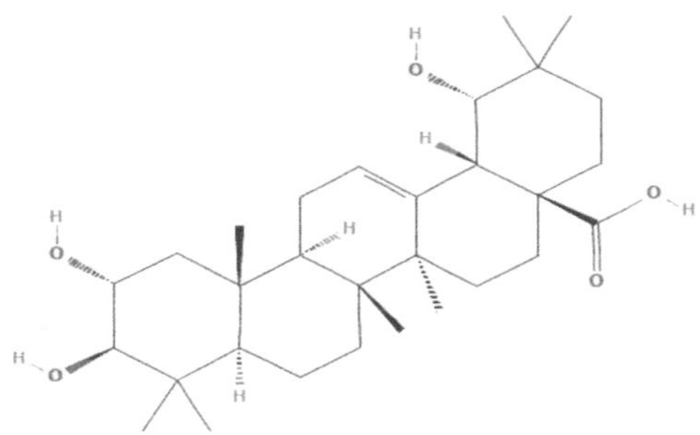	Cardioprotective nature against cyclosporin A–induced toxicity	Mythili et al., 2012
		c. Arjungenin		Free radical scavenging activity	Pawar and Bhutani, 2005 Nakayama et al., 2021

		d. Arjunglucoside-I	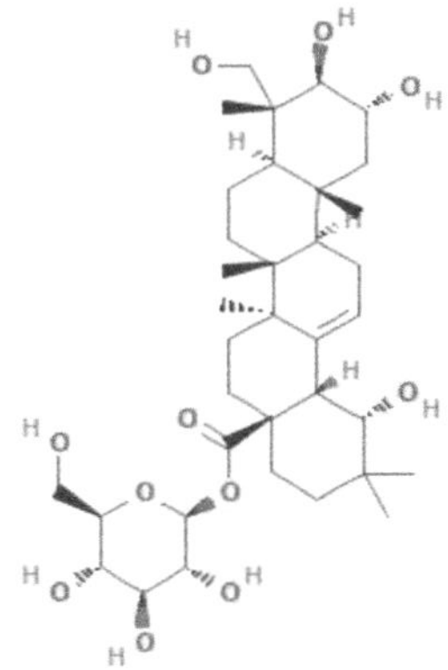	–	–
		e. Arjunglucoside-II	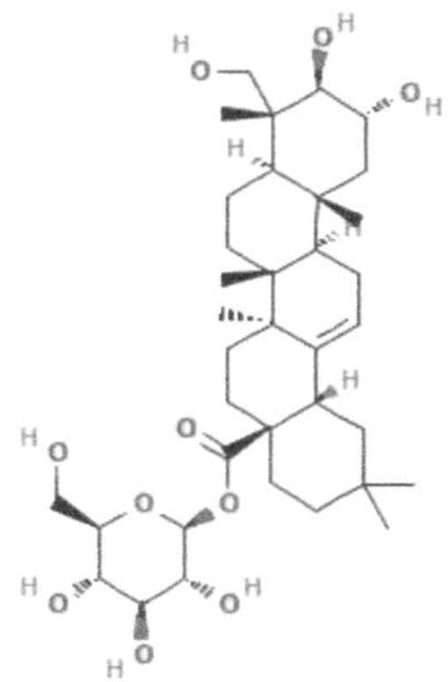	–	–

(*Continued*)

TABLE 3.2 (*Continued*)
Chemical Structures and Uses of Some Important Bioactive Molecules Isolated from Medicinal Tree Species

S. No.	Species	Phytochemicals	Chemical Structure	Uses	References
5.	*Eucalyptus* sp	a. Eucalyptol	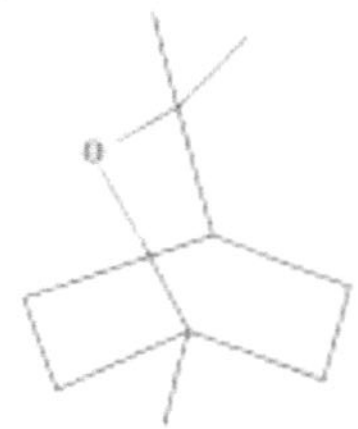	Respiratory problems, antiviral activity, antifungal growth, anticancer activity against HCT-116, MCF-7 and HepG-2	Galan et al., 2020 Tolba et al., 2015 Astani et al., 2010 Vimalanathan and Hudson, 2014 Ghareeb et al., 2016
		b. D-limonene	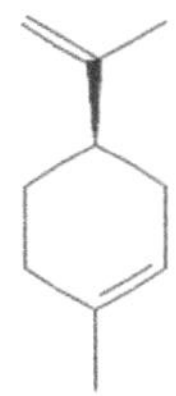	Antimicrobial acitivity with minimal adverse effects	Salvatori et al., 2023
		c. Globulol	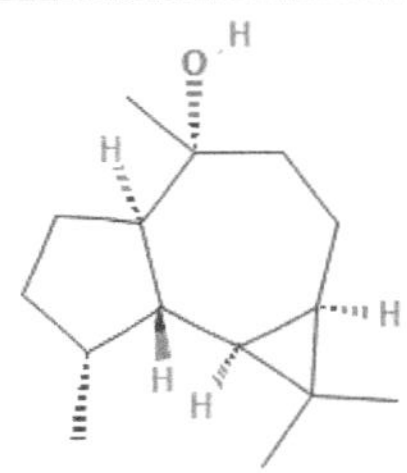	Antibacterial activity on *Bacillus subtilis* and other bacteria	Chaves et al., 2018

6.	*Rhododendron arboreum*	a. Rutin	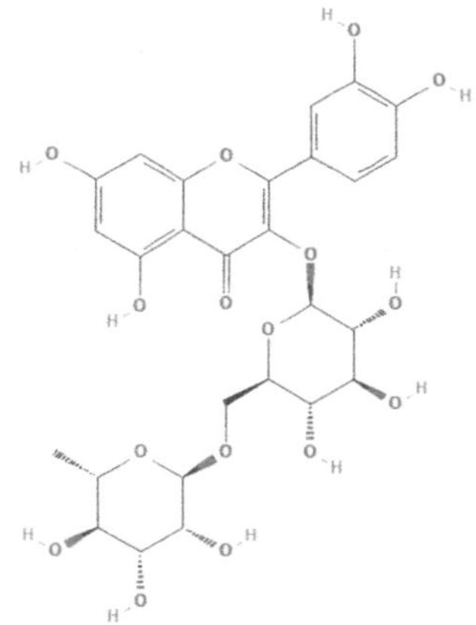	Antitumour activity	Sonar et al., 2013
		b. Quercetin	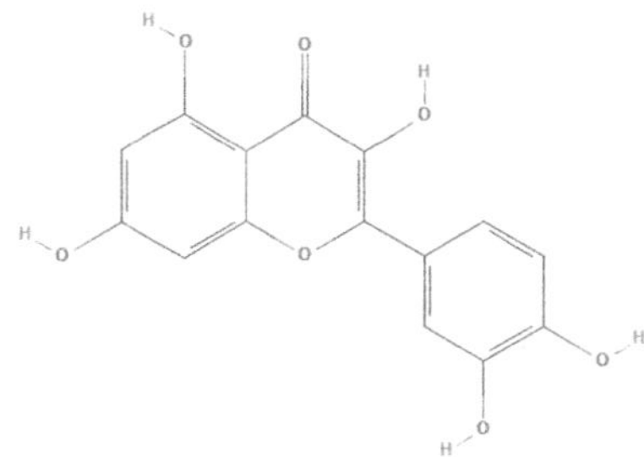	Antimicrobial activity, DPPH and ABTS scavenging activity	Sahu et al., 2013 Dede et al., 2019

(*Continued*)

TABLE 3.2 *(Continued)*
Chemical Structures and Uses of Some Important Bioactive Molecules Isolated from Medicinal Tree Species

S. No.	Species	Phytochemicals	Chemical Structure	Uses	References
		c. Hyperin	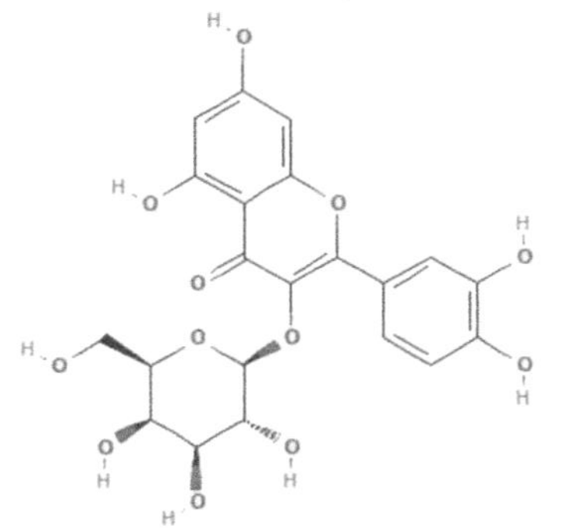	Antidibaetic activity, hypoglycaemic nature	Verma et al., 2012
		d. Betulinic acid	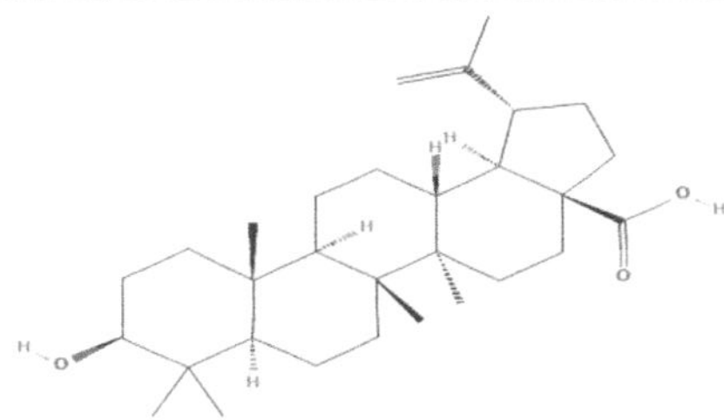	–	–

7.	*Ficus carica*	a. Luteolin	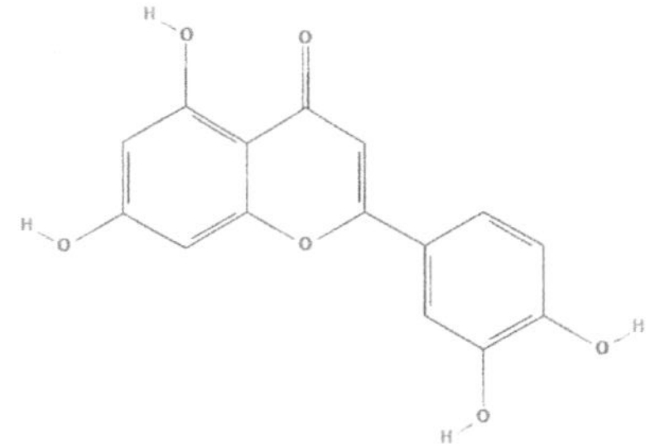	–	–
		b. Biochanin A	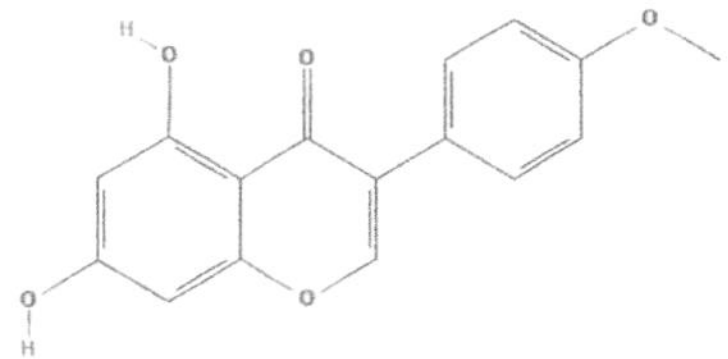	Cardioprotective nature, hypoglycaemic and antiobesity activity	Harini et al., 2012
		c. Psoralm	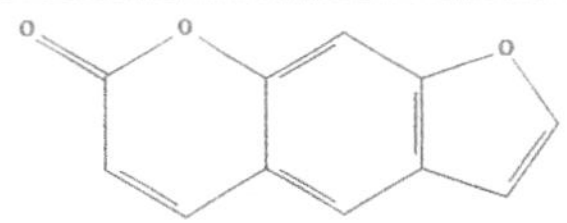	–	–
		d. Quercetin	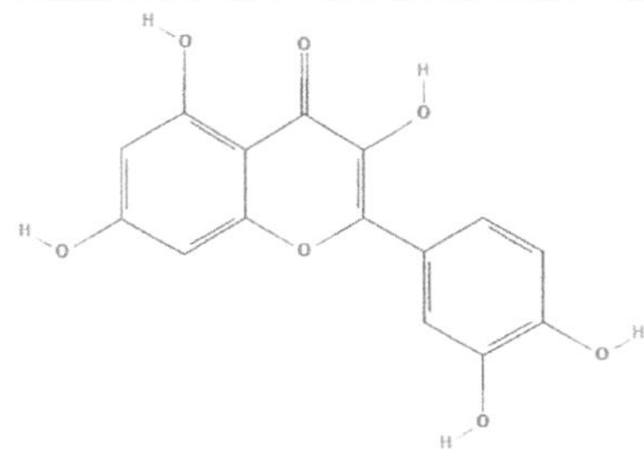	Antioxidant and DPPH radical inhibtion activity	Ude et al., 2013

(*Continued*)

TABLE 3.2 *(Continued)*

Chemical Structures and Uses of Some Important Bioactive Molecules Isolated from Medicinal Tree Species

S. No.	Species	Phytochemicals	Chemical Structure	Uses	References
8.	*Salix* sp.	a. Salicin	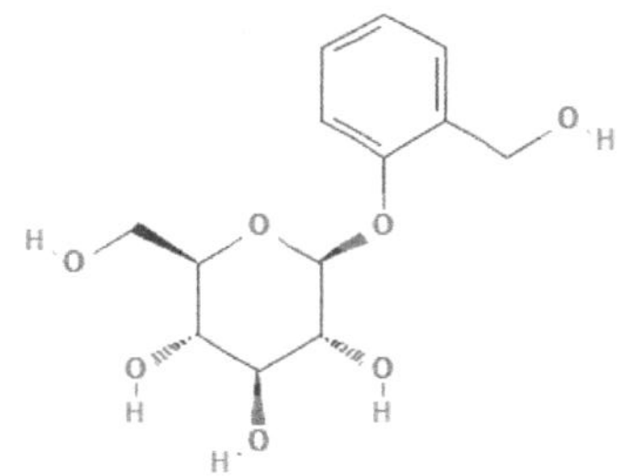	Inhibits cyclooxygenase (COX I, II)	Fiebich and Chrubasik, 2004 Antoniadou et al., 2021
		b. Luteolin	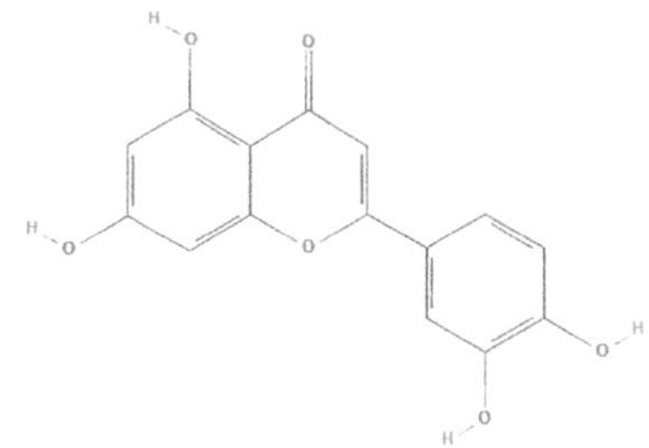	–	–
		c. Rutin	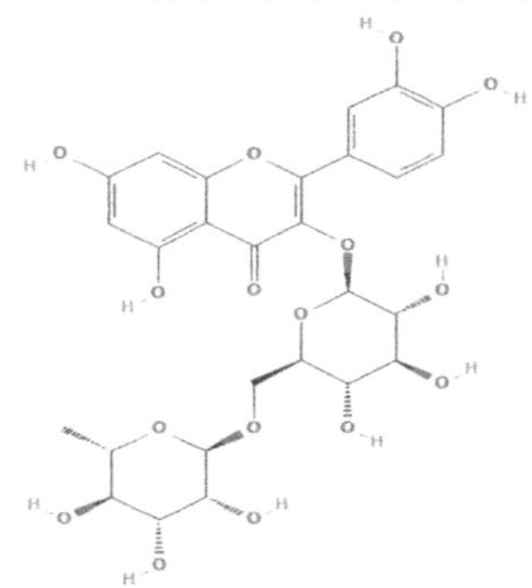	–	–

9.	*Olea europea*	a. Oleuropein	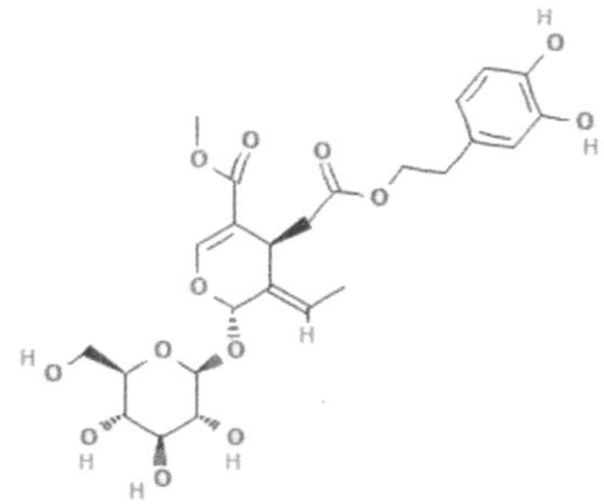	Inhibition of the LDL oxidation, neuroprotective nature with anti-inflammatory and scavenging activity	Omar, 2010
		b. Elenolic acid	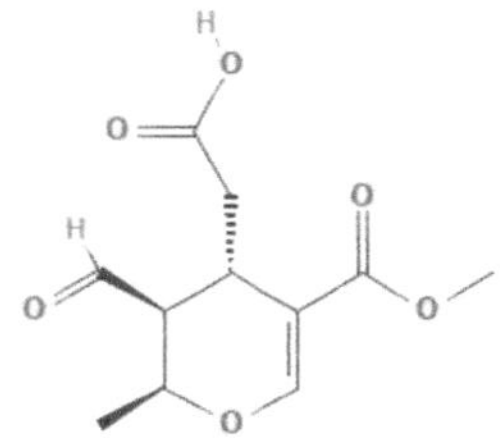	–	–
		c. Quercetin	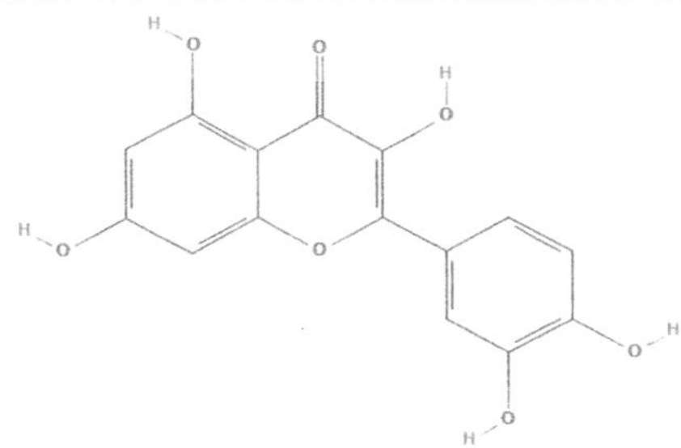	–	–

(*Continued*)

TABLE 3.2 (*Continued*)
Chemical Structures and Uses of Some Important Bioactive Molecules Isolated from Medicinal Tree Species

S. No.	Species	Phytochemicals	Chemical Structure	Uses	References
10.	*Taxus* sp.	a. Paclitaxel	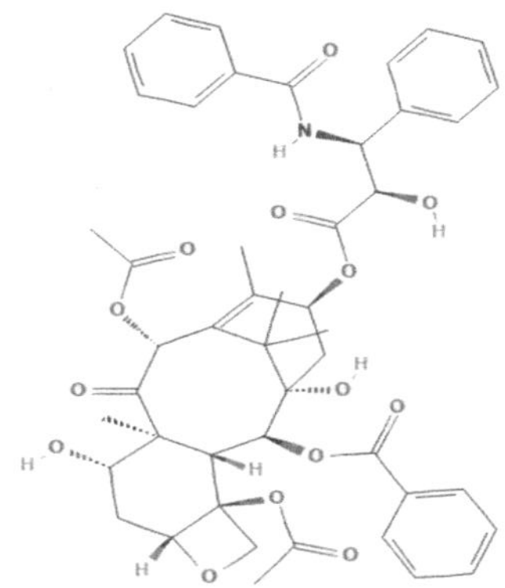	Anticancer properties	Suffness and Wall, 2021 Yan-hua et al., 2020
		b. Taxine B	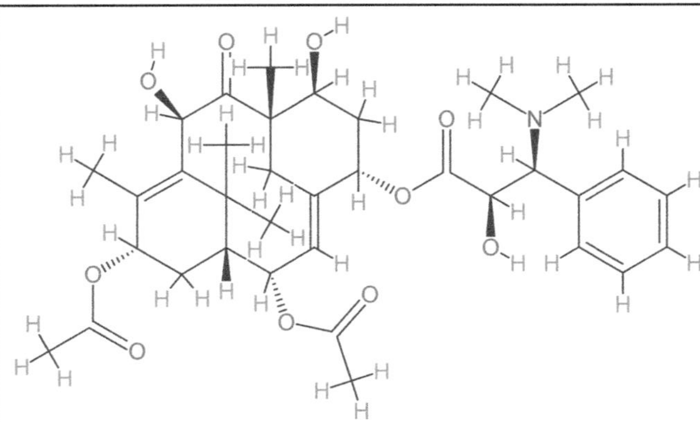	–	–

11	*Cedrus deodara*	a. Taxifolin	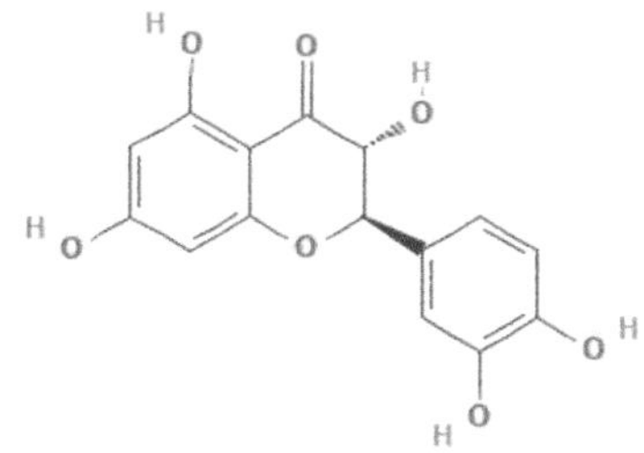	Antihypertensive effects, antioxidative nature, prevents neurotoxicity	Bernatova and Liskova, 2021 Kuang et al., 2017 Zhang et al., 2019
		b. Cedrin	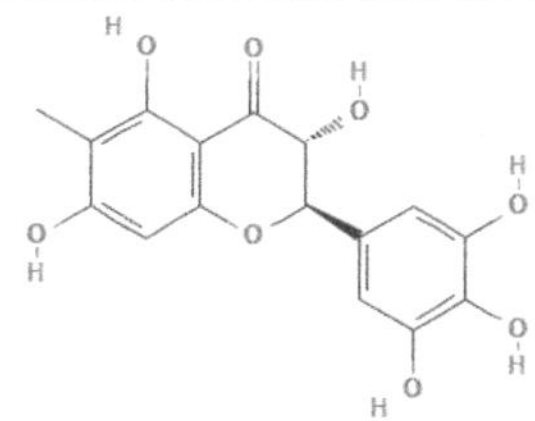	Improves mitochondrial membrane potential and permeability, increases cell viability with neuroprotective nature	Zhao et al., 2018
		c. Matairesinol	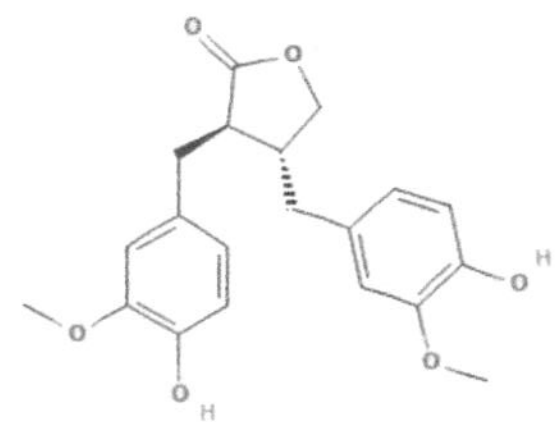	Supresses reactive oxygen species	Lee et al., 2012

(*Continued*)

TABLE 3.2 (*Continued*)
Chemical Structures and Uses of Some Important Bioactive Molecules Isolated from Medicinal Tree Species

S. No.	Species	Phytochemicals	Chemical Structure	Uses	References
12.	*Acacia auriculiformis*	Acaciaside B	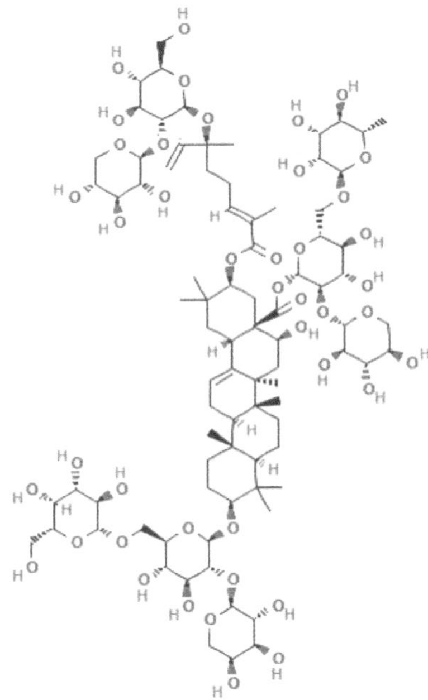	–	–
		Pedunculagin	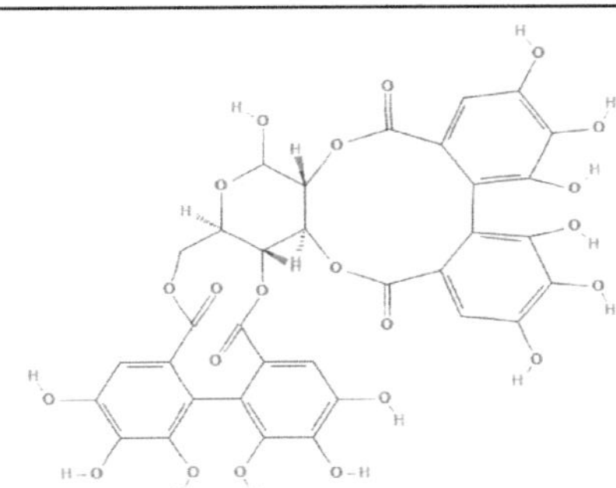	–	–

(2.5 mg/L) and KIN (0.5 mg/L), while IBA (1 mg/L) works best for rooting (Al-Mizory and Mayi, 2012). The main problem with *J. regia* propagation is the browning of the plantlets. In order to combat this, studies conducted by Bhat et al. (2022) revealed that the best outcomes were obtained when the plantlets were treated with ascorbic acid (550 mg/L) in DKW (Driver and Kuniyuki Walnut; Driver and Kuniyuki, 1984) medium (Bhat et al., 2022; Table 3.1). *J. regia* tissue culture is effective in increasing the amount of phytochemicals present in it using a variety of techniques. In a same vein, its medical qualities encourage further investigation of its facets.

3.2.3 *Azadirachta indica* A. Juss.

Azadirachta indica is a tree of the mahogany family, Meliaceae. The Persian language is the source of the scientific name. It is also known as Indian lilac, margosa and neem. This tree is tropical or subtropical, meaning it can withstand extreme high temperatures. In addition to southeast Asian countries like China, Bangladesh, Malaysia, Vietnam and Africa, it is indigenous to India. The temperature range for *A. indica*, an evergreen tree, is 22–35 °C. The tree can grow to a height of 20 metres and forms fruits known as drupe (Hashmat et al., 2012).

It is a rich source of phytochemicals which treat several health ailments and have had medical applications since the pre-historic times in the Indus valley civilization. Parts of the neem tree have been used by the Indians since time immemorial in their daily lives as a tonic for boosting immunity, blood decoction and gut health. Small branches of the tree are used as toothbrushes. Leaves are eaten and are often boiled to be used in bathing for eliminating skin problems. Neem oils are extracted from the tree and used for the same (Kumar and Navaratnam, 2013). The extracts are obtained from leaves, seeds, bark and its flowers. The fractionated extracts from various parts of the neem tree have been studied and found to have potential antimicrobial, anti-inflammatory, antioxidant and insecticidal properties, as well as being effective in digestive and skin ailments, etc (Faujdar et al., 2020).

Limonoids such as azadirachtin and its derivatives and terpenes such as nimbidin, nimbocinolide, nimbin, nimbolide, nimbilic acid, etc., are among the phytochemicals that were extracted. Azadirachtin-A to G is the primary phytochemical that is extracted and has a variety of uses (Saleem et al., 2018). *Escherichia coli* and *Staphylococcus aureus* are used to demonstrate the bactericidal effect of neem using its ethanolic extract (Pandey et al., 2014). Spermicidal activity of neem extracts has been demonstrated in rats (Idu et al., 2023). Glutathione levels, lipid profiles and glycaemic index can all be improved by neem extracts (Pingali et al., 2021). According to Chatterjee et al. (2011), clinical research has demonstrated that mouthwash containing neem extracts and chlorohexidine had similar effects on gingivitis and plaque (Table 3.2).

Leaflets have used for the successful embryo formation using MS media with KIN (6.9 μM) and indole-3-acetic acid (IAA 8.6 μM; Shekhawat et al., 2009). Nodal stem segments in MS media with BAP (1.11μM) and IAA (1.4μM) and adenine hemi-sulphate (81.43 μM) provided maximum multiple shoot induction. For *in vitro* rooting, MS media was used with IBA (0.2 mg/L). The survival rate was 100%, and the genetic fidelity and azadirachtin content were on par with the mother tree, determining this protocol as safe enough for the *in vitro* cloning of the neem tree (Arora et al., 2010). Rafiq and Dahot (2010) cultured different parts of the tree and found the best inoculation was done with immature flowers in MS media with 2,4-dichlorophenoxyacetic acid (2,4-D 1 mg/L), BAP (1 mg/L) and 1-naphthaleneacetic acid (NAA 0.2 mg/L). Suspension cultures were incubated in shakers at 25 ± 2°C at 100 rpm for 16 hours of light at 2500 lux. These cell suspension cultures of neem showed increased concentration of azadirachtin due to ammonium nitrate, 2,4-D and BAP (Rafiq and Dahot, 2010). Dhandapani et al. cultured various parts of the tree and observed the effects on the propagation. In ½ MS, nodes and cotyledons were utilized for shoot bud sprouting with BAP (3 mg/L), NAA (0.5 mg/L) and casein hydrosylate (1 g/L). It was shown that sub-culturing the cotyledons produced the greatest results. On the other hand, there have been observations of the fast multiple shoot induction but not plantlet regeneration at the end. Nonetheless, the cotyledons' effective rooting in IBA (2 mg/L) in MS media indicates that cotyledon *in vitro* culturing can be a successful technique for the tree's mass propagation (Dhandapani et al., 2021). The lateral and terminal buds have been cultured by Bello et al. (2022) by sterilizing them in mercuric chloride. They observed the growth using BAP and IBA using WPM, MS and Linsmaier and Skoog (LS) medium. To determine the optimal procedure, it was all done at various concentrations and combinations up until the hardening stage. The highest rooting was obtained with LS media with IBA (4 mg/L) and pulsing, whereas MS media containing BAP (0.7 mg/L) was optimal for shooting. At a room temperature of 23–24°C, hardening was carried out in a 2:1 ratio using peat and perlite soil (Bello et al., 2022). In MS medium, nodal segments used as explants have demonstrated the best and greatest BAP values of 08.88 μM and 4.44 μM, respectively. In MS medium, tryptophan at 146.89 μM was the highest concentration for rooting (Gehlot et al., 2014; Table 3.1).

3.2.4 *Taxus baccata* Zucc.

Taxus baccata Zucc. is an evergreen tree species of Gymnosperm belonging to the Taxaceae family. It is commonly called yew. *Taxus* is the genus, which is identified more on the basis of its geographic location due to the uncanny resemblance in its sub-species as and when classified. For instance, *T. baccata* is found in European countries and so is called European yew. *Taxus wallichiana* is found in Asia in the Himalayan regions and therefore is called Himalayan yew. In a nutshell, the *Taxus* genus is found in the Northern Hemisphere. Narrowing our readings to *T. baccata* Zucc, it is native to Europe but now is spread worldwide (Sharma and Garg, 2015). The tree size ranges from medium to large and can reach up to the height of 20 m.

Mention of the uses of *Taxus* has been made by the Greeks, Romans, medieval Europeans mythology and Italy along with Unani and Ayurveda TM systems. Its uses have been mentioned as cardiotonic, carminative, nerve problems, epilepsy, stomach problems and many more. The extracts of its leaves, wood, roots, etc., have been studied, and the leaves were found to have the maximum content of the phytochemical useful for humans (Akbar and Akbar, 2020). *Taxus* has been reported for various chemotherapeutic properties. The drug paclitaxel (Taxol®) extracted from the bark of the tree has shown its value as an anti-tumour drug (Suffness and Wall, 2021). It works by preventing the depolymerization of microtubules, which are responsible for the cell cycle, and thus arresting the cycle at the G2/M phase. This ultimately prevents the growth of cancerous cells (Sah et al., 2020). It has given positive effects in the treatment of breast, ovarian and lung cancer during clinical trials. The bark and needles are used as the source of paclitaxel extraction, and around 10,000 kg of bark is used to produce 1 kg of Taxol (Ishtiyak and Puni, 2017). This brings us to the need of cultivating more trees of *Taxus* to prevent exploitation of its natural forests and quantifying the chemical at the cellular level. Doing both can be a better option to meet the needs (Table 3.2).

Tissue culture of *Taxus* has been done for the obvious reasons owing to its contribution to the medical industry of treating cancer. The most effective media for the *in vitro* multiplication of *Taxus* from different explants has been determined to be MS. *Taxus* has been propagated in nutritional media. Branched twigs that have been incubated with AC, BAP and vitamins in MS medium have been used as explants to induce multiple shoots. Excellent rooting outcomes (acclimatization) were obtained by using ½ MS in conjunction with IBA for the rooting process (Abbasin et al., 2010). As explants, apical buds were cultivated in MS medium supplemented with 1 mg/L of KIN and 1 mg/L of IBA to induce numerous shoots. KIN (2 mg/L) and IBA (1 mg/L) were added to MS media for rooting (Sahari Moghadam et al., 2022). Additionally, leaf and shoot tips have been utilized as explants for growth on MS medium supplemented with KIN (3 mg/L) and AC for optimum leaf proliferation and shoot induction. Nonetheless, the propagated branches were rooted in WPM medium. Baharak et al. (2014) conducted a study on *Taxus* embryo cultivation that was successfully carried out on 100-year-old seeds, demonstrating that the most effective medium was ½ MS with AC

(2 g/L; Davarpanah et al., 2014). Analysis of the production of taxanes in various culture media has revealed that DKW, when combined with plant hormones, is the most effective medium (Table 3.1). Using the standards Taxol, Baccatin III and 10-DEBA (deacetylbaccatin III), HPLC was used to analyse its efficacy (Kajani et al., 2012). According to Kolewe et al. (2012), it has been reported that paclitaxel synthesis increases when *Taxus* suspension cultures are formed into smaller aggregates. Tissue culture has led to an effort to strike a compromise between the need to prevent tree exploitation and the necessity to produce enough Taxol to meet demand.

3.2.5 *Terminalia arjuna* Roxb. Wight and Arn.

The angiosperm *Terminalia arjuna* Roxb. Wight and Arn. is a member of the Combretaceae family. It is indigenous to the subcontinent of India. But it has since expanded to Malayasia, Bangladesh, Pakistan, Afghanistan, Thailand, China and Indonesia, among other South Asian nations. Arjuna is its colloquial name. It is 20–30 meters tall and evergreen. The tree's inner bark is red, which is most likely why it has a high concentration of phytochemicals. The tree bears drupe fruit and has a trunk with buttresses (Dwivedi and Chopra, 2014). Ayurveda makes reference to the tree's medicinal value. Its bark has been traditionally used to treat ulcers, fractures, spasmodic disorders, hypertension, cardiotonic effects, antimicrobial and antioxidant properties and nerve regulation (Amalraj and Gopi, 2017).

T. arjuna is a treasure trove of many phytochemicals with therapeutic qualities. These phytochemicals were detected by various chromatographic techniques, including HPLC, silica column chromatography, thin layer chromatography (TLC), reverse phase-HPLC (RP-HPLC) and liquid chromatography–mass spectrometry (LC-MS), as well as sterols, triterpenoids, saponins, carbohydrates, lactones, alkaloids, phenols, proteins and glycosides. All together, they have a cardioprotective quality since they can lower blood pressure, inflammation, oxidative stress, cholesterol and vascular problems. The bark and roots are used to extract these compounds. Their chemical makeup and side chains have allowed for their identification and naming. Root extracts contain triterpene carboxylic acid, whereas terpenoids and glycosides are isolated from the bark as terminic acid, arjunoside-III and arjunoside-IV. Glycosides called termiarjunoside I, II and arjunophthanoloside, an antioxidant and nitric acid inhibitor, are taken from the bark of the arjuna plant (Table 3.2). Tannins contained within the bark include punicalagin, pyrocatechols and casuarin. Flavonoids such as kaempferol, arjunone, gallic acid, ellagic acid and flavones are also extracted (Kapoor et al., 2014; Amalraj and Gopi, 2017).

T. arjuna has been cultured *in vitro* using MS medium. According to Kumar et al. (2010), apical buds and leaves were employed as an explant in the MS media along with KIN, 2,4-D and gibberellic acid (GA_3) as PGRs demonstrating multiple shoot induction. 2,4-D was also used in the MS media for rooting. Multiple shoot induction in the explants is caused by nodal segments in the modified MS media containing BAP (0.5 mg/L) and NAA (0.1 mg/L). Rooting is carried out in modified MS (mMS) media with IBA (250 mg/L) and 1-naphthoxyacetic acid (NOA 250 mg/L) in the *ex vitro* settings to shorten the regeneration time. The genetic fidelity of the plantlets was maintained with respect to the mother tree. Additionally, the goal of creating artificial seeds with an 80% plantlet survival rate was accomplished (Gupta et al., 2014). When BAP (4.44 μM) and NAA (0.54 μM) were added to axillary buds on MS medium, multiple shoot induction was observed. In ½ MS media with IBA (984 μM) and AC (100 mg/L), effective rooting took place. When producing trees in large quantities, this methodology works well. It is evident that the tree's chopped portions provide a better tissue culture candidate due to their effective sterilization (Choudhary et al., 2015). Nodal segments in MS media containing 3.9 μM BAP and 4.4 μM phloroglucinol resulted in shooting, while 2.5 μM of IBA in MS media led to roots in the propagules that had been shooted (Table 3.1). The protocol's effectiveness is demonstrated by the fact that the *in vitro* plantlets' genetic fidelity test yielded results that were similar to those of the mother tree (Dangi et al., 2014).

3.2.6 *Eucalyptus* L'Hér

Eucalyptus L'Hér, collectively called gum trees, are a group of flowering trees that belong to the Myrtaceae family, which includes over 700 species. The Greek words "eu" (meaning "well") and "calyptos" (meaning "covered") are the origin of the word "eucalyptus," which refers to certain species of the plant that have hooded blossoms. It is indigenous to Australia, where it grows to form pastures and forests. Their habitat is found in valleys, hill slopes, mixed forests and semi-arid to sub-alpine areas. In a nutshell, they can withstand stressful situations since they are adapted to a variety of habitats, including tropical, subtropical, dry and fire-prone areas of the planet. The southwest of Australia is where the eucalyptus diversifies. *E. deglupta* is indigenous to the Philippines, Indonesia, Australia and New Guinea. Different species have different common names with respect to the area of localization and characteristic features: *E. regnans* (mountain ash), *E. diversicolor* (karri), *E. deglupta* (rainbow gum), *E. alba* (white gum), *E. grandis* (red gum), *E. camadulensis* (river red gum), *E. viminalis* (manna gum), *E. globulus* (safeda, maiden's blue gum) and *E. melliodora* (yellow box). The species spans up to 10 metres, from shrubs to trees. In Australia, shrubby ones are collectively referred to as marlock. They have a single stem (*E. platypus*). Trees with narrow trunks are called mallet trees (*E. astringens*). The Eucalyptus multistem species (*E. gamophylla* and *E. polybractea*) are referred to as mallee species. Additionally, they contain epicorm and lignotubers, which are missing in marlock and mallet. The bark varies from species to species and might be smooth or rough, ribbon-like, deeply furrowed, stringy or iron-like. One characteristic of eucalyptus trees is its lignotubers. When damaged by fire or burned, they readily emerge again from latent buds (Salehi et al., 2019).

Eucalyptus oil has been used by Australians for various purposes as insect repellents and insect bites, mouthwash, body pains, fever, cold and sinus problems. Leaf, flower and bark extracts are abundant in phytochemicals that may have medical applications for humans. Antimicrobial, antiviral, antifungal, anti-inflammatory and insect repellent properties are among them. Additionally, it helps treat sinusitis, the common cold, congestion, respiratory conditions and even certain viral illnesses like influenza. The essential oil eucalyptol (1,8-cineole) from the leaves possesses antimicrobial and antifilm properties against *Candida* and gram-negative bacteria. When eucalyptus leaf powder was tested on *Bacillus subtilis* and *Escherichia coli*, the results showed a diversion of the zone of inhibition of 17 mm and 14 mm, respectively (Mumtaz et al., 2022). In 2012, Bachir and Benali reported similar positive outcomes with the gram-positive bacteria *Staphylococcus aureus* (Bachir and Benali, 2012). The most potent efficacy against *S. aureus*, *Streptococcus pyogenes*, *Streptococcus agalactiae* and *Streptococcus pneumoniae* has been demonstrated by *E. odorata*. *Pseudomonas aeruginosa* has demonstrated the greatest resistance (Elaissi et al., 2011). Depending on the oil content, pre-treating viruses with *E. biostata* and *E. astringens* has demonstrated antiviral effectiveness. *E. globulus* has shown positive effects in the light of the objective to decrease the drug-resistant mutant of the anti-influenza drug (Vimalanathan and Hudson, 2014). Astani and Reichling's (2010) tests using the HSV-1 viral protein have yielded positive findings (Astani et al., 2010). Using molecular docking to conduct an *in silico* investigation of SARS-CoV2, it was demonstrated that 1,8-cineole, -pinene and monoterpenes block the virus's membrane protein (Sharma, 2020; Panikar et al., 2021).

The heat treatment of *E. globulus* essential oil (EGEO) has demonstrated its efficacy as a natural beverage preservative by effectively inhibiting the growth of *Saccharomyces cerevisiae* in the juice of the genuine Orangina brand (Boukhatem et al., 2020). The radial growth of *Helminthosporium oryzae* (250 ppm), *Rhizoctonia solani* (250 ppm) and *Fusarium oxysporum* (500 ppm) has been inhibited by 500 ppm of *E. tereticornis* oil (Kaur et al., 2012). *Magnaporthe oryzae*, a rice blast fungus, has had its mycelial growth inhibited by oil derived from the hybrid of *E. grandis* and *E. urophylla*. It has been demonstrated that altering the products of expressed genes can reduce growth and cause a functional abnormality, which in turn impairs mycelial growth (Zhou et al., 2016), making it an effective biopesticide. *Fusarium* sp. isolates' mycelial growth has been suppressed by oil derived from *E. camaldulensis* leaves (Gakuubi et al., 2017). *E. citridora* has also been demonstrated to exhibit anti-biofilm properties against *Candida*. According to Maaloul et al. (2019), *E. camaldulensis* has high root metabolism properties and can manage waste water. It will help lower the prevalence of water-borne illnesses like dysentery, typhoid and diarrhoea.

Epicormic shoots of the eucalyptus have been used in WPM media with BA (0.60 mg/L) for shooting and NAA (0.10 mg/L) for rooting (Silva de Oliviera et al., 2015). Nodal segments have been used in both WPM and MS media with BAP (0.1 mg/L) and NAA (0.05 mg/L) and BAP (0.50 mg/L), respectively, for bud formation by Brondani et al., 2012 and Navroski et al., 2014. Cotyledons have been used in WPM media with TDZ + 4.4 mM BAP and 0.89 mM BAP + 0.05 mM NAA in mini-incubators leading to the growth of adventitious buds (de Oliviera et al., 2022). *In vitro* rooting of the eucalyptus has not shown efficient results. So, the rooting is done in the natural conditions to increase the efficiency of the mature plants (Table 3.1).

It is impossible to discount the therapeutic efficacy of eucalyptus, and the generation of extracts during tree growth can be measured using tissue culture techniques. When dicamba and picloram (4.1 μM) were present as PGRs, the cotyledons of *E. urophylla* and *E. grandis* explants demonstrated more somatic pro-embryo induction (Moura et al., 2017). To summarize, this is a species that is endemic to Australia and varies in its ability to adapt to different geographical and climatic situations. Its ethnobotanical significance contributes to its medicinal usefulness. As such, it cannot be disregarded in light of possible future opportunities in the research and development sector.

3.2.7 *Rhododendron* Sp.

Rhododendron arboreum Sm. is an evergreen tree belonging to the family Ericaceae. The word rhododendron is derived from the Greek words "rhodo" meaning rose and "dendron" meaning tree, as the tree bears red-coloured flowers. It is found over the height of 1500–3600 asl in the temperate forests. However, there are several species of *Rhododendron* growing all over the world. *Rhododendron* is found in the Indian sub-continent and other South Asian countries which includes Malaysia, Burma, Sumatra, Japan, Thailand and China and in Europe. In India, it is distributed from the east of Kashmir to Nagaland. It is capable of reaching up to the height of 15 metres. Its fruit is a capsule, and flowers vary in shades of red to pink and white blooming in a single truss. The bark is soft, and its flowering occurs from March to July. The significance of the tree its diverse uses, from timber to medicinal, keystone nature and cultural value (Iqbal et al., 2023).

The petals of the flowers are used in making of jellies, jams and juices which are of medicinal importance. Fresh flowers are used in curing diarrhoea and blood dysentery. Leaves are useful in reducing fever and headaches. Several research studies have been done on the animal models from the extracts of *R. arboreum*. The antinociceptive effect and anti-inflammatory effect of the extract has been evident in mice. This has been due to the presence of flavonoids (hyperin) or other phytochemicals in the ethyl acetate extract (Table 3.2). `Ethyl acetate extract of the *Rhododendron* has shown an antihepatic nature. Several tests on animal models with the *R. arboreum* plant (RAP) extracts have led to fruitful results as well. CCl_4-induced liver damage in curative and preventive model of the rats was tested upon to check the effects of the extract. The treatment with this extract has caused the decrement of glutathione and other serum levels like SGOT, bilirubin, SALP, etc. (Verma et al., 2010, 2011). Magnesium- and castor oil–induced diarrhoea in mice has been regulated by the dose-dependent treatment of *Rhododendron* extract (Verma et al., 2011). The aqueous methanolic extract has inhibited the α-glucosidase level in the intestine of the rats. Also, the ethyl acetate extract of α- glucosidase inhibitor is determined as potent antidiabetic medicine. *In vivo* studies have shown *R. arboreum* flower as a good replacement for synthetic antioxidants available in the market. The extract has shown hypolipidemic and hypercholesteraemic effects, determined by the experiments by Verma et al. (2011). So, it is good replacement of synthetic antioxidants. Lipopolysaccharide (LPS) causes oxidative damage through ROS by causing organ injury, especially in the liver and brain. Other than this, LPS-induced problems in the body include renal issues causing kidney damage, hyperbilirubinemia and abnormal levels of albumin in the body. *R. arboreum* polysaccharides extracts in appropriate dosage amounts have a potential antioxidant nature which combats the ROS and improves the condition of the aforementioned issues. RAP treatment helps in restoring the levels of SOD and CAT enzymes, which are helpful in inhibiting ROS. LPS-induced inflammation and injury due to gram-negative bacteria causes septic shock in the body which is directly caused by reduced tumour necrosis factor alpha (TNF-α) levels in the body. So, RAP extracts help in neutralizing the antibodies, which prevents the TNF-α levels from going down and thus prevents organ failure deaths. However, this is only efficient when the RAP treatment is given in the short-term pre/post after stimulus (Ahmad et al., 2020). The cardiac protection property of *Rhododendron* flower has efficacy due to the presence of flavonoids present in it, as has been shown in rabbits by Yuan et al. (2006). Alcoholic extracts of the *R. arboreum* leaves have shown ability to suppress humoral immune response and cell-mediated immunity in Levamisole mice (Sonar et al., 2013).

Hypolipidemic and hypocholesteraemic effects were determined by the administration of *Hypophaerhaminoides* leaves and *R. arboreum* flower juice in the ratio of 1:4 (Murty et al., 2010). Diet-fed rabbits have shown an increase in high-density lipoprotein (HDL) and decrease in low-density lipoprotein (LDL) and other parameters by the consumption of *R. arboreum* flower (Thangaraj, 2013).

Traditional methods of propagation are even today for its growth. The problem arises due to the phenolic substances that inhibit the root initiation. Ahmad and Puni successfully initiated rooting with 60% efficiency using IAA (1000 ppm) and NAA (2500 ppm) and modifying the air layering method to prevent the phenolic from inhibiting the rooting; they called it the FRI-wire technique (Ahmad and Puni, 2010). *In vitro* culturing of *Rhododendron* has been done by its flowers with the ovary and pedicle in Anderson's media using isopentyl adenine causing multiple shoot induction and, for rooting IBA was used along with Anderson's media. (Mao et al., 2011). Nodal segments of *R. fortunei* were used to do *in vitro* propagation in two different media: Economu and Read media (ER) and WPM along with two plant growth regulators: 6-purine (4 mg/L) or zeatin and NAA (1 mg/L). Axillary shoots were induced. Those axillary shoots were used further for the rooting in both the media along with the auxin IBA (1 mg/L; Wei et al., 2018). Sandeepani et al. did *in vivo* and *in vitro* propagation of *R. arboreum* subsp. *zeylanicum*. They were able to successfully propagate *R. arboreum* seeds *in vivo* using coir dust:sand in the ratio of 1:1. In the *in vitro* propagation of the *Rhododendron* seeds they observed that sterilization with 20% Clorox with few drops of Teepol and shaking for 10 minutes is better for the seeds. Regarding the media, it was observed that Anderson's media was found to be more effective followed by ½ MS media. Full-strength MS media showed a high contamination percentage (Sandeepani et al., 2021; Table 3.1). These results can be used for further research in the propagation of the *R. arboreum*. Protocols have also been developed successfully for the other species of the genus *Rhododendron*. There is further scope of discovering a novel protocol for the *in vitro* propagation of *Rhododendron arboreum*.

3.2.8 *Ficus carica* L.

Ficus carica L., commonly called fig tree, belongs to the family of mulberry, Moraceae. It is a genus of 800+ species of shrubs, vines and trees distributed pantropically. It is distributed from Turkey to India and the Mediterranean region. Around 36 species are found in Nepal. The characteristic features of fig are their method of pollination, which depicts mutualism with the fig wasps and their special inflorescence, syconium. Different species have different common names based on the morphological characteristic and their cultural/aesthetic values (Crisosto et al., 2011): *F. religiosa* (peepal), *F. benghalensis* (banyan tree), *F. carica* (common fig), *F. benjamina* (weeping fig) and *F. palmata* (Punjab fig or bedu).

The extracts of the tree were used to treat inflammation-related diseases, ulcers, skin ailments of various kinds from the microbial or insecticidal sources and swelling. *Ficus* sp. is rich in minerals like Ca, K, P and Fe and phytochemicals like flavonoids, luteolin, quercetin, biochanin A, bergapten, psoralen, phytosterol, anthocyanin amino acids, phenols, fatty acids, etc. (Table 3.2). They are extracted from the leaves, bark and roots of the plant (Lansky et al., 2008). Owing to its richness of phytochemicals, there is richness in pharmacological uses too. Patil et al. (Patil and Patil, 2010) demonstrated antipyretic and anti- inflammatory effects of the *Ficus* extracts comparable to the standard drug. The ethanolic extract of the leaves showed reduction in the body temperature in a dose-dependent manner. Its efficacy was comparable to the standard drug paracetamol. The anti-inflammatory effects were demonstrated by the hydroalcoholic extract of *F. carica* leaves in carrageenan0induced inflammation in paws of rats, which were comparable to the standard drug indomethacin (Ali et al., 2012). The antioxidant property has been demonstrated by Allahyari et al. (2014) through methanolic extract of *F. carica* leaves. It showed a reduction in ischaemia and reperfusion-induced cardiac arrhythmia and infarction in rats. Bey and Louaileche (2015) showed more antioxidant efficacy in the extracts of the dark variety of *F. carica*. The antioxidative stress action is decreased due to dried fig consumption, and this is seen in ethanol-induced liver damage, which causes hepatotoxicity and oxidative stress. A total of 10% dried fig (DF) consumption in such rats caused fewer lesions and lower levels of liver damage serum marker enzymes like lactate dehydrogenase (LDH), aspartate aminotransferase (AST) and alanine aminotransferase (ALT) levels (Turan and Celik, 2016). The antioxidant activity is due to the presence of flavonoids, saponins and polyphenolic compounds. *F. carica* leaf extracts have shown cytotoxic effects and apoptosis of cancer cell lines of MCF7 (breast cancer), HepG2 (hepatocellular carcinoma), CaCo-2 (colon cancer) and Hep-2 (human laryngeal carcinoma). Hep-2 and HepG2 have shown the most inhibition percentage (Abdel-Rahman et al., 2021). Huh7it liver cancer cell lines have shown apoptosis by activating p53 and p21 genes (Purnamasari et al., 2019). Improvement in the motor coordination, spatial memory and cognitive deficits have been shown as the result of the consumption of fig in a transgenic mice model. Hexane extracts of *F. carica* leaf extracts have also shown significant results on the memory and behavioural models in Swiss albino mice (Vasundhara et al., 2013). Methanolic extracts of *F. carica* leaves have shown strong nematicidal activity against *Caenorhabditis elegans, Bursaphelenchus xylophilus* and *Panagrellus redivivus* (Liu et al., 2011). The antihelminthic property was also evident from *F. carica* extract in *Pheretimaposthuma*, and it was again comparable to the standard drug mebendazole (Patil and Patil, 2010).

During the tissue culture of *F. carica*, Dhage et al. (2012) used *in vitro* leaf explant in MS media, and callusing was observed with the addition of NAA (0.5 mg/L) and TDZ (7 mg/L). Rooting was successful in ½ MS media with IBA (1 mg/L) and AC (2 g/L). In 2013, Taha et al. conducted tissue culture of *F. carica* using shoot tip as an explant. Shooting was done in mMS media with BAP (0.5 mg/L) and root induction in half strength of MS media with NAA (0.1 mg/L) and BAP (0.5 mg/L; Taha et al., 2013). Danial et al. used leaf discs as explants in MS media with BAP (0.3 mg/L) for shoot induction and IBA (0.5 mg/L) and NAA (0.5 mg/L) for root induction in the same media (Danial et al., 2014). Moniruzzaman et al., 2015 did shoot induction of *F. carica* by shoot tip using MS media with BAP (0.5 mg/L) and NAA (0.2 mg/L; Moniruzzaman et al., 2015). Using a nodal as an explant was used by Shahcheraghi and Shekafandeh (2016) in the Dehdez variety of *F. carica* and Dessoky et al. (2016) in *F. carica*. MS media was used by both with BA (0.5 mg/L) and 2iP (0.2 mg/L) by the former and BAP (3 mg/L) and KIN (0.1 mg/L) by the latter for shoot induction. For rooting, IBA (1.5 mg/L) at the same strength of MS media was used by the former and half-strength MS media with IAA (2 mg/L), IBA (0.1 mg/L) and AC (2 g/L) by the latter (Shahcheraghi and Shekafandeh, 2016; Dessoky et al., 2016). Ling et al., 2022 has used shoot tip in MS media with BAP (0.5 mg/L) for shoot induction and WPM media with IBA (3 mg/L) for rooting (Ling et al., 2022; Table 3.1).

3.2.9 *Acacia auriculiformis* A. Cunn ex. Benth.

Acacia auriculiformis A. Cunn ex. Benth. is an evergreen tree belonging to the Fabaceae family. The word *Acacia* is derived from the Greek word "akis" meaning pointed/barb, and *auriculiformis* is derived from two Latin words, "auricula," meaning external ear of an animal and "forma," meaning figure, referring to its pod shape. It is commonly called "Australian wattle" in English and "babool" in Hindi. It is native to Australia, Indonesia and New Guinea and now is found worldwide due to its tolerance to abiotic stresses of water, soil, season, etc. It is found in all states of India except Jammu and Kashmir, Sikkim and Andhra Pradesh. The height of the tree can reach up to 30 meters, and the fruit is an oblong-shaped pod (Haque et al., 2021).

The ethnomedicinal importance of the tree includes its treatment in several ailments like eye-sore, rheumatism, antimalarial nature, etc. This has been due to the presence of several phytochemicals like flavonoids, saponins, tannins and anthocyanidins. The pharmacological value of the tree has been proved by various researchers in their studies on the biological activity of these phytochemicals (Rangra et al., 2019).

The antioxidant activity has been evident by DPPH radical scavenging activity using ethyl acetate extract of the bark (Urmi et al., 2013). The methanolic fruit extract has shown 78.9% antioxidant activity and 73.1% phenolic activity using the same method (Prakash et al., 2011). Antifungal and antibacterial activity has been demonstrated by Mandal et al. (2005) by acaciaside A and B (saponins). They inhibited the growth of fungi, *Aspergillus ochraceous* and *Curvularia lunata*, and bacteria, *Bacillus megaterium, P. aeruginosa* and *Salmonella typhimurium*. Ethanolic leaf extract has shown significant results in *Plasmodium bergheii*–infected mice in dose-dependent manner and that too in comparison to the standard drug chloroquine (Okokon et al., 2010). Microfilaria infection has shown to reduce to more than 80% by the intake of 100 mg/kg of the extract (Chakraborty et al., 1995). The presence of tannins, flavonoids and phenolic compounds have a potent wound healing capability due to the hydroxyproline content and others (Singh and Sharma, 2014; Table 3.2). The ethyl acetate–soluble portion of the extract has CNS depressant capacity due to the presence of auriculosides (Hussain et al., 2016). The extracts of the bark and empty pods have shown promising results in diabetes and in restoring the normal levels of liver damage serum markers. The results were comparable to the standard drug silymarin for liver injury (Sathya and Siddhuraju, 2013). Also, the extract has shown *in vitro* spermicidal activity more than the standard drug Triton X-100 with no mutagenic potential (Sharma et al., 2016).

Nodal explants were cultured in MS media with BAP (2 mg/L) and NAA (0.1 mg/L) for shoot induction, and for rooting MS media was used (Girishankar, 2011). Nodal segments were used by Ismail et al. (2012, 2016) as well. In the previous year they did two cultures, one with BAP (0.5 mg/L) alone and the other with BAP (2 mg/L) and KIN (2 mg/L) together to multiply shoot induction. In the later year, they used MS media with BAP (0.1 mg/l) for shooting. Banerjee (2013) cultured cotyledons in MS media with 2iP (inorganic phosphate; 2 mg/L) for shoot induction and NAA (1 mg/L) for root induction. Shoots as explants were also cultured in MS media with KIN (2 mg/L) and IAA (0.5 mg/L) for shoot induction and IAA (0.1 mg/L) for root induction (Yadav et al., 2016; Table 3.1).

3.2.10 *Salix* sp.

Salix sp., commonly called willow, belongs to the Poplar family Saliaceae. The genus includes approximately 500 species of shrubs and trees distributed in the Northern Hemisphere in Asia, Africa, North America and Europe. The height of the tree reaches up to 10 metres. The flowers are dioecious and in catkin inflorescence (Wagner et al., 2021).

Different species of *Salix* have different common names with respect to their morphological features like *S. alba* (white willow), *S. terasperma* (Indian willow), *S. egyptiaca* (musk willow) and *S. babylonica* (weeping willow). Phytochemicals include flavonoids, glycosides, cyanidins, terpenes etc., extracted from the leaves, roots and bark of the tree. Salicinoids, derivatives of salicin, are the characteristic phytochemicals derived from the *Salix* sp. *Salix* was used traditionally for gastrointestinal problems, inflammation and fever. *S. egyptiaca* is used in anaemia, vertigo and as a cardiotonic agent (Asgarpanah, 2012; Table 3.2). *S. alba* is used in fever, acute inflammation, pain and infection. *S. tetrasperma* is used in epilepsy, piles, diabetes, rheumatism and swelling stones (Prashith Kekuda et al., 2017).

The antibacterial property of the bark extract of *S. alba* has been shown against the dental biofilm of *Streptococcus mutans* and *Lactobacillus* (Fayaz and Sivakumaar, 2014). *S. babylonica* shows its activity against gram-negative organisms as compared to the broad-spectrum antibiotic thiamphenicol. *S. tetrasperma* has shown anti-quorum-sensing activity against *P. aeruginosa* (Mostafa et al., 2020). *S. egyptiaca* extract has an antiretroviral effect (Eftekhari et al., 2014). The hydroalcoholic extract of *S. tetrasperma* has an anti-inflammatory effect in a carrageenan-induced rat paw model due to the presence of phenolic glycosides: salicin, flavonoids, luteolin, quercetin and rutin (Kishore et al., 2014). *S. canariensis* decreases writhing moderately in formalin-induced pain (Gutiérrez et al., 2017). It has also been reported as effective with dose-dependent cytotoxic effects against human colon and lung cancer (Hostanska et al., 2007). The extracts of *Salix* have also shown neural stimulating properties capable of affecting the muscles and thus locomotion (Virupaksha et al., 2016). The extract has an antioxidant property along with hepatoprotective ability.

Tissue culture aspects have improved in *Salix* with increased biotechnological interventions. Regueira et al. (2018) used temporary immersion systems for the *in vitro* propagation of the *Salix* sp. They used nodal segments of the *S. vimnalis* in liquid MS media with half-strength nitrates and BAP (0.22 μM) in the bioreactors with periodic aeration. The root induction was spontaneous after the shoot induction with successful acclimatization rate. This experiment clearly stated the applications of Plantform™ for shoot culturing and RITA® vessel for yielding in 4–6 weeks. Khan and Anis (2012) used nodal segments of *S. tetrasperma* in WPM media with TDZ (2.5 μM), BA (1.0 μM) and NAA (0.5 μM). Rooting was done in half-strength WPM media with IBA (0.5μM). Nodal segments or shoots of *S. retusa* and *S. alpina* gave multiple shoot induction in the presence of MS media with KIN (0.5 mg/L) and AC (2 g/L). Also, the addition of several antioxidants favoured multiple shoots. (Chornobrov et al., 2019). Parzymies et al. have been working on the *Salix* sp. for several years now. The shoots of *S. lapponnum* used as explants showed optimal shoot induction in MS media with KIN (mg/dm^3) and IAA (0.5 mg/dm^3; Parzymies et al., 2023). Later, Parzymies et al. (2023) used shoots again for *in vitro* culture of *S. myritilloides* in MS media (Table 3.1). To their surprise, the control setup showed better response with the use of any plant growth regulators.

3.2.11 *Olea europaea* L.

Olea europaea L. is a deciduous tree belonging to the family Oleaceae. It is commonly called olive tree. It is native to the Mediterranean countries in the Europe and some Asian and African countries. Now it is cultivated in Australia, New Zealand and North and South America with a variety of cultivars. Its height reaches up to 15 meters, and fruit is called drupe. The cultivars vary in the morphology and quality of the olive (Khan et al., 2007). The major commercial use of olives is in the manufacturing of olive oil. The cultivars available have variations in olive oil quality.

The traditional uses of olive tree include treating gallstones, diuretic problems, diarrhoea, hair loss, gastrointestinal problems, inflammation, diabetes and immunity booster (Khan et al., 2007).

Phytochemicals are present in the bark, leaves, fruits and seeds of the tree. It includes flavonoids, secoiridoids, phenols, glucosides, tyrosol and hydroxytyrosols, lignans and triterpenoids. One of the most potent of all phytochemicals present in it is oleuropein (Table 3.2). It has multiple pharmacological activities as an antioxidant, antimicrobial, antibacterial, neuroprotective, immunoprotective, prevention of LDL oxidation, thyroid, antidiabetic and platelet aggregation, among others. Olive extract helps in preventing the oxidation of LDL, which is the most harming state of cholesterol for the heart, and also helps in dilation of the arteries, reduction in angiotensin-II levels, biomarkers for oxidative stress and lowering the blood pressure. This has been demonstrated in *in vivo* (Vazquez et al., 2019) and in human trials (Hermans et al., 2020). Since the major health issues arise due to the presence of ROS production in the body, the extracts of leaves, fruits and seeds have high antioxidant activity comparable to Trolox (Silva et al., 2006). This has been due to the presence of oleuropein, tyrosol and hydroxytyrosol. The presence of maslinic acid along with phenols and ascorbic acid also has the capability to increase the antioxidant activity. The extract of leaves has also shown strong activity against bacteria and viruses. It helps in controlling the thyroid by interfering with the feedback mechanism (Al-Qarawi et al., 2002).

The reasons for the tissue culture of the olive are obvious owing to its rich phytochemical constituent. Mangal et al. (2014) used nodal segments in MS media with IBA (1 mg/L) and KIN (1 mg/L) for shoot induction, and for the shoot establishment half-strength MS media with BAP (2 mg/L) is suitable, along with IBA (0.2 mg/L) or NAA (1 mg/L) with AC (1.5 g/L) for root induction. Lambardi et al. (2012) gave the protocol for the micro propagation of the olive using nodal segments with the olive medium (OM). They used zeatin (1–4 mg/L) for shoot induction. For root induction, half-strength of OM is used either with IBA (2–3 mg/L) or NAA (1 mg/L). Another protocol used by Rostami and Shahsava (2012) was done with nodal segments, again in the MS media, with BAP (2.1 mg/L) and GA_3 (0.6 mg/L) for shooting and half-strength MS with IBA (4 mg/L) for rooting. There have been several new methods developed for the micropropagation. This includes the use of Plantform™, a temporary immersion system which uses liquid culture and immerses the explants periodically in the culture with a proper aeration time period. This is proven to be an effective method (Benelli and De Carlo, 2018). Khalid et al. (2022) used OM media on nodal segments and successfully induced shoots with zeatin (9.12 μM) and BAP (17.7 μM; Table 3.2). These tissue culture methods not only aim to develop the *in vitro* culturing but also improve the biomass yield and thus prepare a highly valued plant.

3.2.12 *Cedrus deodara* Roxb.

Cedrus deodara Roxb. is a conifer of the Pinaceae family native to the Western Himalayas. It is commonly called deodar or Himalayan cedar (Pijut et al., 2008). It is an evergreen tree with a height reaching up to 85 metres. The *Cedrus* genus is generally distributed in tropical and sub-tropical climates. *C. deodara* is found in the Himalayan moist temperate region (Champion and Seth, 1968) within the altitude of 1200–3030 metres in well-drained, acidic soil. The leaves are needle shaped and triquetrous. The tree is monoecious, and flowering occurs in September to October. It has a life span of 600 years (Gupta et al., 2011).

Deodar, being native to the Indian sub-continent, has its name originating from the Sanskrit word "devadaru." The word "deva" means divine or deity, and "daru" means tree. It thus refers to the tree of God (Sharma et al., 2018). The deodar forests were the favourite place for the meditation, aroma and healing among the sages. Other than this, there are three popular species of *Cedrus* in the world, namely *C. atlantica*, *C. brevifolia* and *C. libani* (Gupta et al., 2011).

India, with its ancient medicinal practice, has mentioned the medicinal properties of cedar. Since ancient times, its oil is used as an analgesic for pain in joints and muscles, insomnia and mental health treatment, a decoction for skin and blood purification, ulcers, etc. The bark of the tree treats gut health issues. The incline of the consumers towards plant-based medicines has led to the exploration of the uses of the tree (Gupta et al., 2011; Chaudhary et al., 2011).

Important phytochemicals extracted from *C. deodara* have been sesquiterpene alcohols like himachalol, allohimachalol, himadarol, isocentradol and centradol. Sesquiterpeneα and β- himachalane are the main constituents of the oil along with himachalone, cedeodarin, cedrin and cedrinoside extracted from the wood; α- and γ-atlantone have also been extracted. (-)-Wikstromal, (-)-mataresinol and dibenzylbutyrolactol has been extracted and isolated from the lignan of stem wood powder (Gupta et al., 2011; Chaudhary et al., 2011; Singh et al., 2007; Table 3.2).

Various *in vitro* and *in vivo* models have been studied to analyse the pharmacological use of deodar. They have shown several properties of the extracts, namely anti-inflammatory, antispasmodic, hypoglycaemic, analgesic, antiapoptotic, antimicrobial, anti-insecticidal, molluscidal, anticancer and even anxiolytic and anticonvulsant properties. Volatile oil extract has shown the stopping of adjuvant-induced arthritis in mice, indicating its anti-inflammatory nature (Shinde et al., 1999b;

Chandur et al., 2011). The analgesic property of the wood oil is comparable to aspirin and morphine in a dose-dependent manner (Shinde et al., 1999a; Sharma et al., 2018). Also, it has the capability to decrease neutrophil production that will affect the phagocytosis action and inhibit humoral-mediated immunity (HMI) and cell-mediated immunity (CMI) response (Shinde et al., 1999a; Chaudhary et al., 2011). The himachalol extract of the cedar has shown antispasmodic effects on the isolated smooth muscles of the rabbit (jejunum), rat's uterus and ileum and seminal vesicles of the guinea pig. Its effects are comparable to papaverine, and there is faster action in some cases (Faris et al., 2023). Ethanolic extracts have been shown to control glucose levels in the blood in diabetic mice induced with streptozotocin along with some antioxidant activity (Singh et al., 2013; Jain et al., 2014). The anticancer effects of mataresinol and wikstromal were studied *in vivo* in colon carcinoma in mice and *in vitro* human cancer cell lines (Sharma et al., 2008; Bisht et al., 2021). (-)-Wikstromal showed positive effects against leukaemia and HIV in *in vitro* conditions (Shashi et al., 2006). Aqueous extracts of the leaf have antibacterial effects. Essential oil of cedar has potent antifungal activity. An experiment has been done and showed activity against *Capsicum annum* L., *Aspergillus fumigatus, Candida albicans, Fusarium oxysporum* and *Alternaria porri* (Gupta et al., 2011). The most significant use of the cedar extracts that can be explored more is the use of heart wood extracts in anxiety and convulsions. The effects of the extract have been observed in the brain tissues of rats induced with pentylenetetrazole. The effects included reduced fear and anxiety in mice. Improved cerebral functions have also been observed (Viswanatha et al., 2009).

Tissue culture of the tree has been done using all possible explants. Zygotic embryos and stem segments were used by Tamta and Palni (2004). They used LP media with BAP and KIN 1 μM each for the first 28 days and then shifted the culture to LP solely. Buds were cultivated with 93% rate, but acclimatization failed. Again, they put the de-coated embryos in the LP media with BAP 20 μM for 10 days and then shifted to LP solely. The cultivation rate was less than 50%, but the rooting was fully successful in the micro-shoots in half-strength LP media. The same protocol was followed for the stem explants with 5 μM BAP in LP media, but efficiency was not on par with efforts. The results clearly demonstrated that acclimatization efficiency is more important than culture cultivation rate (Tamta and Palni, 2004). Ahmad et al. (2013) inoculated the seed for 2–3 weeks in the MS media and tried to influence the germination with a photoperiod along with BAP (1 mg/L) and GA_3 (0.5 mg/L) at the pH of 5.6–5.8. A light photoperiod of 16 hours gave the best shooting response and followed by a dark photoperiod of 16 hours gave the best rooting response without the PGRs in the media. The antioxidant activity measured by them suggested good concentrations of it for to be used for medicinal purposes. The main point to be taken care of the surface sterilization of the seeds is to be done by $HgCl_2$ at 0.1% (Ahmad et al., 2013). Rafi and Salehi worked on four genotypes of cedar to get an insight in the factors affecting the *in vitro* culture and gave the first successful report on the culturing of the *C. deodara*. They identified that the defoliated shoot tips produced the maximum healthy shoots in all the basal media with BAP. The next factor they determined was the presence of TDZ 0.8 μM with WPM media giving the maximum axillary shoots. The final factor they determined in their experiment was the addition of BAP (2.5 μM) with TDZ 0.8 μM gives the best results. The experiment explained the details in a simplified manner, paving the way to culture the conifers through explants (Rafi and Salehi, 2018). Later in 2022, Chauhan et al. used nodal segments of the cedar in MS media with BAP and 2,4-D. They manipulated the cytokinin and auxin levels and the method of sterilization of the explant. They found that maximum shooting response was when $HgCl_2$ (0.01 mg/L) was used along with BAP (1 mg/L) and 2,4-D (0.2 mg/L; Chauhan et al., 2022; Table 3.1).

3.3 CHALLENGES ASSOCIATED WITH MICROPROPAGATION OF MEDICINAL TREES

The major limitation associated with *in vitro* propagation of medicinal trees is the cost associated with the technique as compared with macropropagation. The cost of plant tissue culture medium, plant growth regulators and the laboratory setup required for *in vitro* establishment of cultures is often a roadblock in micropropagation of plants. However, cost-effective and cheap medium formulation and alternative plant growth regulators have often been described as economical ways of micropropagation. However, when it comes to saving a tree species which is rare, endangered or threatened, the cost associated with *in vitro* germplasm conservation and its propagation can easily be justified. Another limitation associated with micropropagation is the somaclonal variations which arise due to *in vitro* culturing of plants. However, somaclonal variations can easily be circumvented by not following the indirect regeneration protocols. Direct regeneration is the most preferred way for clonal propagation and obtaining true-to-type plants similar to the mother plants. Moreover, molecular makers such as SCoT, RAPD and ISSR can be utilized to check the genetic homogeneity of the *in vitro* raised plantlets. Secondary metabolites which are produced in the plants for their self-defence is another impediment in the micropropagation of tree species when compared to herbaceous plants. Tree species, owing to their long-life cycle, often have a more robust defence mechanism against the invading pathogens, and these metabolites are released into the culture medium upon inoculation. These metabolites hamper the growth of *in vitro* cultures. Many additives such as AC, casein hydrolysate, ascorbic acid, polyvinylpyrrolidone and citric acid have often been used to overcome this problem. Another constraint in the micropropagation of trees is the poor success rate in culture response of the tissue taken from mature adult trees. Juvenile tissue of the plants during its seasonal growth can be used to establish cultures *in vitro* (Giri et al., 2004; Abdalla et al., 2022).

3.4 CONCLUSION AND FUTURE PROSPECTS

The chapter described the importance of many medicinal tree species both in TM systems and alternative medicine systems. A renewed interest in plant-based natural medicine has resulted in 40% or more pharmaceuticals being directly or indirectly derived from natural resources like trees. This has put the natural population of such trees in grave danger due to over-exploitation. The chapter also discussed the significance and justifications for the necessity of large-scale medicinal tree micropropagation. However, exploring the depths of these tree species is still limited. The path is not without its risks and obstacles, but it also offers the chance for sustainable improvement and growth. Both the outside world and technological breakthroughs are undergoing rapid change. Protocols for *in vitro* propagation of many medicinally important tree species have not yet been developed, and therefore the science enthusiasts might use tissue culture techniques and biotechnological interventions to delve deeper and learn more about such trees.

REFERENCES

Abbasin, Z., Zamani, S., Movahedi, S., Khaksar, G., & Tabatabaei, B. S. (2010). *In vitro* micropropagation of yew (*Taxus baccata*) and production of plantlets. *Biotechnology*, *9*(1), 48–54.

Abdalla, N., El-Ramady, H., Seliem, M. K., El-Mahrouk, M. E., Taha, N., Bayoumi, Y., … & Dobránszki, J. (2022). An academic and technical overview on plant micropropagation challenges. *Horticulturae*, *8*(8), 677.

Abdel-Rahman, R., Ghoneimy, E., Abdel-Wahab, A., Eldeeb, N., Salem, M., Salama, E., & Ahmed, T. (2021). The therapeutic effects of *Ficus carica* extract as antioxidant and anticancer agent. *South African Journal of Botany*, *141*, 273–277.

Ahmad, A., Wali, A. F., Rehman, M. U., Khan, A., Raish, M., Kazi, M., & GM Rao, P. (2020). Therapeutic potential of *Rhododendron arboreum* polysaccharides in an animal model of lipopolysaccharide-inflicted oxidative stress and systemic inflammation. *Molecules*, *25*(24), 6045.

Ahmad, N., Fazal, H., Abbasi, B. H., & Ali, M. (2013). Factors influencing *in vitro* seeds germination and correlation of antioxidant activity with tissue development in *Cedrus deodara*. *Forest Systems*, *22*(3), 559–563.

Ahmad, P. I., & Lokho Puni, L. P. (2010). Vegetative propagation of *Rhododendron arboreum* Sm. using FRI-Wire technique.

Akbar, S., & Akbar, S. (2020). *Taxus baccata* L. (Taxaceae) (Syns.: *T. adpressa* Carrière; *T. aurea* K. Koch). In I. A. Rather & S. Akbar (Eds.), *Handbook of 200 Medicinal Plants: A Comprehensive Review of Their Traditional Medical Uses and Scientific Justifications* (pp. 1753–1761). Springer.

Ali, B., Mujeeb, M., Aeri, V., Mir, S. R., Faiyazuddin, M., & Shakeel, F. (2012). Anti-inflammatory and anti-oxidant activity of *Ficuscarica* Linn. leaves. *Natural Product Research*, *26*(5), 460–465.

Allahyari, S., Delazar, A., & Najafi, M. (2014). Evaluation of general toxicity, anti-oxidant activity and effects of *Ficuscarica* leaves extract on ischemia/reperfusion injuries in isolated heart of rat. *Advanced Pharmaceutical Bulletin*, *4*(Suppl 2), 577.

Al-Mizory, L. S. M., & Mayi, A. A. (2012). *In vitro* propagation of walnut (*Juglans regia*) by nodal explants. *Journal of Agricultural Science and Technology. B*, *2*(6B), 665.

Al-Qarawi, A. A., Al-Damegh, M. A., & El-Mougy, S. A. (2002). Effect of freeze dried extract of *Olea europaea* on the pituitary–thyroid axis in rats. *Phytotherapy Research*, *16*(3), 286–287.

Alshatwi, A. (2011). Anti-proliferative effects of organic extracts from root bark of *Juglans regia* L.(RBJR) on MDA-MB-231 human breast cancer cells: Role of Bcl-2/Bax, caspases and Tp53. *Asian Pac J Cancer Prev*, *12*(2), 525–530.

Alshatwi, A. A., Hasan, T. N., Shafi, G., Syed, N. A., Al-Assaf, A. H., Alamri, M. S., & Al-Khalifa, A. S. (2012). Research article validation of the antiproliferative effects of organic extracts from the green husk of *Juglans regia* L. on PC-3 human prostate cancer cells by assessment of apoptosis-related genes.

Al-Snafi, A. E. (2018). Chemical constituents, nutritional, pharmacological and therapeutic importance of *Juglans regia* - A review. *IOSR Journal of Pharmacy*, *8*(11), 1–21.

Amalraj, A., & Gopi, S. (2017). Medicinal properties of *Terminaliaarjuna* (Roxb.) Wight &Arn.: A review. *Journal of Traditional and Complementary Medicine*, *7*(1), 65–78.

Amiri, M. E., & Gharati, S. (2012). Influence of medium composition on multiplication of walnut (*Juglans regia* L.) growth. *J Med Plants Res*, *6*(8), 1482–1485.

Ana-Maria, R., & Radu, T. (2011). Preliminaries results concerning the *in vitro* multiplication of *Ginkgo biloba* species. *Journal of Horticulture, Forestry and Biotechnology*, *15*(1), 227–232.

Antoniadou, K., Herz, C., Le, N. P. K., Mittermeier-Kleßinger, V. K., Förster, N., Zander, M., … & Lamy, E. (2021). Identification of salicylates in willow bark (*Salix* cortex) for targeting peripheral inflammation. *International Journal of Molecular Sciences*, *22*(20), 11138.

Arora, K., Sharma, M., Srivastava, J., Ranade, S. A., & Sharma, A. K. (2010). Rapid *in vitro* cloning of a 40-year-old tree of *Azadirachtaindica* A. Juss.(Neem) employing nodal stem segments. *Agroforestry Systems*, *78*, 53–63.

Arora, K., Sharma, M., Srivastava, J., Ranade, S. A., & Sharma, A. K. (2011). In vitro cloning of *Azadirachta indica* from root explants. *Biologia Plantarum*, *55*, 164–168.

Asgarpanah, J. (2012). Phytopharmacology and medicinal properties of *Salix aegyptiaca* L. *African Journal of Biotechnology*, *11*(28), 7145–7150.

Asgary, S., Parkhideh, S., Solhpour, A., Madani, H., Mahzouni, P., & Rahimi, P. (2008). Effect of ethanolic extract of *Juglans regia* L. on blood sugar in diabetes-induced rats. *Journal of Medicinal Food*, *11*(3), 533–538.

Ashrafuzzaman, M., Haque, M. S., & Luna, L. N. (2008). In vitro clonal propagation of the neem tree (*Azadirachta indica* A. Juss.). *African Journal of Biotechnology*, 7(4).

Astani, A., Reichling, J., & Schnitzler, P. (2010). Comparative study on the antiviral activity of selected monoterpenes derived from essential oils. *Phytotherapy Research: An International Journal Devoted to Pharmacological and Toxicological Evaluation of Natural Product Derivatives*, *24*(5), 673–679.

Bachir, R. G., & Benali, M. (2012). Antibacterial activity of the essential oils from the leaves of *Eucalyptus globulus* against *Escherichia coli* and *Staphylococcus aureus*. *Asian Pacific Journal of Tropical Biomedicine*, *2*(9), 739–742.

Baharak, B. S., Mansoor, O., Reza, N. M., Farhad, H. A., Sepideh, K. J., Mehdi, S., & Mohammad, S. (2014). Effect of explants, salts concentration medium and hormone treatments on *Taxus baccata in vitro* culture. *Int J Biosci*, *5*, 1–9.

Banerjee, P. (2013). Rapid *in vitro* propagation of *Acacia auriculiformis* on solid and liquid media: Role of organic additive, antioxidant and plant growth regulators. *Cibtech Journal of Bio-Protocols*, *2*(1), 39–49.

Bello, A. M., Kamali Aliabad, K., Saravi, A. T., & SodaeiZade, H. (2022). Determination of the best culture medium and plant growth regulators for micropropagation of neem tree (*Azadirachtaindica* A. Juss). *International Journal of Horticultural Science and Technology, 9*(2), 237–245.

Benelli, C., & De Carlo, A. (2018). *In vitro* multiplication and growth improvement of *Olea europaea* L. cv Canino with temporary immersion system (Plantform™). *3 Biotech, 8*(7), 317.

Bernatova, I., & Liskova, S. (2021). Mechanisms modified by (−)-epicatechin and taxifolin relevant for the treatment of hypertension and viral infection: Knowledge from preclinical studies. *Antioxidants, 10*(3), 467.

Bey, M. B., & Louaileche, H. (2015). A comparative study of phytochemical profile and *in vitro* antioxidant activities of dark and light dried fig (*Ficuscarica* L.) varieties. *The Journal of Phytopharmacology, 4*(1), 41–48.

Bhat, S. N., Khalil, A., Nazir, N., Mir, M. A., Khan, I., Mubashir, S. S., & Hossain, M. A. (2022). *In vitro* prevention of browning in Persian Walnut (*Juglans regia* L.) cv. Sulaiman. *International Journal of Plant Biology, 13*(3), 330–342.

Bisht, A., Jain, S., Misra, A., Dwivedi, J., Paliwal, S., & Sharma, S. (2021). *Cedrus deodara* (Roxb. ex D. Don) G. Don: A review of traditional use, phytochemical composition and pharmacology. *Journal of Ethnopharmacology, 279*, 114361.

Boukhatem, M. N., Boumaiza, A., Nada, H. G., Rajabi, M., & Mousa, S. A. (2020). *Eucalyptus globulus* essential oil as a natural food preservative: Antioxidant, antibacterial and antifungal properties *in vitro* and in a real food matrix (Orangina fruit juice). *Applied Sciences, 10*(16), 5581.

Brondani, G. E., de Wit Ondas, H. W., Baccarin, F. J. B., Gonçalves, A. N., & de Almeida, M. (2012). Micropropagation of *Eucalyptus benthamii* to form a clonal micro-garden. *In Vitro Cellular & Developmental Biology-Plant, 48*, 478–487.

Capuana, M., Nissim, W. G., & Klein, J. D. (2022). Protocol for in vitro propagation of *Salix acmophylla* (Boiss.) studies on three ecotypes. *Forests, 13*(7), 1124.

Chakraborty, T., Sinhababu, S. P., & Sukul, N. C. (1995). Antifilarial effect of a plant *Acacia auriculiformis* on canine dirofilariasis. *Trop Med, 37*(1), 35–37.

Champion, H. G., & Seth, S. K. (1968). *A Revised Survey of the Forest Types of India.* Manager of Publications.

Chandur, U., Shashidhar, S., Chandrasekar, S. B., & Rao, M. N. (2011). *Cedrusdeodara. Research Journal of Pharmaceutical, Biological and Chemical Sciences, 2*(3), 654–659.

Chatterjee, A., Saluja, M., Singh, N., & Kandwal, A. (2011). To evaluate the antigingivitis and antipalque effect of an *Azadirachtaindica* (neem) mouthrinse on plaque induced gingivitis: A double-blind, randomized, controlled trial. *Journal of Indian Society of Periodontology, 15*(4), 399.

Chaudhary, A. K., Ahmad, S., & Mazumder, A. (2011). *Cedrus deodara* (Roxb.) Loud.: A review on its ethnobotany, phytochemical and pharmacological profile. *Pharmacognosy Journal, 3*(23), 12–17.

Chaudhary, S. P., Mishra, A., Singh, A. K., Dwivedi, K. N., & Ram, B. (2015). A FT-IR spectroscopic study of phytoconstituents of prepared formulation of arjuna (*Terminalia arjuna* lin.) and shilajatu. *International Journal of Science and Research* (IJSR), *5*, 12.

Chauhan, A., Lal, S., & Chandra, V. (2022). Investigation on the effects of growth hormones for the micropropagation of *Cedrusdeodara. Plant Archives, 22*(2), 335–338.

Chaves, T. P., Pinheiro, R. E. E., Melo, E. S., Soares, M. J. D. S., Souza, J. S. N., de Andrade, T. B., … & Coutinho, H. D. (2018). Essential oil of *Eucalyptus camaldulensis* Dehn potentiates β-lactam activity against *Staphylococcus aureus* and *Escherichia coli* resistant strains. *Industrial Crops and Products, 112*, 70–74.

Cheng, S., Zhang, W., Nannan, S. U. N., Feng, X. U., Linling, L. I., Yongling, L. I. A. O., & Cheng, H. (2014). Production of flavonoids and terpene lactones from optimized *Ginkgo biloba* tissue culture. *NotulaeBotanicaeHortiAgrobotanici Cluj-Napoca, 42*(1), 88–93.

Chornobrov, O., Bilous, S., Chornobrov, O., & Manko, M. (2019). Peculiarities of morphogenesis of the endangered species of willow (*Salix* spp.) *in vitro. Biologija, 65*(1).

Choudhary, M., Jaiswal, S., Singh, R., Arya, I. D., & Arya, S. (2015). A micropropagation protocol for mass multiplication of *Terminalia arjuna*-a valuable medicinal tree. *Advances in Forestry Science, 2*(1), 1–6.

Collard, B. C. Y., & Mackill, D. J. (2009). Start codon targeted (SCoT) polymorphism: A simple, novel DNA marker technique for generating gene-targeted markers in plants. *Plant Mol. Biol. Rep, 27*, 86–93.

Crisosto, H., Ferguson, L., Bremer, V., Stover, E., & Colelli, G. (2011). Fig (*Ficus carica* L.). In *Postharvest Biology and Technology of Tropical and Subtropical Fruits* (pp. 134–160). Woodhead Publishing.

Dhage, S. S., Pawar, B. D., Chimote, V. P., Jadhav, A. S., & Kale, A. A. (2012). In vitro callus induction and plantlet regeneration in fig (*Ficus carica* L.). *Journal of Cell and Tissue Research, 12*, 3395–3400.

Dangi, B., Khurana-Kaul, V., Kothari, S. L., & Kachhwaha, S. (2014). Micropropagtion of *Terminalia bellerica* from nodal explants of mature tree and assessment of genetic fidelity using ISSR and RAPD markers. *Physiology and Molecular Biology of Plants, 20*, 509–516.

Danial, G. H., Ibrahim, D. A., Brkat, S. A., & Khalil, B. M. (2014). Multiple shoots production from shoot tips of fig tree (*Ficus carica* L.) and callus induction from leaf segments. *International Journal of Pure and Applied Sciences and Technology, 20*(1), 117.

Davarpanah, S. J., Lahouti, M., & Karimian, R. (2014). Micropropagation of common yew using embryo culture. *Journal of Applied Biotechnology Reports, 1*(2), 77–80.

Davis, P., Valacchi, G., Pagnin, E., Shao, Q., Gross, H. B., Calo, L., & Yokoyama, W. (2006). Walnuts reduce aortic ET-1 mRNA levels in hamsters fed a high-fat, atherogenic diet. *The Journal of Nutrition, 136*(2), 428–432.

de Oliveira, A. F., Menezes, G. L., Goncalves, L. C., de Araujo, V. E., Ramirez, M. A., Júnior, R. G., … & Lana, A. M. Q. (2022). Pasture traits and cattle performance in silvopastoral systems with *Eucalyptus* and *Urochloa*: Systematic review and meta-analysis. *Livestock Science*, 262, 104973.

Dede, E., Genc, N., Elmastas, M., Aksit, H., & Erenler, R. (2019). Chemical constituents isolated from *Rhododendron ungernii* with antioxidant profile. *The Natural Products Journal, 9*(3), 238–243.

Dessoky, E. D. S., Attia, A. O., & Mohamed, E. A. A. (2016). An efficient protocol for *in vitro* propagation of fig (*Ficuscarica* sp.) and evaluation of genetic fidelity using RAPD and ISSR markers. *Journal of Applied Biology and Biotechnology, 4*(4), 057–063.

Dhandapani, B., Ramasamy, G., Natesan, S., & Kalyanasundaram, K. (2021). Regeneration potential of different explants during micropropagation of neem tree (*Azadirachtaindica* A. Juss.). *Journal of Phytology, 13*, 171–177.

Driver, J. A., & Kuniyuki, A. H. (1984). *In vitro* propagation of Paradox walnut rootstock. *HortScience, 19*(4), 507–509.

Dwivedi, S., & Chopra, D. (2014). Revisiting *Terminalia arjuna*–an ancient cardiovascular drug. *Journal of Traditional and Complementary Medicine, 4*(4), 224–231.

Eftekhari, Y., Rustaiyan, A., Monajjemi, M., & Khavarinejad, R. A. (2014). Study of anti-retroviral effects of Salix aegyptiaca L. herbal extract on HIV-1 in-vitro. *International Journal of Molecular and Clinical Microbiology, 1*, 398–405.

Elaissi, A., Salah, K. H., Mabrouk, S., Larbi, K. M., Chemli, R., & Harzallah-Skhiri, F. (2011). Antibacterial activity and chemical composition of 20 Eucalyptus species' essential oils. *Food Chemistry, 129*(4), 1427–1434.

El-Boghdady, N. A. (2013). Increased cardiac endothelin-1 and nitric oxide in adriamycin-induced acute cardiotoxicity: Protective effect of *Ginkgobiloba* extract.

Faris, A., Edder, Y., Louchachha, I., Lahcen, I. A., Azzaoui, K., Hammouti, B., …&Jodeh, S. (2023). From himachalenes to trans-himachalol: Unveiling bioactivity through hemisynthesis and molecular docking analysis. *Scientific Reports, 13*(1), 17653.

Faujdar, S. S., Bisht, D., & Sharma, A. (2020). Antibacterial potential of neem (*Azadirachtaindica*) against uropathogens producing beta-lactamase enzymes: A clue to future antibacterial agent?. *Biomedical and Biotechnology Research Journal (BBRJ)*, 4(3), 232–238.

Fayaz, M., & Sivakumaar, P. K. (2014). Phytochemical analysis and antimicrobial activity of *Salix alba* against dental biofilm forming bacteria. *International Journal of Pharmaceutical & Biological Archives, 5*, 137–140.

Fenning, T. M., & Gershenzon, J. (2002). Where will the wood come from? Plantation forests and role of biotechnology. *Trends Biotechnol, 20*, 291–296.

Fiebich, B. L., & Chrubasik, S. (2004). Effects of an ethanolic *Salix* extract on the release of selected inflammatory mediators *in vitro*. *Phytomedicine, 11*(2–3), 135–138.

Gakuubi, M. M., Maina, A. W., & Wagacha, J. M. (2017). Antifungal activity of essential oil of *Eucalyptus camaldulensis* Dehnh. against selected *Fusarium spp. International Journal of Microbiology, 2017*, 8761610.

Galan, D. M., Ezeudu, N. E., Garcia, J., Geronimo, C. A., Berry, N. M., & Malcolm, B. J. (2020). Eucalyptol (1, 8-cineole): An underutilized ally in respiratory disorders. *Journal of Essential Oil Research, 32*(2), 103–110.

Gehlot, A., Gupta, R. K., Tripathi, A., Arya, I. D., & Arya, S. (2014). Vegetative propagation of *Azadirachtaindica*: effect of auxin and rooting media on adventitious root induction in mini-cuttings. *Advances in Forestry Science, 1*(1), 1–9.

Ghareeb, M. A., Refahy, L. A., Saad, A. M., & Ahmed, W. S. (2016). Chemical composition, antioxidant and anticancer activities of the essential oil from *Eucalyptus citriodora* (Hook.) leaves. *Der Pharma Chem*, *8*, 192–200.

Giri, C. C., Shyamkumar, B., & Anjaneyulu, C. (2004). Progress in tissue culture, genetic transformation and applications of biotechnology to trees: An overview. *Trees*, *18*, 115–135.

Girijashankar, V. (2011). Micropropagation of multipurpose medicinal tree *Acacia auriculiformis*. *Journal of Medicinal Plants Research*, *5*(3), 462–466.

Gu, X., Hou, Y. K., Li, D., Wang, J. Z., & Wang, F. J. (2015). Separation, purification, and identification of angiotensin I–converting enzyme inhibitory peptides from walnut (*Juglans regia* L.) hydrolyzate. *International Journal of Food Properties*, *18*(2), 266–276.

Gupta, A. K., Harish, Rai, M. K., Phulwaria, M., Agarwal, T., & Shekhawat, N. S. (2014). *In vitro* propagation, encapsulation, and genetic fidelity analysis of *Terminalia arjuna*: A cardioprotective medicinal tree. *Applied Biochemistry and Biotechnology*, *173*, 1481–1494.

Gupta, S., Walia, A., & Malan, R. (2011). Phytochemistry and pharmacology of *Cedrusdeodara*: An overview. *International Journal of Pharmaceutical Sciences and Research*, *2*(8), 2010.

Gutiérrez, S. D., Kuri, S. A., & Martín-Herrera, D. (2017). The bioguided fractionation and pharmacological activity of an endemic *Salix canariensis* species. *Actapharmaceutica*, *67*(2), 265–273.

Haque, M. M., Ni, Y., Akon, A. J. U., Quaiyyum, M. A., & Jahan, M. S. (2021). A review on *Acacia auriculiformis*: Importance as pulpwood planted in social forestry. *International Wood Products Journal*, *12*(3), 194–205.

Harini, R., Ezhumalai, M., & Pugalendi, K. V. (2012). Antihyperglycemic effect of biochanin A, A soy isoflavone, on streptozotocin-diabetic rats. *European Journal of Pharmacology*, *676*(1–3), 89–94.

Hashmat, I., Azad, H., & Ahmed, A. (2012). Neem (*Azadirachtaindica* A. Juss)-A nature's drugstore: An overview. *Int Res J BiolSci*, *1*(6), 76–9.

Hayes, D., Angove, M. J., Tucci, J., & Dennis, C. (2016). Walnuts (*Juglansregia*) chemical composition and research in human health. *Critical Reviews in Food Science and Nutrition*, *56*(8), 1231–1241.

Hermans, M. P., Lempereur, P., Salembier, J. P., Maes, N., Albert, A., Jansen, O., & Pincemail, J. (2020). Supplementation effect of a combination of olive (*Oleaeuropea* L.) leaf and fruit extracts in the clinical management of hypertension and metabolic syndrome. *Antioxidants*, *9*(9), 872.

Hosseini, S., Huseini, H. F., Larijani, B., Mohammad, K., Najmizadeh, A., Nourijelyani, K., & Jamshidi, L. (2014). The hypoglycemic effect of *Juglansregia* leaves aqueous extract in diabetic patients: A first human trial. *DARU Journal of Pharmaceutical Sciences*, *22*, 1–5.

Hostanska, K., Jürgenliemk, G., Abel, G., Nahrstedt, A., & Saller, R. (2007). Willow bark extract (BNO1455) and its fractions suppress growth and induce apoptosis in human colon and lung cancer cells. *Cancer Detection and Prevention*, *31*(2), 129–139.

Hussain, F., Poddar, S. K., Ganguly, A., & Rahman, S. A. (2016). Investigation of CNS depressant, antidiarrheal and cytotoxic activities of crude methanolic extracts of *Acacia nilotica* and *Justicia adhatoda* root. *Indo Am J Pharm Res*, *6*(1), 3954–3961.

Idu, M., Ogedegbe-George, S., Oriarewo, P. E., & Gabriel, B. O. (2023). Spermicidal, antifertility and contraceptive effect of *Azadirachtaindica* A. Juss. seed extract in female and male Wistar rats. *Biology, Medicine, & Natural Product Chemistry*, *12*(1), 79–88.

Iqbal, K., Negi, A. K., Pala, N. A., & Todaria, N. P. (2023). Seedling recruitment of *Rhododendron arboreum*: An important NTFP species of North-Western Himalaya, India. *Ecological Questions*, *34*(3), 1–19.

Ishtiyak, P., & Puni, L. (2017). Standardization of in-situ propagation technique for *Taxus baccata* Linn. an endangered medicinal plant of Western Himalayas. *Studies on Ethno-Medicine*, *11*(4), 332–340.

Ismail, H., Kumar, S. M., Aziah, M. Y., Hasnida, N. H., & Nor Aini, A. S. (2016). *In vitro* micropropagation of *Acacia auriculiformis* from selected juvenile sources. *Dendrobiology*, *75*, 157–165.

Ismail, H., Noraini, A. S., Aziah, M. Y., Hassan, N. H., Zainudin, F., Abdullah, N., & Rahman, S. S. A. (2012). *In vitro* shoot induction of *Acacia auriculiformis* from juvenile and mature sources. *Journal of Biotechnology and Pharmaceutical Research*, *3*(5), 88–93.

Jain, S., Jain, A., Malviya, N., Kumar, D., Jain, V., & Jain, S. (2014). Antidiabetic activity of *Cedrus deodara* aqueous extract and its relationship with its antioxidant properties. *Journal of Pharmaceutical Sciences and Pharmacology*, *1*(3), 187–194.

Kajani, A. A., Moghim, S., & Mofid, M. R. (2012). Optimization of the basal medium for improving production and secretion of taxanes from suspension cell culture of *Taxus baccata* L. *DARU Journal of Pharmaceutical Sciences*, *20*, 1–6.

Kang, J. W., Kim, J. H., Song, K., Kim, S. H., Yoon, J. H., & Kim, K. S. (2010). Kaempferol and quercetin, components of *Ginkgo biloba* extract (EGb 761), induce caspase-3-dependent apoptosis in oral cavity cancer cells. *Phytotherapy Research*, *24*(S1), S77–S82.

Kapoor, D., Vijayvergiya, R., & Dhawan, V. (2014). *Terminalia arjuna* in coronary artery disease: Ethnopharmacology, pre-clinical, clinical & safety evaluation. *Journal of Ethnopharmacology, 155*(2), 1029–1045.

Kaur, S., Pal Singh, H., R Batish, D., & Kumar Kohli, R. (2012). Role of monoterpenes in *Eucalyptus* communities. *Current Bioactive Compounds, 8*(1), 101–107.

Kepenek, K., & Kolağasi, Z. (2016). Micropropagation of walnut (*Juglans regia* L.). *ActaPhysicaPolonica A, 130*(1), 150–156.

Khalid, L., Jamil, I., Sajid, Z. A., & Aftab, F. (2022). A rapid plant regeneration system by using stem and leaf explants of *Olea europea* L. cv. Koroneiki; a multipurpose bioenergy tree. *Research Square*. doi: 10.21203/rs.3.rs-2211677/v1.

Khan, Y., Panchal, S., Vyas, N., Butani, A., & Kumar, V. (2007). *Oleaeuropaea*: A phyto-pharmacological review. *Pharmacognosy Reviews, 1*(1), 114–118.

Khan, M. I., & Anis, M. (2012). Modulation of in vitro morphogenesis in nodal segments of *Salix tetrasperma* Roxb. through the use of TDZ, different media types and culture regimes. *Agroforestry Systems, 86*, 95–103.

Kishore, R. N., Anjaneyulu, N., Abhinayani, G., & Sravya, N. (2014). Investigation of anti-inflammatory and invitro antioxidant activities of hydroalcoholic extract of bark of *Salix tetrasperma* Roxb. *International Journal of Pharmaceutics and Drug Analysis, 2*(6), 506–509.

Kobus-Cisowska, J., Flaczyk, E., Rudzińska, M., & Kmiecik, D. (2014). Antioxidant properties of extracts from *Ginkgo biloba* leaves in meatballs. *Meat Science, 97*(2), 174–180.

Kolewe, M. E., Roberts, S. C., & Henson, M. A. (2012). A population balance equation model of aggregation dynamics in *Taxus* suspension cell cultures. *Biotechnology and Bioengineering, 109*(2), 472–482.

Kuang, H., Tang, Z., Zhang, C., Wang, Z., Li, W., Yang, C., …& Kong, A. N. (2017). Taxifolin activates the Nrf2 anti-oxidative stress pathway in mouse skin epidermal JB6 P+ cells through epigenetic modifications. *International Journal of Molecular Sciences, 18*(7), 1546.

Kumar, C., Singh, A. K., & Nehar, S. (2010). In-vitro propagation and callus formation from seedling of *Terminalia arjuna* Roxb. *An International Quarterly Journal of Life Sciences, 2*, 423–429.

Kumar, V. S., & Navaratnam, V. (2013). Neem (*Azadirachtaindica*): Prehistory to contemporary medicinal uses to humankind. *Asian Pacific Journal of Tropical Biomedicine, 3*(7), 505–514.

Lambardi, M., Ozudogru, E. A., & Roncasaglia, R. (2013). In vitro propagation of olive (*Olea europaea* L.) by nodal segmentation of elongated shoots. *Methods in Molecular Biology (Clifton, N.J.), 11013*, 33–44.

Lansky, E. P., Paavilainen, H. M., Pawlus, A. D., & Newman, R. A. (2008). *Ficus* spp. (fig): Ethnobotany and potential as anticancer and anti-inflammatory agents. *Journal of Ethnopharmacology, 119*(2), 195–213.

Lee, B., Kim, K. H., Jung, H. J., & Kwon, H. J. (2012). Matairesinol inhibits angiogenesis via suppression of mitochondrial reactive oxygen species. *Biochemical and Biophysical Research Communications, 421*(1), 76–80.

Ling, W. T., Tan, L. V., Khor, S. P., Sriskanda, D., Subramaniam, S., & Chew, B. L. (2022). Rapid *in vitro* propagation of fig (*Ficus carica* L.) 'Violette de Solliès' supported by molecular and microscopy analyses. *Horticulturae, 8*(11), 1025.

Liu, F., Yang, Z., Zheng, X., Luo, S., Zhang, K., & Li, G. (2011). Nematicidal coumarin from *Ficus carica* L. *Journal of Asia-Pacific Entomology, 14*(1), 79–81.

Maaloul, A., Michalet, S., Saadaoui, E., Ghzel, N., Bekir, J., Romdhane, C. B., & Romdhane, M. (2019). Effect of treated wastewater on growth and secondary metabolites production of two Eucalyptus species. *Agricultural Water Management, 211*, 1–9.

Majeed, M., Nagabhushanam, K., Prakasan, P., & Mundkur, L. (2023). The pursuit of natural medicine—A current perspective. In *Nutrition Science, Marketing Nutrition, Health Claims, and Public Policy* (pp. 173–192). Academic Press.

Mandal, P., Babu, S. S., & Mandal, N. C. (2005). Antimicrobial activity of saponins from A*cacia auriculiformis*. *Fitoterapia, 76*(5), 462–465.

Mangal, M., Sharma, D., Sharma, M., & Kumar, S. (2014). In vitro regeneration in olive (Olea europaea L.) cv, 'Frontio' from nodal segments. *Indian Journal of Experimental Biology, 52*, 912–914.

Mantovani, N. C., Grando, M. F., Xavier, A., & Otoni, W. C. (2013). *In vitro* shoot induction and multiplication from nodal segments of adult Ginkgo biloba plants. *Horticultura Brasileira, 31*, 184–189.

Mao, A. A., Kaliamoorthy, S., Ranyaphi, R. A., Das, J., Gupta, S., Athili, J., & Chanu, L. I. (2011). *In vitro* micropropagation of three rare, endangered, and endemic *Rhododendron* species of Northeast India. *In Vitro Cellular & Developmental Biology-Plant, 47*, 674–681.

Maurer, K., Ihl, R., Dierks, T., & Frölich, L. (1997). Clinical efficacy of *Ginkgo biloba* special extract EGb 761 in dementia of the Alzheimer type. *Journal of Psychiatric Research, 31*(6), 645–655.

Moniruzzaman, M., Yaakob, Z., & Taha, R. A. (2015, August). *In vitro* production of fig (*Ficus carica* L.) plantlets. *V International Symposium on* Fig *1173* (pp. 231–236).

Mostafa, I., Abbas, H. A., Ashour, M. L., Yasri, A., El-Shazly, A. M., Wink, M., & Sobeh, M. (2020). Polyphenols from *Salix tetrasperma* impair virulence and inhibit quorum sensing of *Pseudomonas aeruginosa*. *Molecules*, *25*(6), 1341.

Moura, L. C. D., Xavier, A., Cruz, A. C. F. D., Gallo, R., Gatti, K. C., Miranda, N. A., & Otoni, W. C. (2017). Effects of explant type, culture media and picloram and dicamba growth regulators on induction and proliferation of somatic embryos in *Eucalyptus grandis* x *E. urophylla*. *RevistaÁrvore*, *41*, e410502.

Mumtaz, R., Zubair, M., Khan, M. A., Muzammil, S., & Siddique, M. H. (2022). Extracts of *Eucalyptus alba* promote diabetic wound healing by inhibiting α-glucosidase and stimulating cell proliferation. *Evidence-Based Complementary and Alternative Medicine*, *2022*, 4953105.

Murashige, T., & Skoog, F. (1962). A revised medium for rapid growth and bio assays with tobacco tissue cultures. *Physiologia Plantarum*, *15*(3).

Murty, D., Rajesh, E., Raghava, D., Raghavan, T. V., & Surulivel, M. K. M. (2010). Hypolipidemic effect of arborium plus in experimentally induced hypercholestermic rabbits. *YakugakuZasshi*, *130*(6), 841–846.

Mythili, P., Parameswari, C. S., & Dayana, J. (2012). Phytochemical analysis of the bark extract of *Terminalia arjuna* and its cardioprotective effect. *Indian J Innov Dev*, *1*, 40–2.

Nagini, S., Nivetha, R., Palrasu, M., & Mishra, R. (2021). Nimbolide, a neem limonoid, is a promising candidate for the anticancer drug arsenal. *Journal of Medicinal Chemistry*, *64*(7), 3560–3577.

Nakayama, T., Honda, R., Kuwata, K., Usui, S., & Uno, B. (2021). Electrochemical and mechanistic study of reactivities of α-, β-, γ-, and δ-tocopherol toward electrogenerated superoxide in N, N-dimethylformamide through proton-coupled electron transfer. *Antioxidants*, *11*(1), 9.

Navroski, M. C., Reiniger, L. R. S., Araújo, M. M., Curti, A. R., & Pereira, M. D. O. (2014). *In vitro* establishment and multiplication of genotypes of *Eucalyptus dunnii* maiden. *Cerne*, *20*, 139–146.

Oh, J., Hwang, I. H., Hong, C. E., Lyu, S. Y., & Na, M. (2013). Inhibition of fatty acid synthase by ginkgolic acids from the leaves of *Ginkgo biloba* and their cytotoxic activity. *Journal of Enzyme Inhibition and Medicinal Chemistry*, *28*(3), 565–568.

Okokon, J. E., Obot, J., Opara, K. N., & Etim, E. (2010). *In vivo* antimalarial activity of ethanolic leaf extract of Acacia auriculiformis. *International Journal of Drug Development & Research*, *2*(3), 482–487.

Omar, S. H. (2010). Oleuropein in olive and its pharmacological effects. *Scientiapharmaceutica*, *78*(2), 133–154.

Pandey, G., Verma, K. K., & Singh, M. (2014). Evaluation of phytochemical, antibacterial and free radical scavenging properties of *Azadirachtaindica* (neem) leaves. *Int J Pharm PharmSci*, *6*(2), 444–447.

Panikar, S., Shoba, G., Arun, M., Sahayarayan, J. J., Nanthini, A. U. R., Chinnathambi, A., & Kim, H. J. (2021). Essential oils as an effective alternative for the treatment of COVID-19: Molecular interaction analysis of protease (Mpro) with pharmacokinetics and toxicological properties. *Journal of Infection and Public Health*, *14*(5), 601–610.

Panth, N., Paudel, K. R., & Karki, R. (2016). Phytochemical profile and biological activity of *Juglans regia*. *Journal of Integrative Medicine*, *14*(5), 359–373.

Parzymies, M., Pogorzelec, M., Głębocka, K., & Śliwińska, E. (2020). Genetic stability of the endangered species *Salix lapponum* L. regenerated in vitro during the reintroduction process. *Biology*, *9*(11), 378.

Parzymies, M., Pogorzelec, M., Głębocka, K., & Sliwinska, E. (2023). Micropropagation protocol and genetic stability of the *Salix myrtilloides* plants cultivated *in vitro*. *Biology*, *12*(2), 168.

Patil, V. V., & Patil, V. R. (2010). A comparative evaluation of anti-inflammatory activity of the bark of *Ficus bengalensis* in plants of different age. *Journal of Basic and Clinical Pharmacy*, *1*(2), 107.

Pawar, R. S., & Bhutani, K. K. (2005). Effect of oleanane triterpenoids from *Terminaliaarjuna*—a cardioprotective drug on the process of respiratory oxyburst. *Phytomedicine*, *12*(5), 391–393.

Pijut, Paula M. 2008. Cedrus Trew. In F. T. Bonner & R.P. Karrfalt (eds.), *Woody Plant Seed Manual. Agric. Handb. 727.* Washington, DC: U.S. Department of Agriculture, Forest Service: 357–362.

Pingali, U., Vuppalanchi, B., Nutalapati, C., & Gundagani, S. (2021). Aqueous *Azadirachtaindica* (neem) extract attenuates insulin resistance to improve glycemic control and endothelial function in subjects with metabolic syndrome. *Journal of Medicinal Food*, *24*(11), 1135–1144.

Prakash, D., Upadhyay, G., Pushpangadan, P., & Gupta, C. (2011). Antioxidant and free radical scavenging activities of some fruits. *Journal of Complementary and Integrative Medicine*, *8*(1), 1–16.

Prashith Kekuda, T. R., Vinayaka, K. S., & Praveen Kumar, S. K. (2017). Antimicrobial activity of *Salix tetrasperma* Roxb. (Salicaceae). *International Journal of Herbal Medicine*, *5*(5), 192–195.

Purnamasari, R., Winarni, D., Permanasari, A. A., Agustina, E., Hayaza, S., & Darmanto, W. (2019). Anticancer activity of methanol extract of *Ficuscarica* leaves and fruits against proliferation, apoptosis, and necrosis in Huh7it cells. *Cancer Informatics*, *18*, 1176935119842576.

Qaderi, A., Mehrafarin, A., Rezazadeh, S., & Zarinpanjeh, N. (2021). The efficient method for *in vitro* micropropagation of *Ginkgo biloba* L. *Journal of Medicinal Plants*, *20*(78), 78–89.

Rafi, Z. N., & Salehi, H. (2018). Factors affecting *in vitro* propagation of some genotypes of Himalayan cedar [*Cedrus deodara* (Roxb. ex Lamb) G. Don.]. *Advances in Horticultural Science*, *32*(4), 479–486.

Rafiq, M., & Dahot, M. U. (2010). Callus and azadirachtin related limonoids production through *in vitro* culture of neem (*Azadirachtaindica* A. Juss). *African Journal of Biotechnology*, *9*(4).

Rahimi, S., Bayati, M., Kordrostami, M., & Ghasemi-Soloklui, A. A. (2023). Tissue culture techniques to conserve endangered medicinal plants with antimicrobial and antiviral activities. In S. Jha & M. Halder (eds.), *Medicinal Plants: Biodiversity, Biotechnology and Conservation* (pp. 675–710). Springer Nature.

Raju, S., Sivanesan, S. K., Gudemalla, K., Mundugaru, R., & Swaminathan, M. (2019). Effect of *Ginkgo biloba* extract on hematological and biochemical alterations in fluoride toxicity in Wistar rats. *Research Journal of Pharmacy and Technology*, *12*(8), 3915–3918.

Rangra, N. K., Samanta, S., & Pradhan, K. K. (2019). A comprehensive review on phytopharmacological investigations of *Acacia auriculiformis* A. Cunn. ex Benth. *Asian Pacific Journal of Tropical Biomedicine*, *9*(1), 1–11.

Regueira, M., Rial, E., Blanco, B., Bogo, B., Aldrey, A., Correa, B., …& Vidal, N. (2018). Micropropagation of axillary shoots of *Salix viminalis* using a temporary immersion system. *Trees*, *32*, 61–71.

Rigney, U., Kimber, S., & Hindmarch, I. (1999). The effects of acute doses of standardized Ginkgo biloba extract on memory and psychomotor performance in volunteers. *Phytotherapy Research*, *13*(5), 408–415.

Rodriguez-Enriquez, F., Torres, I., & Vina, D. (2017). Natural products for the treatment of Alzheimer's disease: Present and future expectations. *Front. Clin. Drug Res.-CNS Neurol. Disord*, *5*, 74–170.

Rostami, A. A., & Shahsava, A. R. (2012). In vitro micropropagation of olive (*Olea europaea* L.) 'Mission' by nodal segments. *Journal of Biological and Environmental Sciences*, *6*(17).

Sahari Moghadam, A., Kaviani, B., Mohammadi Torkashvand, A., Abdossi, V., & Eslami, A. (2022). Micropropagation of English yew, an ornamental-medicinal tree. *Journal of Ornamental Plants*, *12*(2), 91–99.

Sah, B., Kumari, M., Subban, K., & Chelliah, J. (2020). Evaluation of the anticancer activity of enzymatically synthesized Baccatin III: An intermediate precursor of Taxol®. *3 Biotech*, *10*, 1–10.

Sahu, S., Saraf, S., Kaur C D., & Saraf, S. (2013). Biocompatible nanoparticles for sustained topical delivery of anticancer phytoconstituent quercetin. *Pakistan Journal of Biological Sciences*, *16*, 601–609.

Saleem, S., Muhammad, G., Hussain, M. A., & Bukhari, S. N. A. (2018). A comprehensive review of phytochemical profile, bioactives for pharmaceuticals, and pharmacological attributes of *Azadirachtaindica*. *Phytotherapy Research*, *32*(7), 1241–1272.

Salehi, B., Sharifi-Rad, J., Quispe, C., Llaique, H., Villalobos, M., Smeriglio, A., …& Martins, N. (2019). Insights into *Eucalyptus* genus chemical constituents, biological activities and health-promoting effects. *Trends in Food Science & Technology*, *91*, 609–624.

Salvatori, E. S., Morgan, L. V., Ferrarini, S., Zilli, G. A., Rosina, A., Almeida, M. O., & Dal Magro, J. (2023). Anti-inflammatory and antimicrobial effects of *Eucalyptus* spp. essential oils: A potential valuable use for an industry byproduct. *Evidence-Based Complementary and Alternative Medicine*, *2023*.

Sandeepani, S. H., Senaratna, M. M. D. J., Weerasinghe, P. A., & Haluwana, H. M. I. G. N. K. (2021). In vivo and In vitro seed germination of an endemic rhododendron species (Rhododendron arboreum subsp. zeylanicum) in Sri Lanka. *Cross Current International Journal of Agriculture and Veterinary Sciences*, *3*(4):25–39.

Sarhan, A. Z., Abou Dahab, T., El-Bana, A. E., & Arafa, R. N. (2010). Micropropagation of ginkgo (*Ginkgo biloba* L.) Plant by tissue culture. *Egyptian Journal of Agricultural Sciences*, *61*(3), 309–315.

Sathya, A., & Siddhuraju, P. (2013). Protective effect of bark and empty pod extracts from *Acacia auriculiformis* against paracetamol intoxicated liver injury and alloxan induced type II diabetes. *Food and Chemical Toxicology*, *56*, 162–170.

Shahcheraghi, S. T., & Shekafandeh, A. (2016). Micropropagation of three endemic and endangered fig (*Ficus carica* L.) genotypes. *Advances in Horticultural Science*, *30*(3), 129–134.

Shahrajabian, M. H., Sun, W., & Cheng, Q. (2022). *Ginkgo biloba*: A famous living fossil tree and an ancient herbal traditional Chinese medicine. *Current Nutrition & Food Science*, *18*(3), 259–264.

Sharma, A., Prashar, B., & Arora, P. (2018). *Cedrusdeodara*: A medicinal herb. *International Journal of Current Research*, *10*(02), 65758–65762.

Sharma, A. D. (2020). Eucalyptol (1, 8 cineole) from Eucalyptus essential oil a potential inhibitor of COVID 19 corona virus infection by molecular docking studies. *Preprints*, *2020*, 2020030455.

Sharma, H., & Garg, M. (2015). A review of traditional use, phytoconstituents and biological activities of Himalayan yew, *Taxuswallichiana*. *Journal of Integrative Medicine*, *13*(2), 80–90.

Sharma, N., Singh, S., & Singh, S. K. (2016). Review on phytopharmacological properties of *Acacia auriculiformis* A. Cunn. ex. Benth. *Planta Activa*, *1*, 1–6.

Sharma, P. R., Shanmugavel, M., Saxena, A. K., & Qazi, G. N. (2008). Induction of apoptosis by a synergistic lignan composition from *Cedrus deodara* in human cancer cells. *Phytotherapy Research: An International Journal Devoted to Pharmacological and Toxicological Evaluation of Natural Product Derivatives*, *22*(12), 1587–1594.

Shashi, B., Jaswant, S., Madhusudana, R. J., Kumar, S. A., & Nabi, Q. G. (2006). A novel lignan composition from *Cedrus deodara* induces apoptosis and early nitric oxide generation in human leukemia Molt-4 and HL-60 cells. *Nitric Oxide*, 14(1), 72–88.

Shekhawat, G. S., Mathur, S., & Batra, A. (2009). Role of phytohormones and nitrogen in somatic embryogenesis induction in cell culture derived from leaflets of *Azadirachtaindica*. *Biologiaplantarum*, *53*, 707–710.

Shengule, S., Mishra, S. A. N. J. A. Y., & Bodhale, S. (2018). Inhibitory effect of a standardized hydroethanolic extract of *Terminalia arjuna* bark on alpha-amylase enzyme. *Asian J. Pharm. Clin. Res*, *11*, 366–369.

Shinde, U. A., Phadke, A. S., Nair, A. M., Mungantiwar, A. A., Dikshit, V. J., & Saraf, M. N. (1999a). Preliminary studies on the immunomodulatory activity of *Cedrus deodara* wood oil. *Fitoterapia*, *70*(4), 333–339.

Shinde, U. A., Phadke, A. S., Nair, A. M., Mungantiwar, A. A., Dikshit, V. J., & Saraf, M. N. (1999b). Studies on the anti-inflammatory and analgesic activity of *Cedrus deodara* (Roxb.) Loud. wood oil. *Journal of Ethnopharmacology*, *65*(1), 21–27.

Shiyong, Y., Yijia, X., Peng, Z., Chong, L., Wuming, H., Linchun, L., & Chunlai, Z. (2015). *Ginkgo biloba* extract inhibits platelet activation via inhibition of Akt. *Integrative Medicine International*, *1*(4), 234–242.

Silva, S., Gomes, L., Leitao, F., Coelho, A. V., & Boas, L. V. (2006). Phenolic compounds and antioxidant activity of *Olea europaea* L. fruits and leaves. *Food Science and Technology International*, *12*(5), 385–395.

Silva de Oliveira, L., Brondani, G. E., Batagin-Piotto, K. D., Calsavara, R., Gonçalves, A. N., & de Almeida, M. (2015). Micropropagation of *Eucalyptus cloeziana* mature trees. *Australian Forestry*, *78*(4), 219–231.

Singh, P., Khosa, R. L., & Mishra, G. (2013). Evaluation of antidiabetic activity of ethanolic extract of *Cedrus deodara* (Pinaceae) stem bark in streptozotocin induced diabetes in mice. *Nigerian Journal of Experimental and Clinical Biosciences*, *1*(1&2), 33–38.

Singh, S. K., Shanmugavel, M., Kampasi, H., Singh, R., Mondhe, D. M., Rao, J. M., & Qazi, G. N. (2007). Chemically standardized isolates from *Cedrusdeodara* stem wood having anticancer activity. *Planta Medica*, *73*(06), 519–526.

Singh, S., & Sharma, N. (2014). Evaluation of wound healing activity of *Acacia auriculiformis* A. Cunn. Stem bark. *Asian J Pharm Clin Res*, *7*(2), 204–207.

Smith, M. A. L., & McCown, B. H. (1983). A comparison of source tissue for protoplast isolation from three woody plant species. *Plant Science Letters*, *28*(2), 149–156.

Sonar, P. K., Singh, R., Verma, A., & Saraf, S. K. (2013). *Rhododendron arboreum* (Ericaceae): Immunomodulatory and related toxicity studies. *Oriental Pharmacy and Experimental Medicine*, *13*, 127–131.

Soto-Maldonado, C., Vergara-Castro, M., Jara-Quezada, J., Caballero-Valdés, E., Müller-Pavez, A., Zúñiga-Hansen, M. E., & Altamirano, C. (2019). Polyphenolic extracts of walnut (*Juglansregia*) green husk containing juglone inhibit the growth of HL-60 cells and induce apoptosis. *Electronic Journal of Biotechnology*, *39*, 1–7.

Suffness, M., & Wall, M. E. (2021). Discovery and Development of Taxol. In *Taxol* (pp. 3–26). CRC press.

Taha, R. A., Mustafa, N. S., & Hassan, S. A. (2013). Protocol for micropropagation of two *Ficuscarica* cultivars. *World Journal of Agricultural Sciences*, *9*(5), 383–388.

Talreja, S., & Tiwari, S. (2023). An in depth exploration of *Ginkgo biloba*: A review. *International Journal of Pharmaceutical Sciences*, *1*(07), 1–1.

Tamta, S., & Palni, L. M. S. (2004). Studies on in vitro propagation of Himalayan cedar (Cedrus deodara) using zygotic embryos and stem segments. *Indian Journal of Biotechnology*, *3*(2), 209–215.

Thangaraj, V. (2013). Hypolipidemic effect of *Rhododendron arboreum* Sm. linn flower juice in experimentally induced hypercholestermic rabbits. *International Journal of Pharmacy & Biomedical Research*, *4*(1), 46–49.

Tolba, H., Moghrani, H., Benelmouffok, A., Kellou, D., & Maachi, R. (2015). Essential oil of Algerian *Eucalyptus citriodora*: Chemical composition, antifungal activity. *Journal de mycologiemedicale*, *25*(4), e128–e133.

Toosi, S., & Dilmagani, K. (2010). Proliferation of *Juglans regia* L. by *in vitro* embryo culture. *J. Cell Biol. Genet*, *1*(1), 12–19.

Turan, A., & Celik, I. (2016). Antioxidant and hepatoprotective properties of dried fig against oxidative stress and hepatotoxicity in rats. *International Journal of Biological Macromolecules*, *91*, 554–559.

Ude, C., Schubert-Zsilavecz, M., & Wurglics, M. (2013). *Ginkgo biloba* extracts: A review of the pharmacokinetics of the active ingredients. *Clinical Pharmacokinetics*, *52*, 727–749.

Urmi, K. F., Chowdhury, S. Y., Hossain, M. K., Bhusal, P., & Hamid, K. (2013). Comparative antioxidant activity and brine shrimp lethality bioassay of different parts of the plant-*Acacia auriculiformis*. *International Journal of Pharmaceutical Sciences and Research*, *4*(2), 875.

Vasundhara, S., Hafsa, A., & Rajiv, G. (2013). Memory enhancing effects of *Ficus carica* leaves in hexane extract on interoceptive behavioral models. *Asian J Pharm Clin Res*, *6*(3), 109–113.

Vazquez, A., Sanchez-Rodriguez, E., Vargas, F., Montoro-Molina, S., Romero, M., Espejo-Calvo, J. A., & Mesa, M. D. (2019). Cardioprotective effect of a virgin olive oil enriched with bioactive compounds in spontaneously hypertensive rats. *Nutrients*, *11*(8), 1728.

Verma, N., Amresh, G., Sahu, P. K., Rao, C. V., & Singh, A. P. (2012). Antihyperglycemic and antihyperlipidemic activity of ethyl acetate fraction of *Rhododendron arboreum* Smith flowers in streptozotocin induced diabetic rats and its role in regulating carbohydrate metabolism. *Asian Pacific Journal of Tropical Biomedicine*, *2*(9), 696–701.

Verma, N., Singh, A. P., Amresh, G., Sahu, P. K., & Rao, C. V. (2010). Anti-inflammatory and anti-nociceptive activity of *Rhododendron arboreum*. *J Pharm Res*, *3*(6), 1376–80.

Verma, N., Singh, A. P., Amresh, G., Sahu, P. K., & Rao, C. V. (2011). Protective effect of ethyl acetate fraction of *Rhododendron arboreum* flowers against carbon tetrachloride-induced hepatotoxicity in experimental models. *Indian Journal of Pharmacology*, *43*(3), 291.

Vimalanathan, S., & Hudson, J. (2014). Anti-influenza virus activity of essential oils and vapors. *American Journal of Essential Oils and Natural Products*, *2*(1), 47–53.

Virupaksha, J. H., Nadendla, R. R., Kumar, M. S., & Kavya, S. (2016). Effect of *Salix tetrasperma* Roxburgh leaf extracts on central nervous system activities. *Research Journal of Pharmaceutical Biological and Chemical Sciences*, *7*(4), 2060–2064.

Viswanatha, G. L., Nandakumar, K., Shylaja, H., Ramesh, C., Rajesh, S., & Srinath, R. (2009). Anxiolytic and anticonvulsant activity of alcoholic extract of heart wood of *Cedrus deodara* Roxb. in rodents. *J Pharm Res Health Care*, *1*(2), 217–239.

Wagner, N. D., He, L., & Hörandl, E. (2021). The evolutionary history, diversity, and ecology of willows (*Salix* L.) in the European Alps. *Diversity*, *13*(4), 146.

Wang, F. J., Yin, X. Y., Regenstein, J. M., & Wang, J. Z. (2016). Separation and purification of angiotensin-I-converting enzyme (ACE) inhibitory peptides from walnuts (*Juglans regia* L.) meal. *European Food Research and Technology*, *242*, 911–918.

Wei, X., Chen, J., Zhang, C., & Wang, Z. (2018). *In vitro* shoot culture of *Rhododendron fortunei*: An important plant for bioactive phytochemicals. *Industrial Crops and Products*, *126*, 459–465.

WHO (2022). World Health Organization. WHO establishes the Global Centre for Traditional Medicine in India. https://www.who.int/news/item/25-03-2022-who-establishes-the-global-centre-for-traditional-medicine-in-india#:~:text=Around%2080%25%20of%20the%20world's,traditional%20medicine%20practices%20and%20products. Accessed on 08.01.2024.

Wu, H., Ao, Q., Li, H., & Long, F. (2023). Rapid and efficient regeneration of *Rhododendron decorum* from flower buds. *Horticulturae*, *9*(2), 264.

Yadav, R., Yadav, N., & Kumar, S. (2016). An improved micropropagation and assessment of genetic fidelity in multipurpose medicinal tree, *Acacia auriculiformis*. *Proceedings of the National Academy of Sciences, India Section B: Biological Sciences*, 86, 921–929.

Yang, X. M., Wang, Y. F., Li, Y. Y., & Ma, H. L. (2014). Thermal stability of ginkgolic acids from *Ginkgo biloba* and the effects of ginkgol C17: 1 on the apoptosis and migration of SMMC7721 cells. *Fitoterapia*, *98*, 66–76.

Yan-Hua, Y. A. N. G., Jia-Wang, M. A. O., & Xiao-Li, T. A. N. (2020). Research progress on the source, production, and anti-cancer mechanisms of paclitaxel. *Chinese Journal of Natural Medicines*, *18*(12), 890–897.

Yuan, L. P., Chen, Z. W., Li, F., Dong, L. Y., & Chen, F. H. (2006). Protective effect of total flavones of rhododendra on ischemic myocardial injury in rabbits. *The American Journal of Chinese Medicine*, *34*(03), 483–492.

Zhang, H. Q., Wang, Y. J., Yang, G. T., Gao, Q. L., & Tang, M. X. (2019). Taxifolin inhibits receptor activator of NF-κB ligand-induced osteoclastogenesis of human bone marrow-derived macrophages *in vitro* and prevents lipopolysaccharide-induced bone loss *in vivo*. *Pharmacology*, *103*(1–2), 101–109.

Zhang, X. B., Zou, C. L., Duan, Y. X., Wu, F., & Li, G. (2015). Activity guided isolation and modification of juglone from *Juglans regia* as potent cytotoxic agent against lung cancer cell lines. *BMC Complementary and Alternative Medicine*, *15*, 1–8.

Zhang, X. B., Zou, C. L., Duan, Y. X., Wu, F., & Li, G. (2015). Activity guided isolation and modification of juglone from *Juglans regia* as potent cytotoxic agent against lung cancer cell lines. *BMC Complementary and Alternative Medicine*, *15*, 1–8.

Zhang, Y., Chen, A. Y., Li, M., Chen, C., & Yao, Q. (2008). Ginkgo biloba extract kaempferol inhibits cell proliferation and induces apoptosis in pancreatic cancer cells. *Journal of Surgical Research, 148*(1), 17–23.

Zhao, Y., Huang, W., Wang, J., Chen, Y., Huang, W., & Zhu, Y. (2018). Taxifolin attenuates diabetic nephropathy in streptozotocin-induced diabetic rats. *American Journal of Translational Research*, *10*(4), 1205.

Zhou, S. X., Medlyn, B. E., & Prentice, I. C. (2016). Long-term water stress leads to acclimation of drought sensitivity of photosynthetic capacity in xeric but not riparian *Eucalyptus* species. *Annals of Botany*, *117*(1), 133–144.

4 Advances in Micropropagation of *Cannabis sativa* L.

A Multipurpose Medicinal Plant

Hemant Lata, Suman Chandra, Ikhlas A. Khan, and Mahmoud A. ElSohly

4.1 INTRODUCTION

Since ancient times, humans have utilized cannabis for a variety of purposes, including as a recreational narcotic, sustenance, fiber, medicine, and in religious rituals. Throughout human history, it has been among the earliest cultivated plants. Originally from the temperate area of Asia, this plant gradually expanded its reach to rest of the world. Presently cultivated in regions spanning alpine foothills of the Himalayas to sea level in the tropics, it is now widely dispersed across the globe and inhabits a variety of natural habitats. Having historical records dating as far back as the 6th century BC, cannabis boasts an extensive medicinal usage history in both the Middle East and Asia. The species was first integrated into Western Europe's medicinal practices during the early 19th century AD.

The cannabis plant synthesizes a diverse blend of aromatic and medicinal compounds. Over 550 distinct compounds have been derived from cannabis so far (Radwan et al., 2017; ElSohly and Slade, 2005). These compounds correspond to a wide variety of chemical classes including cannabinoids, which is the most significant and distinctive category of compounds. To date, over 120 cannabinoids have been documented in cannabis (ElSohly et al., 2017). Primarily synthesized in the glandular trichomes of the plant, Δ^9-tetrahydrocannabinol (Δ^9-THC) is the most significant and psychologically potent compound among them (Hammond and Mahlberg, 1977).

Originally, Δ^9-THC is found in the plant in its acidic form, Δ^9-THCA. Over time or when subjected to heat, Δ^9-THCA decarboxylates into Δ^9-THC (Turner et al., 1980; Pertwee, 2006). The structure of Δ^9-THC was fully elucidated in 1964 by Gaoni and Mechoulam (Gaoni and Mechoulam, 1964). Numerous studies have documented the pharmacological and therapeutic efficacy of cannabis and its principal active constituent, Δ^9-THC (Mechoulam, 1986; Mattes et al., 1994; Brenneisen et al., 1996; Pryce and Baker, 2005; Abrams et al., 2007; Sirikantaramas et al., 2007; Cascio et al., 2017).

Cannabidiol (CBD) is a non-psychoactive compound in cannabis which exhibits pharmacological and therapeutic potential against pediatric epilepsy. CBD was initially isolated from cannabis plants in 1940 by Adams et al., and its structure was determined in 1963 by Mechoulam and Shvo (Adams et al., 1940; Mechoulam and Shvo, 1963). Other major cannabinoids in cannabis are tetrahydrocannabivarin (THCV), cannabigerol (CBG), cannabichromene (CBC), and cannabinol (CBN).

4.2 PHYTOCANNABINOID AND TERPENE BIOSYNTHESIS

Figures 4.1 and 4.2 illustrate the chemical structures of major phytocannabinoids and terpenes, respectively. Among the seed-raised plants, female cannabis plants exhibit a higher concentration of phytocannabinoids in comparison to their male counterparts. Within the female plants, cannabinoids

DOI: 10.1201/b23374-4

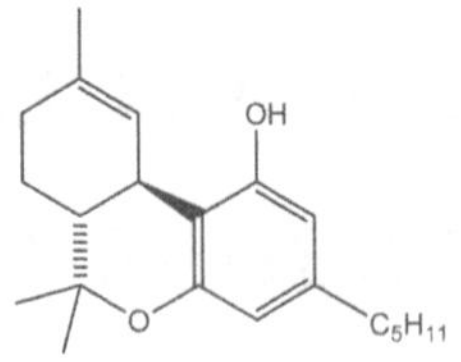

Δ^9-Tetrahydrocannabinol

(Δ^9-THC)

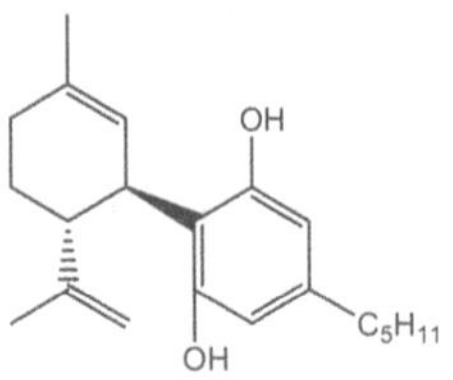

Cannabidiol

(CBD)

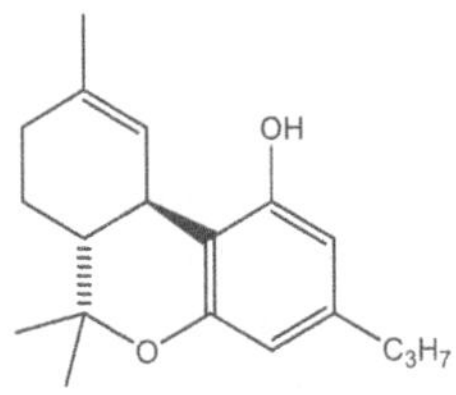

Tetrahydrocannabivarin

(THCV)

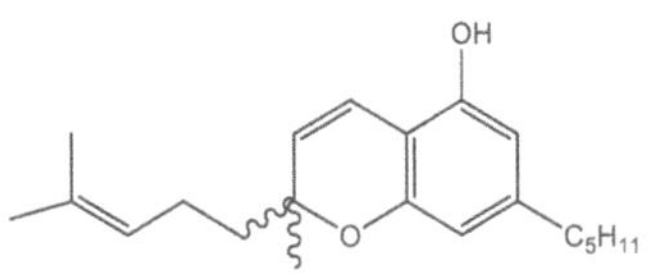

Cannabichromene

(CBC)

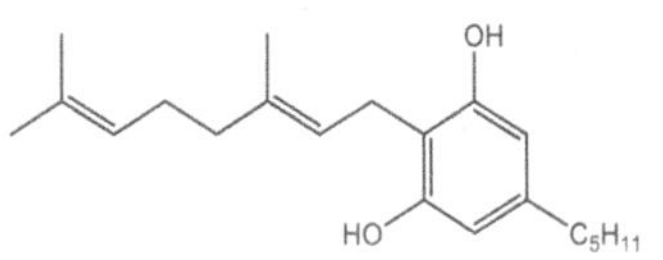

Cannabigerol

(CBG)

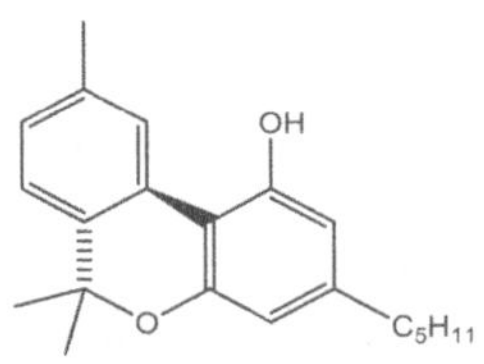

Cannabinol

(CBN)

FIGURE 4.1 Chemical structures of major phytocannabinoids commonly found in *Cannabis sativa.*

and terpenes are predominantly synthesized and accumulated in the glandular trichomes of mature flower buds, followed by young flowers and leaves.

Figure 4.3 illustrates the biosynthetic pathway of cannabinoids. It has been reported that the plastidial methylerythritol phosphate (MEP) pathway is primarily responsible for terpenoid biosynthesis. Olivetolic acid (OLA) is generated as the primary product in the biosynthetic pathway of cannabinoids. OLA and geranyl diphosphate (GPP) are generated through the polyketide and deoxyxylulose phosphate (DOXP)/MEP pathways, respectively. Numerous researchers have conducted comprehensive reviews on the biosynthesis of cannabinoids in *Cannabis sativa* (Shoyama et al., 1975; Kajima and Piraux, 1982; Fellermeier and Zenk, 1998; Sirikantaramas et al., 2004; Sirikantaramas et al., 2007).

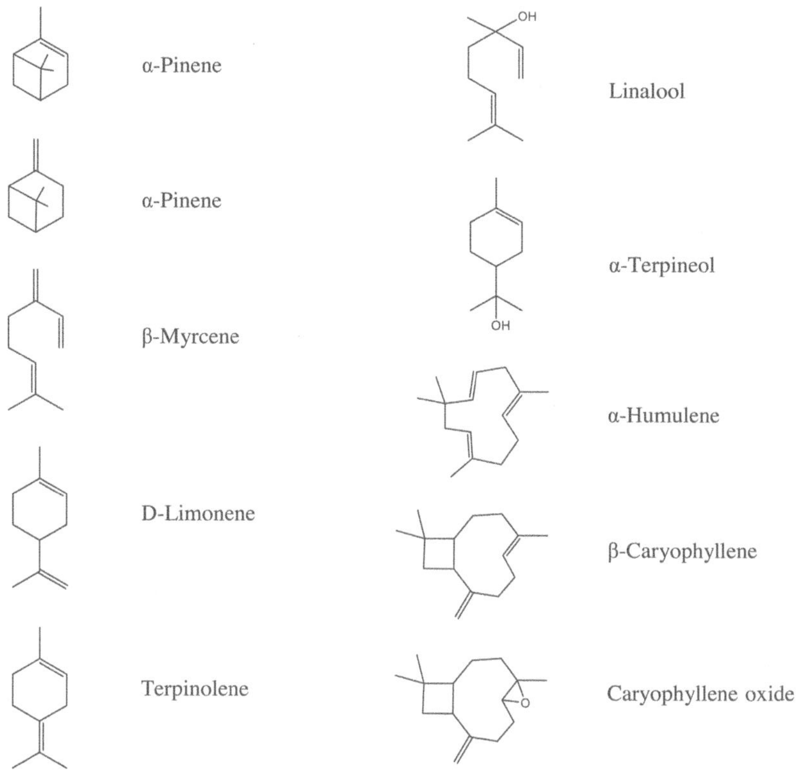

FIGURE 4.2 Chemical structures of major terpenes in the cannabis plant.

4.3 CANNABIS CONSTITUENTS: CANNABINOIDS, PHENOLICS, TERPENES, AND ALKALOIDS

Cannabis contains a wide range of secondary metabolites, which encompass both cannabinoid and non-cannabinoid components. As of the present moment, *C. sativa* has yielded over 500 documented compounds, 125 of which are cannabinoids (Radwan et al., 2021). Non-cannabinoid components consist of flavonoids, terpenes, alkaloids, non-cannabinoid phenols, and others. Cannabinoids comprise a class of compounds distinguished by their C21 terpenophenolic structure. The aforementioned terminology is applicable to cannabinoid derivatives, transformation products, and progenitor cannabinoids. Cannabis contains 125 cannabinoids that occur naturally; these cannabinoids are further categorized into 11 subclasses (Table 4.1). The primary non-cannabinoid components fall into the following four categories: (1) Alkaloids, (2) flavonoids, (3) terpenes, and (4) non-cannabinoid phenols. Forty-two flavonoids, 34 non-cannabinoid phenols, 120 terpenes, and 2 alkaloids have been identified in *C. sativa* to date.

4.4 CANNABIS NOMENCLATURE AND CLASSIFICATION: SIMPLE YET COMPLICATED

C. sativa plants that are typically cultivated both indoors and outdoors are illustrated in Figures 4.4a and 4.4b, respectively. Cannabis is a taxonomically complex plant. It is a herbaceous, annual plant that belongs to genus *Cannabis* (family Cannabaceae).

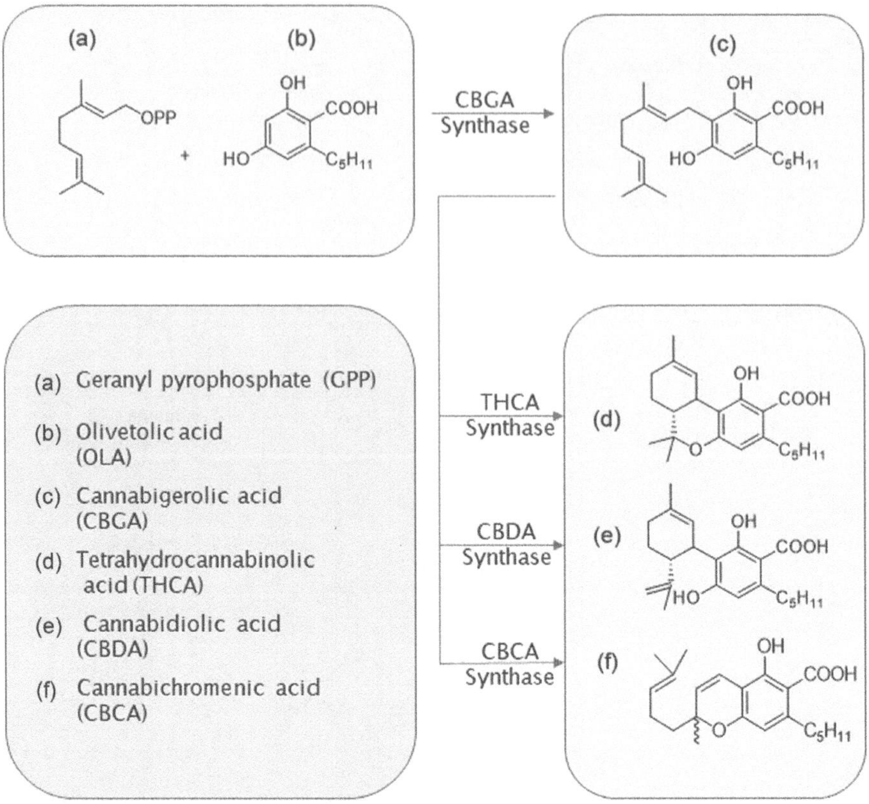

FIGURE 4.3 Biosynthetic pathway of cannabinoids.

TABLE 4.1
Number of Cannabinoids in Different Chemical Sub-classes in *Cannabis sativa* L.

Cannabinoids Sub-classes	Number of Cannabinoids
CBG type	16
CBC type	9
CBD type	10
Δ^9-THC type	25
Δ^8-THC type	5
CBL type	3
CBE type	5
CBN type	11
CBND type	2
CBT type	9
Miscellaneous type	30
Total	125

FIGURE 4.4 Indoor (a) and outdoor (b) cultivation of cannabis.

Botanical Nomenclature:

Kingdom: Plantae (plants)
Subkingdom: Tracheobionta (vascular plants)
Superdivision: Spermatophyta (seed plants)
Division: Magnoliophyta (flowering plants)
Class: Magnoliopsida (dicotyledons)
Sub-class: Hamamelididae
Order: Urticales
Family: Cannabaceae
Genus: *Cannabis*
Species: *Cannabis sativa* L.

The number of species in cannabis has been a subject of ongoing discussion for a long period of time. Whether the genus consists of one or more species is a matter of divided opinion. The subject of taxonomic dispute pertains to the allocation of scientific nomenclature to distinct cannabis strains characterized by variations in appearance and chemical composition, with particular emphasis on contemporary hybrid cultivars.

The Swedish taxonomist Carl Linnaeus was the first who classified cannabis as a monomorphic genus having a single species, *Cannabis sativa* L. (Linnaeus, 1737). However, Lamarck described the 'Indian cannabis strain' taxonomically different from 'European hemp' and gave it a specific name: '*Cannabis indica*' (Lamarck, 1785).

FIGURE 4.5 Drug (a) and fiber (b) types of cannabis plants.

In recent reports, cannabis is considered to belong to two (or three) species (*Cannabis sativa, Cannabis indica*, and *Cannabis ruderalis*) by some botanists, while others treat cannabis as one extremely polygamous species, *Cannabis sativa* L. (Clarke and Merlin, 2013; Small, 2017). Nevertheless, in the present day, *Cannabis* is considered to have a single species (*Cannabis sativa* L.), with *sativa, indica*, and *ruderalis* being its varieties.

On the other hand, it is straightforward to divide cannabis into three distinct chemotypes according to chemical profiles: (1) Drug types (THC >0.30%), intermediate (THC ~ CBD), and fiber type (THC <0.30%) (Figure 4.5). The chemotherapeutic classification of cannabis varieties are being utilized more frequently and gaining widespread acceptance in situations where the objective is to develop phytopharmaceuticals through clinical studies utilizing cannabis biomass (Russo, 2016, Figure 4.6).

4.5 SELECTION OF ELITE CANNABIS CHEMOVARS FOR VEGETATIVE AND MICROPROPAGATION

The cannabis plant is classified as dioecious due to the presence of distinct inflorescences on each male and female branch. Nevertheless, cannabis can occasionally exhibit a monoecious characteristic. Male and female plants are exceedingly difficult to distinguish during the early vegetative or juvenile stages. However, the distinct characteristics of male flowers (Figure 4.7a) and that they emerge prior to female flowers (Figure 4.7b) make them easily recognizable.

As a result of the cross-pollination tendency of cannabis, plants grown from seeds are extremely heterozygous. A considerable degree of variability is evident in the profile and concentration of cannabinoids among individual plants. For the cannabis production in terms of drug development, sinsemilla or seedless plants are preferred over male plants because (1) female plants contain greater quantities of cannabinoids than male plants; (2) when male plants are present, female flowers produce seeds, which results in a reduction in the drug content; and (3) cannabis, being an allogamous plant that is pollinated by wind, presents a challenge in maintaining the chemical profile of a specific variety when cultivated in conjunction with other varieties via seeds.

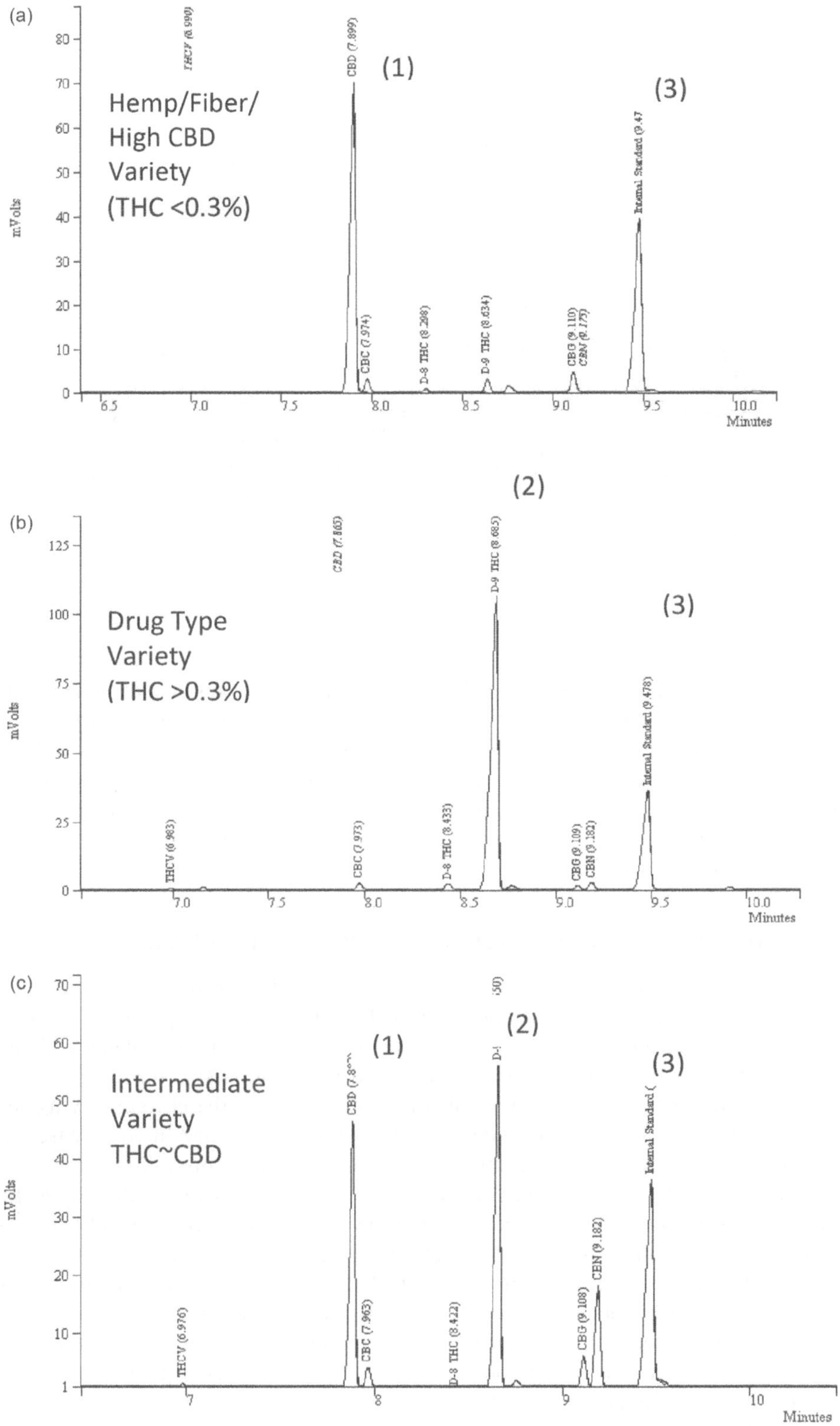

FIGURE 4.6 (a) Fiber-type cannabis variety (CBD > THC). (b) Drug-type cannabis (THC > CBD). (c) Intermediate-type cannabis variety (THC ~ CBD). Peak (1), cannabidiol, CBD; peak (2), Δ^9-tetrahydrocannabinol, THC and peak 3, internal standard, IC.

FIGURE 4.7 Flowering buds of male (a) and female (b) cannabis plants.

In case of several varieties growing together from seeds and in presence of male plants, cross pollination influences the biomass products, leading their cannabinoids profile to differ from that of the original varieties. However, a consistency in the cannabinoids profile is necessary if the biomass is to be grown for clinical research. This can be accomplished through the selection of maternal clones that possess the desired chemical profile, their preservation for subsequent use, and their propagation via vegetative cuttings and/or micropropagation, which yield biomass products that maintain the desired chemical profile.

4.6 CANNABIS MICROPROPAGATION: EFFORTS AND UPDATES

Micropropagation is a highly recognized and widely employed plant biotechnological technique utilized to preserve, maintain, and mass-produce medicinally significant plants. It is a distinctive method of asexual reproduction that allows for the large-scale production of plants with an identical chemical profile. Micropropagation-based advanced propagation systems are not only useful for producing true-to-type, identical individuals in a timely manner, thereby conserving space and resources, but they can also be implemented routinely to ensure the availability of disease-free stock material. Figure 4.8 provides a representative depiction of the micropropagation process employed to cultivate *Cannabis sativa* L., as well as various phases of plant growth and development. As the fact that cannabis is an allogamous species, plants grown from seeds are extremely heterozygous; thus, *in vitro* propagation and screening of exceptional female clones provide advantages over conventional methods. Various botanical components, including leaves, branches, and roots, have been utilized to cultivate cannabis plants *in vitro*. The establishment of dependable *in vitro* regeneration methodologies is anticipated to have a significant impact on forthcoming manufacturing systems. A compilation of research endeavours conducted and published regarding cannabis micropropagation over the past few decades has been provided in Table 4.2. However, compiling an entirely current list of publications is exceedingly challenging, if not impossible. Nevertheless, we earnestly present this table as a means of showcasing the advancements achieved in the field of cannabis micropropagation.

FIGURE 4.8 Representative illustrations of micropropagation of *Cannabis sativa* L. with different stages of plant growth and development. (a) Development of shoots, (b–d) well rooted plants, (e) rooted plants during the acclimatization and (f) well grown cannabis plants.

4.6.1 Propagation Through Nodal Segments (Direct Organogenesis)

With the objective of developing a dependable supply of pharmacologically active elite chemotypes that exhibit consistency in cannabinoid content, our goal was to devise an efficient and dependable method for cultivating *C. sativa* plants *in vitro*. In that regard, a two-step direct organogenesis protocol was developed for cannabis using apical nodal explants obtained from a high-Δ^9-THC-yielding, indoor-grown cannabis chemovar. The best response for shoot initiation and multiplication was achieved with MS + 0.5 μM TDZ medium. For root development, fully developed plants were subsequently transferred to activated charcoal with MS medium and different concentrations cytokinins such as IBA, NAA, and IAA. Out of different combinations tried, micropropagated plants utilizing 1/2 MS salts + 500 mg/L activated charcoal with 2.5 μM IBA exhibited the maximum rooting percentage (Lata et al., 2009a).

However, a one-step regeneration protocol utilizing the aromatic cytokinin meta-topolin (*m*T) and adventitious shoot induction for efficient root development has been developed further (Lata et al., 2016).

Among all the combinations of growth regulators tested, it was noted that the explants that showed the highest rate of shoot initiation and proliferation responded most positively to MS medium + 2 μM *m*T. Four to six weeks after multiplying on the same medium, two subcultures of shoots were capable of producing robust roots. The utilization of an auxin-containing medium was unnecessary for root induction. Tissue culture–raised plants were transferred, effectively acclimatized, and grown to maturity under the outdoor conditions.

4.6.2 Propagation through Leaf Tissues (Indirect Organogenesis)

Additionally, an *in vitro* propagation protocol utilizing a callus derived from leaf explants of *C. sativa* was developed (Lata et al., 2010). The regulation of indirect shoot organogenesis in cultures of *C. sativa* is predominantly determined by the combination and concentration of plant growth regulators. Leaf explants were inoculated with MS medium supplemented with different amounts of IAA, IBA, NAA, 2,4-D (individually, ranging from 0.5 μM to 2.0 μM), and TDZ (1.0 μM) to form

TABLE 4.2
Advances in Micropropagation of *Cannabis sativa* L. in the Last Two Decades

Author	Explant(s)	Media and PGR	Response
Stephen et al., 2023	Shoot tips	MS, LS, WPM, DKW medium BA, 2iP and TDZ (0.1, 5.0, 10.0 mM) IBA, NAA (0.2, 0.5, and 2.5 mM)	Shoot multiplication and *in vitro* roots
Hesami et al., 2023	Nodal segments	DKW medium	Shoot multiplication
Rico et al., 2022	Apical shoots	MS with half-strength nitrates + Vitamins formula β-H and β-A, no vitamins both medium with 2 mM	Shoot multiplication
Zhang et al., 2021	Immature embryo hypocotyls, true leaves, cotyledons and hypocotyls	*Callus induction medium*: MS+ 1 mg/L nicotinic acid + 1 mg/L pyridoxine-HCl + 10 mg/L thiamine-HCl + 0.1 g/L myo-inositol + 3% sucrose + 2.5 g/L phytagel + 1 mg/L 2,4-D + 0.25 mg/L Kin + 100 mg/L casein hydrolysate *Regeneration medium*: Half-strength MS + 1.5% sucrose + 3.5 g/L phytagel + 0.5 mg/L TDZ + 0.3 mg/L 6-BA + 0.2 mg/L NAA + 0.2 mg/L IAA *Rooting medium*: Half-strength MS + 1.5% sucrose + 3.5 g/L phytagel + 0.2 mg/L NAA + 0.5 mg/L IBA + 0.01 mg/L Zea^{RIB}	Indirect shoot organogenesis
Zarei et al., 2021	Stem segments	Safari Flower (SF) vegetative fertilizer solution + Aigma-Aldrich Canada + Ethanesulfonic acid + 5-mM MES (2-(N-morpholino)	Shoot growth and development
Page et al., 2021	Stem segments with two nodes	MS, DKW, WPM, B5, BABI media + TDZ (0.5 μM), 2,4-D (10, 20, 30 μM)	Callogenesis, shoot proliferation
Murphy and Adelberg, 2021	Apical shoot tip and single node	DKW without PGRs	Shoot proliferation
Lubell-Brand et al., 2021	Stem tips	MS, MS + Mesos components, 2.5× MS with vitamins, MS with vitamins + Added Mesos, MS with vitamins + Added vitamins, MS with vitamins + Added Mesos and vitamins MS with vitamins + Added Mesos	Shoot proliferation
Wróbel et al., 2022	Nodal segments containing axillary buds, shoot tips	*Shoot regeneration*: MS+ BAP (0.5–2 mg/L), TDZ (0.1–0.5 mg/L), mT (0.1–1 mg/L) *In vitro rooting*: 1/2 MS + (0.25, 0.5, 0.75 mg/L) of IBA and IAA	Shoot regeneration and *in vitro* rooting
Mestinšek-Mubi et al., 2020	Axillary buds	MS + TDZ (0.011, 0.1, 0.11, 0.22, 0.44, 0.88, 1.76 mg/L), mT (0.012, 0.12, 0.24, 0.48, 0.5, 0.96, 1.93 mg/L), BAP (1, 2.5, 5 mg/L), IAA (0.1 mg/L)	Shoot regeneration

(*Continued*)

TABLE 4.2 (*Continued*)
Advances in Micropropagation of *Cannabis sativa* L. in the Last Two Decades

Author	Explant(s)	Media and PGR	Response
Monthony et al., 2021a	Inflorescences (single florets vs. pairs of florets)	DKW + BAP (0.0, 0.01, 0.1, 1.0, and 10 μM) for both cultivars DKW + mT (0.0, 0.01, 0.1, 1.0, and 10 μM) for U91 cultivar	Direct shoot regeneration
Monthony et al., 2021b	Leaf	MS and DKW + NAA (0.5 μM), TDZ (0.5 and 1 μM)	Callogenesis
Ioannidis et al., 2020	Axillary buds	*Shoot regeneration medium*: Full- or half-strength MS + BAP (1.0, 2.0, 4.0 and 8.0 μM), TDZ (1.0, 2.0, 4.0 and 8.0 μM) *In vitro rooting medium*: Full- or half-strength MS + IBA (1.0, 2.0, 4.0 and 8.0 μM), NAA (1.0, 2.0, 4.0 and 8.0 μM)	Shoot regeneration and rooting
Galán-Ávila et al., 2020	Leaves, hypocotyl, and cotyledon	MS+ BAP (0.5, 1, 2 μM), TDZ (0.4, 1 μM), NAA (0.02, 0.2 μM), IBA (0.5 μM), 2,4-D (0.1 μM), 4-CPPU (1.0 μM), ZTRIB (1, 2), BAP^{RIB} (1 μM)	Direct shoot regeneration
Codesido et al., 2018	Axillary buds	MS, B_5 with vitamins (Formula βA), and B_5 without MS vitamins (Formula βH) + 2 μM mT, 2 μM IBA NAA, 2 μM IBA	Shoot regeneration
Smýkalová et al., 2019	Cotyledonary node, epicotyl with first node, hypocotyl, epicotyl with first and second node, shoot apical meristem, and shoot apex	MS + 9.31 μg/L NAA + 0.23 mg/L BAP + mT (1–5), BAP9THP (1–5), PEO-IAA (10 μmol/l)	Shoot regeneration
Piunno et al., 2019	Immature and mature inflorescences	*Shoot organogenesis medium*: MS + TDZ (0.1, 2, 5, 10 μM)	Shoot regeneration
Grulichova et al., 2017	Shoot tips (SM)	MS + 0.54 μM NAA + 1.78 μM	Shoot regeneration
Movahedi et al., 2015	Cotyledon and epicotyl	*Indirect shoot organogenesis medium*: MS + BAP (0.1, 0.2, 0.5, 1, 2, 3 mg/L), IBA (0.5 mg/L), TDZ (0.1, 0.2, 0.5, 1, 2, 3 mg/L), IAA (0.5 mg/L) *In vitro rooting medium*: MS + IBA (0.1, 0.2, 0.5, 1 mg/L) + NAA (0.1, 0.2, 0.5, 1 mg/L)	Indirect shoot organogenesis, and *in vitro* rooting
Movahedi et al., 2016a	Leaf and hypocotyl	*Callogenesis and shoot regeneration medium*: MS + 2,4-D (0.1, 0.2, 0.5, 1 mg/L), NAA (0.5, 1, 2, 3 mg/L), BAP (0.5 mg/L) *In vitro rooting*: MS+ (0.1, 0.2, 0.5, 1 mg/L) IBA and NAA	Callogenesis, indirect shoot organogenesis, and *in vitro* rooting
Movahedi et al., 2016b	Leaf and hypocotyl	*Callogenesis and shoot regeneration medium*: MS + BAP (0.1, 0.2, 0.5, 1, 2, 3 mg/L), TDZ (0.1, 0.2, 0.5, 1, 2, 3 mg/L), IBA (0.5 mg/L)	Callogenesis

(*Continued*)

TABLE 4.2 (*Continued*)
Advances in Micropropagation of *Cannabis sativa* L. in the Last Two Decades

Author	Explant(s)	Media and PGR	Response
Lata et al., 2016	Nodal segments containing axillary buds	*Shoot regeneration medium*: MS + 500 mg/L activated charcoal + TDZ (0.05, 0.50, 1, 2, 3, 4, and 5 µM), mT (0.05, 0.50, 1, 2, 3, 4, and 5 µM) *In vitro rooting medium*: ½ MS + 500 mg/L activated charcoal + IBA (0.05, 0.50, 1, 2, 3, 4, and 5 µM)	Shoot proliferation and *in vitro* rooting
Chaohua et al., 2016	Cotyledons	*Callogenesis medium*: MS + BAP (4, 6, 8 mg/L), ZT (0.5, 1, 1.5 mg/L), TDZ (0.1, 0.2, 0.4 mg/L), NAA (0.2, 0.4, 0.6 mg/L) *Shoot organogenesis medium*: MS + TDZ (0.1, 0.2, 0.3, 0.4, 0.5 mg/L), NAA (0.2, 0.4, 0.6 mg/L) *In vitro rooting*: ½ MS + IBA (0.2, 0.5, 1, 2 mg/L)	Callogenesis, shoot regeneration, and *in vitro* rooting
Movahedi et al., 2015	Cotyledons and epicotyls (callogenesis + shoot regeneration, rooting)	*Callogenesis/Shoot regeneration*: MS + 8.88 µM BAP + 2.46 µM IBA *Rooting medium*: MS + 0.49 µM IBA	Callogenesis
Jiang et al., 2015	Internodes	MS + 1 mg/l BAP + 0.5 mg/L NAA	Callus culture
Farag and Kayser, 2015	Seedling	B5 + 4 mg/L NAA	Hairy root cultures from callus
Wahby et al., 2013	Hypocotyl	*A. rhizogenes* and *A. tumefaciens* strains	Hairy root cultures
Lata et al., 2012	Nodal segments with axillary buds	MS + 0.5µM TDZ MS + 2.5 µM IBA	Direct organogenesis; shoot and root regeneration
Lata et al., 2010	Leaf	*Callogenesis medium*: MS + 1.0 µM TDZ + (0.5, 1.0, 1.5, and 2.0 µM) of IAA, NAA, IBA *Shoot organogenesis*: MS + (0.5, 1.0, 2.5, 5.0, and 10.0 µM) of BAP, KIN, TDZ *In vitro rooting medium*: ½ MS + (0.5, 1.0, 2.5, 5.0, and 10.0 µM) of IAA, IBA, and NAA	Callogenesis, shoot organogenesis, and *in vitro* rooting

(*Continued*)

TABLE 4.2 (*Continued*)
Advances in Micropropagation of *Cannabis sativa* L. in the Last Two Decades

Author	Explant(s)	Media and PGR	Response
Wang et al., 2009	Shoot tips	*Shoot regeneration medium*: MS + BAP (1.0, 2.0, 5.0 mg/L), KIN (1.0, 2.0, 5.0 mg/L), TDZ (0.1, 0.2, 0.5 mg/L), NAA (0.05, 0.1, 0.5 mg/L) *Rooting medium*: ½ MS, MS, B5 or NN + NAA (0.05, 0.25 mg/L), IAA (0.05, 0.25 mg/L), IBA (0.1, 0.5 mg/L)	Shoot proliferation and *in vitro* rooting
Lata et al., 2009	Nodal segments containing axillary buds	*Shoot regeneration medium*: MS + BAP (0.5–9 μM), KIN (0.5–9 μM), TDZ (0.5–9 μM), GA (0.7 μM) *In vitro rooting medium*: ½ MS + 500 mg/L activated charcoal + IAA (2.5, 5 μM), IBA (2.5, 5 μM), NAA (2.5, 5 μM)	Shoot proliferation and *in vitro* rooting
Lata et al., 2009	Nodal segments containing axillary buds	*Shoot regeneration medium*: MS + BAP (0.5–9 μM), KIN (0.5–9 μM), TDZ (0.5–9 μM), GA (0.7 μM) *In vitro rooting medium*: ½ MS + 500 mg/L activated charcoal + IAA (2.5, 5 μM), IBA (2.5, 5 μM), NAA (2.5, 5 μM)	Shoot proliferation and *in vitro* rooting
Flores-Sanchez et al., 2009	Leaves (callus suspension cultures and somatic embryogenesis)	*Suspension culture medium*: MS + B5 vitamins + 4.52 μM 2,4-D + 4.65 μM kinetin *Somatic embryogenesis*: Media composition N.S	Somatic embryogenesis
Wielgus et al., 2008	Cotyledons, stems, and roots	Knopp's medium + KIN (1), BAP (0.2), NAA (0.03–0.05), IAA (2)	Callogenesis, shoot regeneration, and *in vitro* rooting
Casano and Grassi, 2009	Cotyledon, stem root	Formula β–based medium	Shoot regeneration
Bing et al., 2007	Lateral buds	*Shoot regeneration*: MS + TDZ (0.1–0.5) + NAA (0.05–0.3) *Rooting*: ½ MS or MS + IBA (0.01–0.5) + NAA (0.01–0.25)	Shoot regeneration and *in vitro* rooting
Michał Pławuszewski and Wielgus, 2006	Stems, roots, and adventitious shoots	Knapp's medium + BAP (NR) + NAA (NR) + IAA (NR)	Direct organogenesis and indirect embryogenesis
Wahby et al., 2006	Hypocotyl	*A. rhizogenes* and *A. tumefaciens* strains	Hairy root cultures

(*Continued*)

TABLE 4.2 (*Continued*)
Advances in Micropropagation of *Cannabis sativa* L. in the Last Two Decades

Author	Explant(s)	Media and PGR	Response
Raharjo et al., 2006	Leaves, flowers, and seedling roots, stems, and shoots (callogenesis, callus suspension cultures)	*Callogenesis medium*: MS + 0.56 mM mesoinositol + 29.65 μM thiamine diHCl + 4.86 pyridoxine HCl + 8.12 μM nicotinic acid + 4.52 μM 2,4-D *Suspension culture (2 steps)*: *Step 1*: MS (as noted earlier, aqueous; 2 weeks) *Step 2*: B5 media + 9.05 μM 2,4-D + 2.85 μM IAA + 2.69 μM NAA + 5.12 μM potassium	Callogenesis and suspension culture
Plawuszewski et al., 2006	Axillary nodes (direct organogenesis), stems, and roots (indirect somatic embryogenesis)	*Callus induction medium*: DARIAind + NAA + BAP (concentrations N.S.) *Shoot proliferation*: DARIApro + NAA + BAP (concentrations N.S.) *Somatic embryogenesis medium*: DARIApro + NAA + BAP (concentrations N.S.) *Rooting medium*: DARIAroot + IAA (concentrations N.S.)	Callogenesis, shoot proliferation, embryogenesis and rooting
Slusarkiewicz-Jarzina et al., 2005	Leaves, petioles, internodes, and axillary buds	MS + 2,4-D (2 and 4), DIC (2 and 3), NAA (0.5, 1 and 2), and KIN (1, 2, and 4)	Callogenesis, shoot regeneration, and *in vitro* rooting
Feeney and Punja, 2003	Stem and leaf segment from seedling	MS + B5 vitamins + 5 μM 2,4 D + 1 μM KIN	Callus culture; agrobacterium-mediated transformation
Mandolino and Ranalli, 1999	Leaf, hypocotyl, cotyledon, and root	MS salts + B5 vitamins +2,4-D (3–10) + BAP (0.01–1)	Callogenesis and shoot regeneration
Jekkel et al., 1989	Anthers	MS + B5 vitamins + 3 mg/L 2,4,5-T B5 + 0.5 mg/L KIN + 1 mg/L 2,4 –D 10% DMSO	Cell suspension cultures, cryopreservation
Braemer and Paris. 1987	Leaf	B5 medium + 1 mg/L 2,4 -D + 0.5 mg/L KIN + 3% glucose	Cell suspension cultures
Fisse and Andres, 1985	Stem, cotyledon, root	MS + NAA	Callus formation
Richez-Dumanois et al., 1986	Apical and axillary buds	MS + IBA (0–20) + BAP (0.45) + 3% glucose + 1% sucrose + Charcoal (0–2 g/L)	Shoot organogenesis and *in vitro* rooting
Loh et al., 1983	Leaf, roots, and stem	MS salts + B5 vitamins medium + 2,4-D (0–5), 2,4,5-T (0–5), NAA (0–5), KIN (0–5), 2iP (0–5), and BAP (0–5)	Callogenesis and cell suspension cultures
Hartsel et al., 1983	Different parts of seedling	MS salts + B5 vitamins medium + 2,4-D (0.1) + KIN (0.5)	Callogenesis and cell suspension cultures

(*Continued*)

TABLE 4.2 (*Continued*)
Advances in Micropropagation of *Cannabis sativa* L. in the Last Two Decades

Author	Explant(s)	Media and PGR	Response
Hemphill et al., 1978	Bracts, calyx	Miller's medium + Murashige's iron source + IAA (0.25, 1) + NAA (0.1, 0.25) + 2,4-D (0.2) + KIN (1, 1.5, 2) + Casein hydrolysate (1)	Callogenesis
Itokawa et al., 1975	Leaves, hypocotyl, root, and female and male floral parts	MS + 2,4-D (1) + KIN (0.01–0.1)	Callogenesis
Veliky and Genest, 1972	Roots derived from *in vitro* grown seedling	*Gamborg's medium*: (67-V), 2,4-D (1.5 mg/L) + NAA (0.1 mg/L) + IAA (1 mg/L) + KIN (0.25 mg/L) + casein hydrolysate (1 mg/L)	Cell suspension cultures

calluses. The highest callus growth was achieved in 1.0 μM TDZ and 0.5 μM NAA. Subsequently, out of different cytokinins and auxins supplemented in different concentrations with MS, the highest number of shoots were achieved in MS medium + 0.5 μM TDZ, whereas rooting was found to be the best in ½ MS + 2.5 μM IBA for cannabis. Nevertheless, it has been demonstrated that regeneration through a callus develops a greater degree of somaclonal variations in plants (Lata et al., 2002; Faisal and Anis, 2007). This renders it impracticable to maintain true-to-type cannabis plants, as regeneration through the callus raises the possibility of genetically unstable variants.

4.7 *IN VITRO* GERMPLASM CONSERVATION FOR FUTURE USE

Micropropagation methods function to preserve a superior germplasm for subsequent applications. Explants derived from maternal clones that have been screened and selected according to their chemical profile can be preserved through *in vitro* propagation under delayed growth conditions, cryopreservation, or synthetic seed (encapsulation) technology.

4.7.1 Encapsulation Technology

Synthetic seed encapsulation technology has been effectively employed in *C. sativa* to preserve the germplasm *in vitro*. These seeds exhibit efficient regrowth, root development, and plantlet formation, making them suitable for both *in vitro* and *ex vitro* applications. To enhance a micropropagation protocol that was previously documented (Lata et al., 2009a), our research group has developed an effective approach to preserve and propagate *C. sativa in vitro* through explant encapsulation (synthetic seed) technology (Lata et al., 2009b). Nodal explants acquired from *in vitro*–raised plants were used for the encapsulation in beads formed with calcium alginate gel. The best bead formation was achieved by combining 5% sodium alginate with 50 μM $CaCl_2$. Under *in vitro* conditions, the highest regrowth was achieved in MS + 0.5 μM TDZ + 0.075% PPM. Regenerated plantlets from encapsulated explants were effectively hardened and subsequently transplanted into the soil.

4.7.2 Slow Growth

The micropropagation process, on the other hand, requires frequent subcultures, which increases its labour intensity and cost. In order to optimize efficiency and minimize labour requirements, the frequency of periodic subcultures is diminished during a specific time period. The extension of the subculture interval can be achieved through various methods: Reducing the temperature of the

TABLE 4.3
Advances in Cryopreservation in *Cannabis sativa* L.

Author	Explant(s)	Media and PGR	Response
Downey et al., 2021	*In vitro* nodal explants	MS, GA3	Recovery (26.7–66%)
Lata et al., 2019	Axillary meristem	V-cryoplate droplet vitrification, MS	Regrowth (42–47%)
Uchendu et al., 2019	Apical shoot tips	Droplet vitrification, MS, mT	Regrowth (55–63%
Jekkel et al., 1989	Cell suspension cultures	Nutritive medium with 0.75 μM ABA (10% DMSO)	Viability (46%)

culture, adjusting the culture media to include osmotic agent supplements, employing growth suppressing compounds, or eliminating growth boosters (Dodds and Roberts 1985). However, in order to ensure the most effective storage of the germplasm, it is crucial that the protocols account for chemical and genetic stability, maximum survival rate, and minimal subculture frequency. Limited research has been conducted on the conservation of *C. sativa* germplasm. A laboratory protocol has been devised to effectively preserve *C. sativa* at a temperature of 15 °C for a maximum of 16 weeks, as described by Lata et al. (2012). Additionally, this system facilitated the establishment of an effective preservation protocol for *C. sativa*, which has resulted in the development of cannabis plants that are genetically stable and homogeneous.

4.7.3 Cryopreservation

Cryopreservation is a valuable method for storing or conserving living plant tissues over an extended period of time at an extremely low temperature (approximately –196 °C, employing liquid nitrogen), while eliminating any modifications to their metabolic or genetic composition. To ensure the long-term viability of plant tissue, various cryopreservation techniques are utilized, such as dehydration, programmed freezing, vitrification, and V-cryo-plate. Table 4.3 presents an inventory of the cryopreservation techniques that have been devised for cannabis germplasm. Initial application of cryopreservation in the cannabis industry involved the preservation of hemp germplasm collections via cell suspension cultures (Jekkel et al., 1989). Out of the various cryoprotectants (glycerol, proline, DMSO), the one utilizing 10% DMSO and a –10 °C temperature transfer achieved the highest reported viability (58%). For the cryopreservation of cannabis, our laboratory has developed techniques utilizing axillary meristem (Uchendu et al., 2019) and apical shoot ends (Lata et al., 2019). The type of regenerated plants obtained from cryopreserved explants was determined to be superior to that of the donor plants. More recently, Downey et al. (2021) reported cryopreservation of 13 commercial *C. sativa* genotypes using *in vitro* nodal explants in another study.

4.8 CONCLUSION

These methodologies serve as instruments for the *in vitro* propagation, preservation, and large-scale production of true-to-type exceptional cannabis genotypes, thereby contributing significantly to the progression of cannabis research in the future.

ACKNOWLEDGMENT

This work is supported in part by the National Institute on Drug Abuse (NIDA), National Institutes of Health (NIH), and Department of Health and Human Services, USA, under contract no. N01DA-15-7793.

REFERENCES

Abrams, D. I., Jay, C. A., Shade, S. B., Vizoso, H., Reda, H., Press, S., Kelly, M. E., Rowbotham, M. C., & Petersen, K. L. (2007). Cannabis in painful HIV-associated sensory neuropathy: A randomized placebo-controlled trial. *Neurology, 68*(7), 515–521.

Adams, R., Baker, B. R., & Wearn, R. B. (1940). Structure of Cannabinol. III. Synthesis of Cannabinol, 1-Hydroxy-3-n-amyl-6, 6, 9-trimethyl-6-dibenzopyran1. *Journal of the American Chemical Society, 62*(8), 2204–2207.

Bing, X., Ning, L., Jinfeng, T., & Nan, G. (2007). Rapid tissue culture method of Cannabis sativa for industrial uses. CN, 1887043, 9.

Braemer, R., & Paris, M. (1987). Biotransformation of cannabinoids by a cell suspension culture of *Cannabis sativa* L. *Plant Cell Reports, 6*(2), 150–152.

Brenneisen, R., Egli, A., Elsohly, M. A., Henn, V., & Spiess, Y. (1996). The effect of orally and rectally administered delta 9-tetrahydrocannabinol on spasticity: A pilot study with 2 patients. *International Journal of Clinical Pharmacology and Therapeutics, 34*(10), 446–452.

Casano, S., & Grassi, G. (2009). Evaluation of media for hemp (*Cannabis sativa* L.) *in vitro* propagation. *Italus Hortus, 16*(2), 109–112.

Cascio, M. G., Pertwee, R. G., & Marini, P. (2017). The pharmacology and therapeutic potential of plant cannabinoids. In: Chandra, S., Lata, H., ElSohly, M. (eds.), *Cannabis sativa L. – Botany and Biotechnology* (pp. 207–225). Cham: Springer.

Chaohua, C., Gonggu, Z., Lining, Z., Chunsheng, G., Qing, T., Jianhua, C., Xinbo, G., Dingxiang, P., & Jianguang, S. (2016). A rapid shoot regeneration protocol from the cotyledons of hemp (*Cannabis sativa* L.). *Industrial Crops and Products, 83*, 61–65.

Clarke, R. C., & Merlin, M. D. (2013). *Cannabis: Evolution and Ethnobotany*. University of California Press.

Codesido, V., Meyer, S., & Casano, S. (2018). Influence of media composition and genotype for successful *Cannabis sativa* L. *in vitro* introduction. In *XXX International Horticultural Congress IHC2018: II International Symposium on Micropropagation and In vitro Techniques 1285* (pp. 75–80).

Dodds, J. H., & Roberts, L. W. (1985). *Experiments in Plant Tissue Culture*. International Potato Center.

Downey, C. D., Golenia, G., Boudko, E. A., & Jones, A. M. P. (2021). Cryopreservation of 13 commercial *Cannabis sativa* genotypes using *in vitro* nodal explants. *Plants, 10*(9), 1794.

ElSohly, M. A., & Slade, D. (2005). Chemical constituents of marijuana: The complex mixture of natural cannabinoids. *Life Sciences, 78*(5), 539–548.

ElSohly, M. A., Radwan, M. M., Gul, W., Chandra, S., & Galal, A. (2017). Phytochemistry of *Cannabis sativa* L. In: Kinghorn, A., Falk, H., Gibbons, S., Kobayashi, J. (eds.), *Phytocannabinoids: Progress in the Chemistry of Organic Natural Products, 103*. Cham: Springer. https://doi.org/10.1007/978-3-319-45541-9_1

Faisal, M., & Anis, M. (2007). Regeneration of plants from alginate-encapsulated shoots of *Tylophora indica* (Burm. f.) Merrill, an endangered medicinal plant. *The Journal of Horticultural Science and Biotechnology, 82*(3), 351–354.

Farag, S., & Kayser, O. (2015). Cannabinoids production by hairy root cultures of *Cannabis sativa* L. *American Journal of Plant Sciences, 6*(11), 1874.

Feeney, M., & Punja, Z. K. (2003). Tissue culture and Agrobacterium-mediated transformation of hemp (*Cannabis sativa* L.). *In vitro Cellular & Developmental Biology-Plant, 39*, 578–585.

Fellermeier, M., & Zenk, M. H. (1998). Prenylation of olivetolate by a hemp transferase yields cannabigerolic acid, the precursor of tetrahydrocannabinol. *FEBS Letters, 427*(2), 283–285.

Fisse, J., & Andres, J. (1985). Organogenesis and biosynthesis in an *in vitro* culture of *Cannabis sativa* L. *Boletin da Academia Galega de Ciencias, 4*, 57–67.

Flores-Sanchez, I. J., Peč, J., Fei, J., Choi, Y. H., Dušek, J., & Verpoorte, R. (2009). Elicitation studies in cell suspension cultures of *Cannabis sativa* L. *Journal of Biotechnology, 143*(2), 157–168.

Galán-Ávila, A., García-Fortea, E., Prohens, J., & Herraiz, F. J. (2020). Development of a direct *in vitro* plant regeneration protocol from *Cannabis sativa* L. seedling explants: Developmental morphology of shoot regeneration and ploidy level of regenerated plants. *Frontiers in Plant Science, 11*, 645.

Gaoni, Y., & Mechoulam, R. (1964). Isolation, structure, and partial synthesis of an active constituent of hashish. *Journal of the American Chemical Society, 86*(8), 1646–1647.

Grulichova, M. A., Mendel, P. E., Lalge, A. B., Slamova, N. I., Trojan, V. A., Vyhnanek, T. O., Winkler, J., Vaverková, M. D., Adamcova, D. A., & Đorđevic, B. (2017). Effect of different phytohormones on growth and development of micropropagated *Cannabis sativa* L. *Mendel Net, 2017*, 618–623.

Hammond, C. T., & Mahlberg, P. G. (1977). Morphogenesis of capitate glandular hairs of *Cannabis sativa* (Cannabaceae). *American Journal of Botany, 64*(8), 1023–1031.

Hartsel, S. C., Loh, W. T., & Robertson, L. W. (1983). Biotransformation of cannabidiol to cannabielsoin by suspension cultures of *Cannabis sativa* and *Saccharum officinarum*. *Planta Medica*, *48*(05), 17–19.

Hesami, M., Adamek, K., Pepe, M., & Jones, A. M. P. (2023). Effect of explant source on phenotypic changes of *in vitro* grown Cannabis plantlets over multiple subcultures. *Biology*, *12*(3), 443.

Ioannidis, K., Dadiotis, E., Mitsis, V., Melliou, E., & Magiatis, P. (2020). Biotechnological approaches on two high CBD and CBG *Cannabis sativa* L. (Cannabaceae) varieties: *In vitro* regeneration and phytochemical consistency evaluation of micropropagated plants using quantitative 1H-NMR. *Molecules*, *25*(24), 5928.

Itokawa, H., Takeya, K., & Akasu, M. (1975). Studies on the constituents isolated from the callus of *Cannabis sativa*. *Shoyakugaku Zasshi*, *29*(2), 106–112.

Jekkel, Z., Heszky, L. E., & Ali, A. H. (1989). Effect of different cryoprotectants and transfer temperatures on the survival rate of hemp (*Cannabis sativa* L.) cell suspension in deep freezing. *Acta Biologica Hungarica*, *40*(1–2), 127–136.

Jiang, Y., Zunmin, X. I. A., Yan, T. A. N. G., Qiang, H. A. N., & Chengwei, H. A. N. (2015). Preliminary studies on the tissue culture of *Cannabis sativa* L. (Industrial hemp). *Agricultural Science & Technology*, *16*(5).

Kajima, M., & Piraux, M. (1982). The biogenesis of cannabinoids in Cannabis sativa. *Phytochemistry*, *21*(1), 67–69.

Lamarck, J. B. (1785). Encyclopédie méthodique. Botanique. 1 part 2, Panckoucke, Paris, pp. 694–695.

Lata, H., Bedir, E., Hosick, A., Ganzera, M., Khan, I., & Moraes, R. M. (2002). *In vitro* plant regeneration from leaf-derived callus of *Cimicifuga racemosa*. *Planta Medica*, *68*(10), 912–915.

Lata, H., Chandra, S., Khan, I. A., & ElSohly, M. A. (2010). High frequency plant regeneration from leaf derived callus of high Δ9-tetrahydrocannabinol yielding *Cannabis sativa* L. *Planta Medica*, *76*(14), 1629–1633.

Lata, H., Chandra, S., Khan, I. A., & ElSohly, M. A. (2009a). Propagation through alginate encapsulation of axillary buds of *Cannabis sativa* L.—An important medicinal plant. *Physiology and Molecular Biology of Plants*, *15*, 79–86.

Lata, H., Chandra, S., Khan, I., & ElSohly, M. A. (2009b). Thidiazuron-induced high-frequency direct shoot organogenesis of *Cannabis sativa* L. *In vitro Cellular & Developmental Biology-Plant*, *45*, 12–19.

Lata, H., Chandra, S., Mehmedic, Z., Khan, I. A., & ElSohly, M. A. (2012). *In vitro* germplasm conservation of high Δ 9-tetrahydrocannabinol yielding elite clones of *Cannabis sativa* L. under slow growth conditions. *Acta Physiologiae Plantarum*, *34*, 743–750.

Lata, H., Chandra, S., Techen, N., Khan, I. A., & ElSohly, M. A. (2016). *In vitro* mass propagation of *Cannabis sativa* L.: A protocol refinement using novel aromatic cytokinin meta-topolin and the assessment of eco-physiological, biochemical and genetic fidelity of micropropagated plants. *Journal of Applied Research on Medicinal and Aromatic Plants*, *3*(1), 18–26.

Lata, H., Uchendu, E., Chandra, S., Majumdar, C. G., Khan, I. A., & ElSohly, M. A. (2019). Cryopreservation of axillary buds of *Cannabis sativa* L. by V-cryoplate droplet-vitrification: The critical role of sucrose preculture. *Cryoletters*, *40*(5), 291–298.

Linnaeus, S. C. (1737) Species Plantarum (ed. 1): 1027.

Loh, W. T., Hartsel, S. C., & Robertson, L. W. (1983). Tissue culture of *Cannabis sativa* L. and *in vitro* biotransformation of phenolics. *Zeitschrift für Pflanzenphysiologie*, *111*(5), 395–400.

Lubell-Brand, J. D., Kurtz, L. E., & Brand, M. H. (2021). An *in vitro*–ex vitro micropropagation system for hemp. *HortTechnology*, *31*(2), 199–207.

Mattes, R. D., Engelman, K., Shaw, L. M., & Elsohly, M. A. (1994). Cannabinoids and appetite stimulation. *Pharmacology Biochemistry and Behavior*, *49*(1), 187–195.

Mechoulam, R. (1986). The pharmaco history of *Cannabis sativa*. In: Mechoulam R. (ed.), *Cannabinoids as Therapeutic Agents* (pp. 1–19). CRC Press.

Mechoulam, R., & Shvo, Y. (1963). The structure of cannabidiol. *Tetrahedron*, *19*, 2073–2078.

Mestinšek-Mubi, Š, Svetik, S., Flajšman, M., & Murovec, J. (2020). *In vitro* tissue culture and genetic analysis of two high-CBD medical cannabis (*Cannabis sativa* L.) breeding lines. *Genetika*, *52*(3), 925–941.

Michał Pławuszewski, W. L., & Wielgus, K. (2006). Regeneration of Polish cultivars of monoecious hemp (*Cannabis sativa* L.) grown *in vitro*. *Renewable Resources and Plant Biotechnology*, 149.

Monthony, A. S., Bagheri, S., Zheng, Y., & Jones, A. M. P. (2021a). Flower power: Floral reversion as a viable alternative to nodal micropropagation in *Cannabis sativa*. *In vitro Cellular & Developmental Biology-Plant*, *57*(6), 1018–1030.

Monthony, A. S., Kyne, S. T., Grainger, C. M., & Jones, A. M. P. (2021b). Recalcitrance of *Cannabis sativa* to de novo regeneration; a multi-genotype replication study. *PloS One*, *16*(8), e0235525.

Movahedi, M., Ghasemiomran, V., & Torabi, S. (2016a). Effect of explants type and plant growth regulators on *in vitro* callus induction and shoot regeneration of *Cannabis sativa* L. *Iranian Journal of Medicinal and Aromatic Plants Research, 32*(1), 83–97.

Movahedi, M., Ghasemiomran, V., & Torabi, S. (2016b). *In vitro* callus induction and regeneration of medicinal plant *Cannabis sativa* L. *Iranian Journal of Medicinal and Aromatic Plants Research, 32*(5), 758–769.

Movahedi, M., Ghasemi-Omran, V., & Torabi, S. (2015). The effect of different concentrations of TDZ and BA on *in vitro* regeneration of Iranian cannabis (*Cannabis sativa*) using cotyledon and epicotyl explants. *Journal of Plant Molecular Breeding, 3*(2), 20–27.

Murphy, R., & Adelberg, J. (2021). Physical factors increased quantity and quality of micropropagated shoots of *Cannabis sativa* L. in a repeated harvest system with ex vitro rooting. *In vitro Cellular & Developmental Biology-Plant*, 1–9.

Page, S. R., Monthony, A. S., & Jones, A. M. P. (2021). DKW basal salts improve micropropagation and callogenesis compared with MS basal salts in multiple commercial cultivars of *Cannabis sativa. Botany, 99*(5), 269–279.

Pertwee, R. G. (2006). Cannabinoid pharmacology: The first 66 years. *British Journal of Pharmacology, 147*(S1), S163–S171.

Piunno, K. F., Golenia, G., Boudko, E. A., Downey, C., & Jones, A. M. P. (2019). Regeneration of shoots from immature and mature inflorescences of *Cannabis sativa. Canadian Journal of Plant Science, 99*(4), 556–559.

Pryce, G., & Baker, D. (2005). Emerging properties of cannabinoid medicines in management of multiple sclerosis. *TRENDS in Neurosciences, 28*(5), 272–276.

Radwan, M. M., Chandra, S., Gul, S., & ElSohly, M. A. (2021). Cannabinoids, phenolics, terpenes and alkaloids of cannabis. *Molecules, 26*(9), 2774.

Radwan, M. M., Wanas, A. S., Chandra, S., & ElSohly, M. A. (2017). Natural cannabinoids of cannabis and methods of analysis. In: Chandra, S., Lata, H., ElSohly, M. (eds.), *Cannabis sativa L. - Botany and Biotechnology* (pp. 161–182). Cham: Springer.

Raharjo, T. J., Eucharia, O., Chang, W. T., & Verpoorte, R. (2006). Callus induction and phytochemical characterization of *Cannabis sativa* cell suspension cultures. *Indonesian Journal of Chemistry, 6*(1), 70–74.

Richez-Dumanois, C., Braut-Boucher, F., Cosson, L., & Paris, M. (1986). Multiplication végétative *in vitro* Du Chanvre (*Cannabis sativa* L.). Application à la conservation des clones sélectionnés. *Agronomie, 6*(5), 487–495.

Rico, S., Garrido, J., Sánchez, C., Ferreiro-Vera, C., Codesido, V., & Vidal, N. (2022). A temporary immersion system to improve *Cannabis sativa* micropropagation. *Frontiers in Plant Science, 13*, 895971.

Russo, E. B. (2016). Current therapeutic cannabis controversies and clinical trial design issues. *Frontiers in Pharmacology, 7*, 309.

Shoyama, Y., Hirano, H., Oda, M., Somehara, T., & Nishioka, I. (1975). Cannabichromevarin and cannabigerovarin, two new propyl homologues of cannabichromene and cannabigerol. *Chemical and Pharmaceutical Bulletin, 23*(8), 1894–1895.

Sirikantaramas, S., Morimoto, S., Shoyama, Y., Ishikawa, Y., Wada, Y., Shoyama, Y., & Taura, F. (2004). The gene controlling marijuana psychoactivity: Molecular cloning and heterologous expression of Δ1-tetrahydrocannabinolic acid synthase from *Cannabis sativa* L. *Journal of Biological Chemistry, 279*(38), 39767–39774.

Sirikantaramas, S., Taura, F., Morimoto, S., & Shoyama, Y. (2007). Recent advances in *Cannabis sativa* research: Biosynthetic studies and its potential in biotechnology. *Current Pharmaceutical Biotechnology, 8*(4), 237–243.

Slusarkiewicz-Jarzina, A. U. R. E. L. I. A., Ponitka, A., & Kaczmarek, Z. (2005). Influence of cultivar, explant source and plant growth regulator on callus induction and plant regeneration of *Cannabis sativa* L. *Acta biologica Cracoviensia. Series, 47*(2005), 145–151.

Small, E. (2017). Classification of *Cannabis sativa* L. in relation to agricultural, biotechnological, medical and recreational utilization. In: Chandra, S., Lata, H., ElSohly, M. (eds.), *Cannabis sativa L. - Botany and Biotechnology* (pp. 1–62). Cham: Springer.

Smýkalová, I., Vrbová, M., Cvečková, M., Plačková, L., Žukauskaitė, A., Zatloukal, M., ... & Griga, M. (2019). The effects of novel synthetic cytokinin derivatives and endogenous cytokinins on the *in vitro* growth responses of hemp (*Cannabis sativa* L.) explants. *Plant Cell, Tissue and Organ Culture (PCTOC), 139*, 381–394.

Stephen, C., Zayas, V. A., Galic, A., & Bridgen, M. P. (2023). Micropropagation of hemp (*Cannabis sativa* L.). *HortScience, 58*(3), 307–316.

Turner, C. E., Elsohly, M. A., & Boeren, E. G. (1980). Constituents of *Cannabis sativa* L. XVII. A review of the natural constituents. *Journal of Natural Products, 43*(2), 169–234.

Uchendu, E., Lata, H., Chandra, S., Khan, I. A., & ElSohly, M. A. (2019). Cryopreservation of shoot tips of elite cultivars of *Cannabis sativa* L. by droplet vitrification. *Medical Cannabis and Cannabinoids*, *2*(1), 29–34.

Veliky, I. A., & Genest, K. (1972). Growth and metabolites of *Cannabis sativa* cell suspension cultures. *Lloydia*, *35*(4), 450–456.

Wahby, I., Caba, J. M., & Ligero, F. (2013). Agrobacterium infection of hemp (*Cannabis sativa* L.): Establishment of hairy root cultures. *Journal of Plant Interactions*, *8*(4), 312–320.

Wang, R., He, L. S., Xia, B., Tong, J. F., Li, N., & Peng, F. (2009). A micropropagation system for cloning of hemp (*Cannabis sativa* L.) by shoot tip culture. *Pakistan Journal of Botany*, *41*(2), 603–608.

Wielgus, K., Luwanska, A., Lassocinski, W., & Kaczmarek, Z. (2008). Estimation of *Cannabis sativa* L. tissue culture conditions essential for callus induction and plant regeneration. *Journal of Natural Fibers*, *5*(3), 199–207.

Wróbel, T., Dreger, M., Wielgus, K., & Słomski, R. (2022). Modified nodal cuttings and shoot tips protocol for rapid regeneration of *Cannabis sativa* L. *Journal of Natural Fibers*, *19*(2), 536–545.

Zarei, A., Behdarvandi, B., Tavakouli Dinani, E., & Maccarone, J. (2021). *Cannabis sativa* L. photoautotrophic micropropagation: A powerful tool for industrial scale *in vitro* propagation. *In vitro Cellular & Developmental Biology-Plant*, *57*(6), 932–941.

Zhang, X., Xu, G., Cheng, C., Lei, L., Sun, J., Xu, Y., … & Su, J. (2021). Establishment of an Agrobacterium-mediated genetic transformation and CRISPR/Cas9-mediated targeted mutagenesis in Hemp (*Cannabis Sativa* L.). *Plant Biotechnology Journal*, *19*(10), 1979–1987.

5 Physicochemical Factors and Their Significance in the Propagation of Medicinal Plants

Sushma Pandey, Chandra Bahadur Thapa, and Bijaya Pant

5.1 INTRODUCTION

Plants and plant products have been a source of food and medicines from the dawn of human civilization. In traditional healthcare practices, medicinal plants play a great role. The importance of medicinal plants provides clues to new areas of research and biodiversity conservation. Products from plants have attracted scientists around the world due to their huge impact on human health. Medicinal plants have been reported to contain numerous types of bioactive compounds of high therapeutic properties, which include all the pharmacological activities like anti-inflammatory, antiviral, anti-tumor, anti-malarial, anti-diabetic properties, and analgesic properties (Aye et al., 2019). The World Health Organization (WHO) reported approximately 80% of the developing population in the world depends on traditional medicines for their primary healthcare needs (Geneva, 2002; Pant, 2013), whereas in developed nations, more than 25% of prescribed medications are made from wild plant species (Hamilton, 2004). The survival of medicinal plants in the wild, however, is under threat from discriminatory heavy exploitation, habitat loss, illegal collection, trade, and global climate change. Therefore, *ex situ* and *in situ* conservation technologies are used to propagate rare, threatened, and valuable medicinal plants to conserve them for the benefit of mankind. The micropropagation approach, which uses even a small part of the plant (explant) for the production of a large number of plants which can be carried out in any season of the year in a short time, may be a trustworthy means of the *ex situ* conservation of medicinal plants. Micropropagation is an alternative approach to vegetative propagation that aids in germplasm storage and the conservation of rare and endangered species. Because they act as an "insurance policy" against extinction, *ex situ* conservation strategies can be used in conjunction with *in situ* conservation techniques (Pant, 2013; Pant, 2020). Although micropropagation is an *in vitro* culture technique for the quick propagation of disease-free and high-quality plants, a few physicochemical parameters exist in the culture system that might be considered for the proper growth, proliferation, and establishment of a culture. Some detrimental effects such as contamination of the culture, loss of totipotency, browning of the tissue and media, and recalcitrance of explants/tissues due to cellular oxidation on *in vitro* cultured explants may be the result of physical and chemical factors interacting with one another or depending on one another. Moreover, the most effective tissue culture approach requires the fulfillment of a few important environmental requirements or physicochemical environmental growth factors such as light, temperature, oxygen, pH, humidity, and salt concentration in growth media at optimum levels. Several physicochemical factors should be taken into account when growing medicinal plants *in vitro*.

5.2 EFFECT OF LIGHT, TEMPERATURE, AND HUMIDITY

Temperature, humidity, and light levels all have a big influence on plant tissue culture in the microenvironment. The synthesis of bioactive compounds, the fresh weight of plantlets, shoot regeneration,

DOI: 10.1201/b23374-5

and even the quantity of chlorophyll in plant cells and tissues are all influenced by the kind, quality, and intensity of light. In tissue culture plants, the amount of light influences both morphogenesis and organogenesis. The kind of plant and stage of growth determine the optimal light intensity for tissue culture (Bhojwani and Dantu, 2013). Depending on the type of plant and the goals of the culture, a typical tissue culture plant needs a photosynthetic photon flux density (PPFD) of 10–100 μmol/m^2/second, or 740–7400 lux, and a photoperiod of 16 hours. Starting with a modest light intensity is often preferable to high intensity, as too much light can easily burn the tissues of plants. A specific temperature is required, depending on the type of tissue being cultured. For best growth, a tissue culture lab's temperature should normally be between 22°C and 28°C, with a temperature variation of less than 0.5°C, and usually 26°C is recommended. The development of plant tissues may be significantly slowed by a temperature rise brought on by an increase in light intensity. Due to increased metabolic activity at higher temperatures, nutrients and water become scarce in nutrient media, which slows down plant growth and development (Cavallaro and Muleo, 2022). Furthermore, the culture room or growth chamber is kept at 50–60% relative humidity for optimum growth. Conditions with high ethylene concentrations, steady temperatures, low CO_2 in light and high CO_2 in darkness, and high relative humidity are excellent for the proper development of plant tissue culture (Kozai et al., 1996). High CO_2 levels during the light period slow plantlet growth and cause senescence (Shimazu and Kurata, 1999). According to studies conducted by (Thapa et al., 2022) and (Pandey et al., 2020b), the most effective photoperiod for micropropagating *Piper longum* and *Valeriana jatamansi*, two valuable medicinal plants, was 16 hours, with a light intensity of 30 μmol/m^2/second, temperature of 25°C $\pm$ 2°C, and relative humidity of 50–60% (Figure 5.1a, b).

FIGURE 5.1 Photograph showing micropropagated plants at effective temperature of 25°C $\pm$ 2°C of (a) *Piper longum* and (b) *Valeriana jatamansi*. (From Pandey et al., 2020b.)

When plants are exposed to low temperatures that are not below freezing, the chilling effect occurs. Certain plants, such as *Prunus tenella*, apricot cultivar "Bebecou," and Rubus, can have their *in vitro* shoot elongation and proliferation impacted by cold. (Koubouris and Vasilakakis, 2006) observed that the apricot cultivar "Bebecou" had the highest number of young micro shoots (18.7) in a culture medium enhanced with 2.2 μM BA + 0.57 μM IAA after 300 hours of chilling. Pre-treatment chilling at 4°C for 6 weeks significantly enhanced *in vitro* Rubus initiation culture, and the success rates for initiation cultures following chilling reached as high as 65%, according to (Wu et al., 2009). Similarly, it was found by (Bouza et al., 1992) that the *in vitro*–grown shoots may be chilled at 5°C for 30 days to restore the *P. tenella* growth rate.

5.3 EFFECT OF MICROBES AND PATHOGENS

The process of tissue culture must be carried out in an entirely aseptic environment, as the culture may become contaminated with viruses and viroids, fungi including yeast, bacteria, and bacteria-like organisms that could harm the cultivated plants. It affects the time, money, and objectives of the experiment. Tissue culture can be contaminated by a variety of sources, including growth rooms, culture laboratories, and endophytes. Because of inappropriate handling techniques, the culture may get contaminated by non-fastidious environmental microorganisms, including fungi, bacteria, and micro-arthropods, that are present in the growing chamber and laboratory. Most of them thrive on culture medium, but some might not because of strong osmotic pressure, certain media components, pH, or they might not develop quickly enough to be noticeable at the temperature of the culture (George, 1993). Once plants are subcultured in a fresh medium, dormant microbes may begin to develop. The ambient microbial inoculum is ubiquitous, although in certain regions it may display seasonal patterns (Gregory, 1973). Endophytes found in the tissues of the explant are another source of contamination in culture. Perennial plants and vegetatively propagated plants might be argued to be endophytically contaminated locally, if not systemically (Bacon and White, 2000). According to (Cassells and Tahmatsidou, 1997), these endophytes are strains of widely occurring microbes that are not harmful. On the other hand, plants endophytically colonized by non-pathogenic bacteria may also have adverse effects on growth (Long et al., 1988). Accordingly, contamination in tissue culture plants results from the introduction of endophytic bacteria, which are responsible for plant diseases, or surface sterilization-resistant microorganisms, which are mostly ambient (Cassells, 2001). Cell death rises in plant tissue cultures containing these microbes, and latent infections can also result in irregular growth, tissue necrosis, reduced shoot multiplication, and impaired roots (Patra, 2020). However, many plant growth–promoting microbes such as *Azotobacter*, *Trichoderma*, *Pseudomonas*, and *Piriformospora indica* are capable of synthesizing plant hormones, and a wide range of other organic chemicals that can enhance plant growth, shoot elongation, and rooting may aid in the acclimatization phase in tissue culture (Chennappa et al., 2019; Shah et al., 2019; Soumare et al., 2021).

5.4 EFFECT OF TYPE OF EXPLANT USED

Explants are tiny pieces of donor plants that are utilized as the initial material for micropropagation in tissue culture. Leaf, root, shoot tip, meristem, nodal segments, ovary, anther, cotyledon, hypocotyl, etc., are a few examples of explants. Certain plants also employ the transverse thin cell layer (tTCL) and the longitudinal thin cell layer (lTCL) as explants (Tran Thanh Van, 1998). Although the fact that any plant tissue may be utilized to regenerate the entire plant is widely recognized, there are a few limitations. Root tips are rarely used as explants for tissue culture because they are difficult to isolate and may be infected with microorganisms. Explants for tissue culture are selected based on the purpose of the study; however, they should be disease-free and immature tissues with dividing totipotent cells. The ability of a single cell to undergo dedifferentiation and redifferentiation and thereby regenerate the entire organism is known as totipotency. The capacity to regenerate explants into plantlets by micropropagation is dependent on several criteria, including age, size, source, genotype, explant location, and the donor plant's physiological stage (Yildiz, 2012). Larger explants have a greater capacity for regeneration than smaller explants because they contain more reserve nutrients and are less dependent on the culture medium; younger plants have a greater capacity for regeneration because they contain more actively dividing cells and relate to greater physiological action than older parts (Pierik, 1987). Explants obtained from the lower sections of donor plants showed a larger potential for regeneration than those removed from the higher areas; healthy plants with high cell division rates have a greater capacity for regeneration than sick ones (Evers, 1984). Moreover, the multiplication of an explant regenerating multiple shoots in cytokinin-rich media is essential for micropropagation, and that depends on the types of explant and plant species used in tissue culture. In the case of *P. longum*, nodal explants regenerate multiple shoots that can be used for micropropagation (Figure 5.1).

5.5 EFFECT OF GAS ACCUMULATION AND AERATION IN CULTURE VESSELS

Plant tissue culture is carried out in various-sized culture tubes and jars for micropropagation. In general, 15-mL culture tubes with a surface area of 16×125 mm^2 and 300-mL culture jars with a surface area of 55×130 mm^2 are used; which one, however, depends on the type of plant growth and its stage of development. The microenvironment of culture vessels affects the growth and development of *in vitro* plantlets. However, the microenvironment of culture containers can be affected by the aeration or gas exchange between the interior and exterior air (Walli et al., 2019). On the other hand, the aeration process of culture vessels can be affected by light, temperature, and relative humidity (Gupta and Ibaraki, 2006). Gases including CO_2, O_2, and C_2H_4 are released by developing tissues and cells and collected in the culture containers. A large accumulation of gas in culture jars or tubes can impact plant growth and development in plant tissue culture. The plant hormone ethylene is released by explants grown *in vitro* and may accumulate in large amounts in the culture vessels, particularly from rapidly dividing non-differentiated callus, meristem, or suspension cultures. As a result, ethylene is likely to have both positive and negative effects on secondary metabolite synthesis, stress response, and growth and development (Biddington, 1992). Inadequate aeration can also cause ethylene and carbon dioxide buildup, oxygen deprivation, and affect explant development in culture vessels (Jackson et al., 1991). The efficacy of the vessel's aeration system and the tissue's capacity for photosynthetic activity determine the CO_2 concentration, whereas the concentrations of ethylene and water vapor (relative humidity) in culture vessels are higher than those in an outer environment (Gupta and Ibaraki, 2006). If the growth room is polluted with environmental microorganisms, aeration between the interior of the culture vessel and the outside environment could contaminate tissues grown *in vitro*.

5.6 EFFECT OF THE PHYSICAL PROPERTY OF CULTURE VESSELS

For tissue culture to operate efficiently, vessels are crucial. The growth and development of plantlets in tissue culture are also influenced by the physical characteristics of the culture vessel, such as the quality of the vessel materials (glass, polypropylene, polycarbonate, or polystyrene) and the size of the vessel (height and width). Culture vessels made of high-quality glass allow for the homogeneous and efficient transmission of light to developing tissues inside for photosynthesis and other metabolic processes (Abd Malik et al., 2022). To exchange gases, minimize water vapor, and maintain ethylene concentration, culture vessels must be the right size based on the size of the growing explant. The materials used to make the lid or cap that covers the opening of the vessel can play a significant role in maintaining the developing tissues inside the vessel. The gaseous exchange between the interior and outside of the vessel is generally affected when the mouth of the vessel is closed with a cotton plug, aluminum foil, or plastic cap (Huang and Chen, 2005).

5.7 EFFECT OF pH OF NUTRIENT MEDIA

Murashige and Skoog (1962) first conducted their experiment in *Nicotiana tabacum* using acidic media (pH 5.7–5.8). However, the majority of known plant tissues need a pH 5.2–5.8 for enough nutrient uptake by growing tissues and for consistency or proper gelling of media (Skirvin et al., 1986). In addition, a pH of less than 5.0 inhibits the proper gelling of the medium, whereas a pH of greater than 6.0 produces hard media. Similar to this, because absorbable nutrients for growing tissues in the medium at lower and higher pH are not accessible, plant cells in cultures cease to develop when the pH is more than 7.0 and less than 4.5. For wood-sedge soils, it is thought that nutrients are accessible to plant roots in the pH range of 5.5–5.8 (Lucas and Davis, 1961). For tissue culture, many different kinds of nutritional media are available. These media include White's media (pH 5.8), MS media (pH 5.8), Gamborg (B5) media (pH 5.5), Chu (N6) media (pH 5.8), Nitsch's media (pH 5.8), and others. Each of these media has a specific pH need. All varieties of media have almost the same pH requirements; however, to promote healthy plant growth and development, the pH level must be maintained within each type of media's specified range (Sudheer et al., 2022).

5.8 EFFECT OF GELLING AGENT

Depending on the purpose of the study, nutritional media are usually used in tissue culture in either a liquid or semi-solid state. When using cell suspension culture to produce secondary metabolites, liquid media with frequent agitation of the medium are preferred. The tissue may be submerged and get insufficient oxygen if a liquid medium is used, which might result in the tissue's death. Liquid or suspension culture may be used for somatic embryogenesis in some species (Ozyigit et al., 2023). However, a semi-solid medium is used in tissue culture in most cases for micropropagation utilizing different kinds of gelling agents. A widely used gelling agent is agar, which is a polysaccharide obtained from seaweeds. Agar is inert at culture temperature, unaffected by plant enzymes, and unaffected by interactions with other media components (Kacar et al., 2010). Other gelling agents are agarose, Gelrite, phytagel, xhantam gum, gelcarin, and alginic acid. Because of their capacity to form gels, they offer semi-solid or solid surfaces for structural support and aeration on which plants or microorganisms can grow. Gelatin can gel at a concentration of 10%, agar can gel at a concentration of 0.5–1%, and other artificial gelling agents can gel at concentrations as low as 1.0–2.5 g/L (Das et al., 2015).

5.9 EFFECT OF METABOLITE/PHENOLIC RELEASED IN CULTURE MEDIA

Several therapeutic plant explants release metabolites into the surrounding plant tissue and culture media, especially phenolics, which build up and oxidize (Jones and Saxena, 2013). As a result, the culture medium and plant material turn brown. Depending on whether or not enzymes are involved in the browning processes, the term "enzymatic" or "non-enzymatic" tissue browning might be used (Wang et al., 2021). However, enzymatic oxidation in the presence of oxygen is the primary cause of explant browning, and this is a serious concern in plant tissue culture, especially with woody and perennial plants. Enzymatic browning is a natural process that occurs when explants are sliced because enzymes oxidize released phenols to quinones. (Permadi et al., 2023). Together, catechol oxidase, polyphenol oxidase (PPO), and other enzymes chemically change natural phenols into benzoquinone and melanin. Explants that show browning show impediments to regenerating and dividing, which might lead to their death or render them inactive in the culture media. To reduce enzymatic browning, antioxidants such as citric acid, ascorbic acid, polyvinyl pyrrolidone (PVP), or activated charcoal are added to the culture medium. Ascorbic acid or citric acid added to the medium at a dose of 10–150 mg/L has been shown to minimize tissue browning (Strosse et al., 2004). Activated charcoal's large surface area and tiny pore network allow it to be used for many different purposes, such as lowering phenolic oxidation buildup, adjusting the medium's pH to the

FIGURE 5.2 Tissue browning in micropropagated plants of *Piper longum*.

perfect level, and establishing a darkened environment that replicates soil conditions (North et al., 2012). It was reported that in the Grande Naine cultivar, activated charcoal, either at 1.5 g/L by itself or in conjunction with 150 mg/L of ascorbic acid or citric acid, inhibited phenolic component oxidation and promoted plantlet growth (Safwat et al., 2015). After some time, the transplants can be subcultured, cut using a sharp knife, submerged in water for a while, and maintained vertically to help avoid tissue browning in tissue culture. Tissue and media became brown due to the release of phenolics into the medium by *P. longum* nodal explants (Figure 5.2). While this effect was reduced by using 100 mg/L of activated charcoal, *Withania somnifera* did not exhibit medium browning during nodal explant culture (Chin et al., 2019).

5.10 EFFECT OF PLANT GROWTH REGULATORS IN MEDIA

Phytohormones, also known as plant hormones or plant growth regulators (PGRs), are organic compounds that can regulate the biochemical, metabolic, and morphological processes in higher plants. They can be synthetic or natural. Five major categories may be used to group the PGRs: Auxin, cytokinins, gibberellins (GAs), abscisic acid (ABA), and ethylene. Apart from these, numerous synthetic and natural derivative chemicals serve as growth regulators for plants. Auxins, cytokinins, and auxin-cytokinin interactions are thought to be the most significant regulators of growth and development in plant tissue culture since these two categories of hormones are often required (Gaspar et al., 1996). Even though they are less prevalent, other PGRs, including ethylene, ABA, salicylic acid, jasmonate acid, etc., are nevertheless important in plant tissue culture. Depending on the objectives of the research, PGRs are employed in plant tissue culture for many functions, such as metabolite augmentation, callus induction, organogenesis, morphogenesis, root initiation, shoot multiplication, and somatic embryogenesis. Plant development, differentiation, and growth in general are influenced by the kind of PGR and their concentrations employed in tissue culture conditions. Typically, two or more growth regulators from different classes must be used in combination, either simultaneously or sequentially (Evans et al., 1981; Pandey et al., 2022). Auxins cause cell growth expansion, cell division, callus induction, root initiation, and vascular differentiation, whereas cytokinins stimulate cell division, often in association with auxins, and shoot initiation in tissue culture (Gaspar et al., 1996). It has been found that cytokinins and auxins interact to control several aspects of organogenesis, differentiation, and cell proliferation in tissue culture. The ratio of auxin to cytokinins influences the morphogenesis in callus culture. For instance, a high ratio of auxin to low cytokinins leads to the development of roots, a high ratio of cytokinin to low auxins results in the development of shoots, and a roughly equal ratio of auxin to cytokinins leads to the development of callus. The necessary concentration of each PGR might vary greatly depending on the type of tissue being grown, the type of PGR being used, and the surrounding conditions (Su et al., 2011). In rare cases, vitrification can be caused by a high cytokinin level in conjunction with other media components. Vitrification is a physiological disturbance that impairs tissue culture plants' ability to grow and develop, giving them a glassy appearance and causing their leaves and stems to become brittle and thick, which is observed in *W. somnifera* (Figure 5.3). Similarly, among the approximately 80 gibberellins found in plants, GA_3 and $GA_4 + 7$ are commonly used in plant tissue culture. Depending on the kind of plant, GAs may or may not decrease or prevent the beginning of roots, shoots, or somatic embryos in tissue culture, but it is employed in the process of embryogenesis, namely in the maturation and germination phases (Gaspar et al., 1996; Kodad et al., 2021). Recent studies on *V. jatamansi* (Pandey et al., 2020a) have demonstrated the advantages of exogenous nitrous oxide (NO) and/or sodium nitroprusside (SNP) supplementation on *in vitro* proliferation (Figure 5.4).

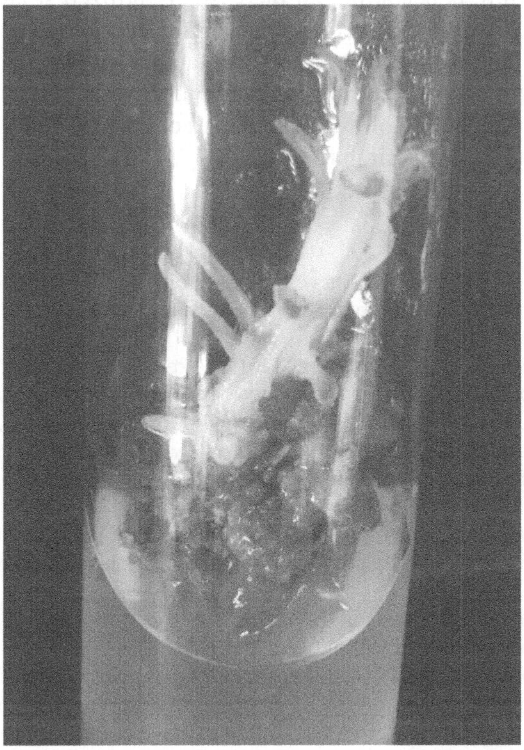

FIGURE 5.3 Vitrification in *Withania somnifera* at high concentration of cytokinins.

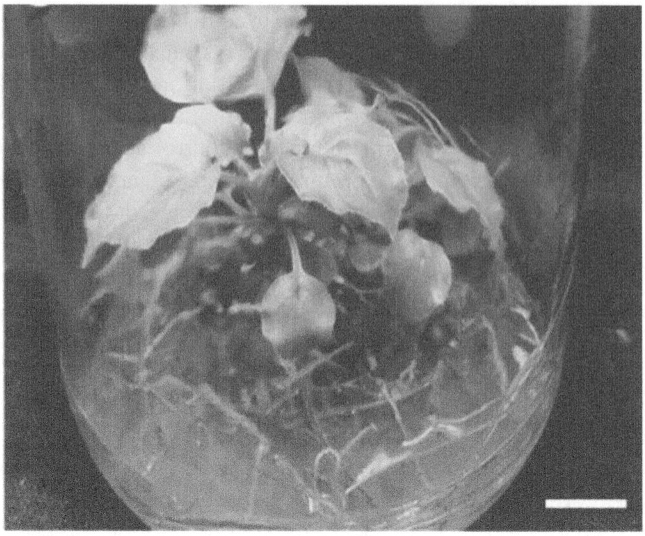

FIGURE 5.4 Positive effects of sodium nitroprusside (SNP) supplementation in *Valeriana jatamansi.*

5.11 CONCLUSION

In this chapter, we have systematically summarized the role and parameters of different physiochemical factors which are very essential in plant tissue culture and micropropagation. Growing plants *in vitro* requires correct parameters and environmental conditions, as every plant species requires different growth factors including seasonal, physical, and chemical variations. The reintroduction of plants in nature is becoming an increasingly utilized strategy in plant conservation and protected areas, which is possible through plant tissue culture technology. Besides, nowadays numerous phytochemical and bioactive secondary metabolite production from plants can be produced by this technology if correct parameters can be applied to the plants while growing them *in vitro*. However, given the paucity of available data, conducting a more thorough study on this topic is both important and highly desirable.

REFERENCES

Abd Malik, A. N., Uddain, J., Chin, C. K., Chew, B. L., Antony, J. J. J., Parab, A. R., & Subramaniam, S. (2022). Impact of culture vessels on micro-morphological features of *in vitro Dendrobium sabin* blue orchid. *BioTechnologia, 103*(1), 41–52.

Aye, M. M., Aung, H. T., Sein, M. M., & Armijos, C. (2019). A review on the phytochemistry, medicinal properties and pharmacological activities of 15 selected Myanmar medicinal plants. *Molecules, 24*(2), 293.

Bacon, C. W., & White, J. F. Jr. (2000). *Microbial Endophytes.* Roca Baton: CRC.

Bhojwani, S. S., & Dantu, P. K. (2013). *Plant Tissue Culture: An Introductory Text* (Vol. 318). India: Springer.

Biddington, N. L. (1992). The influence of ethylene in plant tissue culture. *Plant Growth Regulation, 11*, 173–187.

Bouza, L., Jacques, M., Mazière, Y., & Arnaud, Y. (1992). *In vitro* propagation of *Prunus tenella* Batsch. cv. 'Firehill': Control of vitrification; Increase of the multiplication rate and growth by chilling. *Scientia Horticulturae, 52*(1–2), 143–155.

Cassells, A. C. (2001). Contamination and its impact in tissue culture. *Acta Horticulturae, 560*, 353–359. https://doi.org/10.17660/actahortic.2001.560

Cassells, A. C., & Tahmatsidou, V. (1997). The influence of local plant growth conditions on non-fastidious bacterial contamination of meristem-tips of Hydrangea cultured *in vitro. Plant Cell Tissue Organ Culture, 47*, 15–26.

Cavallaro, V., & Muleo, R. (2022). The effects of LED light spectra and intensities on plant growth. *Plants, 11*(15), 1911.

Chennappa, G., Naik, M. K., Udaykumar, N., Vidya, M., Sreenivasa, M. Y., Amaresh, Y. S., & Mathad, P. F. (2019). Plant growth promoting microbes: A future trend for environmental sustainability. *New and Future Developments in Microbial Biotechnology and Bioengineering*, 163–178. https://doi.org/10.1016/b978-0-12-818258-1.00010-8

Chin, C. K., Lee, Z. H., Mubbarakh, S. A., Antony, J. J. J., Chew, B. L., & Subramaniam, S. (2019). Effects of plant growth regulators and activated charcoal on somaclonal variations of protocorm-like bodies (PLBs) of *Dendrobium sabin* blue orchid. *Biocatalysis and Agricultural Biotechnology*, *22*, 101426.

Das, N., Tripathi, N., Basu, S., Bose, C., Maitra, S., & Khurana, S. (2015). Progress in the development of gelling agents for improved culturability of microorganisms. *Frontiers in Microbiology*, *6*, 698.

Evans, D. A., Sharp, W. R., & Flick, C. F. (1981). Growth and Behavior of Cell Cultures: Embryogenesis and Organogenesis. In T. A. Thorpe (eds.), *Plant Cell Culture: Methods and Applications in Agriculture* (pp. 45–113). New York: Academic Press.

Evers, P. W. (1984). *Growth and Morphogenesis of Shoot initials of Douglas Fir, Pseudotsuga mensiesii* (Mirb.) Franco, *in vitro* (pp. 1–6). The Netherlands: Diss. Agric. Univ. Wageningen.

Gaspar, T., Kevers, C., Penel, C., Greppin, H., Reid, D. M., & Thorpe, T. A. (1996). Plant hormones and plant growth regulators in plant tissue culture. *In Vitro Cellular & Developmental Biology - Plant*, *32*(4), 272–289. https://doi.org/10.1007/bf02822700

Geneva, S. (2002). Traditional Medicine Growing Needs and Potential–WHO Policy Perspectives on Medicines.

George, E. F. (1993). *Plant Propagation by Tissue Culture, Part 1, the Technology.* 2nd Edition, Edington: Exegetics Limited.

Gregory, P. H. (1973). *The Microbiology of Atmosphere.* Leonard Hill: Aylesbury.

Gupta, S. D., & Ibaraki, Y. (Eds.). (2006). *Plant Tissue Culture Engineering* (Vol. 6, pp. 313–327). Netherlands: Springer.

Hamilton, A. C. (2004). Medicinal plants, conservation and livelihoods. *Biodiversity & Conservation*, *13*, 1477–1517.

Huang, C., & Chen, C. (2005). Physical properties of culture vessels for plant tissue culture. *Biosystems Engineering*, *91*(4), 501–511.

Jackson, M. B., Abbott, A. J., Belcher, A. R., Hall, K. C., Butler, R., & Cameron, J. (1991). Ventilation in plant tissue cultures and effects of poor aeration on ethylene and carbon dioxide accumulation, oxygen depletion and explant development. *Annals of Botany*, *67*(3), 229–237.

Jones, A. M., & Saxena, P. K. (2013). Inhibition of phenylpropanoid biosynthesis in *Artemisia annua* L.: A novel approach to reduce oxidative browning in plant tissue culture. *PLoS One*, *8*(10), e76802. https://doi.org/10.1371/journal.pone.0076802

Kacar, Y. A., Biçen, B., Varol, I., Mendi, Y. Y., Serçe, S., & Çetiner, S. (2010). Gelling agents and culture vessels affect *in vitro* multiplication of banana plantlets. *Genetics and Molecular Research*, *9*(1), 416–424.

Kodad, S., Melhaoui, R., Hano, C., Addi, M., Sahib, N., Elamrani, A., & Mihamou, A. (2021). Effect of culture media and plant growth regulators on shoot proliferation and rooting of internode explants from Moroccan native Almond (*Prunus dulcis* Mill.) genotypes. *International Journal of Agronomy*, *2021*, 1–10.

Koubouris, G., & Vasilakakis, M. (2006). Improvement of *in vitro* propagation of apricot cultivar 'Bebecou. *Plant Cell, Tissue and Organ Culture*, *85*, 173–180.

Kozai, T., Fujiwara, K., & Kitaya, Y. (1996). Modeling, measurement, and control in plant tissue culture. *Acta Horticulturae*, *393*, 63–73.

Long, R. D., Curtin, T. F., & Cassells, A. C. (1988). An investigation of the effects of bacterial contaminants on potato nodal cultures. *Acta Horticulturae*, *225*, 83–92.

Lucas, R. E., & Davis, J. F. (1961). Relationships between pH values of organic soils and availabilities of 12 plant nutrients. *Soil Science*, *92*, 171–182.

Murashige, T., & Skoog, F. (1962). A revised medium for rapid growth and bio assays with tobacco tissue cultures. *Physiologia Plantarum*, *15*(3).

North, J. J., Ndakidemi, P. A., & Laubscher, C. P. (2012). Effects of antioxidants, plant growth regulators and wounding on phenolic compound excretion during micropropagation of *Strelitzia reginae. International Journal of Physical Sciences*, *7*(4), 638–646.

Ozyigit, I. I., Dogan, I., Hocaoglu-Ozyigit, A., Yalcin, B., Erdogan, A., Yalcin, I. E., & Kaya, Y. (2023). Production of secondary metabolites using tissue culture-based biotechnological applications. *Frontiers in Plant Science*, *14*, 1132555.

Pandey, S., Sundararajan, S., Ramalingam, S., & Pant, B. (2020a). Effects of sodium nitroprusside and growth regulators on callus, multiple shoot induction and tissue browning in commercially important *Valeriana jatamansi* Jones. *Plant Cell, Tissue and Organ Culture (PCTOC)*, *142*, 653–660.

Pandey, S., Sundararajan, S., Ramalingam, S., Baniya, M. K., & Pant, B. (2020b). Rapid clonal propagation and valepotriates accumulation in cultures of *Valeriana jatamansi* Jones, a high-value medicinal plant. *Journal of Applied Botany & Food Quality*, *93*, 177–185.

Pandey, S., Sundararajan, S., Ramalingam, S., & Pant, B. (2022). Elicitation and plant growth hormone-mediated adventitious root cultures for enhanced valepotriates accumulation in commercially important medicinal plant *Valeriana jatamansi* Jones. *Acta Physiologiae Plantarum*, *44*, 1–13.

Pant, B. (2013). Medicinal orchids and their uses: Tissue culture a potential alternative for conservation. *African Journal of Plant Science*, *7*(10), 448–467.

Pant, B. (2020). Biotechnology for Plant Conservation. In M. Siwakoti, P. K. Jha, S. Rajbhandary, S. K. Rai (eds.), *Plant Diversity in Nepal, 2020* (pp. 237–251). Kathmandu: Botanical Society of Nepal.

Patra, S. (2020). Microbial organisms in tissue culture. *Biochemistry and Physiology*, *9*, 281.

Permadi, N., Nurzaman, M., Alhasnawi, A. N., Doni, F., & Julaeha, E. (2023). Managing lethal browning and microbial contamination in *Musa spp.* tissue culture: Synthesis and perspectives. *Horticulturae*, *9*(4), 453.

Pierik, R. L. M. (1987). *In Vitro Culture of Higher Plants, Handbook.* Martinus Nijhoff.

Safwat, G., Abdul-Rahman, F., & Sharbasy, A. S. (2015). The effect of some antioxidants on blackening and growth of *in vitro* culture of banana (*Musa* spp. cv. Grand Naine). *Journal of Genetics and Cytology*, *44*, 47–59.

Shah, S., Thapa, B. B., Chand, K., Pradhan, S., Singh, A., Varma, A., & Pant, B. (2019). Piriformospora indica promotes the growth of the *in-vitro*-raised *Cymbidium aloifolium* plantlet and their acclimatization. *Plant Signaling & Behavior*, *14*(6), 1596716.

Shimazu, T., & Kurata, K. (1999). Relationship between the production of carrot somatic embryos and dissolved oxygen concentration in liquid culture. *Plant Cell, Tissue and Organ Culture*, *57*(1), 29–38.

Skirvin, R. M., Chu, M. C., Mann, M. L., Young, H., Sullivan, J., & Fermanian, T. (1986). Stability of tissue culture medium pH as a function of autoclaving, time, and cultured plant material. *Plant Cell Reports*, *5*(4), 292–294. https://doi.org/10.1007/bf00269825

Soumare, A., Diedhiou, A. G., Arora, N. K., Al-Ani, L. K. T., Ngom, M., Fall, S., Hafidi, M., Ouhdouch, Y., Kouisni, L., & Sy, M. O. (2021). Potential role of utilization of plant growth promoting microbes in plant tissue culture. *Frontiers in Microbiology*, *12*, 649878.

Strosse, H., Van den Houwe, I., & Panis, B. (2004). Banana Cell and Tissue Culture-Review. In *Banana Improvement: Cellular, Molecular Biology, and Induced Mutations* (pp. 1–12). https://doi.org/10.3389/fmicb.2021.649878

Su, Y. H., Liu, Y. B., & Zhang, X. S. (2011). Auxin–cytokinin interaction regulates meristem development. *Molecular Plant*, *4*(4), 616–625.

Sudheer, W. N., Praveen, N., Al-Khayri, J. M., & Jain, S. M. (2022). Role of Plant Tissue Culture Medium Components. In *Advances in Plant Tissue Culture* (pp. 51–83). Academic Press.

Tran Thanh Van, K. (1998). Control of morphogenesis in *in vitro* cultures. *Annual Review of Plant Physiology*, *32*, 291–311.

Walli, M. H., Jasim, H. M., & Ali, F. H. (2019). Effect of exchange of gases for tissue culture vessels to produce the meristem of solanum tuberosum *in vitro*. *International Journal of Drug Delivery*, *10*(09754415), 1.

Wang, C., Zhang, X., Gao, Y., Han, Y., & Wu, X. (2021). Path analysis of non-enzymatic browning in *Dongbei suancai* during storage caused by different fermentation conditions. *Food Chem*, *335*, 127620.

Wu, J. H., Miller, S. A., Hall, H. K., & Mooney, P. A. (2009). Factors affecting the efficiency of micropropagation from lateral buds and shoot tips of *Rubus. Plant Cell, Tissue and Organ Culture (PCTOC)*, *99*, 17–25.

Yildiz, M. (2012). The prerequisite of the success in plant tissue culture: High-frequency shoot regeneration. *Recent Advances in Plant in Vitro Culture.* https://doi.org/10.5772/51097

6 Nutrient Medium and Its Fortification for *In Vitro* Cultivation of Medicinal Plants

Gunjan Garg, Alok Bharadwaj, Shweta Chaudhary, and Sanjay Kataria

6.1 INTRODUCTION

Medicinal plants contain medicinal and therapeutical properties that have been utilized by societies around the world as natural remedies for their healthcare. However, many of these plants are often slow-growing, susceptible to diseases, and influenced by environmental factors. *In vitro* cultivation addresses these challenges by providing a reliable and efficient method for the mass production of medicinal plant material (Porwal *et al.*, 2020). *In vitro* cultivation, often referred to as plant tissue culture or micropropagation, is a valuable technique for the controlled and efficient propagation of medicinal plants. This method allows for the cultivation of plant cells, tissues, or organs in a sterile and controlled environment. It has several advantages in the cultivation of medicinal plants, including the rapid production of disease-free, genetically identical plant material, which is crucial for consistent and high-quality medicinal compounds. *In vitro* cultivation involves the aseptic (sterile) culture of plant cells, tissues, or organs in a nutrient-rich medium (Bhattacharjee *et al.*, 2020). This medium contains essential nutrients, vitamins, and growth regulators that support the growth and development of plant cells. The process typically takes place in a controlled environment with regulated temperature, light, and humidity. The cultivation of medicinal plants can be highly rewarding, but it also presents several challenges that need to be addressed for successful and sustainable production. Medicinal plants often have specific climate and environmental requirements. Changes in temperature, precipitation, and other climatic conditions can impact the growth and quality of these plants. Climate change alters the geographical distribution of these plant species and affects their medicinal properties (Ahmadi *et al.*, 2020; Chen *et al.*, 2022). Similarly, variations in soil conditions and genetic diversity can lead to variations in the content of bioactive compounds. Many medicinal plants are wild-harvested, leading to habitat destruction and endangering plant species. Hence, sustainable cultivation practices of these plant species are essential to protect biodiversity and prevent overexploitation. Some medicinal plants are difficult to propagate or cultivate. They may have slow growth rates, low germination rates, or specific propagation requirements. Parallel to all these challenges, *in vitro* cultivation of medicinal plants has its own set of challenges. Maintaining aseptic conditions in the laboratory is crucial for successful *in vitro* culture. *In vitro* culture can sometimes lead to genetic instability, a phenomenon known as somaclonal variation. This can result in unintended genetic changes in the cultured plants, affecting the quality and consistency of medicinal compounds. Cultured plant tissues may not always produce the desired secondary metabolites (medicinal compounds) at levels comparable to those found in whole plants (da Silva *et al.*, 2022). Enhancing the production of these compounds can be a significant challenge. Compliance with regulatory requirements and ethical considerations related to biotechnology, intellectual property rights, and access to genetic resources can be challenging. Addressing these challenges requires a holistic approach that combines sustainable cultivation practices, conservation efforts, scientific research, and collaboration with local communities and traditional knowledge holders.

DOI: 10.1201/b23374-6

6.2 METHODS OF PLANT PROPAGATION AND PROLIFERATION

For the culturing and propagation of plants under controlled conditions, *in vivo* and *in vitro* methods are used. An *in vivo* method of propagation is a traditional type of method for the propagation of plants in field conditions. In this method we removed the apical dominance via decortications and decapitation and used suckers of the plant material for propagation. This is an alternative method of plant propagation through which we availed equivalent results to *in vitro* raised plantlets (Muhie and Teshome, 2023). This method required a minimum infrastructure facility. No sophisticated laboratory for research work and propagation of plantlets is required; hence, there is a low initial establishment investment cost. Materials needed for the propagation are easily available in and around the field and farm. Electricity consumption, contamination risk in propagated plantlets, and skilled manpower requirements are minimized. Therefore, this method of propagation, considered the best alternative method of propagation, resulted in tremendous production potential of producing uniform planting material (Murthy *et al.,* 2023).

In vitro culture of a plant tissue in control conditions, is a familiar practice that has been used from 1902 for the development of elite and disease-free plants within a short duration of time, irrespective of the season. Micropropagation of plants under controlled condition is a useful method for the large-scale production of crop, medicinal, aromatic, ornamental, and other economically important plants (Abbasi Khalaki *et al.*, 2021; Abha Manohar *et al.*, 2022; Singh and Sharma, 2020). This method has been adapted in the fields of agriculture, horticulture, floriculture, and forestry. This is a valuable method for fast and rapid production of elite and disease-free plants. In the micropropagation method, a selected explant or the cultured cell followed totipotency and morphogenetic response with the distinct developmental pathway of de-differentiation and re-differentiation. *In vitro* cultured tissue or plant cells become meristematic under the influence of plant growth regulators (PGRs) and induce organogenesis or somatic embryogenesis. Organogenesis and somatic embryogenesis modes of micropropagation involve four and five definite stages, respectively. Stage 0 of organogenesis involves selection of the explant, where selected specific cells or tissues are cultured on a specific nutrient medium. Stage 1 involves initiation and establishment of the culture. In this stage, explant cells start responding under the influence of PGRs and start division and developmental processes of organogenesis. Stages 2 and 3 involve shooting multiplication and elongation, and roots begin to form. Stage 4 involves the acclimatization process of *in vitro* raised new plantlets and then transferring them to a greenhouse environment (Murthy *et al.*, 2023). In somatic embryogenesis, stages 0 and 1 showed initiation or induction of embryogenesis of the primary cultures or explant in the culture media supplemented with specific growth regulators (Ruta *et al.*, 2022). Stage 2 is similar to the first phase. It showed proliferation of embryogenic cultures on the specified liquid or solid media. The third phase of somatic embryogenesis showed prematuration of somatic embryos in cytokinin lacking PGR medium, which stimulates formation and early development of somatic embryos. The fourth stage of embryo development involves the maturation of somatic embryos by culturing them on a medium supplemented with abscisic acid (ABA) along with the reduction of osmotic potential (Guru *et al.*, 2022). Conversion of somatic embryos into plantlets on a medium lacking a PGR is completed in the fifth stage of the development process (Ruta *et al.*, 2022).

6.3 NEW INSIGHTS OF THE CULTURE MEDIA IN *IN VITRO* PLANT REGENERATION

For the culturing of a plant cell, tissue, or organ in *in vitro* conditions, optimal morphogenesis and organogenesis, attaining a high survival rate and minimizing the physiological disorders of plantlets' mineral nutritional composition of the artificial nutrient media (or basal medium) is an essential component. This artificial nutrient media is a complex mixture of inorganic salts, a carbon source, vitamins, and the phytohormones (growth regulators), which supply all the minimum requirements in terms of energy, nutrients, and water to the explant. Such artificial nutrient media is categorized as synthetic or natural media. When all the chemical components of a medium are well defined, it is referred to as a synthetic-defined medium, whereas media that contains chemically unidentified components like fruit/vegetable extract, plant extract, coconut juice, etc., is considered a natural-undefined medium. In some cases gelling agents like agar-agar are also used for basal medium solidification. On the basis of consistency, media is categorized into three different types: (1) Broth or liquid water based-media do not have a solidifying agent (like agar); (2) semi-solid media having <1% concentration of agar; and (3) solid media containing a high percentage (1.5–2.0%) of agar. In plant tissue culture, a suitable medium can be developed by altering the conditions of PGRs and changing the carbon and inorganic salts. Several types of culture media have been studied to date like Murashige and Skoog's (1962) (MS) medium, Linsmaier and Skoog (1965) (LS) medium, Nitsch and Nitsch (NN) medium, woody plant medium (WPM), White's medium, N6 (Chu) medium, and Gamborg's (B5 medium). The most commonly and widely used nutrient medias are MS and LS, supplemented with the different types of phytohormones like auxins (2,4 D; NAA, IAA), cytokinins (BAP, Kn, zeatin), ABA, gibberellins (GAs), ethylene (gaseous hormone), and other growth regulators, which show similar types of metabolic activity (Muhie and Teshome, 2023). To date, the different types of specific media are well-known in the market for *in vitro* culturing of a specific plant (e.g., orchid media, banana media, citrus media, etc.). In general, the explant shows a response to the PGRs in three stages: (1) Explant cell reorganized phytohormone signaling and commencement of the dedifferentiation; (2) hormonal balance of PGRs and successive formation of specific organs; and (3) the appearance of independent plant morphogenesis. It has been determined that no single medium can be recommended for all types of plant tissues and organs cultured in *in vitro* conditions. Nutrient requirements for optimal growth of a plant tissue/or cell *in vitro* may vary from species to species of the same plant. Even tissues from different parts of the same plant may have different nutrient and PGR requirements for optimal growth. In a preliminary system, it is very essential to work out on medium and the nutrient components that will cover up the minimal requirements of the tissue. In the last 25 years, the need to culture different tissues or organs has led to the development of multiple recipes of culture media for culturing of plantlets *in vitro.* In the plant tissue culture system, root culturing (White's media) and callus culturing (Gautheret's media) were developed from macro and micro nutrient solutions, which were previously used for culturing a whole plant. Subsequently Gautheret's composite is the medium based on Knop's (1865) salt solution, and White developed Uspenski and Uspenskaia's medium (1925) for the culture of algae. Hence, the different types of tissue culture media cover a broad spectrum. Presently, a number of culture mediums are known, which are widely used as culture media in research labs as well as at the commercialization level. Some of them are described here:

- *MS medium*: Most of the plant tissue culture–based research in plant biotechnology still uses MS media, which was described more than 50 years ago by Murashige and Skoog (1962). This medium was invented by two scientists named Toshio Murashige and Folke K. Skoog in 1962 while the two scientists were working on protocol optimization of tobacco callus bioassay for facilitating the study of cytokinins (Table 6.1). This medium is used to induce organogenesis, callus culture, micropropagation, and cell suspension. MS medium provides all the essential macro-elements and microelements. Potassium dihydrogen phosphate serves as a source of phosphate. Microelements like boron, manganese, molybdenum, copper, and zinc play a vital role in plant metabolism (Dadi *et al.*, 2020). Boron plays a key role in carbohydrate metabolism. Thiamine, nicotinic acid, pyridoxine, and inositol act as enzymatic cofactors in universal pathways, including glycolysis and the tricarboxylic acid (TCA) cycle, along with primary and secondary metabolism in the plants. Glycine serves as a source of amino acid. Sometimes we have observed some numbers behind the MS, which indicate the concentration of sucrose in the medium. For example, MS0 indicates sucrose absence and MS10 indicates the presence of 10 g/L sucrose in the medium.

TABLE 6.1
Stock Solution for MS Medium

Constituents	Amount (mg/L)
Stock Solution 1	
NH_4NO_3	33,000
KNO_3	38,000
$CaCl_2.2H_2O$	8800
$MgSO_4.7H_2O$	7400
KH_2PO_4	3400
Stock Solution 2	
KI	166
H_3BO_3	1240
$MnSO_4.4H_2O$	4460
$ZnSO_4.7H_2O$	1720
$Na_2MoO_4.2H_2O$	50
$CuSO_4.5H_2O$	5
$CoCl_2$.6H2O	5
Stock Solution 3	
$FeSO_4.7H_2O$	5560
$Na_2EDTA.2H_2O$	7460
Stock Solution 4	
Inositol	20,000
Nicotinic acid	100
Pyridoxine. HCl	100
Thiamine. HCl	20
Glycine	400

MS medium has comparatively high salt levels compared to White's medium but lower salt concentration than B5 medium.

- *LS medium*: LS medium was developed in 1965 by Linsmaier and Skoog. This was the first media used for the optimization of organic requirements of the tobacco cultures in *in vitro* conditions. The medium has similar basal salts as MS, along with the supplementation of LS vitamins. LS medium is mainly used for *in vitro* organogenesis, callus culture, cell suspension, and micropropagation. The macroelements of this medium are the mixture of inorganic salts, which provide all essential elements to plants for their growth. Potassium dihydrogen phosphate (KH_2PO_4) and magnesium sulfate are the main source of phosphate and magnesium, respectively. Both the macroelements maintain the ionic balance in plants. Magnesium along with sulfur work as precursors in many metabolic pathways. Similarly, potassium nitrate (KNO_3) and ammonium nitrate (NH_4NO_3) are used as a good source of nitrogenous supplement and maintain the pH of the medium.
- *Gamborg (B5) medium*: O. L. Gamborg created this media in 1968. He applied the media for callus and suspension cultures of *Glycine max* L. (Fabaceae family). This medium contains a mixture of vitamins, inorganic salts, and carbohydrates. The main components of B5 basal medium are potassium nitrate and ammonium sulfate, which enhance the cell growth and callus culture via increasing the nitrate ion concentration in the media. Microelements like boron (B), manganese (Mn), molybdenum (Mo), copper (Cu), iron

(Fe), and zinc (Zn) play an essential role in the regulation of plant metabolism. Boron specifically regulates the carbohydrate metabolic pathway of the *in vitro* raised plantlets. B5 media is mainly used for protoplast culturing and *in vitro* culture of *Linum album*, *Eucalyptus camaldulensis*, *Taxus globosa*, and *Splachnum ampullaceum*.

- *NN medium*: J. P. Nitsch created the medium in 1969 in order to establish anther culture of the *Nicotina* plant in *in vitro*. This medium helps in initiation and culture of haploid plantlets from the microspore. NN medium is mainly used for the culturing of haploids or anthers in many plant species. NN medium contains a high concentration of thiamine, biotin, and folic acid that are the essential media components for anther culture.
- *White's medium*: White's medium is one of the earliest plant tissue culture media developed to generate the tomato root cultures in 1963 by Philip R. White. This medium contains a low salt concentration of $MgSO_4$. White's medium is mainly utilized for callus formation and shoot culture, specifically for cultivation of Musa and Daucus plants.

6.4 MEDIA PREPARATION

Plant tissue culture is a set of processes used to obtain cells, tissues, or tissue organs in a nutrient medium of known composition. Nutrient or culture medium is the most important part of tissue culture, and it is highly reliant on the correct formulation of cultural components of the media. Plant tissue culture media usually consist of inorganic and organic compounds, along with support matrices like vitamins and carbon sources. To stimulate cell division and differentiation, the culture media provides the necessary availability of inorganic/organic nutrients along with the PGRs to the growing cultures or the explant, which are normally available in the soil. Solid and liquid media for plant cultures can be prepared either by dissolving the commercially available basal salts or by mixing the self-prepared stock solutions of macronutrients and micronutrients according to the list of components given in Table 6.1 and stored at 4 °C for a week or 15 days only. Generally we prepare concentrated stock solutions (e.g., 10× macroelement stock, 100× microelement stock, 100× Fe stock, and 100× organic stock). In all the concentrated stocks, only the iron stock solution is placed in an aluminum foil–wrapped bottle; the rest can be stored in reagent bottles and labeled properly. All the stock solutions must be stored in the refrigerator (at 4 °C). It has been recommended that organic compounds cannot be stored for more than of 2 weeks in solution form. Heat-labile compounds such as PGRs, antibiotics, or glutamine should be sterilized by the filtration method and added only after the medium has been autoclaved and cooled. Media are commonly available in powder form, which contains all the essential inorganic salts, vitamins, and amino acids and is the easiest way to prepare nutrient media in plant tissue culture. In media preparation, sugar or carbon sources, agar, and other required supplements are mixed with the powder form, which is finally mixed in filtered water to make up the final volume. The medium was autoclaved and finally adjusted to the required pH. These varieties of basic plant tissue culture medium in dry-powered form with their manuals are available in the commercial market for the micropropagation of certain plant species. Two approaches are used for the preparation of media. First, a classic approach is more practical and well-liked, where stock solutions of nutrient are prepared and mixed in proportionate amounts to obtain a desired culture medium. This method of culture media preparation is still in practice in the research laboratories. This method is not only suitable to prepare different culture media using common stocks and solutions but also offers a range for periodic replacement of nutrient stocks as and when required. A second recently used approach for the preparation of culture media is from packaged powder, which is easily available in the market. The drawback of this method is that the powdered form of the media is very hygroscopic in nature. Hence, extra precaution is required to protect it from atmospheric moisture. Generally this type of media should be used immediately after opening (Babich *et al.*, 2020; Thiem *et al.*, 2023). Further preparation of the concentrated form of stocks is not recommended, as precipitations of some salt complexes are quite common, which may affect the shelf life of the culture media. Study of precipitates indicated the high proportion of iron, phosphate, and zinc ions in the medium. The solubility of ferric phosphate increases the precipitation process. The possible cause of the precipitates is the expected oxidation of ferrous ions to ferric ions and the presence of unchelated ferric ions. To avoid all this, we store this type of dry medium in desiccators at 0–5 °C. Prediction of deterioration of this type of powdered medium may be monitored by the changing of the color, granulation or clumping formation, and insolubility of the supplements. The culture media's composition is one of the most important parts for the plant cloning through tissue culture. Therefore, thorough work on a medium's composition is required. This is mainly based on the culture system.

The medium used for the culturing of explants will either be available in liquid or solid form. Agar-agar is one of the most effective and popular solidifying gelling agents used in plant media. It forms a highly stable, clear, non-toxic, metabolism-resistant gel that melts at about 100 °C and solidifies at about 45 °C. For more purity in culture media, we generally use agarose in place of agar. Agarose is noticeably more expensive than agar (Hasnain *et al.*, 2022; Lemma *et al.*, 2020). We can also use Gelrite and Phytagel for the solidification of the medium. But we generally avoid using these two due to the clumping formation and easy contamination at an early stage. The selection of a gelling agent is specific to the plants and plant species, as tissues of some species grow more vigorously on one gelling agent than on another (Sarkar and Banerjee, 2020). Prepared media are maintained in a sterile environment to eliminate the contaminating factors like microbes, fungi, and pathogens. These contaminants interfere with cell reproduction and reduce the growth of cultures. To avoid contamination in the culture media, we can also add plant preservative mixture (PPM) in the medium for both prevention and treatment as preventing contamination measures along with other applicable sterilization methods. Addition of PPM to the medium will help prevent fungi growth and will minimize the microbial contamination. PPM worked as a biocide and a culture preservative. It is a broad-spectrum biocide, prepared specifically for the plant tissue culture procedures. Addition of PGRs (auxins, cytokinins, gibberellins, etc.) in the medium was done in the form of stock solutions at a concentration of 1.0 mg/mL. However, if the growth regulator concentrations in the culture media are expressed in micromoles, it is optional to formulate the calculation of stock solution in millimolar or micromolar strength. This would avoid extensive calculation in medium preparation. Adjusting of pH is an essential part for the addition of agar. At pH below 5.5, the agar will not gel properly, while at pH above 6.0, the agar will become too hard. Also the required pH is needed to hold and support the different stages of growth and differentiation.

6.5 FORTIFICATION OF *IN VITRO* CULTURE MEDIA OF MEDICINAL PLANTS THROUGH ORGANIC GROWTH ADDITIVES

Using fortification strategies of *in vitro* culture media of medicinal plants with the addition of organic growth additives (OGAds) can enhance the growth and development of plant tissues and improve the secondary metabolite synthesis and production rate, which is of therapeutic importance. OGAds are the substances derived from natural sources (Malothu *et al.*, 2023; Pandey *et al.*, 2022; Rat *et al.*, 2021). They are also known as plant growth–promoting compounds. A list of some important common OGAds that are used for fortification of *in vitro* culture media of medicinal plants is given here:

- *Plant hormones*: Via natural plant hormones such as auxins like indole-3-acetic acid (IAA), indole-3-butyric acid (IBA), phenylacetic acid (PAA), indole 3 acetonitrile IAN) and cytokinins such as zeatin, kinetin (Kn) and 6-benzylaminopurine (BAP) regulate plant growth and differentiation in the culture medium (Muhie and Teshome, 2023). IAA is synthesized from the tryptophan amino acid and is transported throughout the plant to regulate plant growth and development. IBA and IAN exist in plants naturally, but in smaller amounts, and have very specific roles in growth and development (Alami *et al.*, 2022;

Wu *et al.*, 2021). These natural auxins influence the rooting or stimulate adventitious root formation Zeatin and kinetin are commonly used in *in vitro* culture media and promote shooting and lateral bud proliferation. Zeatin is mainly used in different forms, including zeatin riboside and their derivatives. These plant hormones can be added to the culture medium at different concentrations depending on the specific requirement of the tissue culture, the plant species, and the growth stage. Further the choice of auxin or the cytokinin and their concentration affect the rate of rooting and shoot proliferation, as well as their branching pattern in the cultivated plants (Alara *et al.*, 2020; Babich *et al.*, 2020; Chamkhi *et al.*, 2021). Hence, it is important to optimize the concentrations of auxins and cytokinin carefully for each species, and it is mainly application based to achieve the desired results. Different types of vitamins, including B vitamins, are essential for growth and metabolism of the plant. Some important vitamins used in culture media are thiamine (vitamin B_1), riboflavin (vitamin B_2), nicotinic acid or niacin (vitamin B_3), pantothenic acid (vitamin B_5), adenine or pyridoxine (vitamin B_6), biotin (vitamin H or vitamin B_7), folic acid or folate (vitamin M or vitamin B_9), cobalamin (vitamin B_{12}), and myo-inositol. The most important vitamins, which are essentially required in culture media, are thiamine and myo-inositol. Other B vitamins that are essential and added in the culture medium as vital elements are riboflavin (B_2), pyridoxine (B_6), and niacin (B_3). Thiamine is involved in carbohydrate metabolism and promotes root and shoot development, whereas riboflavin and pyridoxine provide strength to the photosynthetic machinery, regulate amino acid metabolism, and enhance protein synthesis and overall growth of cultured plantlets (Choudhary *et al.*, 2021; Guru *et al.*, 2022; Hasnain *et al.*, 2022). They act as an important energy transfer source. The specific concentrations and combinations of vitamins and B vitamins in the culture medium may vary depending on the plant species and specific growth goals. Defined optimization of the vitamin content in the medium is necessary for increasing the growth of explants in plant tissue culture.

- *Amino acids and organic extracts*: In plant tissue culture, amino acids and organic extracts can be added to the culture medium to provide essential nutrients and bioactive compounds to support the growth, development, and metabolic processes of cultured plant tissues. Amino acids are a good source of nitrogen in the culture medium (Chandran *et al.*, 2020). They act as building blocks of proteins. They are essential for protein synthesis, enzyme function, and various metabolic processes. They act as metabolic precursors for the biosynthesis of important secondary metabolites, including alkaloids, flavonoids, and phenolics, in medicinal plant species. Amino acids act as stress protectants or stress-responsive molecules and they help the cultured tissues or explants cope with environmental stressors. Amino acids generally added to plant tissue culture media include glutamine, arginine, proline, tryptophan, and asparagine. Proline and tryptophan are the mainly used amino acids, as proline acts as an osmo-protectant and helps to maintain cellular turgor during environmental stresses like drought, salinity, etc., and protect the plant cells from oxidative stress, whereas tryptophan acts as a precursor for indole compounds and regulates auxins synthesis like IAA. Tryptophan also contributes to the production of secondary metabolites like alkaloids and flavonoids (Bharat *et al.*, 2020). To test the effect of organic extracts on the growth enhancement, culture medium are supplemented with a variety of organic substances or extracts including coconut milk, malt extract, banana and pineapple pulp, papaya juice, banana homogenate, orange juice, and tomato juice. Organic extracts contain compounds that stimulate the production of secondary metabolites with pharmaceutical or industrial applications. The choice of organic extracts and their optimized concentrations in the culture medium depends on the specific plant species, and it is very essential to achieve the best results. Researchers frequently conduct experiments to conclude the effective combination and concentration for their particular applications. Wiszniewska *et al.* (2020) reported the successful microculture of explants in the MS media supplemented with organic growth additives, including coconut water and pineapple pulp. Similarly Manawadu *et al.* (2022) used different organic growth additives such as coconut water, coconut milk, ground spinach leaves, ground potato tubers, ground carrot, rice flour, green gram, ground pumpkin, banana fruit, and orange juice for the culture of different plant species in the *in vitro* conditions. Recently aloe vera gel (AvG) has gained attention of researchers because of its biological effectiveness, chemical composition, and antioxidant and antimicrobial properties. Nowadays, it is anticipated as an alternative to additives used in plant micropropagation media. Hamdeni *et al.* (2022) and Haque and Ghosh (2022) reported that AvG served as an organic nutritional supplement for the highest number of formed shoots per explant in MS medium supplemented with 10% AvG. For root initiation, adding AvG (20–30%) to the MS amplified the response to 100% rooting. The study on the *in vitro* micropropagation of *Bacopa chamaedryoides* resulted in 100% rooting response.

- *Essential oils*: Oxidative browning and microbial infectivity are the major problems observed during the *in vitro* cultivation of explants/or tissue. They cause a threat to the growth and development of the plantlets. Browning of the explants in culture media is the major residue of *in vitro* recalcitrance or tissue death. Today, plant extracts of essential oils are used as alternative procedures for disinfection and establishing the aseptic cultures. These essential oils also impede the oxidation of the phenols and prevent browning in the culture media. The effectiveness of *Rosmarinus officinalis* and *Thymus vulgaris* extract in controlling enzymatic browning and contamination of cultures was examined by Hamdeni *et al.* (2022). They reported that *R. officinalis* and *T. vulgaris* essential oil inhibited microbial growth in the culture media and induced the 100% survival rate of explants with no signs of browning after 4 weeks of cultures. Taghizadeh *et al.* (2022) obtained sterile medium conditions by using eugenol, carvacrol, or thymol. They inhibited the growth of fungi and bacteria contaminates with no autoclaving of the medium. Similarly, Jasim *et al.* (2016) observed the inhibitory effect of essential oils from *Mentha piperita*, *T. vulgaris*, and *Cinnamomum camphora* extract against *Alternaria* spp., *Fusarium* spp., and *Aspergillus* spp. of fungi in the cultures of *Phoenix dactylifera*. They also detected the disinfecting properties of thymol and carvacrol against the fungi and bacterial infection in the nodal explants cultures of *Cynodon dactylon*.
- *Phenolic compounds*: Addition of phenolic-rich extracts to the culture medium may be beneficial for some explants, specifically in the case of woody and medicinal plants. Different phenolic compounds like dopamine, catechin, chlorogenic acid, cinnamic acid, hydroxybenzoic, pyrogallic acid, salicylic acid, ferulic acid, vanillin, coumarin, and phenols are generally found in plants (Bryan *et al.*, 2020). Functionally these phenolics may be beneficial for the plant in different ways. They work as antioxidants and provide structural strength to the plant tissues. They help in pollination by attracting pollinators and protect the plant against harmful radiation. Also they improve the plant's immune system against pathogens. Even though these phenolic compounds serve up enormous benefits for the good growth and survival of plants, under *in vitro* conditions they tend to be detrimental. Browning is caused by the oxidation of phenolic compounds. This phenomenon is quite common in the case of woody plant species. This is because woody plants parts are more susceptible to secretion of higher levels of phenolic compounds from the cut or wounded ends of the explants. To prevent the browning of medium and the effects of phenolic compounds, several studies have been conducted by plant scientists. Addition of activated charcoal is the most commonly used method to prevent the browning of the culture media. Activated charcoal traps the toxins and avoids absorption of the phenolic compounds by the cells of the explants. Some polyphenolic inhibitors (like sodium chloride, sulfur dioxide, and 2-aminoindane-2-phosphonic acid) and antioxidants (e.g., ascorbic acid, citric acid, polyvidone, cysteine, silver nitrate) added in the nutrient medium showed significant reductions in visual tissue browning and led to a significant increase in tissue growth and

development. Culturing of the plantlets in dark (i.e., dark culturing) is another way to regulate the phenolic-oxidation pathway. In principle, light promotes the production of phenolic compounds. Thus, culturing the explants in dark (about 72 hours to 10 days) can significantly reduce the chances of browning.

- *Seaweed extracts, plant elicitors, and hormone precursors*: Seaweed extracts are rich in PGRs, vitamins, and minerals. Several species of seaweed are commonly used for their extract and show distinctive properties and applications. Brown seaweed (*Ascophyllum nodosum, Fucus vesiculosus*), *Gracilaria* spp., and *Chondrus crispusare* are mainly used in *in vitro* culture media to enhance the growth of explants and promote the production of bioactive compounds (Jeyasri *et al.,* 2023). Many ongoing research studies continue to uncover new uses and benefits of seaweed in biotechnological research. Natural elicitors, such as jasmonic acid, salicylic acid, and chitosan, can be added to the culture medium to trigger the synthesis of secondary metabolites. Elicitors mimic stress conditions and induce plants to produce bioactive compounds as a defense mechanism. Precursors to natural plant hormones can be used to stimulate the biosynthesis of endogenous hormones within the plant tissues, which can influence growth and development. When incorporating organic growth additives into *in vitro* culture media, it is essential to conduct preliminary trials and optimize the concentrations based on the specific needs of the medicinal plant species and the goals of the culture (Roy *et al.*, 2021). The choice of additives and their concentrations will depend on the particular requirements and desired outcomes of your tissue culture experiment. Additionally, the exact formulation and concentrations of these additives should be based on scientific literature and previous successful protocols for the specific medicinal plant in question.

6.6 CONCLUSION

The results of this chapter conclude that different techniques used to propagate the plants via plant tissue culture can be improved by addition of organic growth additives. However, it is very important to know about the mode of action and relationship of these supplements among each other. Using this strategy for the micropropagation of plant species could be an efficient tool to restore the biodiversity via the large-scale production of medicinal and forest species. However, *in vitro* plant propagation is a costly technique compared to conventional systems. Therefore, implementation of procedures having lower production costs without sacrificing the plant quality is essential. This can be attained by improving the culture media composition. Incorporating organic growth supplementation speeds up the formation and development of cultures and can be considered an economic way to stimulate the plant/tissue growth in culture media.

REFERENCES

Abbasi Khalaki, M., Moameri, M., Asgari Lajayer, B., & Astatkie, T. (2021). Influence of nano-priming on seed germination and plant growth of forage and medicinal plants. *Plant Growth Regulation*, *93*, 13–28.

Abha Manohar, K., Shukla, G., Roy, B., & Chakravarty, S. (2022). Effects of plant growth regulators and growing media on propagation and field establishment of *Stevia rebaudiana*: A medicinal plant of commerce. *CABI Agriculture and Bioscience*, *3*(1), 1–12.

Ahmadi, F., Samadi, A., & Rahimi, A. (2020). Improving growth properties and phytochemical compounds of *Echinacea purpurea* (L.) medicinal plant using novel nitrogen slow release fertilizer under greenhouse conditions. *Scientific Reports*, *10*(1), 13842.

Alami, M. M., Ouyang, Z., Zhang, Y., Shu, S., Yang, G., Mei, Z., & Wang, X. (2022). The current developments in medicinal plant genomics enabled the diversification of secondary Metabolites' biosynthesis. *International Journal of Molecular Sciences*, *23*(24), 15932.

Alara, O. R., Abdurahman, N. H., & Olalere, O. A. (2020). Ethanolic extraction of flavonoids, phenolics and antioxidants from *Vernonia amygdalina* leaf using two-level factorial design. *Journal of King Saud University-Science*, *32*(1), 7–16.

Babich, O., Sukhikh, S., Pungin, A., Ivanova, S., Asyakina, L., & Prosekov, A. (2020). Modern trends in the *in vitro* production and use of callus, suspension cells and root cultures of medicinal plants. *Molecules, 25*(24), 5805.

Bharat, A., Querrey, M., Markov, N. S., Kim, S., Kurihara, C., Garza-Castillon, R., & Budinger, G. S. (2020). Lung transplantation for patients with severe COVID-19. *Science translational medicine, 12*(574), eabe4282.

Bhattacharjee, T., Sen, S., Chakraborty, R., Maurya, P. K., & Chattopadhyay, A. (2020). Cultivation of medicinal plants: Special reference to important medicinal plants of India. In Sen, S., Chakraborty, R. (eds.), Herbal *Medicine* in India: Indigenous *Knowledge, Practice, Innovation* and *Its Value*, p. 105. Springer.

Bryan, C. J., Bryan, A. O., & Baker, J. C. (2020). Associations among state-level physical distancing measures and suicidal thoughts and behaviors among US adults during the early COVID-19 pandemic. *Suicide and Life-Threatening Behavior, 50*(6), 1223–1229.

Chamkhi, I., Benali, T., Aanniz, T., El Menyiy, N., Guaouguaou, F. E., El Omari, N., & Bouyahya, A. (2021). Plant-microbial interaction: The mechanism and the application of microbial elicitor induced secondary metabolites biosynthesis in medicinal plants. *Plant Physiology and Biochemistry, 167*, 269–295.

Chandran, H., Meena, M., Barupal, T., & Sharma, K. (2020). Plant tissue culture as a perpetual source for production of industrially important bioactive compounds. *Biotechnology Reports, 26*, e00450.

Chen, B., Zou, H., Zhang, B., Zhang, X., Jin, X., Wang, C., & Zhang, X. (2022). Distribution pattern and change prediction of *Saposhnikovia divaricata* suitable area in China under climate change. *Ecological Indicators, 143*, 109311.

Choudhary, S., Zehra, A., Mukarram, M., Naeem, M., Khan, M. M. A., Hakeem, K. R., & Aftab, T. (2021). An insight into the role of plant growth regulators in stimulating abiotic stress tolerance in some medicinally important plants. In Aftab, T., Hakeem, K. R. (eds.), Plant Growth Regulators: Signalling under *Stress Conditions* (pp. 75–100). Springer.

Dadi, T. H., Vahjen, W., Zentek, J., Melzig, M. F., Granica, S., & Piwowarski, J. P. (2020). *Lythrum salicaria* L. herb and gut microbiota of healthy post-weaning piglets. Focus on prebiotic properties and formation of postbiotic metabolites in *ex vivo* cultures. *Journal of Ethnopharmacology, 261*, 113073.

da Silva, R. F., Carneiro, C. N., de Sousa, C. B. D. C., Gomez, F. J., Espino, M., Boiteux, J., & Dias, F. D. S. (2022). Sustainable extraction bioactive compounds procedures in medicinal plants based on the principles of green analytical chemistry: A review. *Microchemical Journal, 175*, 107184.

Guru, A., Dwivedi, P., Kaur, P., & Pandey, D. K. (2022). Exploring the role of elicitors in enhancing medicinal values of plants under *in vitro* condition. *South African Journal of Botany, 149*, 1029–1043.

Hamdeni, I., Louhaichi, M., Slim, S., Boulila, A., & Bettaieb, T. (2022). Incorporation of organic growth additives to enhance in vitro tissue culture for producing genetically stable plants. *Plants, 11*(22), 3087.

Haque, M. M., Ghosh, M. K., & Hoyanagi, K. (2022). Coastal development in southwestern Bangladesh: Understanding the interplay between storms and sea level rise. *Progress in Physical Geography: Earth and Environment, 46*(3), 331–356.

Hasnain, A., Naqvi, S. A. H., Ayesha, S. I., Khalid, F., Ellahi, M., Iqbal, S., & Abdelhamid, M. (2022). Plants *in vitro* propagation with its applications in food, pharmaceuticals and cosmetic industries; Current scenario and future approaches. *Frontiers in Plant Science, 13*, 1009395.

Jasmin, K. M., McGettigan, C., Agnew, Z. K., Lavan, N., Josephs, O., Cummins, F., & Scott, S. K. (2016). Cohesion and joint speech: Right hemisphere contributions to synchronized vocal production. *Journal of Neuroscience, 36*(17), 4669–4680.

Jeyasri, R., Muthuramalingam, P., Karthick, K., Shin, H., Choi, S. H., & Ramesh, M. (2023). Methyl jasmonate and salicylic acid as powerful elicitors for enhancing the production of secondary metabolites in medicinal plants: An updated review. *Plant Cell, Tissue and Organ Culture, 153*(3), 447–458.

Lemma, D. T., Banjaw, D. T., & Megersa, H. G. (2020). Micropropagation of medicinal plants. *International Journal of Plant Breeding and Crop Science, 7*(2), 796–802.

Malothu, G. S., Marka, R., & Nanna, R. S. (2023). Effect of plant growth regulators on callus induction in *Solanum torvum* SW: A medicinal plant. *Vegetos*, 1–7, https://doi.org/10.1007/s42535-023-00722-1.

Manawadu, I. N., & Shehara, W. V. T. (2022). Determinants of accounting practice in small and medium enterprises: evidence from Sri Lanka. In *CMA National Management Accounting Conference*.

Muhie, S. H., & Teshome, A. S. (2023). Effects of benzylaminopurine and macro propagation techniques on multiplication of dessert banana (*Musa acuminate* L.). *Journal of Agriculture and Food Research, 11*, 100515.

Murashige, T., & Skoog, F. (1962). A revised medium for rapid growth and bio assays with tobacco tissue cultures. *Physiologia Plantarum, 15*(3).

Murthy, H. N., Joseph, K. S., Paek, K. Y., & Park, S. Y. (2023). Bioreactor systems for micropropagation of plants: Present scenario and future prospects. *Frontiers in Plant Science*, *14*, 1159588.

Pandey, S., Sundararajan, S., Ramalingam, S., & Pant, B. (2022). Elicitation and plant growth hormone-mediated adventitious root cultures for enhanced valepotriates accumulation in commercially important medicinal plant *Valeriana jatamansi* Jones. *Acta Physiologiae Plantarum*, *44*, 1–13.

Porwal, O., Singh, S. K., Patel, D. K., Gupta, S., Tripathi, R., & Katekhaye, S. (2020). Cultivation, collection and processing of medicinal plants. *Bioactive phytochemicals: Drug Discovery to Product Development.*

Rat, A., Naranjo, H. D., Krigas, N., Grigoriadou, K., Maloupa, E., Alonso, A. V., & Willems, A. (2021). Endophytic bacteria from the roots of the medicinal plant *Alkanna tinctoria (Boraginaceae)*: Exploration of plant growth promoting properties and potential role in the production of plant secondary metabolites. *Frontiers in Microbiology*, *12*, 633488.

Roy, A. (2021). Hairy root culture an alternative for bioactive compound production from medicinal plants. *Current Pharmaceutical Biotechnology*, *22*(1), 136–149.

Roy, D., Ghosh, R., Dubey, S., Dubey, M. J., Benito-Leon, J., & Ray, B. K. (2021). Neurological and neuropsychiatric impacts of COVID-19 pandemic. *Canadian Journal of Neurological Sciences*, *48*(1), 9–24.

Ruta, C., Campanelli, A., De Mastro, G., & Blando, F. (2022). *In vitro* propagation by axillary shoots culture and somatic embryogenesis of *Daucus carota* for biodiversity conservation purposes. *Horticulturae*, *8(12)*, 1150.

Sarkar, J., & Banerjee, N. (2020). Influence of different cytokinins on micropropagation of an important medicinal plant, *Solanum erianthum*, and assessment of the genetic fidelity of the regenerants. *In Vitro Cellular & Developmental Biology Plant*, *56*, 480–490.

Singh, B., & Sharma, R. A. (2020). *Secondary Metabolites of Medicinal Plants: Ethno-Pharmacological Properties, Biological Activity and Production Strategies*, Volume 4. John Wiley & Sons.

Taghizadeh, S. F., Karimi, G., Tzatzarakis, M., Tsakiris, I., Ahmadpourmir, H., Azizi, M., & Rezaee, R. (2022). Probabilistic risk assessment of exposure to multiple metals and pesticides through consumption of fruit juice samples collected from Iranian market. *Food and Chemical Toxicology*, *170*, 113493.

Thiem, B., Hermosaningtyas, A. A., Budzianowska, A., & Kikowska, M. (2023). Polish contributions in developing medicinal plant *in vitro* propagation system. *Plant Cell, Tissue and Organ Culture*, *155*(1), 1–28.

Wiszniewska, J., Petecki, Z., Krzemińska, E., Grabarczyk, A., & Demaiffe, D. (2020). A review. Geophysics, petrology, geochronology and isotopic signature. *Geological Quarterly*, *64*(2).

Wu, W., Du, K., Kang, X., & Wei, H. (2021). The diverse roles of cytokinins in regulating leaf development. *Horticulture Research*, *8*, 118, https://doi.org/10.1038/s41438-021-00558-3.

7 Somaclonal Variations for Improvement in Medicinally Active Plant Species

Rambod Abiri

7.1 INTRODUCTION

Medicinal herbs are described as plants which have been traditionally applied for their therapeutic benefits in different parts of the world. The chemical compounds found in the seeds, leaves, shoots, barks, roots, and flowers of medicinal plants serve as valuable natural resources for preventing, alleviating, or even curing the impact of various diseases on the human body (Poswal et al., 2019). The history of using plants for treating diseases and sickness dates back to the dawn of civilisation when early humans realised the influence of plants on sickness in ancient societies such as China, Greece, and Egypt (e.g., opium, garlic, aloe vera, etc.) (Chevallier, 2016). One of the historical examples of using medicinal plants in ancient Egypt is the utilisation of willow bark, which contains salicylic acid, for treating fever and reducing pain. Interestingly, in ancient societies like Greece, the application of medicinal plants such as peppermint, lavender, and chamomile served as another example of soothing remedies used to calm people. Additionally, in Greece, mandrake and henbane were employed to alleviate pain in patients. The history of using medicinal plants in China dates back to 206 BC to 220 AD when liquorice, astragalus, and ginseng root were employed for their therapeutic potential against diseases. The Middle East region, particularly Iran, Iraq, and Turkey, played pivotal roles as enriched origins for the utilisation of herbal medicine in ancient times. As mentioned, the application of these pharmaceutical plants dates back as far as humanity itself, with some of the earliest documents describing their usage dating back 1000–5000 years. For example, the Sumerian clay slab from Nagpur, which is 5000 years old, contains 12 herbal medicine recipes and mentions 250 various plants. Similarly, the Chinese book Pen T'Sao, approximately 2500 years old, describes 350 diverse medicines, some of which are still in use today (Sovilj and Spasojević, 2001; Kurhekar, 2021). Nowadays, we possess a wealth of information about the use of plants in human treatment, particularly from ancient Greek sources. Homer, in *The Iliad* and *The Odyssey*, mentioned 63 herbal medicines used in pharmacotherapy. Herodotus (500 BC) advocated the health benefits of cabbage, mustard, hellebore, and garlic. Hippocrates described more than 300 medicinal plants in his works, while Dioscorides, considered the father of pharmacology, detailed 657 drugs derived from 944 different plants (Wiart, 2007; Petrovska, 2012).

There are additional documents from the time of ancient Rome. For instance, Pliny the Elder (23 AD–79) records more than 1000 therapeutic plants, both wild and domesticated, whereas Galen (131 AD–200) wrote about more than 220 plants, 61 minerals, and 19 animal medicine products in his book (Deming, 2016). Beginning in the early 19th century, advances in chemical methods enabled the discovery and isolation of numerous active compounds, including quinine (1820), ipecacuanha (1817), glycosides, tannins, saponosides, etheric oils, vitamins, and hormones. This was the turning point in the use of medicinal plants (Grzywacz et al., 2014). The invention of synthetic pharmaceuticals during the 20th century temporarily decreased the usage of herbal remedies having a vegetable origin. The idea that herbal remedies are safer and healthier has raised interest in plant species with therapeutic characteristics (Duta-Cornescu et al., 2023).

 DOI: 10.1201/b23374-7

Recently, the outbreak of the coronavirus worldwide, with its high fatality rate and economic damages, has led plant scientists to re-explore the potential of medicinal plants in countering this global threat (Abiri et al., 2021). Nevertheless, the belief that herbal remedies are safer and healthier has sparked renewed interest in plant species with therapeutic characteristics. In the present day, medicinal herbs have gained increasing popularity in different parts of the world for various reasons. For example, medicinal herbs are often affordable, accessible, and easy to grow locally, making them valuable natural resources of medicine that can be accessed by people of any income level, especially in poor communities (Grzywacz et al., 2014). In contrast to synthetic drugs, medicinal plants are natural and generally have fewer side effects or harmful impacts. These valuable natural resources are diverse, containing unique properties that can help cure a wide range of diseases. Medicinal plants are renewable and sustainable natural resources, especially in underdeveloped or developing countries. Furthermore, the traditional knowledge of using medicinal plants is passed down from generation to generation, with information deeply rooted in local communities (George, 2011). Based on human society and nature, different types of medicinal plants have been used in different folk communities. Notwithstanding the application of medicinal plants in the folk medicine systems, many modern medicines have been developed or generated from the available active compounds of these types of plants (Wang et al., 2020).

Depending on the physiological and morphological features of the medicinal plants, different propagation methods have been used to multiply the plants including seed, cuttings, division, layering, and tissue culture. Like other plants, seeds are the most common propagation method for some medicinal plants including fenugreek, ashwagandha, and neem (Faehnrich et al., 2021). Cutting is another well-known propagation method of some medicinal plants including mint, lavender, and rosemary. In this method, a part of the mother plant is cut and promoted to rooting in a suitable condition (Nicola et al., 2003). In the division method, the mature medicinal plants (e.g., comfrey and echinacea) are divided into smaller sections and replanted in suitable media. Layering is another method of medicinal plant propagation. In this method, the branches of some medicinal plants, like elderberry and blackberry, are bent and covered in the suitable soil allowing plants to root and shoot (Namdeo, 2018). Although the mentioned propagation methods have been widely applied to multiply the medicinal plants in different parts of the world, there are always some restrictions that push the plant scientists and growers to find a speedy and cheap way in the area of medicinal plant propagation. To overcome the routine restrictions, the tissue culture method shows a reliable technique for medicinal plant propagation.

Plant tissue culture (PTC) is a laboratory process of germinating, growing, maintaining, and multiplying cells, tissues, and organs of plants under aseptic *in vitro* conditions in basal nutrient media with the necessary plant growth factors (Custódio et al., 2022). This vital technique is used in various aspects of plant sciences, agriculture, horticulture, and forestry. Generally, the PTC process can propagate plants, produce pathogen-free generations, preserve endangered species of plants, and produce genetically modified plants. PTC can be utilised to generate several true-to-type copies of plants from a single cell, known as micropropagation (Abiri et al., 2020). Micropropagation helps plant breeders, plant biotechnologists, and growers to produce identical plants in high numbers, which can be applied for conversation, research, and commercial purposes (Abiri et al., 2017). Like other plants' tissue cultures, the tissue culture of medicinal plants possesses several steps including explant selection, explant sterilisation, initiation and multiplication of cultures, and acclimatisation of plantlets.

When plants are propagated asexually, like cloning or propagation of the plant in tissue culture, a type of beneficial, harmful, or neutral genetic variation may occur due to mutations in different morphological characteristics of plants including changing in the leaves colour, size, or shape; resistance and or susceptibility to diseases; tolerance to abiotic stresses; and growth habit and the plant's development (Bairu et al., 2011). In plant breeding, plant biotechnology, or even agriculture fields, somaclonal variation (SV) is as an important feature. This feature can provide new genetic variation, which can help to improve desirable traits or introduce new varieties. Notwithstanding the mentioned benefit of SV, this phenomenon has some bottlenecks due to the unpredictability and uncontrollability of mutation (Ferreira et al., 2023).

The *in vitro* cultivation technology serves two primary objectives when it comes to medicinal plants: First, it helps to propagate certain valuable genotypes, while preserving their genetic makeup, and second, it aids in obtaining fresh somaclones with the highest possible level of bioactive substances. As a result, it is important to closely observe the genetic and epigenetic changes in the regenerants to ensure the preservation of genetic integrity of the mother plant and to make the most of any beneficial somaclonal changes that arise spontaneously, enhancing the therapeutic qualities of medicinal plants without incurring any extra expenses. Taking into consideration all the mentioned facts, the current chapter aims to provide an overview of the concept of SV in medicinal plants. It delves into the types and sources of SV in detail. Furthermore, the chapter explores the role of pre-existing somatic variation, plant chimeras, aberrations, rearrangements of chromosomes, cell cycle dynamics, and the influence of transposable elements. Additionally, the epigenetic aspects of SV are discussed, including DNA methylation patterns, histone modifications, non-coding RNAs, and transgenerational epigenetic inheritance, within the context of medicinal plants.

7.2 AN INTRODUCTION TO TISSUE CULTURE OF MEDICINAL PLANTS

PTC possesses a crucial role in the biotechnology of medicinal plants, offering a promising solution to those facing low yield and vulnerability to biotic stress (Nhut, 2022). By employing PTC techniques, medicinal plants can be cultivated and propagated under controlled conditions, ensuring a dependable and sustainable supply of diverse beneficial compounds and products (Mofokeng et al., 2022). PTC brings about success in various areas, including micropropagation, *ex situ* and *in situ* conservation, bioreactor applications, polyploidy induction, and metabolite engineering through gene transformation, all contributing to the utilisation and exploration of the untapped genetic potential of medicinal plants (Grzegorczyk-Karolak et al., 2018). PTC enables stable conditions for evaluating the impact of diverse experimental factors on secondary metabolite production in medicinal plants and endogenous hormone metabolism, signalling, and transport (Kumari et al., 2018; Lalaleo et al., 2018). PTC allows studying the effects of diverse PGRs on medicinal plants' physiological properties, including photosynthesis and stomatal function. Additionally, it protects plants, increases the production of active secondary metabolites, and facilitates the development of designed molecules and synthetic biology techniques. As a result, PTC aids in the development of new compounds and customised medicinal plants that are useful in pharmaceutical, food, and other industries (Espinosa-Leal et al., 2018; Rosa et al., 2018). PTC plays a crucial role in enabling various methodologies like medium optimisation, biotransformation, elicitation, Agrobacterium transformation, and scale-up, all of which collectively contribute to enhancing the *in vitro* synthesis of significant plant compounds (Matkowski, 2008). Furthermore, PTC provides a platform for *in vitro* chromosomal doubling and artificial polyploidy induction in medicinal plants, which greatly enhances the plants' quantitative and qualitative performance (Matkowski, 2008; Niazian, 2019). Moreover, PTC is essential for the genes overexpression and regulation, which results in the creation of high-quality biologically active compounds. In addition, PTC facilitates the unique strategy of hairy root cultures, which exhibit greater biosynthetic capability and genetic stability than regular *in vitro* cultures (Niazian, 2019).

7.3 AN OVERVIEW OF SOMACLONAL VARIATION

The term "somaclonal variation" generally refers to a type of morphological and physiological variation that occurs in cell and tissue cultures. The phrase "SV" is frequently used to describe all variations obtained through tissue culture. However, alternative terms like protoclonal, gametoclonal, and mericlonal variation are regularly used to refer to different plants obtained through protoplast, anther, and meristem cultures, respectively (Bairu et al., 2011). Ever since Braun (1959) first reported SV, it has continued to be a significant issue for numerous plants that are grown through tissue culture. The cultivation of plant cells in a laboratory setting and their subsequent development into complete plants is a type of asexual reproduction that only involves mitotic cell division. In theory, this process should not result in any variation. The ideal outcome is the production of identical plants through clonal multiplication, ensuring genetic uniformity (Jain et al., 1998); therefore, the unexpected phenomenon is the appearance of uncontrolled spontaneous variation in the morphological features of plants during the culture process. Despite the aforementioned drawbacks of SV, there is also substantial evidence that it can be used to improve crops by generating new variants. Induced SV has the potential to be extremely useful in altering crops with polygenic features. Moreover, it can be a helpful tool for plant breeding, creating new varieties with disease resistance, higher quality, and greater yields. Therefore, it's crucial to understand and recognise the possible benefits of SV (Bairu et al., 2011).

According to Van Harten's definition (Van Harten, 1998), mutations refer to changes in the DNA sequence that have the potential to be passed down through generations (inherited) and are independent of genetic segregation or recombination processes. Predieri (2001) and Roux (2004) offered comprehensive details on the precise physical and chemical treatments and tissue culture procedures that can be used to create genetic variation in mutants. The term "SV" was used by Anastassopoulos and Keil (1996) to describe mutations in plants that were generated from tissue culture. Interestingly, Yang et al. (2010) have observed that SV results in a higher frequency of dominant expression compared to other forms of mutation. Studies suggest that the interaction of mutagenesis and tissue culture can have favourable outcomes in ornamental and crop species (Wang et al., 2007). The enhanced genetic variability and mutations produced by *in vitro* culture, alone or in combination with mutagenesis, can be used to develop innovative commercial cultivars (Orbović et al., 2008). In the absence of axillary buds, mutagens can be applied to undifferentiated tissues and organs (Predieri, 2001). It is possible that some of the somaclones that arise will exhibit stable, desirable features that will be passed on to future generations (Roy and Mandal, 2005). Tissue culture improves the efficiency of mutagenic treatments, handling of large populations, and rapid cloning of selected variants (Predieri, 2001). The domestication of plants coincided with the appearance of spontaneous mutations in vegetatively propagated plants, resulting in successful cultivars (Hubbard et al., 2002). *In vitro*-generated variations occur more frequently and can be easily detected. Changes in physical and morphological characteristics of undifferentiated callus, *in vitro* organisation and organ formation abilities, variances in the phenotypes of differentiated plants, and chromosomal rearrangements all contribute to SV (Rai, 2022).

7.4 TYPES OF SOMACLONAL VARIATION

As mentioned, the concept of SV has been suggested to refer to the range of differences that arise from *in vitro* multiplication techniques (Côte et al., 2001). SV may exhibit either temporary or permanent differences from the original plant, and the true causes of SV remain only partially understood. The previous studies have suggested that SV is not restricted to nuclear DNA and can affect anything from a single characteristic to the entire plant genome (Duta-Cornescu et al., 2023). The manifestations of this variability can be either somatically or meiotically stable events (Peredo et al., 2006). The phrase "mutation" refers to the variants that are meiotically stable, whereas the term "variation" refers to non-heritable epigenetic SV that can be reversed. In the study of SV, this terminology is commonly employed (Côte et al., 2001). SV must be somatically persistent and noticeable in the new plant products in order to be classified as meiotically heritable. Recessive mutations or other reasons may cause some of the altered traits to only manifest in the progeny. In fact, there are two categories of somatic variation: Heritable and epigenetic. In contrast to epigenetic variation, which may not even be persistent during asexual propagation, heritable variation remains

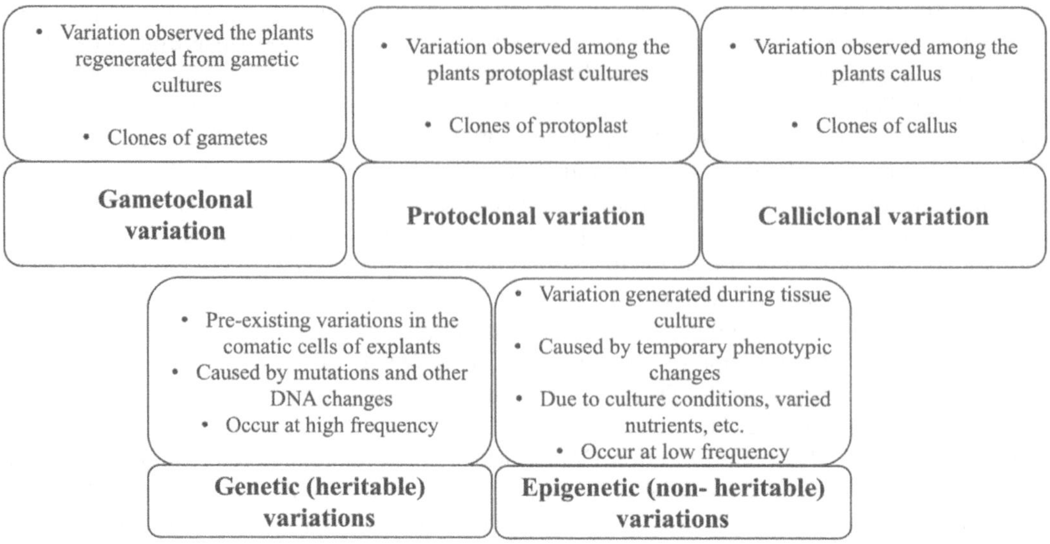

FIGURE 7.1 Types of somaclonal variation.

stable throughout the sexual cycle or recurrent asexual reproduction. The variation can affect one or multiple genes and can result from changes in DNA bases, genes, chromosomes, or even entire sets of chromosomes (Orton, 1984; Krishna et al., 2016). The somatically stable SV includes traits such as habituation of cultures (Pischke et al., 2006) and physiologically induced SV, which can be seen in the regenerants but not their progeny (Kaeppler et al., 2000). Different types of SV are presented in Figure 7.1.

7.5 SOURCES OF SOMACLONAL VARIATION IN TISSUE CULTURE OF MEDICINAL PLANTS

As mentioned earlier, tissue culture is an effective means of asexual propagation, but it often results in the production of regenerants with a range of SVs (Singh, 2021). These variations are typically caused by newly arisen mutations during the tissue culture process (Holme et al., 2019). The occurrence of mutations in tissue culture has been attributed to various stress factors, such as wounding, exposure to a sterilant during sterilisation, the incompleteness of tissues (protoplasts being a severe example), imbalances in media components such as high levels of plant growth regulators (auxins and cytokinins), sugar from nutrient media replacing photosynthesis in leaves, lighting conditions, and a disrupted relationship between high humidity and transpiration (Smulders and De Klerk, 2011). A significant portion of the differences seen in micropropagated plants may be linked to or caused by damage from oxidative stress during their growth in a laboratory setting (Agarwal et al., 2022). Oxidative stress can lead to increased levels of reactive oxygen species (ROS) like hydrogen peroxide and superoxide, which can result in modifications to DNA methylation and changes in chromosome number, strand breaks, and rearrangements (Dong et al., 2022). These changes can then cause mutations in the plant cells while *in vitro*. The genetic variations seen in SV are similar to those caused by induced mutation since they both cause comparable types of DNA alterations. Various factors can influence the likelihood of SV developing while the plants are growing in the laboratory (Capriotti et al., 2022).

The type and frequency of SV may differ depending on the tissue source used for plant regeneration. Highly differentiated tissues, like roots, leaves, and stems, tend to have more variation than tissues with pre-existing meristems, such as axillary buds and shoot tips (Pijut et al., 2012).

The likelihood of variation increases when older or more specialised tissues are used since adventitious shoot regeneration occurs from atypical origins, either directly or through a callus stage (e.g., leaves, petioles, shoot internodes, root segments, anthers, hypocotyls, cotyledons). SV can also arise from somatic mutations present in the donor plant, including chimeras in explants (Krishna et al., 2016). Additionally, initiation and subculture of cultures can cause oxidative stress on explants, which may lead to mutations (Yudoh et al., 2005). Stressful procedures like callus formation and protoplast culture have been shown to impose a greater negative impact on explants, with the extent of stress varying depending on the tissue culture technique used (Smulders and De Klerk, 2011). While the production of plants through axillary branching typically does not result in variant production, cultures that undergo a callus phase are believed to have a higher potential for mutation (Ghosh et al., 2021). Chromosome variability is more likely to occur during the callus phase of tissue culture than in adventitious shoots, suggesting that disturbed genomes lose competence (Desai et al., 2022). This may be due to the different levels of disturbance that cells experience, as normal division patterns occur in adventitious shoots, while callus formation involves a dedifferentiation phase followed by uncontrolled cell division (Simonović et al., 2020). Different tissue culture techniques can cause varying degrees of stress on cultured cells, similar to other stressful situations such as protoplast preparation. Despite the popular belief that unorganised callus growth is necessary for genetic variation, variability can also be observed in plants regenerated from adventitious shoots (Nogueira et al., 2022). *In vitro* regeneration can be achieved through somatic embryogenesis, which is often favoured for propagule generation. Compared to organogenic differentiation, this route is expected to increase the likelihood of producing plants with a homogenous genetic makeup. According to the literature, DNA methylation levels are low during the earliest phases of somatic embryogenesis, which contributes to this increased genetic uniformity (Sahijram et al., 2003).

Several studies suggest that the chance of SV in plants increases with the length of time spent in *in vitro* culture (Kuznetsova et al., 2006; Gao et al., 2010; Farahani et al., 2011; Jevremović et al., 2012; Sun et al., 2013). This is due to the accumulation of different karyotypes in an ageing callus, which increases the likelihood of producing various plants in future subcultures (Zayova et al., 2010). Additionally, the rapid tissue multiplication involved in micropropagation can impact genetic stability, as demonstrated by Khan et al. (2011) who observed an increase in the number of SV and a decrease in the multiplication rate of banana propagules after the eighth subculture. Longer culture durations and subculturing can increase the rate of SVs, particularly in cell suspension and callus cultures (Bairu et al., 2006; Sun et al., 2013). SV is more likely in plants regenerated from long-term cultures, as demonstrated by several studies (Etienne and Bertrand, 2003). Rival et al. (2013) found that *in vitro* proliferation can lead to time-dependent DNA hypermethylation and changes in DNA methylation, which may affect the embryogenic capacity of oil palm during tissue culture.

Inadequate control of the cell cycle *in vitro* is known to cause SV, as growth regulators, temperature, light, osmolarity, and agitation rate can influence the cell cycle in plants (Karp, 1994; Nwauzoma and Jaja, 2013). Disruption of normal cell cycle controls during tissue culture is thought to result in chromosomal breakage and aberrations, such as deletions, duplications, inversions, and translocations (Phillips et al., 1994; Duncan, 1996). Plant growth regulators can also increase the rate of SV by promoting adventitious shoots and increasing the multiplication rate (Gao et al., 2010). It has been suggested that some plant growth regulators may even act as mutagens at certain concentrations or in combination with other factors in the culture medium (D'Amato and Bayliss, 1985).

Certain growth regulators, including 2,4-dichlorophenoxy acetic acid (2,4-D), naphthalene acetic acid (NAA), 6-benzylaminopurine (BAP), and synthetic phenylurea derivatives (4-CPPU, PBU, and 2,3-MDPU), have been linked to genetic variability (Siragusa et al., 2007; Sun et al., 2013; Sales and Butardo, 2014). Long-term exposure to 2,4-D in culture medium can cause higher DNA ploidy levels in callus cells (Silva and Carvalho, 2014), and adding 2,4-D to culture medium can lead to high levels of cytosine methylation, resulting in variations in tissue-cultured plants (Sales and Butardo, 2014). Changes in DNA methylation rate are thought to be responsible for the development of SV in oil palm (Jaligot et al., 2011). Indole-3-acetic acid (IAA) and inositol in the growth medium have also been shown to induce DNA rearrangements and methylation changes in carrot callus cultures (Arnhold-Schmitt, 1993). Growth regulators such as kinetin can cause extensive hypomethylation of DNA in carrot root explants, while auxins such as NAA can cause hypermethylation (LoSchiavo et al., 1989). Changes in chromatin remodelling genes and histone methylation genes can also occur during tissue culture, leading to disruption in the methylation pathway and DNA hypo/hypermethylation patterns that can be transmitted to plants regenerated from these cultures (Shearman et al., 2013). The ratio of different growth regulators in culture media also affects the occurrence of variations *in vitro*. A high auxin/cytokinin ratio can result in the lowest incidence of variant flowering in oil palm, while using media with a high cytokinins/auxin ratio can result in a high incidence of variant flowering (Eeuwens et al., 2002). The role of cytokinin was further confirmed by Ooi et al. (2013), who found that the mantled inflorescences of oil palm contained higher levels of cytokinins than normal inflorescences.

The use of plant growth regulators and culture media *in vitro* morphogenesis varies depending on the plant genotype (Eftekhari et al., 2012). The plant genotype is the main factor affecting SV (Nwauzoma and Jaja, 2013). Eeuwens et al. (2002) identified oil palm clones as low/moderate or high risk for "mantle" flowering based on *in vitro* terminal inflorescence data. High-risk clones had a significantly higher incidence of mantled flowering in the field, suggesting that *in vitro* data can be used to screen for risk of mantled flowering. The differences in genotype and epigenetically inherited changes during the callus initiation and maintenance stages of the culture process may contribute to this result.

Adverse factors during process of PTC can cause SV, but it affects different genomes in varying ways, suggesting that it has a genotypic basis (Weckx et al., 2019). The stability of the variation depends on the genetic makeup of the plant, with some genomes being more susceptible to instability during tissue culture due to differences in repetitive DNA sequences, which vary in quantity and quality across plant species (Ghosh et al., 2021).

7.6 PRE-EXISTING SOMATIC VARIATION

Cellular variation that can be passed down from one generation to the next can occur due to mutations, changes in epigenetics, or a combination of both mechanisms, as explained by Kaeppler et al. (2000). It's important to distinguish between these mechanisms because genetic mutations are usually permanent and can be inherited by offspring, while epigenetic changes are typically not transmitted through sexual reproduction, according to George (1993) and Kaeppler et al. (2000). Various factors, such as chimeric plants, changes in ploidy level, chromosome abnormalities induced by tissue culture, cell cycle regulation mechanisms, and activation of transposable elements, are believed to contribute as the pre-existing somatic variation factors.

7.7 PLANT CHIMERAS

According to George (1993), chimeras can be a source of pre-existing variation *in vitro*. The stability of the chimera depends on how the genetically different tissues are arranged within the plant meristem. For example, McPheeters and Skirvin (1983) found that almost half of the tissue obtained from tissue-cultured chimeral "Thornless Evergreen" blackberries were pure "Thornless Evergreen" dwarfs. In statice plants, protoclones with abnormal flowers and low pollen fertility were linked to chimera formation from leaf segments of meristem-derived plants, as reported by Kunitake et al. (1995). Similarly, Krikorian et al. (1993) showed that the genetic uniformity of the explant source plays a crucial role in regenerating stable variants, such as "Superplantano" from a spontaneously produced useful variant chimeric cv. "Maricongo". Taking more than one explant from a single plant can often cause variation, as explained by Kunitake et al. (1995). These examples highlight the importance of the genetic composition and uniformity of the mother plant used as starting material for tissue culture. Therefore, it is essential to evaluate the entire plant for genetic uniformity before using it for tissue culture, a practice that is often overlooked in most tissue culture experiments.

7.8 ABERRATION AND REARRANGEMENTS OF CHROMOSOMES

A better understanding of SV has been achieved through thorough characterisation and classification of tissue culture–induced chromosome aberrations. Among tissue-cultured somaclones, variation in chromosome number and structure has been observed, as reported by Al-Zahim et al. (1999), Hao and Deng (2002), and Mujib et al. (2007). Detailed studies have shown that structural chromosome changes are the most accurate reflection of the frequency and extent of karyotypic changes, according to Lee and Phillips (1988). Changes in chromosome structure are the predominant type of aberration observed in tissue-cultured cells, with events leading to chromosome breakage, and in some instances subsequent exchange or reunion of fragments being of fundamental importance.

Two possible origins of chromosome rearrangement in tissue culture are late replicating heterochromatin and nucleotide pool imbalance, as proposed by Bryant (1976). The mitotic cell cycle of higher organisms has four phases: G1 (gap), S (synthesis of DNA), G2 (gap), and M (mitosis consisting of prophase, metaphase, anaphase, and telophase), each with a species-specific and cell type–specific duration. Any disruption in the synchrony between chromosome replication during the S phase and cell division is likely to result in chromosome aberration. Heterochromatic regions, which replicate later than euchromatic segments, may be particularly vulnerable to fluctuations in the cycle.

7.9 THE CELL CYCLE

The regulation of the cell cycle can have a direct impact on plant growth and morphogenesis. In order to discuss the importance of the cell cycle in growth, Kaplan (1992) proposed two opposing theories: The "cell theory" and the "organismal theory." The initial perspective considers cells as the fundamental components of an organism, and growth is correlated with an increase in the number of cells, whereas the latter sees cell division as a result of growth rather than its origin. However, Beemster et al. (2003) provided an integrated model of the two theories, emphasising that numerous signalling and growth chemicals synthesised by individuals or groups of cells control cell growth and development.

Since the cell cycle plays such a crucial role in controlling plant development and morphogenesis, mistakes in cell cycle control during tissue culture have the potential to disrupt natural life processes (Larkin and Scowcroft, 1981). For instance, mistakes in microtubule synthesis, spindle formation, spindle orientation, chromatid segregation, and cross-wall development have been observed often in protoplast cultures, resulting in changes in chromosomal number and structure (Karp, 1994). This suggests that any modification to the cell cycle's routine can result in SV.

Somatic crossing over and sister chromatid exchange can generate different types of chromosome rearrangements observed in tissue culture, particularly if the exchanges were symmetric or between non-homologous chromosomes, but evidence supporting the importance of mitotic recombination for producing SV is lacking, with a few exceptions (Larkin and Scowcroft, 1981). SV can also develop as a result of other types of chromosomal rearrangements, such as those that affect the number of copies of a given sequence, the expression of a given gene, the appearance of regenerated plants, the strength of position effects, or the nature of the variation itself.

7.10 TRANSPOSABLE ELEMENTS

DNA sequences known as transposable elements are transportable within a genome and have the ability to alter it genetically and create mutations. These components, which comprise a large part of the genomes of most plants, were initially documented in maize by McClintock (1950). The phenomenon of SV can arise through the activation of formerly dormant transposable elements within the chromosomal structure. According to Peschke et al. (1987), transposable elements in maize have the potential to be activated through chromosome breakage. It is suggested that tissue culture may create a favourable environment for such activation. Recent studies have indicated that introducing transposons into a rice cultivar which was regenerated by tissue culture led to the emergence of SV (Gao et al., 2009). Several researchers have proposed that transpositional occurrences, including the activation of transposable elements and variations in methylation patterns, have a substantial impact on SV (Barret et al., 2006). However, the specific process behind this phenomenon remains unclear (Kaeppler et al., 2000).

7.11 EPIGENETIC ROLE IN SOMACLONAL VARIATION

SV in medicinally active plant species has been studied, and its underlying mechanisms have been analysed, with both genetic and epigenetic variables being considered (Wang et al., 2013; Roostika et al., 2015). More study is needed to determine how much each process contributes to somaclonal variance. The stability of specific qualitative characteristic mutations across sexual generations implies a genetic underpinning, hence showing molecular alterations within the plant's genome. The stable mutations observed in this context might include modifications in particular genes or rearrangements of chromosomes (Jung et al., 2004; Latzel et al., 2012; Udomchalothorn et al., 2014). Reversed insertions of transposable elements are one example of this phenomenon; in this case, the insertion and subsequent removal of transposable elements can lead to changes in gene expression patterns. These epigenetic modifications may not endure during the process of sexual reproduction, hence emphasising the dynamic characteristic of SV (Kaeppler et al., 2000; Pawełkowicz et al., 2021).

In order to fully understand the intricacies of SV, it is imperative to examine the interaction between genetic and epigenetic processes. DNA methylation patterns, histone modifications, and the participation of non-coding RNAs are all examples of epigenetic regulation that have recently come into focus in studies of SV. Understanding the establishment, maintenance, and potential inheritance of these epigenetic markers through generations will shed light on the stability and heritability of SV (Pietsch and Anderson, 2007; Azizi et al., 2020). Furthermore, improvements in high-throughput sequencing technologies and epigenomic profiling methods have made it possible to analyse epigenetic changes brought on by SV in detail. The integration of molecular methods with functional research has the potential to advance our comprehension of the molecular mechanisms behind SV and its implications for enhancing medicinally active plant species (Azizi et al., 2020). This section examines the impact of epigenetics in SV and how it affects the development of medicinally active plant species. The potential mechanism underlying the generation of secondary metabolites in medicinal plants through *in vitro* cultivation, specifically as a consequence of SV, is illustrated in Figure 7.2.

7.12 DNA METHYLATION PATTERNS

The involvement of epigenetic modifications, specifically DNA methylation, has been linked to the occurrence of SV. Studies have shown that SVs differ from their wild-type counterparts in terms of DNA methylation patterns (Borges et al., 2021). For example, a study by Azizi et al. (2020) examined the DNA methylation profiles of SVs in a species of medicinally active plants.

Changes in DNA methylation patterns are seen as a primary source of epigenetic variation, leading to SV and irregular plant development (Kaeppler and Phillips, 1993; Kaeppler et al., 2000; Saze, 2008; Mirouze and Paszkowski, 2011). Throughout plant and organ development or tissue differentiation, the control of DNA methylation patterns and chromatin position is maintained (Fukai et al., 2010; Law and Jacobsen, 2010; Miguel and Marum, 2011). In specific instances, there is a notable rise in global DNA methylation during the initial stage of somatic embryogenesis (SE), followed by a decrease in methylation during the embryo maturation phase, and ultimately, a subsequent

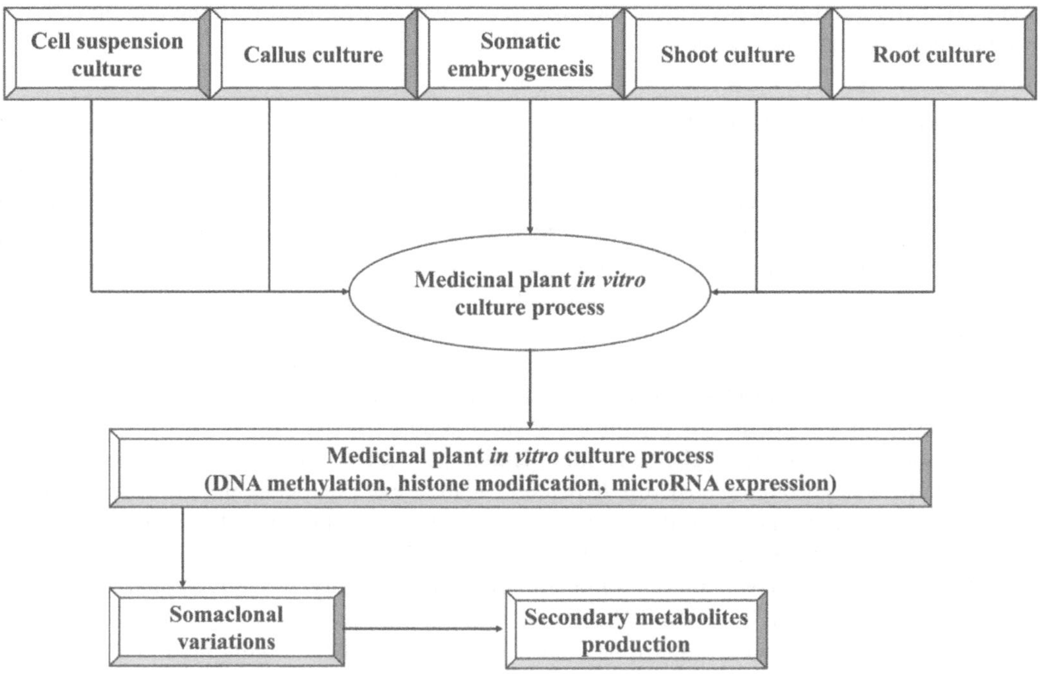

FIGURE 7.2 Effect of somaclonal variations on secondary metabolites production in medicinal plant *in vitro* culture.

increase during the later stages of embryo development. Furthermore, a comprehensive investigation encompassing various phases of embryo development, including somatic embryogenesis and zygotic embryogenesis, has provided evidence of a correlation between DNA methylation and embryogenic differentiation (Landey et al., 2013; Nic-Can et al., 2013). Embryogenic cell suspensions of plants are prone to instability, with a heightened susceptibility to genetic and epigenetic alterations. However, during SE regeneration, only minimal genetic and epigenetic changes have been identified (Landey et al., 2013).

For instance, during *in vitro* culture of different plants, a significant alteration or resetting of the overall methylation level of genomic DNA may be associated with the culture's age, the type of culture, and the concentration of growth regulators (Neelakandan and Wang, 2012; Landey et al., 2013). For example, it has been noted that as the concentration of kinetin (a cytokinin) increases, global methylation levels decrease, whereas increasing concentrations of 2,4-D (an auxin) lead to enhanced methylation levels. The extent of DNA methylation variation in the genome of *in vitro* regenerated plants has been evaluated in different crops, including barley (Bednarek et al., 2007), cocoa (Rodríguez López et al., 2010), pea (Smýkal et al., 2007), grapevine (Schellenbaum et al., 2008), *Plantago* (Kour et al., 2009), *Doritaenopsis* (Park et al., 2009), *Arabidopsis* (Yu et al., 2010), and maize (Yu et al., 2011).

In the oil palm species *Elaeis guineensis* Jacq, the occurrence of mantled floral organs resulting from SV has been the subject of investigation by Jaligot et al. (2004) in both male and female flowers. In cases of mantled male floral organs, stamens and staminodes undergo transformation into pseudocarpels, potentially leading to fruit abortion under severe conditions (Alwee et al., 2006). This anomaly caused a decrease of 5% in regenerated plants through SE and continued to pose a threat to oil palm efficiency and fruit-based oil commercialisation (Rival, 2000). Almost all of these genes belong to the MADS box transcription factor family. Moreover, in this abnormality, the contribution of epigenetic mechanisms, rather than genetic changes, is anticipated due to its frequent occurrence. Consequently, random DNA methylation modifications have been observed among regenerated oil palm plants, but they were not found to be linked with the manifestation of the abnormality (Jaligot et al., 2002).

7.13 HISTONE MODIFICATIONS

Histone modifications, such as acetylation, methylation, and phosphorylation, play a crucial role in regulating gene expression and chromatin structure. Several studies have indicated the involvement of histone modifications in SV (Azizi et al., 2020). For example, a study by Miguel and Marum (2011) investigated the histone modification patterns in SV of a medicinally active plant species. They observed changes in the levels and distribution of specific histone modifications, which correlated with altered gene expression and phenotypic variations in the variants.

In the post-translational stage, modifications such as acetylation, methylation, phosphorylation, and ubiquitination can impact the N-terminal tails of core histone proteins (Kouzarides, 2007). Gene transcription is regulated by a number of internal and external stimuli, such as abiotic and biotic stressors and hormone levels; these modifications collectively play a vital role in governing dynamic changes in chromatin structure (Chen and Tian, 2007; Anzola et al., 2010). As presented in Table 7.1, the separation and proliferation of cells in plant cell and tissue culture are heavily influenced by histone modifications. The capacity of *in vitro* cell culture dedifferentiation to regenerate intact plants is exemplified by plant cell totipotentiality and versatility. Protoplasts in *in vitro* culture have been studied extensively for this behaviour.

The ability of cells to achieve pluripotential competence has been highlighted in this research by focusing on their ability to re-enter the synthesis phase (S) of the cell cycle, which is related to chromatin restructuring (Williams et al., 2003; Avivi et al., 2004). The histone modifications within nucleosomes serve to both orchestrate the chromatin architecture and offer access to the core proteins that are attached to DNA. Gene expression and replication are ultimately controlled by this regulatory process (Choudhuri, 2008). The amino tails of core histones (H2A, H2B, H3, and H4) undergo covalent alterations at multiple terminal sites, characterised by arginine and lysine-rich sequences. These modifications encompass a variety of chemical marks, including methylation,

TABLE 7.1
Several Histone Modifications Occurring with *In Vitro* Plant Cell and Tissue Culture

Histone	Modification	Histone Site	Function	References
H3	Methylation	H3K27 me3	Controlling the flow of the auxin hormone, which plays a role in organ development, cell differentiation, and the creation of somatic embryos.	(Liu et al., 2010; Makarevitch et al., 2013)
H3	Methylation	H3K27 me2	Controls the expression of WUS transcription factor and other somatic embryogenesis–related genes during shoot development.	(Lindroth et al., 2011)
H3	Acetylation	H3K9ac	Activates pRb/E2F target genes and facilitates entry into the S phase of the cell cycle by participating in the redistribution of heterochromatin.	(Williams et al., 2003)
H3	Acetylation	H3K14ac	Redistributes heterochromatin to activate pRb/E2F target genes, permitting cell cycle progression to the S phase.	(Williams et al., 2003)
H3	Methylation/ Demethylation	H3K9	Required for producing mature plants in culture from undifferentiated cells	(Grafi et al., 2007)
HA	Deacetylation	HAP3A	Prevents the embryonic development process	(Uddenberg et al., 2011)

ubiquitination, acetylation, SUMOylation, phosphorylation, adenosine diphosphate (ADP) ribosylation, and biotinylation (Johnson et al., 2004; Zhang et al., 2007). Histone modifications, with their diverse combinations, exert significant effects on precise cellular dedifferentiation, proliferation, cell characteristics, and plant regeneration during cell culture. Additionally, these modifications play a role in SV. These findings suggest that histone modifications contribute to the epigenetic regulation of SV (Ikeuchi et al., 2013).

7.14 NON-CODING RNAs

Non-coding RNAs, including microRNAs (miRNAs) and long non-coding RNAs (lncRNAs), have emerged as important regulators of gene expression and genome stability. Recent studies have implicated non-coding RNAs in the modulation of SV (Duta-Cornescu et al., 2023). For instance, a study by Rodriguez-Enriquez et al. (2011) explored the expression profiles of non-coding RNAs in SV of a medicinally active plant species. They identified differentially expressed miRNAs and lncRNAs, some of which were associated with the regulation of genes involved in secondary metabolite biosynthesis.

In *Eucommia ulmoides*, which serves as a source of natural rubber (Eu-rubber), a total of 29,103 lncRNAs have been identified. Some of these RNAs are associated with subcellular microstructures and cellular processes, including cell division, growth, and cell wall formation. Additionally, they are predicted to be involved in the differentiation and production of laticifers, where Eu-rubber is synthesised. Furthermore, in cotton plants, Lu et al. (2016) reported the influence of lcRNAs in regulating the hormonal pathway in response to drought stress. Furthermore, there is evidence of miRNA expression in cotton ovules, both in the presence and absence of fibre and seed formation These findings suggest that non-coding RNAs play a role in shaping SV and the production of medicinally important compounds (Guan et al., 2014).

7.15 TRANSGENERATIONAL EPIGENETIC INHERITANCE

Epigenetic modifications acquired during SV can be transmitted to subsequent generations, resulting in transgenerational epigenetic inheritance. This phenomenon has implications for the stability and heritability of SV (Verhoeven and Preite, 2014; Alfalahi et al., 2022). A study by Hauser et al. (2011) investigated transgenerational epigenetic changes in SV of a medicinally active plant species. They observed persistent alterations in DNA methylation patterns and histone modifications across multiple generations, indicating the potential for stable epigenetic changes in SV.

Transgenerational epigenetic inheritance necessitates the transmission of epigenetic marks, like DNA methylation, through the germline without being erased by surveillance mechanisms that guarantee the establishment of cellular totipotency during the initiation of ontogenesis (Lange and Schneider, 2010). While the erasure of epigenetic marks in early developmental stages is extensively studied in mammals, its significance for developmental determinations during plant embryogenesis remains less comprehended (Jullien and Berger, 2010). However, the observation that epigenetic alleles (epialleles) are stably passed down over multiple generations in various organisms, including plants, suggests that the resetting to an epigenetic "default state" is likely an imperfect process (Hauser et al., 2011). For a long time, transgenerational epigenetic inheritance mechanisms had limited genetic and molecular evidence because initial experiments were conducted in higher plants, particularly focusing on the mechanisms responsible for pigment variations in corn lines, which could not be explained solely by conventional Mendelian segregation (Brink, 1956).

A well-known example of transgenerational epigenetic inheritance in plants involves a symmetry change in the flowers of *Linaria vulgaris* (toadflax) (Hauser et al, 2011). The shift from bilateral to radial symmetry is associated with increased DNA methylation in the upstream promoter region of the *Lcyc* locus, while somatic reversion back to bilateral flower symmetry is linked to decreased DNA methylation. The question of whether the observed variation in DNA methylation actually causes the transmission of the aforementioned altered epigenetic states remains unresolved (Cubas et al., 1999). Evidence of transgenerational epigenetic inheritance has also been documented in non-plant multicellular organisms (Cavalli and Paro, 1998; Rakyan and Whitelaw, 2003; Jablonka and Raz, 2009). For example, studies on agouti viable yellow and axin fused epialleles in mice have suggested that variations in chromatin modifications of retrotransposons, which insert upstream of the respective loci, drive heritable changes in gene expression (Rakyan and Whitelaw, 2003). These findings highlight the importance of transgenerational epigenetic inheritance in the long-term stability and utilisation of SV (Blewitt et al., 2006).

7.16 CASE STUDIES OF SOMACLONAL VARIATION IN MEDICINALLY ACTIVE PLANT SPECIES

Changes in a plant's genetic makeup due to tissue culture are referred to as SV. Mutations, changes in gene expression, chromosomal abnormalities, and epigenetic modifications are all possible causes of genetic variation in plants. Since the 1980s, scientists have employed SV as a means of crop development in pursuit of better genotypes (Larkin and Scowcroft, 1981). Crop plants including sugarcane, tobacco, rice, maize, and barley have been researched extensively for their somaclonal variety which results from plant tissue culture (Oksman-Caldentey, 1998). Nonetheless, very little is known about the ways in which SV can be used in medicinal plants and their cell cultures (Ngezahayo, 2018). Aside from its potential application in plant breeding, plant cell cultures present an alternative method for extracting useful compounds from plants (Bapat et al., 2023). Using SV, researchers have been able to boost the potency of many medicinal plants, including *Artemisia annua* (Zhang et al., 2022), *Catharanthus roseus*, *Panax ginseng*, *Withania somnifera*, *Taxus baccata*, and *Hypericum perforatum* (Qu et al., 2019; Lee et al., 2020; Alaakel, 2021). However, it is imperative to thoroughly assess the benefits and drawbacks of SV prior to implementing it in plant breeding projects. The purpose of this chapter is to present case examples of SV in species of medicinally active plants, together with an analysis of the genetic mechanisms behind these variations, their implications for plant breeding, and the benefits and drawbacks of this approach.

7.16.1 *ARTEMISIA ANNUA*

Due to its poor agronomic qualities and low yield, cultivation of the plant *A. annua*, which yields the powerful antimalarial medicine artemisinin, is restricted (Zhang et al., 2022). In response to such issues, scientists have used tissue culture to develop high-yielding, high-quality somaclonal variants of *A. annua* (Pandey et al., 2016). So, for example, in a study by Pandey et al. (2016), an SV was chosen because it had a significantly higher artemisinin content than the wild-type plant. The agronomic characteristics of this variant were improved. For example, it branched out more and produced more biomass. The research also aimed to improve the artemisinin content of the plant by creating an SV with a higher trichome density. The resistance of these variations to various diseases and pests was significantly enhanced.

A. annua SVs were tested for resistance to several plant diseases. Disease symptoms and severity were evaluated after inoculating the variations and the wild-type plants with the same pathogens. The discovery of disease-resistant SV demonstrates the potential of this type of variation to boost the plant's immune response and decrease its vulnerability to diseases. The implications of these findings for agricultural practices and crop development techniques are encouraging, as they show the value of SV as a potential approach for enhancing disease resistance in plants (Mesa et al., 2015). A tissue culture experiment on the creation of artemisinin-rich somaclones has shown that these plants are capable of producing much more artemisinin than wild-type plants. In other words, SV has the ability to increase both the quantity and the quality of medicinally important plants, as demonstrated by this study (Baldi and Dixit 2008; Jain and Dixit 2010; Wang et al., 2010; Weathers et al., 2011; Dangash et al., 2014; Xiang et al., 2015; Ikram and Simonsen 2017; Ngezahayo 2018; Al-Khayri et al. 2022). The genetic modifications that occurred during the tissue culture process were responsible for the observed rise in artemisinin concentration. Further investigation indicated a relationship between artemisinin concentration and ploidy level, suggesting polyploidisation as a potential strategy for artemisinin enhancement (Wang et al., 2005).

Artemisinic acid is a precursor compound in the production of artemisinin, hence its presence in *A. annua* SV was investigated. Quantitative analyses of artemisinic acid levels in the variant were conducted using analytical methods such as gas chromatography–mass spectrometry (GC-MS), and the results were compared to those obtained from the corresponding wild-type plants (Huang et al., 2021). Furthermore, the secondary metabolite profile of *A. annua* SV was investigated. Liquid chromatography combined with mass spectrometry (LC-MS) was utilised by the researcher to determine the types and concentration of secondary metabolites in both variant and wild-type plants. Discovering significant differences in secondary metabolite composition between SV, indicating the possibility of SV altering the plant's bioactive compound profile, was the primary result of this investigation (Liu et al., 2022). Moreover, the resistance of *A. annua* SV to abiotic stressors, such as salinity and drought, have also been studied. Researchers treated both variants and the wild-type plants to artificially induced stress and then analysed their physiological and biochemical responses. The study's effectiveness in identifying SV with higher stress tolerance suggests that SV may further boost the plant's resistance to adverse conditions (Mahmood et al., 2012; Pandey et al., 2016).

7.16.2 *Catharanthus roseus*

Vinblastine and vincristine, two alkaloids used in cancer chemotherapy, are biosynthesised by the plant *C. roseus*, which is used to treat a wide range of diseases such as leukaemia (Alam et al., 2017; Qu et al., 2019). Scientists have made substantial progress in the field by using tissue culture techniques to isolate SV containing greater alkaloid content (Lee et al., 2020). Tissue culture techniques were used to cultivate SV lines, which significantly surpassed their wild-type counterparts in terms of alkaloid production. These results highlight the potential of SV as a means for increasing the production of *C. roseus*'s valuable compounds (Perronnet et al., 1988; Vanisree et al., 2004; Ataei-Azimi et al., 2008; Hoda 2013; Lee et al., 2020; Alaakel 2021).

C. roseus's isolated SV displayed intriguing genetic changes. In particular, the variation showed a chromosomal number of 2n = 58 instead of the typical 2n = 52. It is possible that this particular chromosomal variation is responsible for the observed increases in alkaloid synthesis. Furthermore, increased expression of alkaloid biosynthesis genes was also observed in the SV, providing additional evidence of its increased alkaloid content. *Catharanthus roseus*'s SV is better understood due to these genetic mechanisms (Xing et al., 2011; Manzoor et al., 2019). Analysis of the SV effect showed a notable increase in vincristine concentration. Compared to the control plants, the SV showed a surprising 50% rise in vincristine concentration. This discovery demonstrates the possibility of SV as a tool to improve the concentration of the active medicinal component of interest in *C. roseus* (Aslam et al., 2009; Lee et al., 2020).

Researchers compared the SV of *C. roseus* to wild-type plants for a variety of morphological parameters, including plant height, leaf size, flower colour, and branching pattern. The investigation's effectiveness in identifying SVs with favourable morphological attributes suggests that SV could be used to improve the plant's appearance and growth pattern (Noormohammadi et al., 2020). The SV of *C. roseus* was examined in a separate study for its regenerative potential *in vitro*. Under controlled laboratory conditions, the variants and wild-type plants were compared for their capacity to regenerate shoots and roots. SV was found to improve regeneration efficiency, suggesting that it could boost the plant's capacity for *in vitro* multiplication and propagation (Hoda, 2013). Resistance to phytopathogens, such as fungal and bacterial, was also investigated in the SV of *C. roseus*. Specific pathogens were introduced to both the variants and the wild-type plants; afterward disease symptoms and severity were recorded and assessed. SVs with increased resistance to phytopathogens were successfully discovered, demonstrating the potential of SV in boosting the plant's defence mechanisms against diseases (Lee et al., 2020).

7.16.3 *Taxus baccata*

T. baccata, a species renowned for its production of the anticancer compound Taxol, has been extensively studied to investigate the possibility of SV in enhancing Taxol production. Numerous studies have provided insights into the isolation of SVs that display a notable enhancement in Taxol content. These findings have been built upon the foundational research conducted by Malik et al. (2011). An SV with increased Taxol levels was successfully isolated and documented in this work. An increased ploidy level and unique chromosomal alterations, including rearrangements and deletions, were among the variant's most fascinating characteristics. Furthermore, it revealed changes in gene expression patterns related to Taxol biosynthesis, which shed light on the molecular mechanisms responsible for the increased Taxol production in the SV of *T. baccata* (Croteau et al., 2006).

Further research endeavours have been directed towards clarifying the greater genetic and physiological ramifications of SV in *T. baccata*. Studies have examined how different tissue culture circumstances, such as growth regulators and culture media compositions, affect producing SV and boosting Taxol synthesis (Han et al., 1994; Han et al., 2000; Khosroushahi et al., 2006; Malik et al., 2011). Moreover, an investigation has been conducted on the characterisation of key regulatory genes and transcription factors associated with Taxol biosynthesis. The objective of the research is to gain a better understanding of their involvement in SV and potentially develop enhanced Taxol-producing variations by genetic engineering. Future research could combine genomic and transcriptome approaches to better understand the complex regulatory networks involved in Taxol biosynthesis and to identify potential targets for increasing Taxol production in *T. baccata*. Furthermore, it is important to acknowledge that the utilisation of SV in *T. baccata* goes beyond the production of Taxol. The investigation of SV has involved the examination of secondary metabolites, including taxanes, which has potential therapeutic applications. The mentioned investigations have provided evidence of encouraging possibilities in the manipulation of SV for enhancing the synthesis of an extensive range of medically valuable compounds in *T. baccata*.

The growth features of *T. baccata* SVs were analysed; these traits included plant height, leaf size, branch architecture, and biomass production. The experiment was effective in identifying SVs with improved growth attributes, showing the potential of SV to improve the plant's agronomic qualities and biomass yield (Sharma et al., 2022). *T. baccata* SVs were tested for their ability to withstand abiotic stress conditions such as salinity and drought. The experiment involved subjecting both variant and wild-type plants to carefully regulated stress treatments, while closely monitoring their physiological responses and rates of survival. The successful identification of SVs with increased resistance to abiotic stress by the study suggests that SV may have a role in enhancing the plant's resistance to adverse environmental factors (Patel et al., 2020). These investigations have laid the foundation for subsequent research endeavours focused on utilising SV to enhance the synthesis of medicinally important compounds in *T. baccata* and other plant species of pharmacological relevance.

7.16.4 *Panax ginseng*

P. ginseng, a popular plant species recognised for its adaptogenic and immunomodulatory characteristics, has been utilised for a long time in the field of traditional medicine. The therapeutic properties of *P. ginseng* are related to the presence of ginsenosides, a bioactive compound that is found in the roots of the plant. However, consistent production of ginseng has been hindered by the fact that there is a wide range of ginsenoside concentrations even among *P. ginseng* populations. SV has been identified as a useful mechanism for addressing this constraint and enhancing the synthesis of ginsenosides, thus presenting potential opportunities for the pharmaceutical sector (Kim et al., 2009; Kiselev et al., 2011; Chelomina et al., 2020; Lee et al., 2023). Kim et al. (2009) reported isolating a somaclonal form of *P. ginseng* with substantially elevated ginsenoside concentration. SVs were acquired using tissue culture methods, and their ginsenoside profiles were analysed and compared. The study found substantial differences in the relative abundance of particular ginsenosides, suggesting the possibility of selectively boosting the production of compounds with desired therapeutic qualities (Kiselev et al., 2011; Chelomina et al., 2020; Lee et al., 2023). Differential expression of genes involved in ginsenoside production, along with modifications in DNA methylation patterns, were observed in the SV. Researchers used molecular markers and DNA sequencing technology to identify that the SV contained a variety of genetic modifications, such as single nucleotide polymorphisms (SNPs) and insertions/deletions (indels). These results provide insight into the molecular mechanisms responsible for synthesising SV in ginsenoside (Kiselev et al., 2013; Gantait et al., 2020).

Furthermore, *P. ginseng* SVs were also developed and examined for bioactivity. Compared to their wild-type counterparts, the variants showed significantly higher antioxidant activity and enhanced inhibitory effects against cancer cell lines. The results revealed that somaclonal polymorphism in *P. ginseng* could potentially increase ginsenoside synthesis and improve the overall bioactivity of the plant, making it more powerful and effective for medicinal applications (Keum et al., 2000; Shibata, 2001; Gantait et al., 2020). An in-depth research study was conducted to discover more about the physiological shifts related to SV in *P. ginseng*. SVs were compared to their wild-type counterparts in terms of growth, photosynthetic efficiency, and nutrient uptake parameters. The research showed that the SV grew better and exhibited greater photosynthetic activity, which resulted in more ginsenosides production. The results of the study revealed that the biochemical composition and general physiology of *P. ginseng* were both affected by SV (Kiselev et al., 2011; Kim et al., 2013; Chelomina et al., 2020; Lee et al., 2023).

In addition, SVs of *P. ginseng* were observed, and these variants showed increased resistance to fungal infections. Compared to their wild-type counterparts, several SVs were found to have lower disease severity and enhanced defence systems. The results of this study showed that SV in *P. ginseng* has the potential to provide higher resistance to pathogens, thereby decreasing the need for chemical fungicides (Chelomina et al., 2020; Lee et al., 2023). Additional research has shown that SV affects *P. ginseng*'s drought tolerance. The physiological responses of SV were studied when they were treated to drought stress conditions. This included measuring water consumption efficiency, stomatal conductance, and antioxidant enzyme activity. Finding the SV with enhanced drought tolerance in this work demonstrates the potential of SV in the breeding of water-hardy *P. ginseng* cultivation (Mahmood et al., 2012; Udomchalothorn et al., 2014). Shen et al. (2019) conducted a study to assess the SV of *P. ginseng* for enhanced agronomic features. The SVs were compared to their wild-type counterparts on a number of characteristics, including overall size and growth, leaf size, root shape, and flowering time. Thus, the study's findings suggest that SVs could be valuable for breeding programmes aiming to enhance the agronomic performance of *P. ginseng* by selecting traits like greater root biomass and earlier blooming.

7.16.5 *Withania somnifera*

As a medicinal plant species, *W. somnifera* is valuable for its adaptogenic and anti-inflammatory qualities thanks to the presence of the bioactive compound withanolide. The amount of withanolide produced by *W. somnifera* has been boosted through SV (Rana et al., 2012; Dhar et al., 2015; Singh et al., 2017; Yousefian et al., 2018). Two separate research groups, Singh et al. (2016) and Rana et al. (2012), have reported isolating a SV with higher withanolide content. Duplications and deletions were seen on the chromosomes of this variant, as well as modifications in gene expression associated with withanolide production. To determine the relative amounts of withanolide present in the SV and the wild-type plants, the researchers used analytical methods such as high-performance liquid chromatography (HPLC) and mass spectrometry. The study found that some SVs of *W. somnifera* had substantially greater withanolide content than others, indicating that SVs could improve the plant's therapeutic efficacy.

In addition, the abiotic stress resistance of *W. somnifera* SV was assessed. The physiological and biochemical responses of the SV and the wild-type plants were analysed after being exposed to controlled stress conditions. As a result of the analysis, SVs with enhanced stress tolerance were identified. These included plants with higher water retention rates, higher levels of antioxidant activity, and increased growth despite adverse environmental conditions. These results suggest that SV could be used to breed *W. somnifera* variants that are more resistant to environmental stressors (Dhar et al., 2015). Furthermore, the morphological characteristics of *W. somnifera*'s SVs were analysed. This included the variation in stature, leaf size, and branching pattern. Under laboratory conditions, the researchers compared the variants to wild-type plants for these characteristics. The study found SVs with favourable morphological traits, such as taller plants, larger leaves, and improved branching. These results highlight the importance of SV in *W. somnifera* breeding programmes for enhancing the species' agronomic characteristics (Rana et al., 2012). Changes in the secondary metabolite profiles of SV of *W. somnifera* were also investigated. Using GC-MS, a cutting-edge analytical approach, secondary metabolite composition, including alkaloids and flavonoids, was compared between SV plants and wild-type plants. SV has the ability to alter the chemical composition and medicinal traits of *W. somnifera*, as evidenced by the discovery of SVs with altered secondary metabolite profiles (Singh et al., 2017).

7.16.6 *Hypericum perforatum*

The antidepressant effects of *H. perforatum* have made it an effective medicinal plant. The hypericin content of this plant has been enhanced through SV. Several research studies investigated the hypericin content of *H. perforatum*'s SV (Čellárová and Bruňaková, 1996). HPLC analysis was used to evaluate hypericin production in both the variants and the wild-type plants. Researchers discovered that some SVs of *H. perforatum* had significantly greater concentrations of hypericin, indicating that SV might be used to improve the plant's medicinal traits (Chen et al., 2003; Koperdáková et al., 2007; Wang 2010, Nobakht et al., 2022). Furthermore, the essential oil composition of somaclonal varieties of *H. perforatum* was analysed. GC-MS was utilised to evaluate the composition of volatile compounds of essential oils extracted from both the variants and the wild-type plants. The study found that the essential oil profiles varied among the SV, suggesting that SV might have a role in regulating the profile of bioactive compounds in *H. perforatum*.

Additionally the morphological characteristics of the SV of *H. perforatum* were also characterised. Growth factors such plant height, leaf area, and flower features were examined between the variants and wild-type plants. Results showed that SV has the ability to alter the overall plant architecture of *H. perforatum*, as evidenced by the identification of SVs with altered morphological features. The study found SVs with altered morphological features, suggesting that SV has the potential to affect *H. perforatum*'s overall plant structure (He and Wang, 2013). Moreover, the abiotic stress tolerance of *H. perforatum* SV, particularly concerning drought and salinity, has been the subject of multiple research studies. The physiological responses of the variants and wild-type plants, including water status, and antioxidant enzyme activity were assessed after being subjected to controlled stress conditions. Finding SVs with increased resistance to drought and salinity stress in this work demonstrated the potential of SV in enhancing *H. perforatum's* resilience to harsh environmental conditions (Cirak and Radusiene, 2019).

7.16.7 *Cannabis sativa*

Cannabis sativa, an adaptive and historical plant, has been used traditionally for therapeutic, recreational, and industrial purposes. The heightened interest in comprehending this substance's genetic variety and enhancing its active compound production has been driven by recent advancements in its medicinal utilisation. In order to realise the plant's full potential, SV, an important source of genetic variability, has been progressively studied (Adhikary et al., 2021; Ioannidis et al., 2022; Hesami et al., 2023). This phenomenon involves the occurrence of genetic alterations during the propagation of tissue culture *in vitro*. Focusing on the impact of SV in cannabis plants for the production and extraction of therapeutic compounds, this section intends to critically analyse case studies that shed light on such variation's amount and significance.

SV in medical cannabis plants has been the subject of multiple case studies, many of which investigated cannabinoids like tetrahydrocannabinol (THC) and cannabidiol (CBD) in SV. The researchers documented substantial differences in cannabinoid profiles across several somaclones, suggesting the possibility of developing specific varieties with customised therapeutic attributes. However, this research faced difficulties in ensuring stable quantities of cannabinoid levels throughout subsequent generations, emphasising the necessity for meticulous selection and genetic stabilisation (Page, 2021; Adams et al., 2021; Adhikary et al., 2021; Ioannidis et al., 2022). The presence of diversity in SVs of cannabis offers distinct potential for optimisation of medicinally active compound synthesis. The study conducted by Casano et al. (2010) aimed to examine the impact of SV on the production of terpenes, specifically focusing on essential oils that possess medicinal properties. The results of their study revealed that specific somaclones displayed a greater quantity of terpenes in comparison to plants grown by normal methods. Using this information could lead to the creation of cannabis strains with improved therapeutic properties, therefore facilitating the development of more effective medicines for a wider range of conditions. Researchers have documented a case of SV causing unstable morphological features, which impacted plant architecture and development. Stabilising desirable features in the face of such unpredictability requires rigorous selection and breeding processes. Additionally, careful testing is necessary to guarantee the safety and effectiveness of medical cannabis products made from SVs due to the possibility of unexpected alterations in other secondary metabolites (Adhikary et al., 2021; Hesami et al., 2021, 2023).

In addition to these case studies, there are numerous additional instances of SVs seen in plant species with medicinal properties. An example of this phenomenon can be found in *Taxus brevifolia*, which serves as the origin of the anticancer medication Taxol (*T. bracca*). It has been documented that *T. brevifolia* exhibits a greater Taxol production than the plant's wild-type counterpart (Ngezahayo, 2018). Improvements in productivity and quality of several medicinal plants have also been achieved through the utilisation of SV. For example, SV has been employed to increase production of the active ingredient andrographolide in the medicinal plant *Andrographis paniculata*, which is used to treat conditions including fever and dysentery. Similarly, in the medical plant *Stevia rebaudiana*, which is used as a natural sweetener to treat diabetes, SV has been employed to increase the production of steviol glycosides (Gantait and Mukherjee, 2021). The utilisation of SV has been extended to the genetic modification of plant-based biopharmaceuticals, including vaccines, antibodies, and enzymes, in addition to its conventional application in medicinal plant research. For instance, a study revealed the development of a tobacco SV that expresses a human antibody against herpes simplex virus type 2 (Wongsamuth and Doran 1997; Hasnain et al., 2022). The aforementioned findings provide evidence of the possibility of SV in producing plant-based biopharmaceuticals that possess distinct therapeutic characteristics (Hasnain et al., 2022).

7.17 ADVANTAGES AND DISADVANTAGES

SV offers numerous benefits in the enhancement of plant species with therapeutic properties. This process facilitates the selection of favourable characteristics, including enhanced productivity, superior quality, and heightened resilience against both biotic and abiotic pressures. Additionally, it facilitates the production of new variants that possess distinct therapeutic characteristics. Nevertheless, SV does include certain drawbacks. It has the potential to contribute to the loss of advantageous characteristics, such as resistance to diseases, while simultaneously giving rise to unfavourable attributes, such as diminished crop productivity and quality. Additionally, this phenomenon has the potential to cause genetic instability and epigenetic modifications, which could have implications for the plant's long-term stability (Ngezahayo, 2018; Duta-Cornescu et al., 2023).

REFERENCES

Abiri, R., Abdul-Hamid, H., Sytar, O., Abiri, R., Bezerra de Almeida, E. Jr, Sharma, S. K., ... Malik, S. 2021. A Brief Overview of Potential Treatments for Viral Diseases Using Natural Plant Compounds: The Case of SARS-CoV. Molecules, 26(13): 3868.

Abiri, R., Atabaki, N., Abdul-Hamid, H., Sanusi, R., Ab Shukor, N. A., Shaharuddin, N. A., ... Malik, S. 2020. The Prospect of Physiological Events Associated with the Micropropagation of Eucalyptus sp. Forests, 11(11): 1211.

Abiri, R., Maziah, M., Shaharuddin, N. A., Yusof, Z. N. B., Atabaki, N., Hanafi, M. M., ... Valdiani, A. 2017. Enhancing Somatic Embryogenesis of Malaysian Rice Cultivar MR219 Using Adjuvant Materials in A High-Efficiency Protocol. International Journal of Environmental Science and Technology, 14: 1091–1108.

Adams, T. K., Masondo, N. A., Malatsi, P., & Makunga, N. P. 2021. *Cannabis sativa*: From Therapeutic Uses to Micropropagation and Beyond. Plants, 10(10): 2078.

Adhikary, D., Kulkarni, M., El-Mezawy, A., Mobini, S., Elhiti, M., Gjuric, R., ... Bhowmik, P. 2021. Medical cannabis and Industrial Hemp Tissue Culture: Present Status and Future Potential. Frontiers in Plant Science, 12: 627240.

Agarwal, A., Maldonado Rosas, I., Anagnostopoulou, C., Cannarella, R., Boitrelle, F., Munoz, L. V., ... Saleh, R. 2022. Oxidative Stress and Assisted Reproduction: A Comprehensive Review of Its Pathophysiological Role and Strategies for Optimizing Embryo Culture Environment. Antioxidants, 11(3): 477.

Alaakel, S. N. 2021. Improving the Production of Indole Alkaloids from Madagascar Periwinkle (*Catharanthus roseus* L.) Using Stressful Tissue Cultures. p. 184. Damascus University. Faculty of agricultural engineering.

Alam, M. M., Naeem, M., Khan, M. M. A., & Uddin, M. 2017. Vincristine and Vinblastine Anticancer Catharanthus Alkaloids: Pharmacological Applications and Strategies for Yield Improvement. Catharanthus roseus: Current Research and Future Prospects, pp. 277–307. Cham: Springer.

Alfalahi, A. O., Hussein, Z. T., Khalofah, A., Sadder, M. T., Qasem, J. R., Al-Khayri, J. M., ... Almehemdi, A. F. 2022. Epigenetic Variation as a New Plant Breeding Tool: A Review. Journal of King Saud University-Science, 34(8): 102302.

Al-Khayri, J. M., Sudheer, W. N., Lakshmaiah, V. V., Mukherjee, E., Nizam, A., Thiruvengadam, M., ... Attimarad, M. 2022. Biotechnological Approaches for Production of Artemisinin, An Anti-Malarial Drug from *Artemisia annua* L. Molecules, 27(9): 3040.

Alwee, S. S., Van der Linden, C., Van der Schoot, J., De Folter, S., Angenent, G., Cheah, S., & Smulders, M. 2006. Characterization of Oil Palm MADS Box Genes In Relation to The Mantled Flower Abnormality. Plant Cell, Tissue and Organ Culture, 8: 331–344.

Al-Zahim, M. A., Ford-Lloyd, B. V., & Newbury, H. J. 1999. Detection of Somaclonal Variation in Garlic (*Allium sativum* L.) Using RAPD and Cytological Analysis. Plant Cell Reports, 18: 473–477.

Anastassopoulos, E., & Keil, M. 1996. Assessment of Natural and Induced Genetic Variation in Alstroemeria Using Random Amplified Polymorphic DNA (RAPD) Markers. Euphytica, 90(2): 235–244.

Anzola, J. M., Sieberer, T., Ortbauer, M., Butt, H., Korbei, B., Weinhofer, I., Müllner, A. E., & Luschnig, C. 2010. Putative Arabidopsis Transcriptional Adaptor Protein (PROPORZ1) Is Required To Modulate Histone Acetylation In Response To Auxin. Proceedings of the National Academy of Sciences of the United States of America, 107: 10308–10313.

Arnholdt-Schmitt, B. 1993. Rapid Changes in Amplification and Methylation Pattern of Genomic DNA in Cultured Carrot Root Explants (*Daucus carota* L.). Theoretical and Applied Genetics, 85: 793–800.

Aslam, J., Mujib, A., Nasim, S. A., & Sharma, M. P. 2009. Screening of Vincristine Yield in Ex Vitro and *in Vitro* Somatic Embryos Derived Plantlets of *Catharanthus roseus* L.(G) Don. Scientia Horticulturae, 119(3): 325–329.

Ataei-Azimi, A., Hashemloian, B. D., Ebrahimzadeh, H., & Majd, A. 2008. High *in Vitro* Production of Ant-Canceric Indole Alkaloids from Periwinkle (*Catharanthus Roseus*) Tissue Culture. African Journal of Biotechnology, 7(16): 2834–2839.

Avivi, Y., Morad, V., Ben-Meir, H., Zhao, J., Kashkush, K., Tzfira, T., Citovsky, V., & Grafi, G. 2004. Reorganization of Specific Chromosomal Domains and Activation of Silent Genes in Plant Cells Acquiring Pluripotentiality. Developmental Dynamics, 230: 12–22.

Azizi, P., Hanafi, M. M., Sahebi, M., Harikrishna, J. A., Taheri, S., Yassoralipour, A., & Nasehi, A. 2020. Epigenetic Changes and Their Relationship to Somaclonal Variation: A Need to Monitor the Micropropagation of Plantation Crops. Functional Plant Biology, 47(6): 508–523.

Bairu, M. W., Aremu, A. O., & Van Staden, J. 2011. Somaclonal Variation in Plants: Causes and Detection Methods. Plant Growth Regulation, 63: 147–173.

Bairu, M. W., Fennell, C. W., & van Staden, J. 2006. The Effect Of Plant Growth Regulators On Somaclonal Variation in *Cavendish banana* (Musa AAA cv.'Zelig'). Scientia Horticulturae, 108(4): 347–351.

Baldi, A., & Dixit, V. K. 2008. Yield Enhancement Strategies for Artemisinin Production by Suspension Cultures of *Artemisia annua*. Bioresource Technology, 99(11): 4609–4614.

Bapat, V. A., Kavi Kishor, P. B., Jalaja, N., Jain, S. M., & Penna, S. 2023. Plant Cell Cultures: Biofactories for the Production of Bioactive Compounds. Agronomy, 13(3): 858.

Barret, P., Brinkman, M., & Beckert, M. 2006. A Sequence Related to Rice Pong Transposable Element Displays Transcriptional Activation By *In Vitro* Culture and Reveals Somaclonal Variations in Maize. Genome, 49(11): 1399–1407.

Bednarek, P. T., Orłowska, R., Koebner, R. M., & Zimny, J. 2007. Quantification of Thetissue-Cultureinduced Variationin Barley (*Hordeum vulgare L.*). BMC Plant Biology, 7: 10.

Beemster, G. T., Fiorani, F., & Inze, D. 2003. Cell Cycle: The Key to Plant Growth Control? Trends in Plant Science, 8(4): 154–158.

Blewitt, M. E., Vickaryous, N. K., Paldi, A., Koseki, H., & Whitelaw, E. 2006. Dynamic Reprogramming of DNA Methylation at an Epigenetically Sensitive Allele in Mice. PLoS Genetics, 2(4): e49.

Borges, F., Donoghue, M. T., LeBlanc, C., Wear, E. E., Tanurdžić, M., Berube, B., ... Martienssen, R. A. 2021. Loss of Small-RNA-Directed DNA Methylation in the Plant Cell Cycle Promotes Germline Reprogramming and Somaclonal Variation. Current Biology, 31(3): 591–600.

Braun, A. C. 1959. A Demonstration of the Recovery of the Crown-Gall Tumor Cell with the Use of Complex Tumors of Single-Cell Origin. Proceedings of the National Academy of Sciences, 45(7): 932–938.

Brink, R. A. 1956. A Genetic Change Associated with the R Locus In Maize Which Is Directed and Potentially Reversible. Genetics, 41(6): 872.

Bryant, J. A. 1976. Molecular Aspects of Gene Expression in Plants. Experimental Botany. 11.

Capriotti, L., Sabbadini, S., Limera, C., Navacchi, O., & Mezzetti, B. 2022. Somaclonal Variation and Induced Mutagenesis in Several Grapevine Cultivars. In XXXI International Horticultural Congress (IHC2022): International Symposium On *In Vitro* Technology and Micropropagated Plants. ISHS Acta Horticulturae, 1359 (pp. 131–138).

Casano, S., Grassi, G., Martini, V., & Michelozzi, M. 2010. Variations in Terpene Profiles of Different Strains of Cannabis sativa L. In XXVIII International Horticultural Congress on Science and Horticulture for People (IHC2010): A New Look at Medicinal. ISHS Acta Horticulturae, 925 (pp. 115–121).

Cavalli, G., & Paro, R. 1998. The Drosophila Fab-7 Chromosomal Element Conveys Epigenetic Inheritance during Mitosis and Meiosis. Cell, 93(4): 505–518.

Čellárová, E., & Bruňaková, K. 1996. Somaclonal Variation in Hypericum perforatum (St. John's Wort). Somaclonal Variation in Crop Improvement II, pp. 267–279. Berlin, Heidelberg: Springer.

Chelomina, G. N., Rozhkovan, K. V., Burundukova, O. L., Gorpenchenko, T. Y., Khrolenko, Y. A., & Zhuravlev, Y. N. 2020. Age-Dependent and Tissue-Specific Alterations in the rDNA Clusters of the *Panax ginseng* CA Meyer Cultivated Cell Lines. Biomolecules, 10(10): 1410.

Chen, M. L., Huang, L. Q., Yang, B., Fu, M. H., & Luo, Y. M. 2003. Determining The Quantity of Hypericin in Medicinal Materials and Asepsis Seedings of *hypericum perforatum* by HPLC. China Journal of Chinese Materia Medica, *28*(11): 1013–1015.

Chen, Z. J., & Tian, L. 2007. Roles of Dynamic and Reversible Histone Acetylation in Plant Development and Polyploidy. Biochimica et Biophysica Acta (BBA) - Gene Structure and Expression, 1769: 295–307.

Chevallier, A. 2016. Encyclopedia of Herbal Medicine: 550 Herbs and Remedies for Common Ailments. Penguin.

Choudhuri, S. 2008. Epigenetic Regulation of Gene and Genome Expression. In Eds S. Choudhuri, D. B. Carlson, 'Genomics. Fundamentals and Applications, pp. 98–126. Boca Raton, FL, USA: CRC Press.

Cirak, C., & Radusiene, J. 2019. Factors Affecting the Variation of Bioactive Compounds in Hypericum Species. Biologia Futura, 70(3): 198–209.

Côte, F. X., Teisson, C., & Perrier, X. 2001. Somaclonal Variation Rate Evolution in Plant Tissue Culture: Contribution to Understanding through a Statistical Approach. *In Vitro* Cellular & Developmental Biology-Plant, 37: 539–542.

Croteau, R., Ketchum, R. E., Long, R. M., Kaspera, R., & Wildung, M. R. 2006. Taxol Biosynthesis and Molecular Genetics. Phytochemistry Reviews, 5: 75–97.

Cubas, P., Vincent, C., & Coen, E. 1999. An Epigenetic Mutation Responsible for Natural Variation in Floral Symmetry. Nature, 401(6749): 157–161.

Custódio, L., Charles, G., Magné, C., Barba-Espín, G., Piqueras, A., Hernández, J. A., ... Rodrigues, M. J. 2022. Application of *In Vitro* Plant Tissue Culture Techniques to Halophyte Species: A Review. Plants, 12(1): 126.

D'Amato, F., & Bayliss, M. W. 1985. Cytogenetics of Plant Cell and Tissue Cultures and Their Regenerates. Critical Reviews in Plant Sciences, 3(1): 73–112.

Dangash, A. J., Pandya, N., Bharillya, A., Jhala, A., & Dharamchand, J. A. I. N. 2014. Impact of Exogenous Elicitors on Artemisinin Production and Trichome Density in *Artemisia annua* L. Under Subtropical Conditions. Notulae Scientia Biologicae, 6(3): 349–353.

Deming, D. 2016. Science and Technology in World History, Volume 4: The Origin of Chemistry, the Principle of Progress, the Enlightenment and the Industrial Revolution (Vol. 4). McFarland.

Dervendzi, V. 1992. Contemporary Treatment with Medicinal Plants. Skopje: Tabernakul, 5–43.

Desai, P., Desai, S., Rafaliya, R., & Patil, G. 2022. Plant Tissue Culture: Somatic Embryogenesis and Organogenesis. In A. C. Rai, A. Kumar, A. Modi, & M. Singh (Eds.), Advances in Plant Tissue Culture (pp. 109–130). Academic Press.

Dhar, N., Razdan, S., Rana, S., Bhat, W. W., Vishwakarma, R., & Lattoo, S. K. 2015. A Decade of Molecular Understanding of Withanolide Biosynthesis and *In Vitro* Studies in *Withania somnifera* (L.) Dunal: Prospects and Perspectives for Pathway Engineering. Frontiers in Plant Science, 6: 1031.

Dong, C., Fang, W., Yi, Q., & Zhang, J. 2022. A Comprehensive Review on Reactive Oxygen Species (ROS) in Advanced Oxidation Processes (AOPs). Chemosphere, 308: 136205.

Duncan, R. R. 1996. Tissue Culture-Induced Variation and Crop Improvement. Advances in Agronomy, 58: 201–240.

Duta-Cornescu, G., Constantin, N., Pojoga, D. M., Nicuta, D., & Simon-Gruita, A. 2023. Somaclonal Variation—Advantage or Disadvantage in Micropropagation of the Medicinal Plants. International Journal of Molecular Sciences, 24(1): 838.

Eeuwens, C. J., Lord, S., Donough, C. R., Rao, V., Vallejo, G., & Nelson, S. 2002. Effects of Tissue Culture Conditions during Embryoid Multiplication on the Incidence of "Mantled" Flowering in Clonally Propagated Oil Palm. Plant Cell, Tissue and Organ Culture, 70: 311–323.

Eftekhari, M., Alizadeh, M., Mashayekhi, K., & Asghari, H. R. 2012. *In Vitro* Propagation of Four Iranian Grape Varieties: Influence of Genotype and Pretreatment with Arbuscular Mycorrhiza. Vitis, 51(4): 175–182.

Espinosa-Leal, C. A., Puente-Garza, C. A., & García-Lara, S. 2018. *In Vitro* Plant Tissue Culture: Means for Production Of Biological Active Compounds. Planta, 248: 1–18.

Etienne, H., & Bertrand, B. 2003. Somaclonal Variation in *Coffea arabica*: Effects of Genotype and Embryogenic Cell Suspension Age on Frequency and Phenotype of Variants. Tree Physiology, 23(6): 419–426.

Faehnrich, B., Franz, C., Nemaz, P., & Kaul, H. P. 2021. Medicinal Plants and Their Secondary Metabolites—State of The Art and Trends in Breeding, Analytics and Use in Feed Supplementation—With Special Focus on German Chamomile. Journal of Applied Botany and Food Quality, 94: 61–74.

Farahani, F., Yari, R., & Masoud, S. 2011. Somaclonal Variation in Dezful Cultivar of Olive (*Olea europaea* subsp. europaea). Gene Conserve, 10(41).

Ferreira, M. D. S., Rocha, A. D. J., Nascimento, F. D. S., Oliveira, W. D. D. S., Soares, J. M. D. S., Rebouças, T. A., … Amorim, E. P. 2023. The Role of Somaclonal Variation in Plant Genetic Improvement: A Systematic Review. Agronomy, 13(3): 730.

Fukai, E., Umehara, Y., Sato, S., Endo, M., Kouchi, H., Hayashi, M., Stougaard, J., & Hirochika, H. 2010. Derepression of the Plant Chromovirus LORE1 Induces Germline Transposition in Regenerated Plants. PLOS Genetics, 6: e1000868.

Gantait, S., Mitra, M., & Chen, J. T. 2020. Biotechnological Interventions for Ginsenosides Production. Biomolecules, 10(4): 538.

Gantait, S., & Mukherjee, E. 2021. Induced Autopolyploidy—A Promising Approach for Enhanced Biosynthesis of Plant Secondary Metabolites: An Insight. Journal of Genetic Engineering and Biotechnology, 19: 1–13.

Gao, D. Y., Vallejo, V. A., He, B., Gai, Y. C., & Sun, L. H. 2009. Detection of DNA Changes in Somaclonal Mutants of Rice Using SSR Markers and Transposon Display. Plant Cell, Tissue and Organ Culture (PCTOC), 98: 187–196.

Gao, X., Yang, D., Cao, D., Ao, M., Sui, X., Wang, Q., … Wang, L. 2010. *In vitro* Micropropagation of *Freesia hybrida* and the Assessment of Genetic and Epigenetic Stability in Regenerated Plantlets. Journal of Plant Growth Regulation, 29: 257–267.

George, E. F. 1993. Plant Propagation by Tissue Culture. Part 1: The Technology (No. Ed. 2). Exegetics Limited.

George, P. 2011. Concerns Regarding the Safety and Toxicity of Medicinal Plants-An Overview. Journal of Applied Pharmaceutical Science, 1: 40–44.

Ghosh, A., Igamberdiev, A. U., & Debnath, S. C. 2021. Tissue Culture-Induced DNA Methylation in Crop Plants: A Review. Molecular Biology Reports, 48(1): 823–841.

Grafi, G., Ben-Meir, H., Avivi, Y., Moshe, M., Dahan, Y., & Zemach, A. 2007. Histone Methylation Controls Telomerase-Independent Telomere Lengthening in Cells Undergoing Dedifferentiation. Developmental Biology, 306: 838–846.

Grzegorczyk-Karolak, I., Wiktorek-Smagur, A., & Hnatuszko-Konka, K. 2018. An Untapped Resource in the Spotlight of Medicinal Biotechnology: The Genus Scutellaria. Current Pharmaceutical Biotechnology, 19(5): 358–371.

Grzywacz, D., Stevenson, P. C., Mushobozi, W. L., Belmain, S., & Wilson, K. 2014. The Use of Indigenous Ecological Resources for Pest Control in Africa. Food Security, 6: 71–86.

Guan, X., Song, Q., & Chen, Z. J. 2014. Polyploidy and Small RNA Regulation of Cotton Fibber Development. Trends in Plant Science, 19: 516–528.

Han, K. H., Fleming, P., Walker, K., Loper, M., Chilton, W. S., Mocek, U., ... Floss, H. G. 1994. Genetic Transformation of Mature Taxus: An Approach to Genetically Control the *In Vitro* Production of the Anticancer Drug, Taxol. Plant Science, 95(2): 187–196.

Han, K. H., Gordon, M. P., & Floss, H. G. 2000. Genetic Transformation of Taxus (Yew) to Improve Production of Taxol.Eds Y.P.S. Bajaj, Transgenic Trees, pp. 291–306.Berlin, Heidelberg: Springer.

Hao, Y. J., & Deng, X. X. 2002. Occurrence of Chromosomal Variations and Plant Regeneration from Long-Term-Cultured Citrus Callus. *In Vitro* Cellular & Developmental Biology-Plant, 38: 472–476.

Hasnain, A., Naqvi, S. A. H., Ayesha, S. I., Khalid, F., Ellahi, M., Iqbal, S., ... Abdelhamid, M. 2022. Plants *in Vitro* Propagation with Its Applications in Food, Pharmaceuticals and Cosmetic Industries; Current Scenario and Future Approaches. Frontiers in Plant Science, 13: 1009395.

Hauser, M. T., Aufsatz, W., Jonak, C., & Luschnig, C. 2011. Transgenerational Epigenetic Inheritance in Plants. Biochimica et Biophysica Acta (BBA)-Gene Regulatory Mechanisms, 1809(8): 459–468.

He, M., & Wang, Z. 2013. Genetic Diversity of *Hypericum perforatum* Collected from the Qinling Mountains of China. Biochemical Systematics and Ecology, 50: 232–239.

Hesami, M., Adamek, K., Pepe, M., & Jones, A. M. P. 2023. Effect of Explant Source on Phenotypic Changes of *In Vitro* Grown Cannabis Plantlets over Multiple Subcultures. Biology, 12(3): 443.

Hesami, M., Baiton, A., Alizadeh, M., Pepe, M., Torkamaneh, D., & Jones, A. M. P. 2021. Advances and Perspectives in Tissue Culture and Genetic Engineering of Cannabis. International Journal of Molecular Sciences, 22(11): 5671.

Hoda, E. 2013. *In vitro* Regeneration and Somaclonal Variation of *Catharanthus roseus* Don. Using Leaf and Internodal Explants. Alexandria Science Exchange Journal, 34: 452–459.

Holme, I. B., Gregersen, P. L., & Brinch-Pedersen, H. 2019. Induced Genetic Variation in Crop Plants by Random or Targeted Mutagenesis: Convergence and Differences. Frontiers in Plant Science, 10: 1468.

Huang, J. Q., Li, D. M., Tian, X., Lin, J. L., Yang, L., Xu, J. J., & Fang, X. 2021. Side Products of Recombinant Amorpha-4, 11-Diene Synthase and Their Effect on Microbial Artemisinin Production. Journal of Agricultural and Food Chemistry, 69(7): 2168–2178.

Hubbard, L., McSteen, P., Doebley, J., & Hake, S. 2002. Expression Patterns and Mutant Phenotype of Teosinte Branched1 Correlate with Growth Suppression in Maize and Teosinte. Genetics, 162(4): 1927–1935.

Ikeuchi, M., Sugimoto, K., & Iwase, A. 2013. Plant Callus: Mechanisms of Induction and Repression. Plant Cell, 25(113): 116053.

Ikram, N. K., & Simonsen, H. T. 2017. A Review of Biotechnological Artemisinin Production in Plants. Frontiers in Plant Science, 8: 1966.

Ioannidis, K., Tomprou, I., Mitsis, V., & Koropouli, P. 2022. Genetic Evaluation of *in Vitro* Micropropagated and Regenerated Plants of *Cannabis sativa* L. Using SSR Molecular Markers. Plants, 11(19): 2569.

Jablonka, E., & Raz, G. 2009. Transgenerational Epigenetic Inheritance: Prevalence, Mechanisms, and Implications for the Study of Heredity and Evolution. The Quarterly Review of Biology, 84(2): 131–176.

Jain, M., Brar, D. S., Ahloowalia, B. S., & Gichner, T. 1998. Somaclonal Variation and Induced Mutations In Crop Improvement. Biologia Plantarum, 41(3): 414.

Jain, R., & Dixit, V. K. 2010. Effect of Miconazole and Terbinafine on Artemisinin Content of Shooty Teratoma of *Artemisia annua*. Natural Product Communications, 5(2): 1934578X1000500203.

Jaligot, E., Adler, S., Debladis, E., Beulé, T., Richaud, F., Ilbert, P., ... Rival, A. 2011. Epigenetic Imbalance and the Floral Developmental Abnormality of the *in vitro*-Regenerated Oil Palm *Elaeis guineensis*. Annals of Botany, 108(8): 1453–1462.

Jaligot, E., Beulé, T., Baurens, F. C., Billotte, N., & Rival, A. 2004. Search for Methylation-Sensitive Amplification Polymorphisms Associated with the "Mantled" Variant Phenotype in Oil Palm (*Elaeis guineensis* Jacq.). Genome, 47(1): 224–228.

Jaligot, E., Beulé, T., & Rival, A. 2002. Methylation-Sensitive RFLPs: Characterisation of Two Oil Palm Markers Showing Somaclonal Variation-Associated Polymorphism. Theoretical and Applied Genetics, 104: 1263–1269.

Jevremović, S., Subotić, A., Miljković, D., Trifunović, M., Petrić, M., & Cingel, A. 2012. Clonal Fidelity of Chrysanthemum Cultivars after Long Term Micropropagation by Stem Segment Culture. In VII International Symposium on *In Vitro* Culture and Horticultural Breeding 961 (pp. 211–216).

Johnson, L., Mollah, S., Garcia, B. A., Muratore, T. L., Shabanowitz, J., Hunt, D. F., & Jacobsen, S. E. 2004. Mass Spectrometry Analysis of Arabidopsis Histone H3 Reveals Distinct Combinations of Post-Translational Modifications. Nucleic Acids Research, 32: 6511–6518.

Jullien, P. E., & Berger, F. 2010. DNA Methylation Reprogramming During Plant Sexual Reproduction? Trends in Genetics, 26(9): 394–399.

Jung, H., Van, K., Kim, M., & Lee, S. 2004. Identification of DNA Variations Using AFLP and SSR Markers in Soybean Somaclonal Variants. The Korean Journal of Crop Science, 49(1): 69–72.

Kaeppler, S., & Phillips, R. 1993. Tissue Culture-Induced DNA Methylation Variation in Maize. Proceedings of the National Academy of Sciences of the United States of America, 90: 8773–8776.

Kaeppler, S. M., Kaeppler, H. F., & Rhee, Y. 2000. Epigenetic Aspects of Somaclonal Variation in Plants. Plant Gene Silencing, 59–68.

Kaplan, D. R. 1992. The Relationship of Cells to Organisms in Plants: Problem and Implications of an Organismal Perspective. International Journal of Plant Sciences, 153(3, Part 2): S28–S37.

Karp, A. 1994. Origins, Causes and Uses of Variation in Plant Tissue Cultures. Plant Cell and Tissue Culture, 139–151.

Keum, Y. S., Park, K. K., Lee, J. M., Chun, K. S., Park, J. H., Lee, S. K., ... Surh, Y. J. 2000. Antioxidant and Anti-Tumor Promoting Activities of the Methanol Extract of Heat-Processed Ginseng. Cancer Letters, 150(1): 41–48.

Khan, S., Saeed, B., & Kauser, N. 2011. Establishment of Genetic Fidelity of *In Vitro* Raised Banana Plantlets. Pakistan Journal of Botany, 43(1): 233–242.

Khosroushahi, A. Y., Valizadeh, M., Ghasempour, A., Khosrowshahli, M., Naghdibadi, H., Dadpour, M. R., & Omidi, Y. 2006. Improved Taxol Production by Combination of Inducing Factors in Suspension Cell Culture of *Taxus baccata*. Cell Biology International, 30(3): 262–269.

Kim, D. S., Kim, S. Y., Jeong, I. Y., Kim, J. B., Lee, G. J., Kang, S. Y., & Kim, W. 2009. Improvement of Ginsenoside Production by *Panax ginseng* Adventitious Roots Induced by γ-Irradiation. Biologia Plantarum, 53: 408–414.

Kim, D. S., Song, M., Kim, S. H., Jang, D. S., Kim, J. B., Ha, B. K., ... Jeong, I. Y. 2013. The Improvement of Ginsenoside Accumulation in *Panax ginseng* as a Result of γ-Irradiation. Journal of Ginseng Research, 37(3): 332.

Kiselev, K. V., Dubrovina, A. S., & Shumakova, O. A. 2013. DNA Mutagenesis in 2-and 20-yr-old *Panax ginseng* Cell Cultures. *In Vitro* Cellular & Developmental Biology-Plant, 49: 175–182.

Kiselev, K. V., Shumakova, O. A., & Tchernoded, G. K. 2011. Mutation of *Panax ginseng* Genes During Long-Term Cultivation of Ginseng Cell Cultures. Journal of Plant Physiology, 168(11): 1280–1285.

Koperdáková, J., Košuth, J., & Čellárová, E. 2007. Variation in the Content of Hypericins in Four Generations of Seed Progeny of *Hypericum perforatum* Somaclones. Journal of Plant Research, 120: 123–128.

Kour, G., Kour, B., Kaul, S., & Dhar, M. K. 2009. Genetic and Epigenetic Instability of Amplification-Prone Sequences of a Novel B Chromosome Induced by Tissue Culture in *Plantago lagopus* L. Plant Cell Reports, 28: 1857–1867.

Kouzarides, T. 2007. Chromatin Modifications and Their Function. Cell, 128: 693–705.

Krikorian, A. D., Irizarry, H., Cronauer-Mitra, S. S., & Rivera, E. 1993. Clonal Fidelity and Variation in Plantain (Musa AAB) Regenerated from Vegetative Stem and Floral Axis Tips *in vitro*. Annals of Botany, 71(6): 519–535.

Krishna, H., Alizadeh, M., Singh, D., Singh, U., Chauhan, N., Eftekhari, M., & Sadh, R. K. 2016. Somaclonal Variations and Their Applications in Horticultural Crops Improvement. 3 Biotech, 6: 1–18.

Kumari, A., Baskaran, P., Plačková, L., Omámiková, H., Nisler, J., Doležal, K., & Van Staden, J. 2018. Plant Growth Regulator Interactions in Physiological Processes for Controlling Plant Regeneration and *In Vitro* Development of *Tulbaghia simmleri*. Journal of Plant Physiology, 223: 65–71.

Kunitake, H., Koreeda, K., & Mii, M. 1995. Morphological and Cytological Characteristics of Protoplast-Derived Plants of Statice (*Limonium perezii* Hubbard). Scientia Horticulturae, 60(3-4): 305–312.

Kurhekar, J. V. 2021. İn preparation of Phytopharmaceuticals for the Management of Disorders. Ancient and Modern Practices in Phytomedicine.

Kuznetsova, O. I., Ash, O. A., & Gostimsky, S. A. 2006. The Effect of the Duration of Callus Culture on the Accumulation of Genetic Alterations in Pea *Pisum sativum* L. Russian Journal of Genetics, 42: 555–562.

Lalaleo, L., Testillano, P., Risueño, M. C., Cusidó, R. M., Palazon, J., Alcazar, R., & Bonfill, M. 2018. Effect of *in Vitro* Morphogenesis on the Production of Podophyllotoxin Derivatives in Callus Cultures of *Linum album*. Journal of Plant Physiology, 228: 47–58.

Landey, R. B., Cenci, A., Georget, F., Bertrand, B., Camayo, G., Dechamp, E., Herrera, J. C., Santoni, S., Lashermes, P., & Simpson, J. 2013. High Genetic and Epigenetic Stability in *Coffea arabica* Plants Derived from Embryogenic Suspensions and Secondary Embryogenesis as Revealed by AFLP, MSAP and the Phenotypic Variation Rate. PLoS One, 8: e56372.

Lange, U. C., & Schneider, R. 2010. What an Epigenome Remembers. Bioessays, 32(8): 659–668.

Larkin, P. J., & Scowcroft, W. R. 1981. Somaclonal Variation-A Novel Source of Variability from Cell Cultures for Plant Improvement. Theoretical and Applied Genetics, 60: 197–214.

Latzel, V., Zhang, Y., Karlsson Moritz, K., Fischer, M., & Bossdorf, O. 2012. Epigenetic Variation in Plant Responses to Defence Hormones. Annals of Botany, 110(7): 1423–1428.

Law, J. A., & Jacobsen, S. E. 2010. Establishing, Maintaining and Modifying DNA Methylation Patterns in Plants and Animals. Nature Reviews. Genetics, 11: 204.

Lee, J. W., Kim, J. U., Bang, K. H., Kim, D. H., Jo, I. H., & Park, Y. D. 2023. Efficient Micropropagation of Genetically Stable *Panax ginseng* Meyer by Somatic Embryogenesis. Agronomy, 13(4): 1139.

Lee, M., & Phillips, R. L. 1988. The Chromosomal Basis of Somaclonal Variation. Annual Review of Plant Physiology and Plant Molecular Biology, 39(1): 413–437.

Lee, O. N., Ak, G., Zengin, G., Cziáky, Z., Jekő, J., Rengasamy, K. R., … Sivanesan, I. 2020. Phytochemical Composition, Antioxidant Capacity, and Enzyme Inhibitory Activity in Callus, Somaclonal Variant, and Normal Green Shoot Tissues of *Catharanthus roseus* (L) G. Don. Molecules, 25(21): 4945.

Lindroth, A. M., Shultis, D., Jasencakova, Z., Fuchs, J., Johnson, L., Schubert, D., Patnaik, D., Pradhan, S., Goodrich, J., Schubert, I., Jenuwein, T., Khorasanizadeh, S., & Jacobsen, S. E. 2011. Dual Histone H3 Methylation Marks at Lysines 9 and 27 Required For Interaction with CHROMOMETHYLASE3. The EMBO Journal, 30: 1874.

Liu, X., Wang, J., Huang, E., Li, B., Zhang, S., Wang, W., … Lu, H. 2022. Metabolomics Analysis of Three Artemisia Species in the Tibet Autonomous Region of China. BMC Plant Biology, 22(1): 1–14.

LoSchiavo, F., Pitto, L., Giuliano, G., Torti, G., Nuti-Ronchi, V., Marazziti, D., … Terzi, M. 1989. DNA Methylation of Embryogenic Carrot Cell Cultures and Its Variations as Caused by Mutation, Differentiation, Hormones and Hypomethylating Drugs. Theoretical and Applied Genetics, 77: 325–331.

Lu, X., Chen, X., Mu, M., Wang, J., Wang, X., Wang, D., Yin, Z., Fan, W., Wang, S., & Guo, L. 2016. Genome-Wide Analysis of Long Noncoding RNAs and Their Responses to Drought Stress in Cotton (*Gossypium hirsutum L.*). PLoS One, 11: e0156723.

Mahmood, I., Razzaq, A., Ashraf, M., Ahmad Hafiz, I., Kaleem, S., Qayyum, A., & Ahmad, M. 2012. *In Vitro* Selection of Tissue Culture Induced Somaclonal Variants of Wheat for Drought Tolerance. Journal of Agricultural Research (03681157), 50(2).

Makarevitch, I., Eichten, S. R., Briskine, R., Waters, A. J., Danilevskaya, O. N., Meeley, R. B., Myers, C. L., Vaughn, M. W., & Springer, N. M. 2013. Genomic Distribution of Maize Facultative Heterochromatin Marked by Trimethylation of H3K27. The Plant Cell, 25(3): 780–793.

Malik, S., Cusidó, R. M., Mirjalili, M. H., Moyano, E., Palazón, J., & Bonfill, M. 2011. Production of The Anticancer Drug Taxol in *Taxus baccata* Suspension Cultures: A Review. Process Biochemistry, 46(1): 23–34.

Manzoor, A., Ahmad, T., Bashir, M. A., Hafiz, I. A., & Silvestri, C. 2019. Studies on Colchicine Induced Chromosome Doubling for Enhancement of Quality Traits In Ornamental Plants. Plants, 8(7): 194.

Matkowski, A. 2008. Plant *In Vitro* Culture For The Production of Antioxidants-A Review. Biotechnology Advances, 26(6): 548–560.

McClintock, B. 1950. The Origin and Behavior Of Mutable Loci in Maize. Proceedings of the National Academy of Sciences, 36(6): 344–355.

McPheeters, K., & Skirvin, R. M. 1983. Histogenic Layer Manipulation in Chimeral 'Thornless Evergreen' trailing Blackberry. Euphytica, 32(2): 351–360.

Mesa, L. E., Lutgen, P., Velez, I. D., Segura, A. M., & Robledo, S. M. 2015. *Artemisia annua L.*, Potential Source of Molecules with Pharmacological Activity in Human Diseases. American Journal of Phytomedicine and Clinical Therapeutics, 3(5): 436–450.

Miguel, C., & Marum, L. 2011. An Epigenetic View of Plant Cells Cultured *in Vitro*: Somaclonal Variation and Beyond. Journal of Experimental Botany, 62(11): 3713–3725.

Mirouze, M., & Paszkowski, J. 2011. Epigenetic Contribution to Stress Adaptation in Plants. Current Opinion in Plant Biology, 14: 267–274.

Mofokeng, M. M., P du Plooy, C., Araya, H. T., Amoo, S. O., Mokgehle, S. N., Pofu, K. M., & Mashela, P. W. 2022. Medicinal Plant Cultivation for Sustainable Use and Commercialisation of High-Value Crops. South African Journal of Science, 118(7-8): 1–7.

Mujib, A., Banerjee, S., & Dev Ghosh, P. 2007. Callus Induction, Somatic Embryogenesis and Chromosomal Instability in Tissue Culture Raised Hippeastrum (*Hippeastrum hybridum* cv. United Nations). Propag Ornam Plants, 7: 169–174.

Namdeo, A. G. 2018. Cultivation of Medicinal and Aromatic Plants. In Natural Products and Drug Discovery (pp. 525–553). Elsevier.

Neelakandan, A. K., & Wang, K. 2012. Recent Progress in the Understanding of Tissue Culture-Induced Genome Level Changes in Plants and Potential Applications. Plant Cell Reports, 31: 597–620.

Ngezahayo, F. 2018. Somaclonal Variations and Their Applications in Medicinal Plant Improvement. Biotechnological Approaches for Medicinal and Aromatic Plants: Conservation, Genetic Improvement and Utilization, 503–519.

Nhut, D. T. 2022. General Information: Some Aspects of Plant Tissue Culture. In Plant Tissue Culture: New Techniques and Application in Horticultural Species of Tropical Region (pp. 1–23). Singapore: Springer Singapore.

Niazian, M. 2019. Application of Genetics and Biotechnology for Improving Medicinal Plants. Planta, 249: 953–973.

Nic-Can, G. I., López-Torres, A., Barredo-Pool, F., Wrobel, K., Loyola-Vargas, V. M., Rojas-Herrera, R., & De-la-Peña, C. 2013. New Insights into Somatic Embryogenesis: *LEAFY COTYLEDON1*, *BABY BOOM1* and *WUSCHEL-RELATED HOMEOBOX4* are epigenetically regulated in *Coffea canephora*. PLoS One, 8: e72160.

Nicola, S., Fontana, E., & Hoeberechts, J. 2003. Effects of Rooting Products on Medicinal and Aromatic Plant Cuttings. In VI International Symposium on Protected Cultivation in Mild Winter Climate: Product and Process Innovation 614 (pp. 273–278).

Nobakht, S. Z., Akaberi, M., Mohammadpour, A., Moghadam, A. T., & Emami, A. 2022. *Hypericum perforatum*: Traditional Uses, Clinical Trials, and Drug Interactions. Iranian Journal of Basic Medical Sciences, 25(9): 1045.

Nogueira, G. F., Luis, Z. G., Salles, L. A., Pasqual, M., & Scherwinski-Pereira, J. E. 2022. High-Efficiency Organogenesis and Evaluation of the Regenerated Plants by Flow Cytometry of a Broad Range of *Saccharum* spp. Hybrids. Biologia, 77(11): 3265–3278.

Noormohammadi, Z., Asghari-Mooneghi, N., & Farahani, F. 2020. Effect of Gamma Radiation on Morphological and Genetic Variation in Regenerated Plantlets *Catharanthus roseus* L.(G) Don. Genetika, 52(1): 15–28.

Nwauzoma, A. B., & Jaja, E. T. 2013. A Review of Somaclonal Variation in Plantain (Musa spp): Mechanisms and Applications. Journal of Applied Biosciences, 67: 5252–5260.

Oksman-Caldentey, K. M. 1998. Somaclonal Variation in Solanaceous Medicinal Plants. Somaclonal Variation and Induced Mutations in Crop Improvement, 233–251.

Ooi, S. E., Novák, O., Doležal, K., Ishak, Z., & Ong-Abdullah, M. 2013. Cytokinin Differences *in Vitro* Cultures and Inflorescences from Normal and Mantled Oil Palm (*Elaeis guineensis* Jacq.). Journal of Plant Growth Regulation, 32: 865–874.

Orbović, V., Ćalović, M., Viloria, Z., Nielsen, B., Gmitter, F. G., Castle, W. S., & Grosser, J. W. 2008. Analysis of Genetic Variability in Various Tissue Culture-Derived Lemon Plant Populations Using RAPD and Flow Cytometry. Euphytica, 161: 329–335.

Orton, T. J. 1984. Somaclonal Variation: Theoretical and Practical Considerations. In Gene Manipulation in Plant Improvement: 16th Stadler Genetics Symposium (pp. 427–468). Springer US.

Page, S. 2021. A Blunt Re-evaluation of *Cannabis sativa* Micropropagation Practices (Doctoral dissertation, University of Guelph).

Pandey, N., Meena, R. P., Rai, S. K., & Pandey-Rai, S. 2016. *In Vitro* Generation Of High Artemisinin Yielding Salt Tolerant Somaclonal Variant and Development of SCAR Marker in *Artemisia annua* L. Plant Cell, Tissue and Organ Culture (PCTOC), 127: 301–314.

Park, S.-Y., Murthy, H., Chakrabarthy, D., & Paek, K.-Y. 2009. Detection of Epigenetic Variation in Tissue-Culture-Derived Plants of Doritaenopsis by Methylation-Sensitive Amplification Polymorphism (MSAP) Analysis. *In Vitro* Cellular & Developmental Biology. Plant, 45: 104–108.

Patel, M. K., Kumar, M., Li, W., Luo, Y., Burritt, D. J., Alkan, N., & Tran, L. S. P. 2020. Enhancing Salt Tolerance of Plants: From Metabolic Reprogramming to Exogenous Chemical Treatments and Molecular Approaches. Cells, 9(11): 2492.

Pawełkowicz, M. E., Skarzyńska, A., Mróz, T., Bystrzycki, E., & Pląder, W. 2021. Molecular Insight into Somaclonal Variation Phenomena from Transcriptome Profiling of Cucumber (*Cucumis sativus* L.) Lines. Plant Cell, Tissue and Organ Culture (PCTOC), 145: 239–259.

Peredo, E. L., Revilla, M. Á., & Arroyo-García, R. 2006. Assessment of Genetic and Epigenetic Variation in Hop Plants Regenerated from Sequential Subcultures of Organogenic Calli. Journal of Plant Physiology, 163(10): 1071–1079.

Perronnet, C., Denécheau, C., Rideau, M., & Chénieux, J. C. 1988. Somaclonal Variations for Indole Alkaloid Accumulation in Protoclones Derived from an Habituated Strain of *Catharanthus Roseus*. In Progress in Plant Protoplast Research: Proceedings of the 7th International Protoplast Symposium, Wageningen, the Netherlands, December 6–11, 1987 (pp. 397–398). Springer Netherlands.

Peschke, V. M., Phillips, R. L., & Gengenbach, B. G. 1987. Discovery of Transposable Element Activity among Progeny of Tissue Culture–Derived Maize Plants. Science, 238(4828): 804–807.

Petrovska, B. B. 2012. Historical Review of Medicinal Plants' Usage. Pharmacognosy Reviews, 6(11): 1.

Phillips, R. L., Kaeppler, S. M., & Olhoft, P. 1994. Genetic Instability of Plant Tissue Cultures: Breakdown of Normal Controls. Proceedings of the National Academy of Sciences, 91(12): 5222–5226.

Pietsch, G. M., & Anderson, N. O. 2007. Epigenetic Variation in Tissue Cultured *Gaura lindheimeri*. Plant Cell, Tissue and Organ Culture, 89: 91–103.

Pijut, P. M., Beasley, R. R., Lawson, S. S., Palla, K. J., Stevens, M. E., & Wang, Y. 2012. *In Vitro* Propagation of Tropical Hardwood Tree Species-A Review (2001–2011). Propagation of Ornamental Plants, 12(1): 25–51.

Pischke, M. S., Huttlin, E. L., Hegeman, A. D., & Sussman, M. R. 2006. A Transcriptome-Based Characterization of Habituation in Plant Tissue Culture. Plant Physiology, 140(4): 1255–1278.

Poswal, F. S., Russell, G., Mackonochie, M., MacLennan, E., Adukwu, E. C., & Rolfe, V. 2019. Herbal Teas and Their Health Benefits: A Scoping Review. Plant Foods for Human Nutrition, 74: 266–276.

Predieri, S. 2001. Mutation Induction and Tissue Culture in Improving Fruits. Plant Cell, Tissue and Organ Culture, 64(2-3): 185–210.

Qu, Y., Safonova, O., & De Luca, V. 2019. Completion of the Canonical Pathway for Assembly of Anticancer Drugs Vincristine/Vinblastine in *Catharanthus roseus*. The Plant Journal, 97(2): 257–266.

Rai, M. K. 2022. Somaclonal Variation in Improvement of Agricultural Crops: Recent Progress. Agricultural Biotechnology: Latest Research and Trends, 129–146.

Rakyan, V. K., Chong, S., Champ, M. E., Cuthbert, P. C., Morgan, H. D., Luu, K. V., & Whitelaw, E. 2003. Transgenerational Inheritance Of Epigenetic States At The Murine Axin Fu Allele Occurs After Maternal And Paternal Transmission. Proceedings of the National Academy of Sciences, 100(5): 2538–2543.

Rakyan, V. K., & Whitelaw, E. 2003. Transgenerational Epigenetic Inheritance. Current Biology, 13(1): R6.

Rana, S., Dhar, N., Bhat, W. W., Razdan, S., Khan, S., Dhar, R. S., ... & Lattoo, S. K. 2012. A 12-Deoxywithastramonolide-Rich Somaclonal Variant in *Withania somnifera* (L.) Dunal—Molecular Cytogenetic Analysis and Significance as a Chemotypic Resource. *In Vitro* Cellular & Developmental Biology-Plant, 48: 546–554.

Rival, A. 2000. Somatic Embryogenesis in Oil Palm. In Eds S. M. Jain, P. K. Gupta, R. J. Newton, Somatic Embryogenesis in Woody Plants. Forestry Sciences (Vol. 67). pp. 249–290. Dordrecht, Netherlands: Springer.

Rival, A., Ilbert, P., Labeyrie, A., Torres, E., Doulbeau, S., Personne, A., ... Jaligot, E. 2013. Variations in Genomic DNA Methylation during the Long-Term *In Vitro* Proliferation of Oil Palm Embryogenic Suspension Cultures. Plant Cell Reports, 32: 359–68.

Rodriguez-Enriquez, J., Dickinson, H. G., & Grant-Downton, R. T. 2011. MicroRNA Misregulation: An Overlooked Factor Generating Somaclonal Variation? Trends in Plant Science, 16(5): 242–248.

Rodríguez López, C. M., Wetten, A. C., & Wilkinson, M. J. 2010. Progressive Erosion of Genetic and Epigenetic Variation in Callus-Derived Cocoa (*Theobroma cacao*) Plants. New Phytologist, 186: 856–868.

Roostika, I., Khumaida, N., & Ardie, S. W. 2015. RAPD Analysis to Detect Somaclonal Variation of Pineapple *in Vitro* Cultures during Micropropagation. BIOTROPIA-The Southeast Asian Journal of Tropical Biology, 22(2): 109–119.

Rosa, W. S., Martins, J. P. R., Rodrigues, E. S., de Almeida Rodrigues, L. C., Gontijo, A. B. P. L., & Falqueto, A. R. 2018. Photosynthetic Apparatus Performance in Function of the Cytokinins Used during the *In Vitro* Multiplication of *Aechmea blanchetiana* (Bromeliaceae). Plant Cell, Tissue and Organ Culture (PCTOC), 133: 339–350.

Roux, N. S. 2004. Mutation Induction in Musa Review. In Banana Improvement: Cellular, Molecular Biology, and Induced Mutations. Proceedings of a Meeting Held in Leuven, Belgium, 24-28 September 2001 (pp. 23–32). Science Publishers, Inc., USA.

Roy, B., & Mandal, A. B. 2005. Towards Development of Al-Toxicity Tolerant Lines in Indica Rice by Exploiting Somaclonal Variation. Euphytica, 145(3): 221–227.

Sahijram, L., Soneji, J. R., & Bollamma, K. T. 2003. Analyzing Somaclonal Variation in Micropropagated Bananas (Musa spp.). In Vitro Cellular & Developmental Biology-Plant, 39: 551–556.

Sales, E. K., & Butardo, N. G. 2014. Molecular Analysis of Somaclonal Variation in Tissue Culture Derived Bananas Using MSAP and SSR Markers. International Journal of Biotechnology and Bioengineering, 8(6): 615–622.

Saze, H. 2008. Epigenetic Memory Transmission through Mitosis and Meiosis in Plants. Seminars in Cell & Developmental Biology, 19(6): 527–536.

Schellenbaum, P., Mohler, V., Wenzel, G., & Walter, B. 2008. Variation in DNA Methylation Patterns of Grapevine Somaclones (*Vitis vinifera* L.). BMC Plant Biology, 8: 78.

Sharma, A., Sharma, A., Thakur, S., Mutreja, V., & Bhardwaj, G. 2022. A Brief Review on Phytochemistry and Pharmacology of *Taxus baccata* L. Materials Today: Proceedings, 48: 1569–1574.

Shearman, J. R., Jantasuriyarat, C., Sangsrakru, D., Yoocha, T., Vannavichit, A., Tragoonrung, S., & Tangphatsornruang, S. 2013. Transcriptome Analysis of Normal and Mantled Developing Oil Palm Flower and Fruit. Genomics, 101(5): 306–312.

Shen, L., Xu, J., Hu, H. Y., Hu, X., Zhang, Y. K., Li, X. W., … Chen, S. L. 2019. Correlative Analysis of Agronomic Traits and Quality of *Panax ginseng* Saponins Cultivation in Farmland. China Journal of Chinese Materia Medica, 44(23): 5124–5128.

Shibata, S. 2001. Chemistry and Cancer Preventing Activities of Ginseng Saponins and Some Related Triterpenoid Compounds. Journal of Korean Medical Science, 16(Suppl): S28–S37.

Silva, T. C. R., & Carvalho, C. R. 2014. Vertical Heterogeneity of DNA Ploidy Level Assessed by Flow Cytometry in Calli of *Passiflora cincinnata*. In Vitro Cellular & Developmental Biology-Plant, 50: 158–165.

Simonović, A. D., M. Trifunović-Momčilov, M., Filipović, B. K., Marković, M. P., Bogdanović, M. D., & Subotić, A. 2020. Somatic Embryogenesis in *Centaurium erythraea* Rafn—Current Status and Perspectives: A Review. Plants, 10(1): 70.

Singh, M. K. 2021. An Overview on Soma Clonal Variation. ACADEMICIA: An International Multidisciplinary Research Journal, 11(10): 1187–1194.

Singh, P., Guleri, R., Angurala, A., Kaur, K., Kaur, K., Kaul, S. C., … Pati, P. K. 2017. Addressing Challenges to Enhance the Bioactives of Withania Somnifera through Organ, Tissue, and Cell Culture Based Approaches. BioMed Research International, 2017: 3278494.

Singh, G., Tiwari, M., Singh, S. P., Singh, S., Trivedi, P. K., & Misra, P. 2016. Silencing of Sterol Glycosyltransferases Modulates the with Anolide Biosynthesis and Leads to Compromised Basal Immunity of *Withania somnifera*. Scientific Reports, 6(1): 25562.

Siragusa, M., Carra, A., Salvia, L., Puglia, A. M., De Pasquale, F., & Carimi, F. 2007. Genetic Instability in Calamondin (*Citrus madurensis* Lour.) Plants Derived from Somatic Embryogenesis Induced by Diphenylurea Derivatives. Plant Cell Reports, 26: 1289–1296.

Smulders, M. J. M., & De Klerk, G. J. 2011. Epigenetics in Plant Tissue Culture. Plant Growth Regulation, 63: 137–146.

Smýkal, P., Valledor, L., Rodriguez, R., & Griga, M. 2007. Assessment of Genetic and Epigenetic Stability in Long-Term *In Vitro* Shoot Culture of Pea (*Pisum sativum L.*). Plant Cell Reports, 26: 1985–1998.

Sovilj, M., & Spasojević, M. 2001. Production and Application of Essential Oils from the Domestic Medicinal Plant. Časopis za procesnu tehniku i energetiku u poljoprivredi/PTEP, 5(1-2): 34–38.

Sun, S., Zhong, J., Li, S., & Wang, X. 2013. Tissue Culture-Induced Somaclonal Variation of Decreased Pollen Viability in Torenia (*Torenia fournieri* Lind.). Botanical Studies, 54(1): 1–7.

Uddenberg, D., Valladares, S., Abrahamsson, M., Sundström, J. F., SundåsLarsson, A., & von Arnold, S. 2011. Embryogenic Potential and Expression of Embryogenesis-Related Genes in Conifers Are Affected by Treatment with a Histone Deacetylase Inhibitor. BMC Proceedings, 5: P151.

Udomchalothorn, T., Plaimas, K., Comai, L., Buaboocha, T., & Chadchawan, S. 2014. Molecular Karyotyping and Exome Analysis of Salt-Tolerant Rice Mutant from Somaclonal Variation. The Plant Genome, 7(3): plantgenome2014-04.

Van Harten, A. M. 1998. Mutation Breeding: Theory and Practical Applications (Vol. 1). Cambridge University Press.

Vanisree, M., Lee, C. Y., Lo, S. F., Nalawade, S. M., Lin, C. Y., & Tsay, H. S. 2004. Studies on the Production of Some Important Secondary Metabolites from Medicinal Plants by Plant Tissue Cultures. Botanical Bulletin of Academia Sinica, 45(1): 1–22.

Verhoeven, K. J., & Preite, V. 2014. Epigenetic Variation in Asexually Reproducing Organisms. Evolution, 68(3): 644–655.

Wang, H., Ma, C., Li, Z., Ma, L., Wang, H., Ye, H., … Liu, B. 2010. Effects of Exogenous Methyl Jasmonate on Artemisinin Biosynthesis and Secondary Metabolites in *Artemisia annua* L. Industrial Crops and Products, 31(2): 214–218.

Wang, M., Park, C., Wu, Q., & Simon, J. E. 2005. Analysis of Artemisinin in *Artemisia annua* L. by LC-MS with Selected Ion Monitoring. Journal of Agricultural and Food Chemistry, 53(18): 7010–7013.

Wang, X. 2010. Determination of Hypericin Content in *Hypericum perforatum* L. by HPLC. Medicinal Plant, 1(9): 79–84.

Wang, X., Wu, R., Lin, X., Bai, Y., Song, C., Yu, X., ... Liu, B. 2013. Tissue Culture-Induced Genetic and Epigenetic Alterations in Rice Pure-Lines, F1 Hybrids and Polyploids. BMC Plant Biology, 13: 1–12.

Wang, W., Xu, J., Fang, H., Li, Z., & Li, M. 2020. Advances and Challenges In Medicinal Plant Breeding. Plant Science, 298: 110573.

Wang, Y., Wang, F., Zhai, H., & Liu, Q. 2007. Production of a Useful Mutant by Chronic Irradiation in Sweet Potato. Scientia Horticulturae, 111(2): 173–178.

Weathers, P. J., Arsenault, P. R., Covello, P. S., McMickle, A., Teoh, K. H., & Reed, D. W. 2011. Artemisinin Production in *Artemisia annua*: Studies in Planta and Results of a Novel Delivery Method for Treating Malaria and Other Neglected Diseases. Phytochemistry Reviews, 10: 173–183.

Weckx, S., Inzé, D., & Maene, L. 2019. Tissue Culture of Oil Palm: Finding the Balance between Mass Propagation and Somaclonal Variation. Frontiers in Plant Science, 10: 722.

Wiart, C. 2007. Ethnopharmacology of Medicinal Plants: Asia and the Pacific. Springer Science & Business Media.

Williams, L., Zhao, J., Morozova, N., Li, Y., Avivi, Y., & Grafi, G. 2003. Chromatin Reorganization Accompanying Cellular Dedifferentiation Is Associated with Modifications of Histone H3, Redistribution of HP1, and Activation of E2F-Target Genes. Developmental Dynamics, 228: 113–120.

Wongsamuth, R., & Doran, P. M. 1997. Production of Monoclonal Antibodies by Tobacco Hairy Roots. Biotechnology and Bioengineering, 54(5): 401–415.

Xiang, L., Zhu, S., Zhao, T., Zhang, M., Liu, W., Chen, M., ... Liao, Z. 2015. Enhancement of Artemisinin Content and Relative Expression of Genes of Artemisinin Biosynthesis in *Artemisia annua* by Exogenous MeJA Treatment. Plant Growth Regulation, 75: 435–441.

Xing, S. H., Guo, X. B., Wang, Q., Pan, Q. F., Tian, Y. S., Liu, P., ... Tang, K. X. 2011. Induction and Flow Cytometry Identification of Tetraploids From Seed-Derived Explants Through Colchicine Treatments in *Catharanthus roseus* (L.) G. Don. Journal of Biomedicine and Biotechnology, 2011: 793198.

Yang, W. R., Zhang, Q. X., Pan, H. T., & Sun, M. 2010. *In Vitro* Regeneration of *Lilium Tsingtauense* Gilg. and Analysis of Genetic Variability in Micropropagated Plants Using RAPD and ISSR Techniques. Propagation of Ornamental Plants, 10(2): 59–66.

Yousefian, Z., Hosseini, B., Rezadoost, H., Palazón, J., & Mirjalili, M. H. 2018. Production of the Anticancer Compound Withaferin a From Genetically Transformed Hairy Root cultures of *Withania somnifera*. Natural Product Communications, 13(8): 1934578X1801300806.

Yu, X., Li, X., Zhao, X., Jiang, L., Miao, G., Pang, J., Qi, X., & Liu, B. 2011. Tissue Culture-Induced Genomic Alteration in Maize (*Zea mays*) Inbred Lines and F1 hybrids. Annals of Applied Biology, 158: 237–247.

Yu, Z., Haberer, G., Matthes, M., Rattei, T., Mayer, K. F., Gierl, A., & Torres-Ruiz, R. A. 2010. Impact of Natural Genetic Variation on the Transcriptome of Autotetraploid Arabidopsis Thaliana. Proceedings of the National Academy of Sciences of the United States of America, 107: 17809–17814.

Yudoh, K., van Trieu, N., Nakamura, H., Hongo-Masuko, K., Kato, T., & Nishioka, K. 2005. Potential Involvement of Oxidative Stress in Cartilage Senescence and Development of Osteoarthritis: Oxidative Stress Induces Chondrocyte Telomere Instability and Downregulation of Chondrocyte Function. Arthritis Research & Therapy, 7: 1–12.

Zayova, E., Vassilevska-Ivanova, R., Kraptchev, B., & Stoeva, D. 2010. Somaclonal Variations through Indirect Organogenesis in Eggplant (*Solanum melongena* L.). Biological Diversity and Conservation, 3(3): 1–5.

Zhang, K., Sridhar, V. V., Zhu, J., Kapoor, A., & Zhu, J.-K. 2007. Distinctive Core Histone Post-Translational Modification Patterns in Arabidopsis Thaliana. PLoS One, 2: e1210.

Zhang, N., Yang, H., Han, T., Kim, H. S., & Marcelis, L. F. 2022. Towards Greenhouse Cultivation of *Artemisia annua*: The Application of LEDs in Regulating Plant Growth and Secondary Metabolism. Frontiers in Plant Science, 13: 1099713.

8 Plant Tissue Culture as an Effective Tool for Virus-Free Medicinal Plant Production

Sonalika Roy

8.1 INTRODUCTION

The demand for medicinal plants and their derived products has increased globally, thanks to the growing awareness about their therapeutic properties and potential benefits in the healthcare and pharmaceutical industries (Kumar et al., 2011). These plants have been used for centuries to treat various ailments and continue to play a vital role in both traditional and modern medicine (Jamshidi-Kia et al., 2017). However, with the rising demand for these plants come several challenges such as overharvesting, habitat destruction, and contamination with plant viruses, which pose significant risks (Rahimi et al., 2023).

Plant viruses pose a substantial threat to the health and productivity of medicinal plants. They can cause a range of symptoms like stunted growth, leaf deformities, discoloration, and reduced yield. Ultimately, this affects the quality and quantity of the valuable compounds found within these plants. Moreover, the presence of viruses can jeopardize the safety of the final medicinal products by potentially causing adverse effects on consumers. To address these concerns and ensure sustainable production of high-quality medicinal plants, researchers and growers have turned towards plant tissue culture as an effective tool for virus-free propagation (Lal et al., 2015).

Plant tissue culture, also known as *in vitro* culture, is a method that involves growing plant cells, tissues, or organs in a controlled environment without any contamination (Ajay et al., 2020). This is typically done in nutrient-rich medium along with the addition of growth regulators. Although this technique was initially introduced in the early 20th century, it has since evolved and expanded significantly. Plant tissue culture offers several advantages compared to traditional propagation methods which include rapid multiplication, production of uniform plantlets, and the ability to generate virus-free plants (Walkey, 1991; Pant, 2016; Viswanath et al., 2023). The demand for virus-free medicinal plants is growing due to concerns regarding safety and quality. Plant tissue culture offers the production of virus-free medicinal plants along with maintaining genetic stability and preserving the essential phytochemical profile of the regenerated plants. The different strategies to eliminate viruses include thermotherapy, meristem culture, chemotherapy, cryotherapy, and somaclonal variations, to name a few. Besides, it is important to use molecular techniques for virus detection and identification in medicinal plants. Polymerase chain reaction (PCR)–based assays, serological methods, next-generation sequencing, and other methods have proven to be invaluable tools for accurate and efficient virus diagnosis. These techniques enable researchers to screen plant materials effectively and select suitable candidates for tissue culture–based propagation.

The main objective of this chapter is to thoroughly analyse and consolidate the current state of research and applications of plant tissue culture as an effective tool for producing virus-free medicinal plants. It delves into various methodologies, techniques, and challenges associated with this innovative approach while highlighting its potential for sustainable medicinal plant production, supported by successful studies. The comprehensive analysis will contribute to our understanding

DOI: 10.1201/b23374-8

of the challenges of production of virus-free medicinal plants, the significance of adopting tissue culture methods to meet the challenges and ensure a sustainable supply of good-quality medicinal resources that offer safe and standardized herbal products for global use.

8.2 PLANT VIRUSES

Plant viruses primarily infect plants causing harmful effects such as reduced crop yields, quality deterioration, and economic losses (Naidu and Hughes, 2003; Nicaise, 2014; Jones and Naidu, 2019). Similar to viruses that affect animals and humans, plant viruses are parasitic in nature and spread by relying on the hosts' cellular machinery for replication (Gelderblom, 1996). Understanding the characteristics of plant viruses is essential for developing control strategies to ensure global food security (Zaitlin and Palukaitis, 2000). In this exploration, the fundamental aspects of plant viruses are discussed including their structure, replication mechanisms, modes of transmission, impact on plant health, and current management methods.

Plant viruses are small organisms consisting of genetic material enclosed within a protein coat known as a capsid. The genetic material can be either RNA or DNA. The capsid is formed by repeating protein subunits that create a shell around the viral genome. The wide range of sizes and complexities in genomes contributes to the extensive variety of plant viruses observed in different plant species (Lucas and Knipe, 2001; Zanotti and Grinzato, 2021). The replication process of plant viruses involves interactions between the viral genome and the cellular components of the host plant. Once inside the host cell, the viral genome takes control of the machinery to produce viral proteins and replicate its genetic material. This results in the generation of new viral particles, which eventually spread to neighbouring plant cells as well as throughout the entire plant system. The speed and efficiency of replication vary depending on factors such as virus species and characteristics of the host plant (Medina-Puche and Lozano-Duran, 2019; Souza and Carvalho, 2019; Wang, 2021).

Plant viruses have developed methods for infecting new hosts. These transmission modes can be broadly categorized into vector-borne transmission, transmission through seeds, soil-borne transmission, and mechanical transmission (Ertunc, 2020). Vector-borne transmission is through insects like aphids, thrips, and whiteflies, which serve as vectors for transmitting plant viruses during feeding activities and subsequently transmitting viruses to healthy plants. Some plant viruses can spread through seeds, causing new infections in the next generation of plants, while certain plant viruses can survive in the soil for long periods and infect new plants through their root systems. Plant viruses can also be transmitted mechanically when infected parts of a plant come into contact with healthy parts often due to human activities like pruning and grafting.

Infected plants often display symptoms such as mosaic patterns like yellowing (chlorosis), tissue death (necrosis), stunted growth, and other deformations. These symptoms occur because the viral invasion disrupts cellular processes in the plants. In addition, plant viruses weaken the hosts' response, making it more susceptible to secondary infections by other pathogens. Ultimately a buildup of infections can result in reduced plant growth, poor fruit development, and significant economic losses for farmers and growers (Jones, 2021; Jiang and Zhou, 2023).

Multiple factors play a role in the spread and prevalence of plant viruses in the natural environment. These factors include the presence of insects that carry the viruses, the number of host plants available, the genetic resistance of those plants to viruses, and environmental conditions that affect both the insect populations and how quickly the viruses replicate. Additionally, human activities like moving plant material across borders and poor sanitation practices can also play a role in spreading plant viruses to different regions (Islam et al., 2020; Gresikova, 2022). Managing plant viruses effectively is therefore crucial for maintaining crop production and ensuring food security.

Plant viruses pose a challenge to global agriculture as they impact the productivity, quality, and economic value of various medicinal plants. To effectively combat their spread and mitigate their effects on plant health, it is crucial to understand the structure, replication mechanisms, transmission modes, and management of these viruses. Continuous research and technological advancements will play a role in developing sustainable and efficient approaches to protect plant health and maintain the medicinal value of plants for growing populations.

8.3 PHYSIOLOGY AND MOLECULAR BIOLOGY OF PLANT VIRUSES

Plant viruses have a significant impact on the physiological functioning of plants. They disrupt cellular processes and cause diseases that can result in substantial agricultural losses. To effectively manage and control these diseases, it is crucial to understand the physiological and molecular aspects of plant virus interactions. In this section, the intricate relationship between plant viruses and their host plants is explored with a specific focus on the changes induced by viral infections and the molecular mechanisms that govern viral replication, movement, and evasion of host defence mechanisms (Souza et al., 2019).

Plant viruses profoundly impact the physiology of infected plants, leading to an array of symptoms and developmental alterations. The observed physiological changes in infected plants are described next (Zanardo et al., 2019):

- Plant viruses can interfere with chloroplast function, which directly affects photosynthesis. As a result, symptoms such as chlorosis (yellowing) and reduced growth may occur.
- Infections can hinder cell division and elongation processes within plants, ultimately resulting in stunted growth and an overall reduction in plant size.
- Infected plants often display morphological abnormalities like leaf deformation or irregular flower development.
- Plant hormones regulate the growth and development of plants, but when they are infected by viruses, these hormones can be manipulated, leading to changes in their growth patterns.
- One significant effect of infections is the disruption of carbohydrate metabolism within the plant. This disruption affects how energy is produced and stored in plants.
- Additionally, viral infections trigger stress responses within plants, which activate defence mechanisms and cause changes in gene expression.

The replication of plant viruses involves a complex series of molecular events that occur within infected plant cells. The key steps in viral replication (Hyodo and Okuno, 2020; Bwalya and Kim, 2023) are shown in Figure 8.1. First, the virus enters into host cells facilitated by viral proteins that interact with cell surface receptors. Once inside the host cell, the viral RNA or DNA serves as a template for synthesizing proteins using the cellular machinery of the host. Further, the viral genome undergoes replication to produce multiple copies of itself. The replication process involves the activity of viral enzymes and the recruitment of host factors, and newly synthesized viral proteins and replicated genomes assemble to form complete viral particles within infected cells. It should be noted that certain plant viruses can travel from the infected cells to neighbouring cells spreading systematically throughout the entire plant. This movement is made possible by movement proteins that interact with components of plant cells.

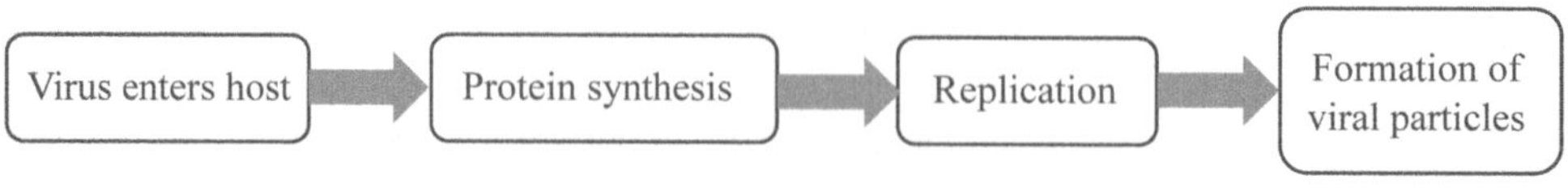

FIGURE 8.1 Key steps of viral replication.

Plant viruses have developed strategies to dodge the immune systems of plants and persist within their hosts. Some ways in which plant viruses counteract host defences include the following (Medina-Puche and Lozano-Duran, 2019; Wang et al., 2022b):

- RNA silencing is a defence mechanism against viruses in plants. Plant viruses produce proteins that can suppress this mechanism, thereby preventing the degradation of RNA and enabling efficient replication.
- Viruses can modify the expression of host genes to their advantage by inhibiting defence-related gene expression while promoting gene expression for their replication.
- Plant hormones play a crucial role in regulating immune responses. Viruses can interfere with hormone signalling pathways to suppress plant defence mechanisms.
- Some viruses can sequester their genomic RNA within specialized compartments, shielding it from host immune sensors.

The interaction between plant viruses and various host factors is a dynamic process that determines the outcome of viral infections and disease development. Key factors involved in these interactions include resistance genes, cell-to-cell movement proteins, RNA binding proteins, and RNA silencing machinery, as enumerated next (Garcia-Ruiz, 2019; Zhao and Li, 2021):

- Resistance (R) genes in plants can recognize viral proteins and initiate defence responses, making it crucial to understand these interactions for breeding virus-resistant plant varieties.
- Plant viruses encode movement proteins that facilitate the movement of particles between cells through plasmodesmata, allowing for systemic infection throughout the entire plant.
- Host RNA binding proteins play a role in viral replication and translation by interacting with viral RNA.
- The RNA silencing machinery in plants, acts as an antiviral defence system. However, viruses have developed proteins that can suppress this system to ensure their own survival.

The study of plant viruses is a captivating and multifaceted subject of research. It is crucial to comprehend the changes caused by viral infections and the intricate molecular mechanisms that drive viral replication and evasion of host defences. This understanding is vital in order to develop strategies for managing and controlling viral diseases in agriculture, horticulture, and forestry. Advancements in this field offer potential for creating medicinal plants that are resistant to viruses by innovative antiviral treatments and sustainable approaches, ensuring global food security. Ongoing research collaboration among plant virologists, molecular biologists, and agronomists will play a role in unravelling the intricate interactions between plant viruses and their host plants.

8.4 TYPES OF PLANT VIRUSES

It is crucial to understand the diversity of plant viruses for effective disease management and the development of strategies that ensure food security and sustainable agriculture (Akutse et al., 2020). The majority of known plant viruses are RNA viruses, while DNA viruses are less common. Double-stranded RNA (dsRNA) plant viruses, possessing a genome composed of double-stranded RNA molecules, also cause a variety of diseases in plants, affecting their growth, yield, and overall health. Some dsRNA plant viruses are of significant economic importance due to the damage they cause to crops. The dominant families of these different viruses affecting plants are listed in Table 8.1.

However, plant viruses can infect a broad spectrum of plant species, although they often display some level of specificity towards certain hosts. While some viruses have a limited range and only infect closely related plant species, others have a wide range and can infect various plant

TABLE 8.1
Families of Different Viruses Affecting Plants

Types of Viruses	Families
RNA viruses	Bromoviridae, Closteroviridae, Potyviridae, and Tombusviridae (Sabharwal, 2019; Ali, 2023).
DNA viruses	Geminiviridae and Caulimoviridae (Clark et al., 2012).
Double-stranded RNA (dsRNA) viruses	Partitiviridae, Totiviruses, and Chrysoviruses (Raza et al., 2023)

families (Saenz et al., 2002; Brodeur, 2012). The diseases associated with these plant viruses can vary greatly, ranging from mild symptoms like mosaic patterns and chlorosis to severe symptoms such as necrosis, stunted growth, and deformed fruits (Jiang and Zhou, 2023). The extensive diversity of plant viruses poses significant challenges for the pharmaceutical industry, global agriculture, and food security.

8.5 RISKS POSED BY VIRUS-INFECTED PLANTS

In recent years, medicinal plants have been widely used in both traditional and modern healthcare systems to address various ailments and improve overall well-being (Shukla, 2023). However, the presence of viruses in these plants can pose significant risks to human health and the environment (Brewer et al., 2018). This section explores the drawbacks associated with virus-infected medicinal plants, shedding light on their potential consequences on human health, agriculture, and biodiversity (Figure 8.2).

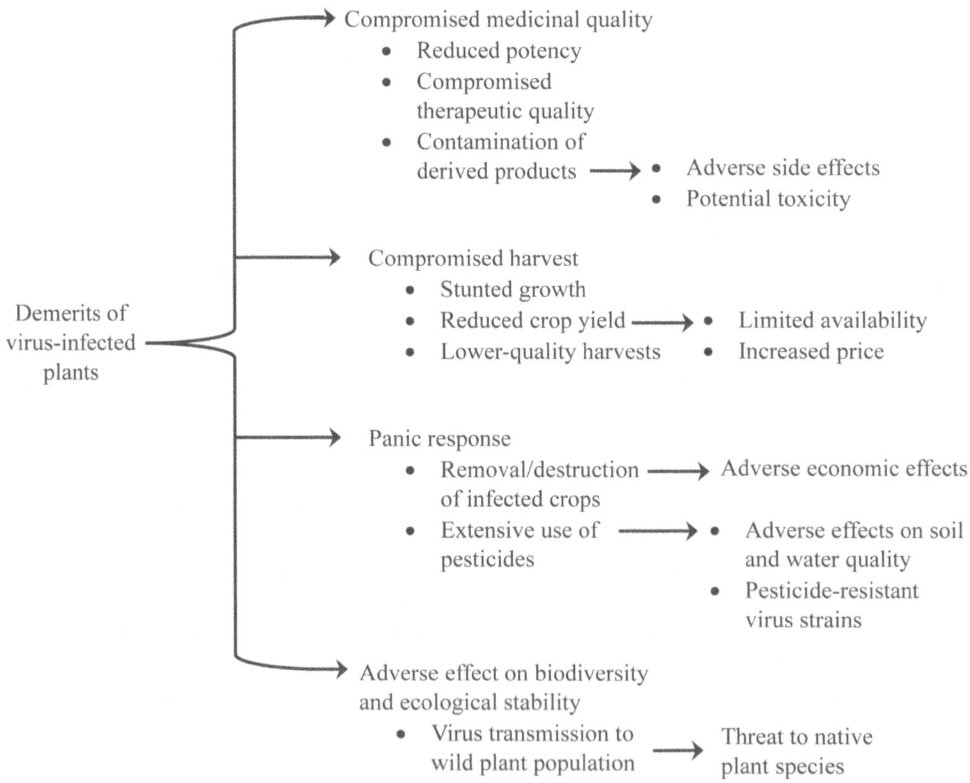

FIGURE 8.2 Potential demerits of virus-infected medicinal plants.

Virus-infected medicinal plants often experience a reduction in potency and compromised therapeutic quality. Viruses can disrupt the synthesis and distribution of critical compounds responsible for the plant's medicinal effects. As a result, traditional remedies and modern pharmaceutical preparations derived from these plants may not exhibit their desired effectiveness (Hull, 2014; Kan et al., 2023). This situation can undermine confidence in herbal medicine among consumers. Additionally, virus-infected medicinal plants can lead to contamination of resulting products. When these infected plants are used to produce herbal remedies or pharmaceuticals, harmful viral particles may be present in the final products. Consuming products that have been contaminated can have negative effects on the health of patients, particularly those with weakened immune systems. Furthermore, viruses can interfere with the desired therapeutic effects of medications, making them less effective or even potentially harmful (Anikina et al., 2023).

Infection with certain plant viruses can change the chemical composition of medicinal plants, resulting in the build-up of toxic substances or alterations in the levels of active compounds. These changes can increase the risk of experiencing adverse side effects and potential toxicity when using or applying these plants for medicinal purposes. People who rely on herbal remedies may unknowingly expose themselves to health risks due to the presence of virus-induced toxic compounds (Ershova et al., 2022). Virus-infected medicinal plants not only affect their medicinal properties but also pose significant threats to agricultural practices and people's livelihood. Viral infections can lead to stunted growth, reduced crop yield, and lower-quality harvests. For farmers who depend on cultivating medicinal plants as a means of income, this can result in financial losses and economic hardships. Moreover, decreased crop yield may strain the availability of medicinal plants, potentially causing price increases and limited access for consumers (Kim et al., 2023).

The transmission of viruses from cultivated medicinal plants to wild plant populations is a concerning issue. Many medicinal plant species are naturally found in specific regions, and when viral infections occur, they can inadvertently introduce foreign viruses into the ecosystem. This introduction of foreign viruses can disrupt the ecological balance and pose a threat to the survival of native plant species, leading to a loss of biodiversity and ecological instability (Roossinck, 2015). Medicinal plants also play a crucial role in providing various ecosystem services such as soil conservation, water purification, and supporting pollinator populations. However, when these plants become infected with viruses, their growth patterns may be altered or their reproductive capacity may get reduced, which can have cascading effects on other organisms. Ultimately, this can lead to further ecological imbalances and reduce overall ecosystem resilience (Charles and Dukes, 2007; Anikina et al., 2023).

It should also be noted that viral infections in medicinal plants can cause a decline in plant populations, posing a threat to the biodiversity of these valuable species. In many cases, infected plants become more vulnerable to other pathogens or environmental stresses, further worsening their decline. The loss of biodiversity not only affects the availability of medicinal resources but also disrupts the ecological balance and jeopardizes the survival of organisms that are dependent on these plants (Mohamed et al., 2022). Detecting viral infections in medicinal plants poses significant challenges. Viruses often remain asymptomatic or exhibit nonspecific symptoms, making early detection difficult. This delayed detection increases the risk of further contamination as the virus spreads to other plants. Furthermore, effectively managing viral infections in large-scale cultivation of medicinal plants presents a complex challenge. Traditional methods of management such as chemical treatments can have negative consequences on the environment and the quality of medicinal plants (Yadav and Khurana, 2016; Mehetre et al., 2021). In an attempt to control the spread of viral infections, farmers might resort to using more pesticides. It is crucial to address these challenges effectively while balancing environmental sustainability and maintaining high-quality medicinal plant cultivation. Excessive use of pesticides can cause detrimental effects on the environment, including negative impacts on soil health, water quality, and non-target organisms. Moreover, prolonged pesticide usage can contribute to the emergence of pesticide-resistant plants, making it more challenging to manage plant infections caused by viruses (Khan et al., 2023).

The presence of viruses in medicinal plants can lead to substantial economic implications for farmers and the agricultural industry. Infected medicinal plants result in financial losses for growers. Furthermore, effectively managing viral infections can be challenging and may involve costly treatments or even the removal and destruction of infected crops, causing additional economic hardships (Hilaire et al., 2022). Virus-infected medicinal plants can disrupt traditional healing practices that rely on specific plants known for their healing properties and cultural significance. The decline of these plants due to viral infections jeopardizes cultural traditions and hampers the transmission of traditional knowledge (Jones, 2009). Furthermore, there may be direct health risks posed to humans by viruses present in medicinal plants in certain cases. Even though plant viruses typically infect plants rather than animals, there is a concern that certain plant viruses could cross species barriers and affect humans. This could potentially lead to the emergence of new diseases or worsen existing health conditions (Wang et al., 2022a).

The harmful effects of virus-infected medicinal plants are extensive and can impact human health, agricultural practices, biodiversity, the pharmaceutical industry, and the global economy. Researchers, policymakers, and stakeholders must invest in managing plant health, detecting infections early on, and implementing preventive measures.

8.6 SIGNIFICANCE OF VIRUS-FREE MEDICINAL PLANTS AND THEIR BENEFITS

For centuries, people have relied on the power of medicinal plants to address various health issues and promote overall well-being. One important aspect that has gained significance is the cultivation of virus-free medicinal plants. This practice plays a crucial role in ensuring that herbal remedies are safe, pure, and effective.

Virus-free medicinal plants demonstrate improved effectiveness compared to their virus-infected counterparts. By eliminating viral interference in biochemical pathways, these plants can synthesize and accumulate a higher concentration of bioactive compounds. As a result, they deliver stronger therapeutic effects. Traditional medicine relies on consistent and reliable herbal formulations, and by utilizing virus-free medicinal plants as raw materials, herbal practitioners can ensure that each batch of remedy maintains uniformity in terms of composition and potency. With virus-free medicinal plants, there is a reduced risk of unintended side effects caused by viral particles or interactions between viral components and active compounds. This promotes the safety and dependability of traditional treatments. The use of virus-free medicinal plants helps preserve the traditional knowledge associated with herbal remedies. These can thrive as long as the medicinal plants remain potent and effective. One way to ensure the sustainability of traditional medicine is by having access to virus-free medicinal plants. When herbal resources are healthy and productive, it reduces the strain on wild populations and promotes responsible harvesting (Okole et al., 2004; Rahimi et al., 2023).

The virus-free plants provide pharmaceutical researchers with a reliable and consistent source of active compounds, which is crucial for reliable drug development. The purity and potency of these compounds are assured. Secondly, cultivating virus-free plants under controlled conditions allows for increased yields of medicinal compounds. This scalability becomes essential for large-scale pharmaceutical production. Furthermore, using such healthy plants in pharmaceutical research helps reduce the risk of research setbacks caused by viral contamination. This leads to cost savings and a more efficient drug development process. Lastly, virus-free medicinal plants enable better quality control in pharmaceutical manufacturing. This ensures that products meet the required standards and specifications (Okole et al., 2004; Shan et al., 2018; Ali et al., 2021).

When using virus-free medicinal plants, healthcare providers can have confidence in the predictable and reproducible treatment outcomes for patients. With their optimal concentration of bioactive compounds, the plants can contribute to enhancing the immune system and promoting overall well-being. Integrating virus-free medicinal plants with modern medicine can offer complementary treatment options, providing patients with a holistic approach to healthcare (Misra and Shukla, 2010; Basavaraju, 2011).

Virus-free cultivation can aid in the conservation of endangered medicinal plant species, ensuring their survival for future generations. Cultivating medicinal plants without viruses reduces the necessity of harvesting them from the wild, which promotes responsible and sustainable sourcing of herbal materials. Introducing such healthy plants back into degraded habitats can aid in restoring ecological balance and thus contribute to habitat conservation efforts. There is an increasing consumer preference for ethically sourced products. Virus-free cultivation ensures that herbal remedies are produced in an environmentally responsible and sustainable manner (Pathak and Abido, 2014; Cao et al., 2023).

Recent studies by Yan et al. (2022), Rani et al. (2022), and Yakhshiboeva and Jumaniyozova (2023) have highlighted progress in agriculture and horticulture where the methods used to produce virus-free medicinal plants have practical applications. Cultivating virus-free medicinal plants encourages the selection and propagation of disease-resistant plant varieties, reducing the reliance on chemical pesticides and promoting ecofriendly farming practices. The availability of virus-free planting material expands the range of cultivated plant species, mitigating the risks associated with monoculture while promoting agrobiodiversity. As discussed by Omiti et al. (2002), utilizing plants that are free from viruses can help farmers reduce losses caused by viral infections. This, in turn, leads to a more stable food supply and improved food security. Moreover, some medicinal plants not only have healing properties but also provide essential nutrients. By cultivating these virus-free plants safely, we can ensure their nutritional value reaches communities without any risks.

The benefits of virus-free medicinal plants extend beyond traditional medicine and pharmaceuticals; they also contribute to human health, biodiversity conservation, and agriculture. By prioritizing the safety, effectiveness, and sustainability of these plant resources through virus-free cultivation practices, we pave the way for a healthier and more secure future. As the demand for natural remedies and pharmaceuticals continues to rise, investing in research and cultivation methods for virus-free medicinal plants becomes increasingly vital for our well-being as well as the environment's.

8.7 PLANT TISSUE CULTURE: AN EFFECTIVE TECHNIQUE FOR PRODUCING GENETICALLY UNIFORM VIRUS-FREE MEDICINAL PLANTS

Plant tissue culture, a revolutionary technique in plant biotechnology, has emerged as a powerful tool for propagating and producing genetically uniform virus-free medicinal plants (Bisht et al., 2010; Rohit, 2014; Pant and Husen, 2022). The method involves the *in vitro* cultivation of plant cells, tissues, or organs under aseptic conditions (Arti et al., 2018). This controlled environment allows for the rapid multiplication of plant material, making it possible to produce large quantities of virus-free plants with consistent genetic characteristics (Haque et al., 2022).

Plant tissue culture is based on the ability of plant cells to dedifferentiate and regain their totipotency, enabling them to regenerate into complete plants under controlled conditions. The basic framework of the process is shown in Figure 8.3 (Phillips and Garda, 2019; Arora et al., 2022;

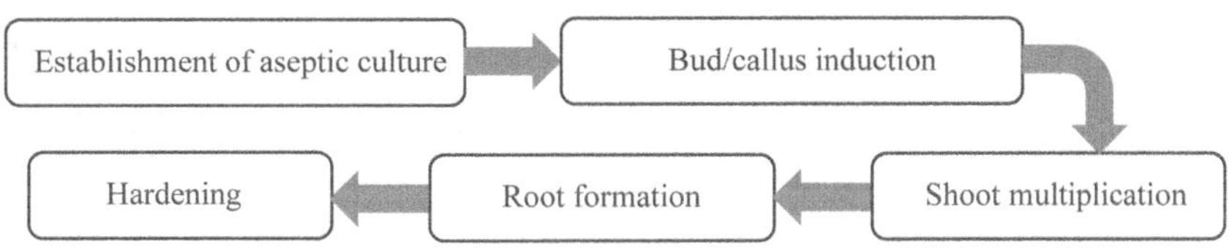

FIGURE 8.3 Basic steps of plant tissue culture.

FIGURE 8.4 Steps of plant tissue culture in *Azadirachta indica:* (a) Bud induction, (b) first multiplication cycle, (c) shoot multiplication, (d) callus, (e) *in vitro* rooting, and (f) shoot taken out for acclimatization.

Pant et al., 2022; Viswanath et al., 2023). The individual steps, with an example of *Azadirachta indica* (neem), are shown in Figure 8.4.

A small piece of plant tissue, known as an explant, is excised from the source plant and sterilized to eliminate contaminants. The explant is placed on a nutrient-rich, gel-like medium containing growth regulators. Here, the formation of a mass comprising undifferentiated cells is observed, known as a callus. Bud induction occurs specifically with the help of nodal explants. The callus/ shoot is transferred to another medium containing specific growth regulators, which induce the

TABLE 8.2
Techniques Used for Producing Virus-Free Medicinal Plants

S.no.	Technique	Description
1	Meristem culture	Meristem, the apical region of a plant shoot, is a region of actively dividing cells with the least likelihood of viral infection. By culturing meristem tips, viruses can be excluded from the regenerated plantlets (Bi et al., 2018).
2	Thermotherapy	Some plant viruses are sensitive to high temperatures, and exposing infected plant material to elevated temperatures can eliminate the viruses without harming the plant (Oves and Gaytova, 2016).
3	Cryotherapy	This involves the storage of plant tissues at extremely low temperatures, such as liquid nitrogen. This process can also help eliminate viruses from plant material before regeneration (Das et al., 2023).
4	Chemotherapy	This method is based on adding broad-spectrum antiviral drugs, such as Virazole, to the nutrient medium where the plant explants are cultivated (Abdelbaset et al., 2023).
5	Micrografting	This method involves the transfer of a small section of healthy plant tissue, known as a micrograft, from a virus-free donor plant onto a virus-infected plant (EL-Zaher et al., 2023).
6	Callus induction	It involves the cultivation of undifferentiated, rapidly dividing plant cells on a nutrient-rich medium (Gambino et al., 2006).
7	Somatic cell hybridization	This technique involves the fusion of two or more different plant cells or protoplasts to create a hybrid plant (Gamborg, 1975).
8	Gene transformation	It involves introducing foreign genes into the genetic material of plant cells to create genetically modified plants (Gantait et al., 2022).
9	Somaclonal variation	It is the phenomenon where plants regenerated from cultured somatic cells (cells other than germ cells, such as egg and sperm cells) exhibit genetic and phenotypic differences from the original parent plant (Koch and Salomon, 1993).
10	Nitrogen-deficient culture medium	Altering nutrient concentrations such as nitrogen content in tissue culture media can have various effects on plant growth, development, and interactions with pathogens, including viruses (Le, 1999).
11	Shoot tip culture	This technique involves the regeneration of whole plants from the axillary buds (Lee et al., 2014).
12	Electrotherapy	It typically refers to treatments that use electrical stimulation for therapeutic purposes (Adil et al., 2022).
13	Antiviral compound-rich medium	An antiviral compound-rich medium in plant tissue culture refers to a growth medium that is supplemented with compounds known to have antiviral properties (Conrad, 1991; Mwangangi et al., 2014).

formation of shoots from the callus cells. The newly formed shoots are transferred to a medium with different growth regulators, stimulating root development. The developed plantlets are transferred to soil and gradually acclimatized to the external environment.

Plant tissue culture offers a reliable and efficient method for producing virus-free medicinal plants. The technique involves multiple approaches to eliminate viral pathogens from the plant material, as listed in Table 8.2.

8.7.1 Advantages of Plant Tissue Culture for Producing Virus-Free Medicinal Plants

Plant tissue culture produces genetically identical clones, ensuring that the propagated plants have the desired traits and characteristics. This consistency is essential for reliable and reproducible medicinal properties (Kumari et al., 2023). It allows for the rapid multiplication of plant material, yielding a large number of virus-free plants within a short period. This scalability is valuable for commercial cultivation and research purposes (Feyisa, 2023). The techniques used in plant tissue culture, such as meristem culture and thermotherapy, to name a few, effectively eliminate viruses from the regenerated plants, ensuring their purity and safety for medicinal use (Xu et al., 2023). Traditional propagation through seeds can result in the transmission of viral pathogens to the next generation. Plant tissue culture bypasses this risk, producing virus-free plants with known genetic backgrounds (Sudheer et al., 2023).

Firstly, the use of genetically uniform virus-free medicinal plants ensures consistent medicinal efficacy, providing herbal practitioners with reliable and predictable remedies. It provides a consistent and reliable source of medicinal compounds for pharmaceutical research and drug development. Secondly, plant tissue culture can help preserve and propagate medicinal plant species that hold cultural and traditional significance in various communities. Furthermore, the mass production of virus-free medicinal plants through tissue culture reduces the pressure on wild populations, contributing to the sustainability of herbal medicine practices. Lastly, traditional medicine often relies on specific plant species, some of which are rare or endangered. Hence, the method can be used for conserving and propagating rare and endangered medicinal plant species, contributing to their preservation and potential reintroduction into the wild (Chandran et al., 2023). Plant tissue culture aids in the conservation and propagation of these valuable species (Pant, 2014; Egbuna et al., 2020).

The ability to rapidly multiply plant material ensures a sufficient supply of medicinal compounds for large-scale pharmaceutical production, and plant tissue culture can be utilized as a platform for the production of biopharmaceuticals, such as vaccines and therapeutic proteins, using genetically engineered plants. The use of virus-free plants in pharmaceutical research and production ensures the safety and purity of the final products, meeting the required quality and standards (Kurz and Constabel, 1979; Bhatia et al., 2015).

By producing virus-free planting material, plant tissue culture contributes to the propagation of disease-resistant plant varieties, enhancing crop productivity and reducing reliance on chemical pesticides. Plant tissue culture techniques can also be used for plant breeding and genetic improvement, leading to the development of new cultivars with desirable traits, including enhanced medicinal properties. Also, for rare or difficult-to-propagate medicinal plant species, tissue culture offers a viable option for conservation and cultivation. Not only this, but it can also be employed to rid valuable crop species of viral diseases, ensuring the production of healthy and productive planting material (Tsay, 1982; Idowu et al., 2009).

Along with the innumerable benefits of plant tissue culture in producing virus-free plants, it has its own limitations. The initial investment in establishing and maintaining a tissue culture laboratory can be substantial, making the technique less accessible to smaller-scale operations. It requires skilled personnel with expertise in aseptic techniques and tissue culture procedures. Also, although tissue culture provides genetic uniformity, some plants may exhibit genetic instability over time, leading to variations in phenotype and performance. A very important step, i.e., hardening, requires extreme care as transitioning plants from the controlled *in vitro* environment to the external field or greenhouse conditions may pose challenges due to differences in growth conditions (Shuler et al., 1984; Birch, 1997). For commercially valuable plant varieties produced through tissue culture, intellectual property rights and plant variety protection are essential for breeders and developers (Boettiger et al., 2004). Tissue culture-derived plants may be subjected to biosafety regulations, especially if they are genetically modified or intended for release into the environment (Alexandrova et al., 2005).

Plant tissue culture is a powerful and versatile technique that has revolutionized the production of genetically uniform virus-free medicinal plants (De Filippis, 2013). Its ability to rapidly propagate plant material and eliminate viral pathogens has significant implications for traditional medicine, pharmaceutical research, and agriculture. By providing a reliable and consistent source of medicinal compounds, virus-free plants contribute to the efficacy and safety of herbal remedies and pharmaceutical products. Moreover, tissue culture aids in the conservation of rare and endangered plant species, promoting biodiversity and sustainable practices in medicinal plant cultivation. Despite challenges, the continued advancement and application of plant tissue culture hold great promise for the future of medicinal plant production and utilization, benefitting both human health and environmental conservation.

8.8 VARIOUS PLANT TISSUE CULTURE METHODS FOR MAINTAINING VIRUS-FREE STOCK

Plant tissue culture techniques have brought about a significant transformation in modern agriculture and horticulture by offering reliable methods to propagate and maintain plant stocks free from viruses. These methods have emerged as effective tools for preserving virus-free plant stocks and safeguarding the genetic purity of valuable plant varieties (Bhojwani and Dantu, 2013).

Meristem culture: Meristem culture stands out as one of the most widely utilized techniques for generating virus-free plant stocks. The meristem refers to the apical region of shoots containing actively dividing cells with minimal chances of viral infection. The process entails isolating and cultivating the meristem tips from infected plants while ensuring that any viruses present in other parts of the plant are kept out of the culture (Lassois et al., 2013; Krishna et al., 2022). The general process of meristem culture is shown in Figure 8.5. Meristem culture exhibits a high success rate in eliminating viruses since isolated meristem tips are less likely to carry viral infections. The resulting plantlets are genetically consistent since they are direct clones of the original meristem tissue, preserving all desired traits and characteristics from the source plant. This technique allows for quick multiplication of virus-free plants, enabling large-scale production within a short time frame. Plantlets produced through meristem culture tend to remain free from viruses throughout their growth stages, ensuring long-term elimination of viruses. Meristem culture is extensively employed for producing disease-free planting materials for various crops, thereby improving overall crop health and yields. It is commonly used to produce virus-free medicinal plants, fruits, and vegetables, ensuring their genetic purity. The technique of meristem culture plays a crucial role in conserving rare and endangered plant species by enabling their propagation and survival in controlled environments (Nehra and Kartha, 1994).

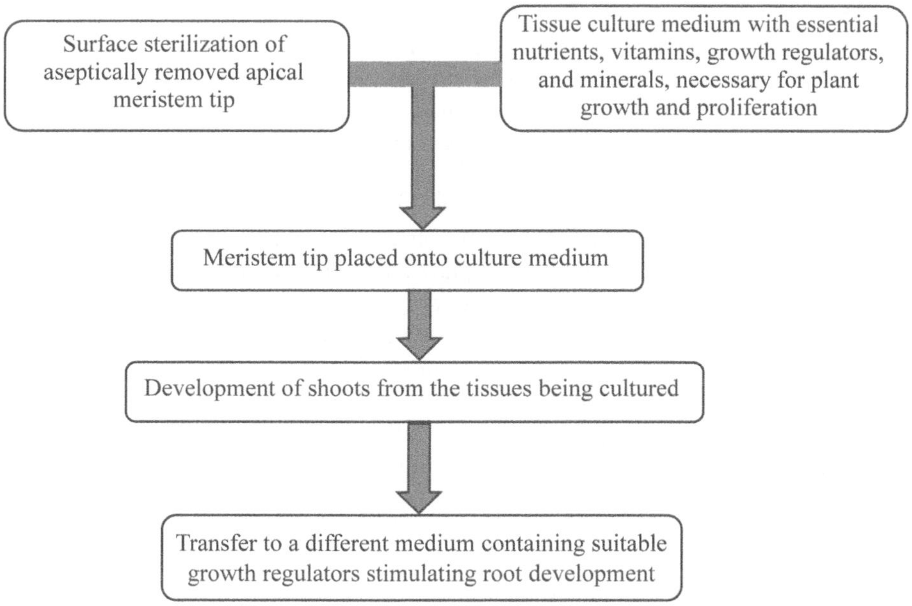

FIGURE 8.5 General procedure for meristem culture.

Thermotherapy: Thermotherapy involves the controlled exposure of virus-infected plant material to elevated temperatures. This method takes precautions to eliminate viruses without causing harm to the plant tissues (Karasawa et al., 2023).

The process includes the following steps:

- *Temperature selection*: Optimal temperature and duration for heat treatment are determined based on the specific virus and the host plant's tolerance to heat stress.
- *Heat treatment*: Infected plant material is subjected to carefully regulated temperatures, either through hot water immersion or controlled environmental conditions.
- *Regeneration*: After heat treatment, the treated plant material is cultured under sterile conditions to regenerate virus-free plantlets.

Thermotherapy is a non-destructive technique that allows for the regeneration of virus-free plantlets from treated plant material. It is quite straightforward to perform the technique and does not require advanced equipment or specialized skills. It can be applied to a large quantity of plant materials at the same time, making it suitable for eliminating viruses on a commercial scale. Thermotherapy is used to treat a wide range of plant species that are infected with various viruses, including those transmitted through seeds or vegetative propagation. The technique can save valuable cultivars and germplasm collections from viral infections, preserving their unique genetic characteristics. It is utilized in producing virus-free fruit tree stocks to enhance fruit quality and yield (Grondeau et al., 1994).

Cryotherapy: Cryotherapy involves using extremely low temperatures, typically liquid nitrogen, to preserve plant tissues and eliminate viral pathogens. This technique is highly effective in eliminating viruses and especially beneficial for plant species that are challenging to propagate using traditional tissue culture methods (Wang et al., 2018a).

Key steps in cryotherapy include:

- *Excision and pre-culture*: A small piece of plant tissue is removed and pre-cultured to initiate callus tissue formation.
- *Cryoprotection*: The callus tissue undergoes a process where it is treated with a solution that enhances its ability to withstand freezing and reduces any harm to the cells.
- *Freezing*: The protected callus tissue is placed in liquid nitrogen for extremely rapid freezing, which helps preserve the genetic material and remove any viruses.
- *Thawing and growth*: The frozen callus tissue is thawed and cultivated in a suitable nutrient medium to regenerate healthy plantlets without any viruses.

Cryotherapy effectively eliminates viruses from infected plant material, resulting in virus-free plantlets. Cryopreserved tissues often display strong genetic stability, ensuring that the desired traits of the original plant material are maintained. It plays a crucial role in preserving the genetic diversity of rare and endangered plant species, safeguarding their survival. Cryopreserved tissues serve as valuable repositories of genetic resources, allowing for long-term storage solutions for valuable plant germplasm. Cryotherapy plays a vital role in producing disease-free stock materials for various crops and fruit tree cultivars (Bettoni et al., 2018).

Chemotherapy: It is a technique in plant tissue culture that involves the use of chemical compounds to eradicate or control viral infections in plants. Chemotherapy entails applying antiviral chemicals to the plant tissues grown in a controlled environment. These chemicals can act through different mechanisms to combat viral infections (Bianculli et al., 2020; Anikina et al., 2023), as listed in Table 8.3.

The primary application of chemotherapy in tissue culture is the production of virus-free plants. Infected plant material is cultured using a medium containing antiviral chemicals, which assists in eliminating or reducing viral load. By producing plants free from viruses, we can ensure that the

TABLE 8.3
Different Chemotherapy Mechanisms to Combat Viral Infections in Plants

Mechanism	Description
Direct virucidal action	Using chemicals (e.g., ribavirin, chitosan) to directly deactivate or eliminate viruses present within plant tissues.
Inhibition of viral replication	Using compounds (e.g., salicylic acid, jasmonic acid, ethylene) to induce plant defence responses that inhibit viral replication and spread, hindering replication and spread of viruses within the plant tissues, effectively suppressing their presence.
Stimulation of plant defence mechanisms	Using chemicals to boost the natural defence mechanisms of plants, enabling them to resist or fight off viral infections more effectively.

genetic diversity of these species is preserved for future generations. We can increase crop yields and reduce the reliance on disease-resistant varieties. Virus-free plants are essential for conducting accurate research on how plants interact with viruses and for breeding programs aimed at developing disease-resistant plant varieties. This holds particular significance in the realms of medicinal plant cultivation and ornament plant landscaping. Chemotherapy also serves as a valuable tool in quarantine and phytosanitary programmes. It aids in preventing the spread of viruses between regions and countries by producing certified planting material that is free from viruses (Panattoni et al., 2013).

Considerations and challenges:

- The choice of chemicals is crucial, as they need to effectively target viruses without harming plant tissues or impacting subsequent plant growth.
- Antiviral chemicals may lose their effectiveness against viruses over time, so it is important to regularly reassess and adjust the chemical treatments to combat this resistance.
- It is crucial to strike a balance between the effectiveness of antiviral chemicals and the health of plant tissues, as certain concentrations can be toxic to them.
- When using antiviral chemicals in plant tissue culture, it is essential to consider their potential impact on non-target organisms and ecosystems in terms of environmental effects.

Chemotherapy in plant tissue culture provides a versatile and efficient approach for producing plants free from viruses, which have various applications. However, careful selection of chemicals, following proper protocols, and ongoing research are necessary to overcome challenges and ensure successful outcomes.

Micrografting: It involves grafting small plant parts like cotyledons, leaves, or micro shoots onto a suitable rootstock. Micrografting is a valuable technique for propagating small or delicate plant materials (Hussain et al., 2014).

When it comes to eliminating viruses, the process involves several steps:

- To ensure that no viral contamination is present, virus-free explants are used as grafting materials.
- Under aseptic conditions, the virus-free explants are grafted onto carefully selected rootstock.
- The micrografted explants are allowed to develop and overgrow the rootstock, effectively replacing any infected tissues.

One of the key advantages of micrografting is its minimally invasive nature, making it suitable for propagating small or sensitive plant materials. This technique enables rapid multiplication of virus-free plantlets, which is highly advantageous for commercial-scale propagation. Micrografting

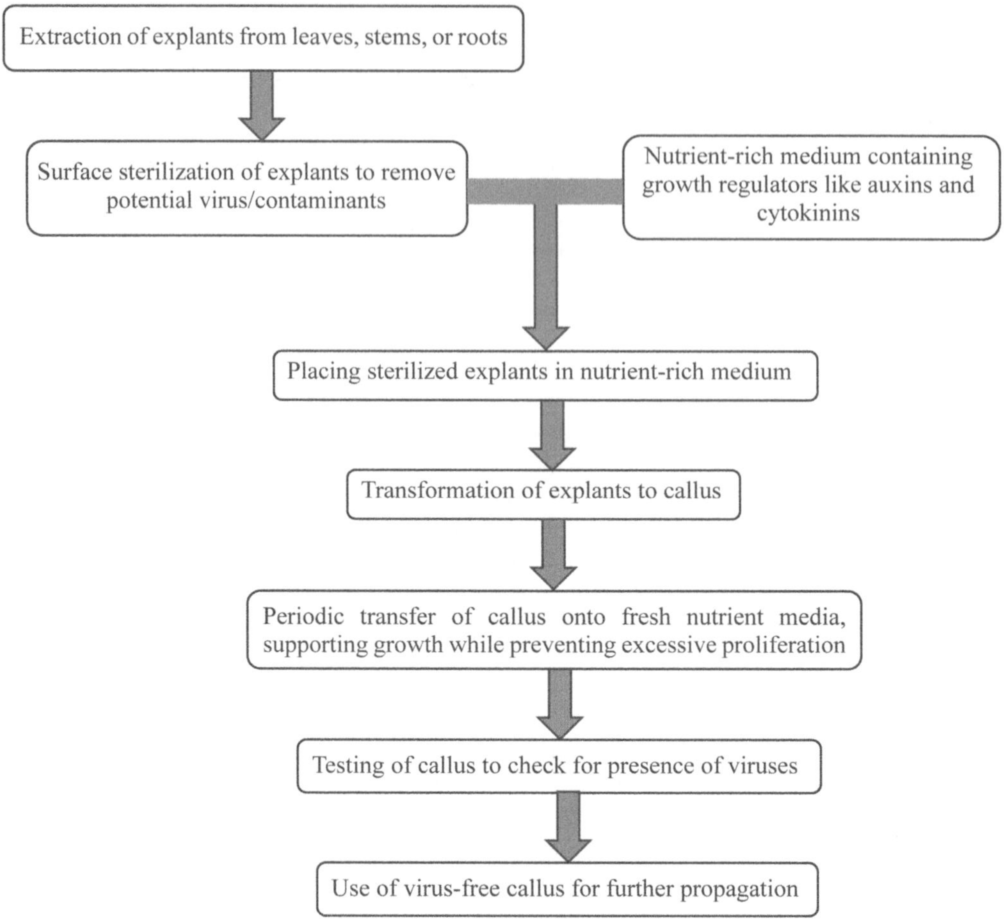

FIGURE 8.6 General procedure of callus induction from explants.

plays a crucial role in eliminating viral infections from plant materials, ensuring the production of healthy and virus-free plants. By aiding in the propagation and conservation efforts, micrografting contributes significantly to preserving rare and endangered plant species that are difficult to propagate using traditional methods.

Callus culture: It is a fundamental technique in plant tissue culture that has applications in producing virus-free plants. It involves the cultivation of undifferentiated, rapidly dividing plant cells on a nutrient-rich medium (Gambino et al., 2006). It starts with small sections of healthy plant tissue, called explants. Figure 8.6 shows the methodology involved in generating virus-free callus from explants using tissue culture technique. Once a virus-free callus is obtained, it can be induced to differentiate into particular plant organs such as shoots or roots, depending on the specific goals and the plant species involved. This step of differentiation of the callus into plant organs plays a crucial role in generating complete plants from the callus culture.

By regenerating plants from virus-free callus, the callus induction process effectively eliminates viral infections. This is especially useful for rescuing valuable plant varieties that have been affected by viral diseases. Callus cultures can serve as a starting point for genetic modification techniques, such as transformation, to introduce specific genes into plant cells. The regenerated plants can then inherit these new traits. Callus cultures enable the rapid multiplication of plant material. Once virus-free callus is established, it can be used to generate multiple plantlets through regeneration. Collecting explants from rare or endangered plant species and initiating callus cultures helps conserve their genetic diversity and prevent extinction. Callus induction plays a role in plant breeding programmes by facilitating the production of disease-free plants with desired traits for agricultural or horticultural purposes. Callus cultures also provide a controlled environment for studying various aspects of plant physiology, development, and interactions with viruses, contributing to scientific knowledge and advancements in the field of plant biology. In general, this process of inducing callus is incredibly important when it comes to creating plants that are free from viruses. This has a wide range of uses in fields like pharmaceuticals, agriculture, horticulture, conservation, and research. It provides a flexible method for producing plant material that is virus-free and can be modified for different purposes (Parmessur et al., 2002).

Somatic hybridization: This method has applications in producing virus-free plants by combining the desirable traits of two parent plants, while also potentially eliminating viral infections. It is based on the use of protoplasts, which are plant cells with their cell walls removed, allowing them to be fused more easily. Such fusion results in the creation of a hybrid cell with combined genetic material from the parent plants. The general procedure (Gamborg, 1975) of using somatic hybridization to produce virus-free plants is shown in Figure 8.7.

Somatic hybridization allows for the combination of desirable traits from different parent plants, including disease resistance and other agronomic qualities. Through protoplast fusion and regeneration, it is possible to obtain virus-free plants if one of the parent plants is virus-free. This provides an avenue for generating plants with improved traits while ensuring their freedom from viral infections. Somatic hybridization can lead to increased genetic diversity by combining the genomes of different parent plants, which can be beneficial for plant breeding and adaptation to changing environments. The technique can be used to introduce new characteristics into crop plants, enhancing

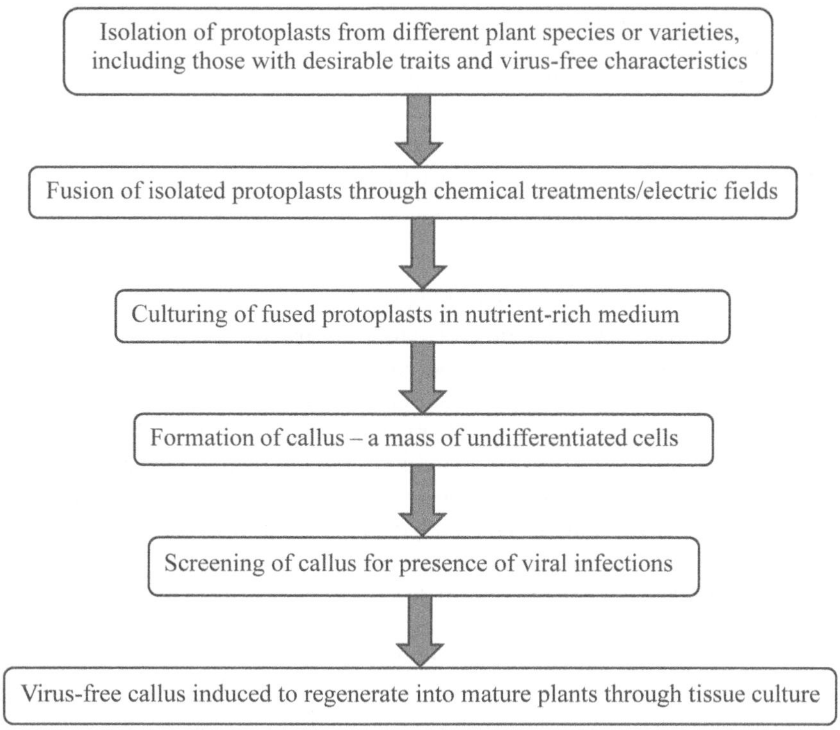

FIGURE 8.7 General process of somatic hybridization to produce virus-free plants.

their productivity, nutritional content, and overall quality. The technique can help conserve rare and endangered plant species by producing hybrid plants that incorporate valuable traits from related species while potentially eliminating viral infections. Also, it can serve as a starting point for generating transgenic plants by introducing specific genes into hybrid cells. It's important to note that somatic hybridization requires technical expertise and optimization of conditions to successfully fuse protoplasts and regenerate plants. The method can offer a powerful means of producing virus-free plants with improved traits, making it a valuable tool in plant breeding, conservation, and biotechnology (Pandey and Pandey, 2018).

Gene transformation: This involves introducing foreign genes into the genetic material of plant cells to create genetically modified plants. This technique has applications in producing virus-free plants by incorporating genes that confer resistance to viral infections. Here's how gene transformation works (Gantait et al., 2022):

- Genes responsible for conferring resistance to viruses are identified and isolated from other sources, such as related plant species, bacteria, or synthetic DNA sequences (Yarra et al., 2019).

The isolated genes are introduced into plant cells in the following way:

- Agrobacterium bacteria are used to transfer the target genes into plant cells.
- Tiny gold or tungsten particles coated with the target genes are shot into plant cells using a gene gun.
- Brief electric pulses create temporary pores in the plant cell membranes, allowing the target genes to enter.
- The transformed cells are cultured under specific conditions that encourage their development into whole plants. This step involves the formation of callus, shoots, and roots from the transformed cells.
- Selective markers, such as antibiotic resistance or fluorescent proteins, are often included along with the target genes during transformation. These markers help identify and select the cells that have successfully incorporated the foreign genes.
- The transformed plants are tested to confirm the presence and expression of the introduced genes. Additionally, the plants can be tested for virus resistance to ensure their virus-free status.

One of the primary applications is to introduce genes that confer resistance to specific viruses, hence protecting plants from viral infections, leading to virus-free or significantly reduced viral loads. Genetically modified plants with enhanced virus resistance can reduce the need for chemical treatments, thus promoting more sustainable and environmentally friendly agricultural practices. Genetic transformation can help protect valuable plant varieties by making them resistant to viruses that could otherwise devastate their populations. Such modified virus-resistant plants can serve as a way to prevent the spread of viral diseases in agricultural settings and natural environments. Also, transgenic plants provide valuable models for studying plant-virus interactions, genetics, and molecular biology. Along with virus resistance, other genes can be introduced to enhance plant stress tolerance, which can indirectly contribute to their overall health and virus resistance. Transgenic plants can also be used to produce proteins of interest, such as antibodies or vaccines, which can be of medical or industrial importance. Genes can be introduced to improve the nutritional content, flavour, shelf life, or other qualities of plants. It's worth noting that genetically modified organisms (GMOs), including transgenic plants, are subject to regulations in many countries due to potential environmental and health concerns. As such, rigorous testing and adherence to regulatory guidelines are essential when developing and deploying gene transformation techniques for producing virus-free plants (Gulzar et al., 2020).

Somaclonal variation: The variation caused by this method can be advantageous in producing virus-free plants with improved characteristics. Here's how somaclonal variation works and its applications in producing virus-free plants (Koch and Salomon, 1993):

- During the tissue culture process, cells dedifferentiate. This process can lead to genetic changes due to factors like mutation, epigenetic modifications, and genetic recombination.
- The regenerated plants may exhibit genetic changes, including mutations or alterations in gene expression patterns. These changes can result from factors such as errors during DNA replication, exposure to stress, or epigenetic modifications.
- Genetic changes can manifest as observable differences in the plants' appearance, growth, and other characteristics. These variations may be subtle or substantial.

Somaclonal variation can lead to the emergence of plants with natural resistance to viral infections. Regenerated plants may develop new genetic or epigenetic traits that confer resistance against specific viruses. The genetic variations that arise can introduce diversity into a population of plants, providing a broader range of genetic material to select from for virus resistance and other desirable traits. This process can be harnessed to select and propagate plants with improved agronomic traits, including virus resistance. It can be more rapid than traditional breeding methods. Variants arising from somaclonal variation may exhibit traits that allow them to thrive in specific environmental conditions, which can be useful in different agricultural or ecological contexts. The variation can provide researchers with a unique opportunity to study the effects of genetic and epigenetic changes on plant traits, contributing to a better understanding of plant biology (Sahijram, 2015).

Challenges and considerations

- While somaclonal variation can be advantageous, the stability of the observed traits needs to be assessed over multiple generations to ensure they persist.
- The specific genetic and phenotypic changes that arise from somaclonal variation can be unpredictable and vary among different plant lines.
- Rigorous testing is required to confirm the virus-free status of the selected variants and to ensure that the introduced variations are beneficial.

Somaclonal variation offers a potential avenue for producing virus-free plants with improved traits, including resistance to viral infections. However, due to the inherent unpredictability of the process, careful selection, testing, and validation are essential steps in harnessing this variation for practical applications.

Nitrogen-deficient culture medium: While the specific concept of using a tissue culture medium deficient in nitrogen content to produce virus-free plants isn't widely known, it's important to note that altering nutrient concentrations in tissue culture media can have various effects on plant growth, development, and interactions with pathogens, including viruses. Adjusting nitrogen levels, among other nutrients, can influence the outcome of tissue culture experiments and potentially impact virus infection (Le, 1999).

Nitrogen deficiency can induce stress responses in plants. It's possible that plants exposed to such stress may trigger their defence mechanisms, potentially leading to resistance against viral infections. This could be applied to create virus-resistant plant varieties. Nitrogen-deficient conditions could alter plant metabolism, influencing the production of secondary metabolites that are involved in plant defence. These metabolites might have antiviral properties that help in producing virus-free plants. Nitrogen limitation might affect the replication and spread of certain viruses, slowing down their progression or preventing their establishment in the tissue culture, and this could contribute to the production of virus-free plants. Also, nutrient stress, including nitrogen deficiency, can lead to genetic and epigenetic changes in plants. This variation might create plant variants with reduced susceptibility to viral infections. Low-nitrogen conditions could potentially enhance the efficiency of micropropagation by influencing shoot proliferation and differentiation, resulting in more uniform and robust plantlets. If low-nitrogen conditions influence the regeneration process favourably, this approach could be used to conserve valuable plant germplasm, including virus-free variants. This approach might be useful in studying plant stress responses and understanding how they impact virus-host interactions, leading to insights into plant defence mechanisms. If nitrogen-deficient conditions induce antiviral effects, this could be explored for biotechnological applications in plant protection and production.

It's important to emphasize that these applications are speculative and hypothetical. The response of plants to altered nutrient conditions is complex and context-dependent. Rigorous research, experimentation, and validation would be required to determine the effectiveness and potential applications of using nitrogen-deficient tissue culture for producing virus-free plants.

Axillary bud culture: Also known as shoot tip culture or axillary shoot proliferation, this is a technique in plant tissue culture that involves the regeneration of whole plants from the axillary buds, which are located in the leaf axils along the stem. Here's how axillary bud culture works and its applications in producing virus-free plants (Lee et al., 2014):

- Healthy plant material with an axillary shoot is selected.
- The explants are subjected to surface sterilization to remove any potential contaminants, including viruses.
- The sterilized explants are placed onto a nutrient-rich tissue culture medium. This medium contains essential nutrients, vitamins, and growth regulators needed for shoot initiation and proliferation.
- The axillary shoots initiate bud growth in the culture medium. Under controlled conditions, these buds give rise to multiple shoots.
- During the proliferation phase, healthy shoots that are free from viral infections are visually selected. Shoots that show no symptoms of infection or viral presence are chosen for further propagation.
- The selected healthy shoots are transferred to a rooting medium to induce root formation. Once roots develop, complete plantlets are obtained.
- The regenerated plantlets are tested for the presence of viruses using diagnostic techniques such as enzyme-linked immunosorbent assay (ELISA) or PCR. Only those plantlets that test negative for viruses are considered virus-free.

Axillary bud explants are commonly used for rapid *in vitro* propagation and are distinguished from meristem tips primarily by the size of the explants (meristem tips are relatively smaller in size (0.1–0.5 mm) (Nehra and Kartha, 1994). Axillary bud culture can be employed to produce virus-free plants for phytosanitary programmes aimed at preventing the spread of viral diseases. This can also be used to preserve genetic diversity by producing virus-free plants from superior germplasm. It provides a controlled environment for plant regeneration and the selection of healthy shoots, ensuring the propagation of plants that are free from viral infections (Wang et al., 2016).

Electrotherapy: Electrotherapy, in the context of plant tissue culture, potentially involves the use of controlled electrical stimuli to induce physiological responses in plants, including the activation of defence mechanisms against viruses. Here are some general insights into how electrical stimulation might be applied in plant tissue culture for producing virus-free plants:

- Controlled electrical pulses could be applied to plant tissues or cells during tissue culture. These pulses might mimic stress responses triggered by viral infections.
- The electrical stimulation could potentially activate the plant's natural defence mechanisms, such as the production of antiviral compounds or changes in gene expression related to immunity.

- The electrical treatment results in enhanced immunity, and it might lead to reduced susceptibility to viral infections. This could contribute to producing virus-free plants.
- Scientists would need to determine the optimal parameters of the electrical stimuli, including pulse frequency, duration, and intensity, to achieve the desired response.
- The resulting plant material would need to be tested to confirm the absence of viruses. Additionally, the long-term stability of any resistance acquired through this method would need to be assessed (Adil et al., 2022).

Antiviral compound-rich medium: This approach aims to create an environment that helps in producing virus-free plants by inhibiting the growth or replication of plant viruses. Antiviral compounds target viral infection by interfering with various stages of the virus life cycle, such as entry, replication, or movement within the plant cells, thus hindering the growth and spread of viruses within the cultured tissue. The general outline of the process (Mwangangi et al., 2014) of using antiviral compound-rich medium to produce virus-free plants within the framework of tissue culture is shown in Figure 8.8.

The primary application of creating an environment that inhibits viral infections is allowing the production of virus-free plants from infected plant material. This technique can be used in research settings to study the effects of specific antiviral compounds on plant-virus interactions and to develop strategies for virus control. It provides a reliable method for producing healthy planting material that is free from viral infections; hence, this approach can rescue valuable plant varieties that have been compromised by viral infections, enabling their propagation and preservation. Regenerated plants obtained through this approach can serve as starting material for breeding programmes to develop new varieties with improved traits, including virus resistance (Conrad, 1991).

It is worth noting that the selection and concentration of compounds as well as their effectiveness can vary depending on the specific plant virus system. Thorough research, experimentation, and

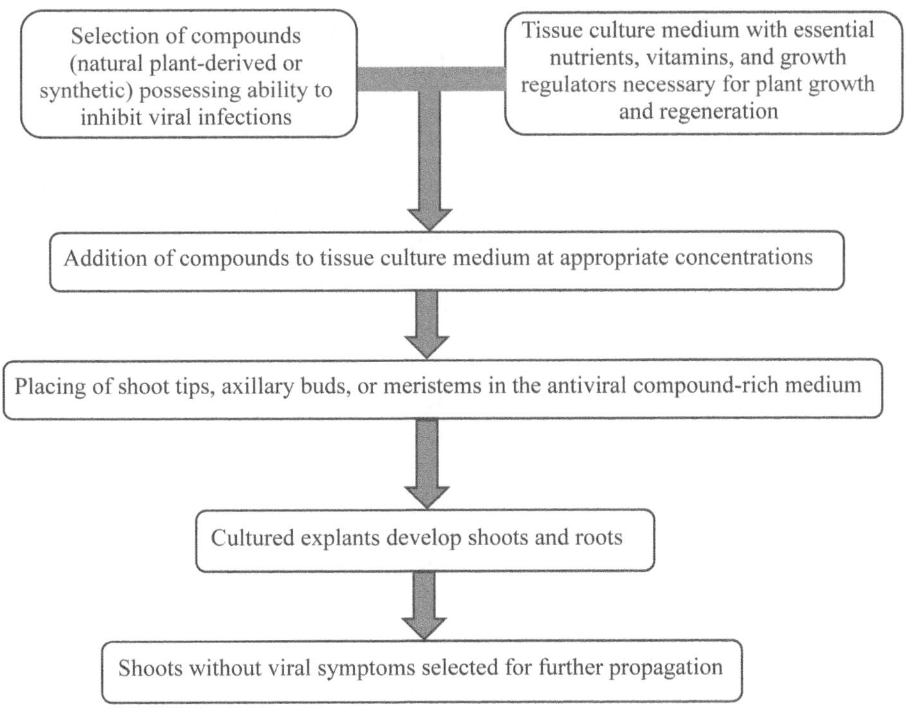

FIGURE 8.8 General procedure of using antiviral compounds in tissue culture medium to produce virus-free plants.

testing are essential to determine the conditions for using a medium rich in antiviral compounds to effectively produce plants free from viruses. Plant tissue culture techniques offer effective ways to maintain virus-free plant stocks. This contributes to agricultural practices, horticulture, and biodiversity conservation.

Through methods like meristem culture, thermotherapy, cryotherapy, micrografting, etc., scientists and growers can eradicate infections and propagate genetically uniform medicinal plants that are free from diseases. The benefits of having virus-free plant stocks extend to medicine, pharmaceutical research, and horticultural practices. This ensures the availability of dependable herbal remedies, pharmaceutical products, and ornamental plants. The ongoing progress in plant tissue culture methods holds immense potential for the future of agriculture, as well as conservation efforts, in sustainable medicinal plant–based industries.

8.9 SOME TECHNIQUES TO DETERMINE THE PRESENCE OF VIRUS IN PLANTS

Detecting plant viruses is of utmost importance in preventing the spread of diseases that can have significant impacts on both the economy and the environment. There are various methods available for identifying the presence of plant viruses in plants. Table 8.4 shows a few commonly used techniques (Naidu and Hughes, 2003; Matthews, 2019; Rubio et al., 2020).

Various methods can be used to identify viruses, even if their genetic material is not easily amplified. It is important to note that each technique has its strengths and limitations, and the selection of a particular method depends on factors such as sensitivity, specificity, cost, and available equipment. In some cases, combining multiple techniques may be employed to enhance the reliability of virus detection.

8.10 SOME RECENT REPORTS FOR THE PRODUCTION OF VIRUS-FREE MEDICINAL PLANTS

The cultivation of plants free from viruses is the key focus in modern agricultural and biotechnological research with far-reaching implications for human health and sustainable farming. Medicinal plants are highly valued for their healing properties. They are vulnerable to viral infections that not only affect their quality and yield but also pose potential risks to consumers. In response to this challenge, scientists have embarked on a mission to develop strategies and techniques that ensure the growth of virus-free medicinal plants. This encompasses a range of approaches in plant tissue culture. Tissue culture methods, as discussed in the earlier sections, have proven successful in producing plantlets which are devoid of viral infections. Thanks to the collaboration between expertise, agricultural knowledge and technological advancements, significant progress has been made in the production of virus-free medicinal plants. As we continue to learn about these advancements through recent reports, this section explores the importance of these efforts by highlighting their contributions towards improving global healthcare, fostering economic growth, and maintaining ecological balance.

In *Hosta capitata*, meristems obtained from sterilized bulbs were cultivated in growth media containing varying concentrations of 1-naphthaleneacetic acid (NAA), along with either benzyl aminopurine (BA) or thidiazuron (TDZ). The combination of 0.1 mg/L NAA and 3.0 mg/L BA yielded the highest number of shoots per plant section. The introduction of nano silver (NAg) particles to the regrowth medium notably enhanced plant development. Genetically stable plants derived from the meristem were confirmed through random amplified polymorphic DNA (RAPD) analysis, demonstrating their comparability to donor plants cultivated in a greenhouse and their freedom from HVX (Hosta virus X) infection (Pe et al., 2020). In another study, the inclusion of polyvinylpyrrolidone (PVP) improved meristem regeneration effectiveness in different *Frageria chiloensis* accessions. Similarly, incorporating 6-benzylaminopurine (BAP) into the culture media increased

TABLE 8.4
Various Methods for Identifying the Presence of Viruses in Plants

Technique	Description
ELISA (Enzyme-linked immunosorbent assay)	ELISA involves utilizing specific antibodies that target the virus in question. These antibodies are tagged with enzymes, and if the virus is present in a plant sample, it will bind to these antibodies. The activity of the enzymes can then be observed, indicating the presence of the virus.
RT-PCR (Reverse transcription polymerase chain reaction)	RT-PCR is a molecular technique that converts viral RNA into complementary DNA (cDNA). This cDNA is then amplified using PCR. By detecting this amplified DNA, we can determine if there is a presence of the virus. Real-time RT-PCR allows for quantitative analysis of viral load.
QPCR (Quantitative polymerase chain reaction)	QPCR is an advanced form of PCR that measures the amount of amplified DNA in real time. It provides valuable quantitative information about viral load within a plant sample.
LAMP (loop-mediated isothermal amplification)	LAMP is another molecular technique that amplifies DNA under constant temperature conditions. These techniques play a vital role in effectively identifying and managing plant viruses to safeguard agriculture and ecological systems. It's actually quite straightforward and can be utilized for swiftly detecting plant viruses out in the field.
Immunoassays	A variety of immunoassay techniques, including lateral flow devices and immunochromatography, rely on antibodies to capture and identify viral antigens. These methods are commonly employed for rapid onsite testing.
Electron microscopy	This technique entails using an electron microscope to directly observe virus particles in plant tissues. It offers a high level of specificity and enables the identification of virus morphology.
Next-generation sequencing (NGS)	NGS technologies enable the sequencing of the complete genetic content of a sample, enabling the detection of both known and unknown viruses. Metagenomic sequencing is helpful in identifying multiple viruses within a single sample.
Serological tests	Serological tests detect virus specific antibodies in plant sap. Techniques such as Western blotting and dot immunobinding assays can reveal the presence of viral proteins.
Nucleic acid hybridization	This method employs labelled probes that are complementary to viral RNA or DNA sequences. If the virus is present, the probe will attach to its genetic material, allowing for detection of a signal.
Immunofluorescence	This approach involves using fluorescently labelled antibodies that bind to viral antigens. The resulting fluorescence can be observed under a microscope.
Mass spectrometry	Mass spectrometry can identify viral proteins present in plant tissues.
Biosensors	These are devices capable of detecting viruses by measuring changes in physical or chemical properties. These changes occur as a result of the interaction between the virus and a specific recognition element.
Thermal imaging camera	This can detect temperature variations caused by certain viruses affecting plant physiology. By identifying these temperature differences, we can infer the presence of a virus.

shoot multiplication rates to an average of three to six shoots per plant (Quiroz et al., 2017). For *Solanum tuberosum* L., a combination of MS + 0.1 mg/L KIN + 0.5 mg/L GA3 was identified as optimal for initial meristem culture establishment, while using 0.5 mg/L BA + 0.5 mg/L GA3 was found most suitable for generating shoots. Potato plantlets grown in a controlled environment without any viruses were successfully obtained using the DAS ELISA test (Azad et al., 2020).

Meristem sections measuring 300 micrometres were taken from Binella potato plants and placed on solidified MS medium for cultivation. The presence of virus infection in the plantlets was determined using DAS ELISA. The results indicated a success rate in producing virus-free plants. The elimination of potato virus Y (PVY) improved when a combination of chemotherapy (using ribavirin at a concentration of 20 mg/L) was applied alongside meristem tips (Almaarri et al., 2023). In *Rehmannia glutinosa*, thermotherapy, chemotherapy, and meristem tip culture (ranging from 0.3 to

0.5 mm) were performed independently. Reverse transcription (RT) PCR tests were employed to detect five viruses, namely 'Rehmannia' mosaic virus, Youcai mosaic virus, broad bean wilt virus 2, *Plantago asiatica* 'mosaic virus', and 'Rehmannia virus 1'. Through chemotherapy using ribavirin (at a concentration of 60 mg/L), the rate of virus elimination reached 55%, while thermotherapy and meristem culture achieved rates of approximately 10% and 37%, respectively (Kwon et al., 2022).

In this next study, *Cnidium officinale* samples were regenerated by placing them on a growth medium supplemented with ribavirin at concentrations of 10, 20, and 40 mg/L for a duration of 4 weeks. After this, these samples were subjected to a thermotherapy treatment ranging from 25 to 37 °C for a period of 15 days. The survival and regrowth of plants were evaluated. The effectiveness of eliminating viruses was determined using RT-PCR with two primer pairs for each virus. Treating the viral infections with ribavirin at a concentration of 40 mg/L resulted in elimination rates of 89.5% and 94.7%, respectively (Kim and Jeong, 2022). To eradicate apple chlorotic leaf spot virus (ACLSV) and apple stem grooving virus (ASGV) in quince through the *in vitro* method, the cultures were subjected to heat treatment at 38 °C for given durations (0, 7, 10, and 15 days) followed by isolating the apical shoot tips. For cryotherapy procedures, the shoot tips were incubated on preculture medium before exposing them to plant vitrification solution for 30 minutes at a temperature of 0 °C before being exposed to liquid nitrogen (LN). The success rate of virus eradication was determined using RT-PCR analysis. The results indicated that droplet vitrification, cryotherapy, and thermotherapy were highly effective in eradicating ACLSV and ASGV from *in vitro* cultures (Farhadi-Tooli et al., 2022).

Potato plants infected with potato leafroll virus (PLRV) and potato spindle tuber viroid (PSTVd) were subjected to currents ranging from 5 to 25 mA for 5–25 minutes. The different parts of the plant such as potato shoot tips, nodal segments, and tuber sprouts were then cultured on a medium. The researchers found that exposing the plant tissues to currents of 5–10 mA for 5–10 minutes did not significantly affect their ability to regenerate. However, when the current was increased to 20–25 mA for 20 minutes, a noticeable number of plants devoid of PLRV and PSTVd were obtained (Singh and Kaur, 2016).

In a study conducted by Guta et al. (2019) grapevine shoot cuttings infected with grapevine fleck virus (GFkV) were treated with an indirect electric current ranging from 40 to 100 mA for 5–20 minutes. Different percentages of grapevines that tested negative for ELISA were obtained, which varied depending on the type of virus infection and the experimental parameters. In another study, the removal of viruses from citrus-infected plants was done by using a combination of micrografting and thermotherapy. Six citrus cultivars were selected. By employing the ELISA technique, plants that tested positive for citrus tristeza virus (CTV) were identified. However, interestingly these plants did not exhibit any symptoms of CTV when they were exposed to high temperatures of 40 °C (Chae et al., 2013). In one of the studies, researchers successfully removed satsuma dwarf virus (SDV), citrus tatter leaf virus (CTLV), and citrus tristeza virus (CTV) from six types of citrus plants by grafting heat-treated shoot tips onto trifoliate orange rootstock. When *in vitro* grafting alone was not sufficient for producing virus-free plants, combining it with antiviral agents and thermotherapy increased the chances of eliminating the viruses (Wang et al., 2018b).

Another research study aimed to explore the use of tepals for embryogenesis (SE) as an alternative technique to induce SE and obtain virus-free plantlets of *Hippeastrum* from infected plants. The main factor for inducing callus was found to be 2,4-dichlorophenoxyacetic acid (2,4 D), while TDZ increased the rate of SE induction. The plantlets derived from SE that germinated *in vitro* successfully tested negative for all viruses analysed (de Almeida et al., 2022). In another study by Ebrahimi et al. (2022), it was discovered that by combining three methods including thermotherapy, meristem culture, and somatic embryogenesis, induction from meristem on MS medium supplemented with 0.5 mg/L 2,4 D and 1 mg/L TDZ proved to be the effective and safe technique for eradicating viruses without encountering challenges related to meristem size. The virus-free samples were grown in immersion bioreactor systems to multiply them. These samples were then preserved for future use in establishing healthy orchards.

The objective of the research conducted by Ahmed et al. (2019) was to develop variations of sugarcane that are resistant to red rot and sugarcane mosaic virus (SCMV) using somaclonal variation. They successfully developed a total of 581 somaclones, out of which 201 survived after the hardening process and 121 after transplantation. These surviving somaclones were then tested for their resistance against rot and SCMV. The researchers evaluated four candidate genes: Catalase (*CAT1*), sucrose phosphate synthase (*SPS*), gibberellin 2 oxidase 4 (GA 2 oxidase 4), and tillering branched 1 (*TB1*) to identify any possible single nucleotide polymorphism (SNP) changes in their exon regions. After analysing the sequences of five somaclones and comparing them with their parental clones, no SNP changes were observed in these candidate genes. This suggests that somaclonal variations can be a genetic improvement method for developing disease-resistant sugarcane without altering the candidate genes through SNPs. A study conducted by Ita et al. (2020) focused was on eliminating yam mosaic virus (YMV) in *Dioscorea rotundata* through cryotherapy of buds. The researchers subjected enlarged buds infected with YMV to freezing in liquid nitrogen for 1 hour followed by rewarming at a temperature of 40 °C and subsequent culturing to regenerate plantlets. The results showed 76.33% plantlet regeneration success rate and complete eradication of YMV when using cryo-treated buds compared to non-treated buds.

Meristem culture, as highlighted by Gao et al. (2023), is widely employed for virus elimination. However, research indicates that the effectiveness of virus elimination can be significantly enhanced by combining different methods. A combination of meristem culture, thermotherapy, and chemotherapy has been proven to be more effective. Studies have also shown that treating infected plants at higher temperatures and incorporating antiviral agents into the medium for meristem regeneration can successfully eliminate various plant viruses (Bhat et al., 2020).

8.11 CONCLUSION

Viral infections pose risks to the quality, effectiveness, and safety of medicinal plants. However, the use of tissue culture techniques has revolutionized the way virus-free medicinal plant stocks can be produced. The technique ensures the preservation of genetic purity, maintenance of medicinal efficacy, and propagation of valuable plant species, hence effectively allowing the production of virus-free medicinal plants in agriculture, horticulture, and pharmaceutical research. By utilizing tissue culture methods such as meristem culture, thermotherapy, cryotherapy, micrografting, etc., researchers and growers have successfully eliminated viral pathogens from infected plant material. These bring about benefits that are vital for human health and environmental well-being. One of the main advantages of using plant tissue culture for producing virus-free medicinal plants is its ability to ensure genetic uniformity. This genetic consistency ensures that each plant in a batch possesses the required medicinal properties. Moreover, tissue culture techniques allow for the multiplication of virus-free medicinal plants, offering a scalable solution for commercial cultivation. Additionally, propagating virus-free plant stocks through tissue culture helps alleviate pressure on wild populations and also helps in conserving rare and endangered medicinal plant species. Furthermore, the technique has reduced the reliance on chemical pesticides for disease resistance, enhanced crop yields, and improved fruit quality. However, continuous training initiatives and knowledge sharing play a major role in the adoption of tissue culture techniques.

By investing in research, education, and collaborative efforts, plant tissue culture will continue to drive advancements in medicine, pharmaceuticals, and environmental conservation.

REFERENCES

Abdelbaset, T. E., Sakr, K. I., Al Shorbagy, A. E., & Elsisi, A. A. (2023). Combination of cold therapy and chemotherapy for eradication of *Citrus exocortis* viroid. *Journal of Plant Protection and Pathology, 14*, 187–194.

Adil, S., Singh, V., Anjum, A., & Quraishi, A. (2022). A mini-review on electrotherapeutic strategy for the plant viral elimination. *Plant Cell, Tissue and Organ Culture (PCTOC), 150*(1), 41–55.

Ahmed, M. S., Gardezi, D. A., Batley, J., Hayashi, S., Zander, M., Javid, A., … Awan, S. I. (2019). Somaclonal variations for red rot and sugarcane mosaic virus resistance and candidate genes integrity assessment in somaclones of selected sugarcane varieties (*Saccharum officinarum* L.). *Pakistan Journal of Agricultural Sciences*, *56*, 15–27.

Ajay, T., Manu, P., & Ankita, L. (2020). Direct and indirect organogenesis in Hippophae salicifolia D. Don., A nutraceutically important plant. *Research Journal of Biotechnology*, *15*, 5.

Akutse, K. S., Subramanian, S., Maniania, N., Dubois, T., & Ekesi, S. (2020). Biopesticide research and product development in Africa for sustainable agriculture and food security–Experiences from the International Centre of Insect Physiology and Ecology (ICIPE). *Frontiers in Sustainable Food Systems*, *4*, 563016.

Alexandrova, N., Georgieva, K., & Atanassov, A. (2005). Biosafety regulations of GMOs: National and international aspects and regional cooperation. *Biotechnology & Biotechnological Equipment*, *19*(sup3), 153–172.

Ali, A. (2023). Single-Stranded RNA Viruses Infecting Cucurbits. In Gaur, R. K., Patil, B. L., Selvarajan, R. (eds.), *Plant RNA Viruses* (39–56). Academic Press.

Ali, S. I., Sheikh, W. M., Rather, M. A., Venkatesalu, V., Muzamil Bashir, S., & Nabi, S. U. (2021). Medicinal plants: Treasure for antiviral drug discovery. *Phytotherapy Research*, *35*(7), 3447–3483.

Almaarri, K., Massa, R., Abouzadan, T., ALSameen, M., & AL-Bisky, F. (2023). Production virus-free plants of grapes and potato by meristem culture from infected plants. *Journal of Agricultural & Marine Sciences (JAMS)*, *28*(2), 64.

Anikina, I., Kamarova, A., Issayeva, K., Issakhanova, S., Mustafayeva, N., Insebayeva, M., … Raposo, A. (2023). Plant protection from virus: A review of different approaches. *Frontiers in Plant Science*, *14*, 1163270.

Arora, K., Rai, M. K., & Sharma, A. K. (2022). Tissue culture mediated biotechnological interventions in medicinal trees: Recent progress. *Plant Cell, Tissue and Organ Culture (PCTOC)*, *150*(2), 267–287.

Arti, K., Ajay, T., & Maqbool, R. M. (2018). *In vitro* plant regeneration from cotyledonary node of *Diploknema butyracea*. *Research Journal of Biotechnology*, *13*(4), 29–36.

Azad, M. A. K., Khatun, Z., Eaton, T. E. J., Hossen, M. I., Haque, M. K., & Soren, E. B. (2020). Generation of virus-free potato plantlets through meristem culture and their field evaluation. *American Journal of Plant Sciences*, *11*(11), 1827–1846.

Basavaraju, R. (2011). Plant tissue culture-agriculture and health of man. *Indian Journal of Science and Technology*, *4*(3), 333–335.

Bettoni, J. C., Dalla Costa, M., Souza, J. A., Volk, G. M., Nickel, O., da Silva, F. N., & Kretzschmar, A. A. (2018). Cryotherapy by encapsulation-dehydration is effective for *in vitro* eradication of latent viruses from 'Marubakaido' apple rootstock. *Journal of Biotechnology*, *269*, 1–7.

Bhat, A. I., Rao, G. P., Bhat, A. I., & Rao, G. P. (2020). *Virus elimination by meristem-tip culture*. In: Characterization of Plant Viruses, edited by Alangar Ishwara Bhat and Govind Pratap Rao. Springer, pp. 465–477.

Bhatia, S., Sharma, K., Dahiya, R., & Bera, T. (2015). *Modern Applications of Plant Biotechnology in Pharmaceutical Sciences*. Academic Press.

Bhojwani, S. S., & Dantu, P. K. (2013). *Plant Tissue Culture: An Introductory Text* (Vol. 318). Springer.

Bi, W. L., Hao, X. Y., Cui, Z. H., Pathirana, R., Volk, G. M., & Wang, Q. C. (2018). Shoot tip cryotherapy for efficient eradication of grapevine leafroll-associated virus-3 from diseased grapevine *in vitro* plants. *Annals of Applied Biology*, *173*(3), 261–270. doi: 10.1111/aab.12459

Bianculli, R. H., Mase, J. D., & Schulz, M. D. (2020). Antiviral polymers: Past approaches and future possibilities. *Macromolecules*, *53*(21), 9158–9186.

Birch, R. G. (1997). Plant transformation: Problems and strategies for practical application. *Annual Review of Plant Biology*, *48*(1), 297–326.

Bisht, P., Pant, M., & Kant, A. (2010). *In vitro* propagation of Gigantochloa atroviolaceae Widjaja through nodal explants. *Journal of American Science*, *6*, 1019–1026.

Boettiger, S., Graff, G., Pardey, P. G., Van Dusen, E., & Wright, B. D. (2004). Intellectual property rights for plant biotechnology: International aspects. *Handbook of Plant Biotechnology*, *1*, 1088–1113.

Brewer, H. C., Hird, D. L., Bailey, A. M., Seal, S. E., & Foster, G. D. (2018). A guide to the contained use of plant virus infectious clones. *Plant Biotechnology Journal*, *16*(4), 832–843.

Brodeur, J. (2012). Host specificity in biological control: Insights from opportunistic pathogens. *Evolutionary Applications*, *5*(5), 470–480.

Bwalya, J., & Kim, K. H. (2023). The crucial role of chloroplast-related proteins in viral genome replication and host defense against positive-sense single-stranded RNA viruses. *The Plant Pathology Journal*, *39*(1), 28.

Cao, H., Han, Y., Peng, J., & Peng, X. (2023). Improving soil moisture content to increase strawberry growth indicators using the hydroponic method. *Water Supply*, *23*, 2927–2939.

Chae, C. W., Yun, S. H., Park, J. H., Hyun, J. W., Koh, S. W., & Lee, D. H. (2013). Micrografting and heat treatment combination for eliminating virus of CTV-infected Citrus. 생명과학회지, *23*(2), 267–272.

Chandran, S., Raghu, A. V., & Mohanan, K. V. (2023). *In Vitro* Conservation of Rare, Endangered, and Threatened Plants. In Sukumaran, S. T., T R, K. (eds.), *Conservation and Sustainable Utilization of Bioresources* (pp. 391–408). Springer Nature Singapore.

Charles, H., & Dukes, J. S. (2007). Impacts of Invasive Species on Ecosystem Services. In Nentwig, W. (ed.), *Biological Invasions* (pp. 217–237). Springer.

Clark, C. A., Davis, J. A., Abad, J. A., Cuellar, W. J., Fuentes, S., Kreuze, J. F., … Valkonen, J. P. (2012). Sweetpotato viruses: 15 years of progress on understanding and managing complex diseases. *Plant Disease*, *96*(2), 168–185.

Conrad, P. L. (1991). Potato virus S-free plants obtained using antiviral compounds and nodal segment culture of potato. *American Potato Journal*, *68*, 507–513.

Das, M. C., Nongsiang, A., & Sanglyne, M. W. (2023). Detection methods and *in vitro* elimination techniques for orchid viruses: A review. *South African Journal of Botany*, *153*, 227–235.

de Almeida, N. V., Rivas, E. B., & Cardoso, J. C. (2022). Somatic embryogenesis from flower tepals of Hippeastrum aiming regeneration of virus-free plants. *Plant Science*, *317*, 111191.

De Filippis, L. F. (2013). Crop improvement through tissue culture. *Improvement of Crops in the Era of Climatic Changes*, *1*, 289–346.

Ebrahimi, M., Habashi, A. A., Emadpour, M., & Kazemi, N. (2022). Recovery of virus-free Almond (*Prunus dulcis*) cultivars by somatic embryogenesis from meristem undergone thermotherapy. *Scientific Reports*, *12*(1), 14948.

Egbuna, C., Mishra, A. P., & Goyal, M. R. (eds.) (2020). *Preparation of Phytopharmaceuticals for the Management of Disorders: The Development of Nutraceuticals and Traditional Medicine*. Academic Press.

El-Zaher, A., Sahar, M. H., Elwahab, M. A., El-Gany, A., El-Sattar, A., & Fathy, H. A. (2023). Differences in morphological, biochemical, and genetic characteristics of grape explants grafted using various micrografting techniques. *Journal of Applied Horticulture*, *25*(1), 42–46.

Ershova, N., Sheshukova, E., Kamarova, K., Arifulin, E., Tashlitsky, V., & Serebryakova, M., & Komarova, T. (2022). *Nicotiana benthamiana* Kunitz peptidase inhibitor-like protein involved in chloroplast-to-nucleus regulatory pathway in plant-virus interaction. *Frontiers in Plant Science*, *13*, 1041867. doi: 10.3389/fpls.2022.1041867

Ertunc, F. (2020). Emerging Plant Viruses. In Ennaji, M. M. (ed.), *Emerging and Reemerging Viral Pathogens* (pp. 1041–1062). Academic Press.

Farhadi-Tooli, S., Ghanbari, A., Kermani, M. J., Zeinalabedini, M., Bettoni, J. C., Naji, A. M., & Kazemi, N. (2022). Droplet-vitrification cryotherapy and thermotherapy as efficient tools for the eradication of apple chlorotic leaf spot virus and apple stem grooving virus from virus-infected quince *in vitro* cultures. *European Journal of Plant Pathology*, *162*(1), 31–43.

Feyisa, A. S. (2023). Protocol optimization for rapid multiplication of cassava (*Manihot esculenta* Crantz) using the nodal segment. *Agricultural Science Digest-A Research Journal*, *43*(1), 18–23.

Gambino, G., Bondaz, J., & Gribaudo, I. (2006). Detection and elimination of viruses in callus, somatic embryos and regenerated plantlets of grapevine. *European Journal of Plant Pathology*, *114*, 397–404.

Gamborg, O. L. (1975). Advances in somatic cell hybridization in higher plants. *Stadler Symp. 7*, 37–46. Univ. Missouri, Columbia.

Gantait, S., Mukherjee, E., Jogam, P., Babu, K. H., Jain, S. M., & Suprasanna, P. (2022). Improving Crops through Transgenic Breeding—Technological Advances and Prospects. In Chandra Rai, A., Kumar, A., Modi, A., Singh, M. (eds.), *Advances in Plant Tissue Culture* (pp. 295–324). Academic Press.

Gao, K., Chen, Q., Pan, B., Sun, Y., Xu, Y., Chen, D., … Huang, C. (2023). Current achievements and future prospects in virus elimination technology for functional Chrysanthemum. *Viruses*, *15*(8), 1770.

Garcia-Ruiz, H. (2019). Host factors against plant viruses. *Molecular Plant Pathology*, *20*(11), 1588–1601.

Gelderblom, H. R. (1996). Structure and classification of viruses. In: Baron S (Ed). *Medical Microbiology*. 4th edition. University of Texas Medical Branch at Galveston.

Gresikova, S. (2022). The transmission of plant viruses. *Agriculture/Pol'nohospodárstvo*, *68*(3), 119–126.

Grondeau, C., Samson, R., & Sands, D. C. (1994). A review of thermotherapy to free plant materials from pathogens, especially seeds from bacteria. *Critical Reviews in Plant Sciences*, *13*(1), 57–75.

Gulzar, B., Mujib, A., Malik, M. Q., Mamgain, J., Syeed, R., & Zafar, N. (2020). Plant Tissue Culture: Agriculture and Industrial Applications. In Kiran, U., Abdin, M. Z., Kamaluddin (eds.), *Transgenic Technology-Based Value Addition in Plant Biotechnology* (pp. 25–49). Academic Press.

Guta, I. C., Buciumeanu, E. C., Tataru, L. D., Oprescu, B., & Topala, C. M. (2019). New approach of electrotherapy for grapevine virus elimination. *Acta Horticulturae, 1242*, 697–702.

Haque, M. I., Singh, P. K., Ghuge, S., Kumar, A., Rai, A. C., Kumar, A., & Modi, A. (2022). A General Introduction to and Background of Plant Tissue Culture: Past, Current, and Future Aspects. In Chandra Rai, A., Kumar, A., Modi, A., Singh, M. (eds.), *Advances in Plant Tissue Culture* (pp. 1–30). Academic Press.

Hilaire, J., Tindale, S., Jones, G., Pingarron-Cardenas, G., Bačnik, K., Ojo, M., & Frewer, L. J. (2022). Risk perception associated with an emerging agri-food risk in Europe: plant viruses in agriculture. *Agriculture & Food Security, 11*(1), 21.

Hull, R. (2014). *Plant Virology*. 5th Edition, Academic Press,

Hussain, G., Wani, M. S., Mir, M. A., Rather, Z. A., & Bhat, K. M. (2014). Micrografting for fruit crop improvement. *African Journal of Biotechnology, 13*(25), 2474–2483.

Hyodo, K., & Okuno, T. (2020). Hijacking of Host Cellular Components as Proviral Factors by Plant-Infecting Viruses. In Carr, J. P., Roossinck, M. J. (eds.), *Advances in Virus Research* (Vol. 107, pp. 37–86). Academic Press.

Idowu, P. E., Ibitoye, D. O., & Ademoyegun, O. T. (2009). Tissue culture as a plant production technique for horticultural crops. *African Journal of Biotechnology, 8*(16), 3782–3788.

Islam, W., Noman, A., Naveed, H., Alamri, S. A., Hashem, M., Huang, Z., & Chen, H. Y. (2020). Plant-insect vector-virus interactions under environmental change. *Science of the Total Environment, 701*, 135044.

Ita, E. E., Uyoh, E. A., Nakamura, I., & Ntui, V. O. (2020). Efficient elimination of Yam mosaic virus (YMV) from white Yam (*Dioscorea rotundata* Poir.) by cryotherapy of axillary buds. *South African Journal of Botany, 130*, 123–129.

Jamshidi-Kia, F., Lorigooini, Z., & Amini-Khoei, H. (2017). Medicinal plants: Past history and future perspective. *Journal of Herbmed Pharmacology, 7*(1), 1–7.

Jiang, T., & Zhou, T. (2023). Unraveling the mechanisms of virus-induced symptom development in plants. *Plants, 12*(15), 2830.

Jones, R. A. (2009). Plant virus emergence and evolution: Origins, new encounter scenarios, factors driving emergence, effects of changing world conditions, and prospects for control. *Virus Res, 141*(2), 113–130. doi: 10.1016/j.virusres.2008.07.028

Jones, R. A. (2021). Global plant virus disease pandemics and epidemics. *Plants, 10*(2), 233.

Jones, R. A., & Naidu, R. A. (2019). Global dimensions of plant virus diseases: Current status and future perspectives. *Annual Review of Virology, 6*, 387–409.

Kan, H., Qu, S., Dong, K., Wang, S., Xu, C., Wang, Y., & Hua, S. (2023). Integrated metabolome and transcriptome analysis unveils the underlying molecular response of *Panax ginseng* plants to the *Phytophthora cactorum* infection. *Agriculture, 13*(2), 509.

Karasawa, M. M. G., Botega, V. T., Pinto, J. E. B. P., Lédo, F. J. S., Vander Pereira, A., & Pinto, J. C. (2023). Effects of thermotherapy and meristem culture on forage production and nutrition value in elephant grass cultivars. *Agronomy Science and Biotechnology, 9*, 1–24.

Khan, B. A., Nadeem, M. A., Nawaz, H., Amin, M. M., Abbasi, G. H., Nadeem, M., … Ayub, M. A. (2023). Pesticides: Impacts on Agriculture Productivity, Environment, and Management Strategies. In Aftab, T. (ed.), *Emerging Contaminants and Plants: Interactions, Adaptations and Remediation Technologies* (pp. 109–134). Springer International Publishing.

Kim, M. H., Jang, S. Y., Choi, J. S., Byun, H. S., Kwak, H. R., Lee, S. H., & Seo, J. K. (2023). Plant viruses and viral host plants newly reported in South Korea in 2020–2022. *Research in Plant Disease, 29*(2), 108–117.

Kim, J. A., & Jeong, H. Y. (2022, August). *In vitro* thermotherapy combined with chemotherapy-based methods for *Cnidium officinale* Makino virus elimination. In *XXXI International Horticultural Congress (IHC2022): International Symposium on Medicinal and Aromatic Plants: Domestication, Acta Horticulturae, 1358* (pp. 79–84).

Koch, M., & Salomon, R. (1993, February). Improvement of garlic via somaclonal variation and virus elimination. In *International Symposium on Alliums for the Tropics, Acta Horticulturae, 358* (pp. 211–214).

Krishna, R., Ansari, W. A., Khandagale, K., Benke, A. P., Soumia, P. S., Manjunathagowda, D. C., … Singh, M. (2022). Meristem Culture: A Potential Technique for *In Vitro* Virus-Free Plants Production in Vegetatively Propagated Crops. In Chandra Rai, A., Kumar, A., Modi, A., Singh, M. (eds.), *Advances in Plant Tissue Culture* (pp. 325–343). Academic Press.

Kumari, N., Gupta, A., Pandey, B. C., Kushwaha, R., & Yaseen, M. (2023). *In Vitro* Cultures: Challenges and Limitations. In Mishra, M. K., Kumari, N. (eds.), *Plants for Immunity and Conservation Strategies* (pp. 371–383). Springer Nature Singapore.

Kumar, S., Kumar, R., & Khan, A. (2011). Medicinal plant resources: Manifestation and prospects of life-sustaining healthcare system. *Continental Journal of Biological Sciences*, *4*(1), 19–29.

Kurz, W. G. W., & Constabel, F. (1979). Plant cell cultures, a potential source of pharmaceuticals. *Advances in Applied Microbiology*, *25*, 209–240.

Kwon, Y. H., Choi, W. I., Kim, H. K., Kim, K. O., Kim, J. H., Huh, Y. S., & Park, W. T. (2022). Efficacy of virus elimination from 'Rehmannia glutinosa' using simultaneous thermotherapy, chemotherapy, and meristem culture. *Plant Omics*, *15*(1), 6–12.

Lal, A., Pant, M., & Rani, A. (2015). The who's who of plant viruses: A cognitive approach. *Asian Journal of Pharmaceutical and Clinical Research*, *8*, 60–68.

Lassois, L., Lepoivre, P., Swennen, R., van den Houwe, I., & Panis, B. (2013). Thermotherapy, Chemotherapy, and Meristem Culture in Banana. In Lambardi, M., Ozudogru, E., Jain, S. (eds.), Protocols for *Micropropagation* of *Selected Economically-Important Horticultural Plants* (pp. 419–433). Humana Press.

Le, C. L. (1999). *In vitro* microtuberization: An evaluation of culture conditions for the production of virus-free seed potatoes. *Potato Research*, *42*, 489–498.

Lee, G., Shin, I. S., Cho, K. H., Kim, S. H., Chun, J. A., Choi, I. M., & Kim, D. H. (2014). Production system of virus-free apple plants using heat treatment and shoot tip culture. 한국원예학회 학술발표요지, 227–227.

Lucas, W., & Knipe, D. M. (2001). Viral Capsids and Envelopes: Structure and Function. eLS. doi: 10.1002/9780470015902.a0001091.pub2

Matthews, R. E. F. (2019). *Diagnosis of Plant Virus Diseases.* CRC Press.

Medina-Puche, L., & Lozano-Duran, R. (2019). Tailoring the cell: A glimpse of how plant viruses manipulate their hosts. *Current Opinion in Plant Biology*, *52*, 164–173.

Mehetre, G. T., Leo, V. V., Singh, G., Sorokan, A., Maksimov, I., Yadav, M. K., … Singh, B. P. (2021). Current developments and challenges in plant viral diagnostics: A systematic review. *Viruses*, *13*(3), 412.

Misra, P., & Shukla, P. K. (2010). *In Vitro Approaches for Mass Production of Medicinal Plants.* International E-Publication.

Mohamed, E., Mustafa, N. A. Z., Sani, H. I., & Abd Aziz, M. H. (2022). Performance of engineering plant virus resistance: miRNA gene silencing. In: *Chemical Process and Sustainability in Agricultural Biotechnology* (pp. 39–50). Penerbit UTHM, UTHM.

Mwangangi, M., Ateka, E., Nyende, A., & Kagundu, A. (2014). Elimination of cassava brown streak virus from infected cassava. *Journal of Biology, Agriculture and Healthcare*, *4*(13), 34–40.

Naidu, R. A., & Hughes, J. D. A. (2003). Methods for the Detection of Plant Virus Diseases. In Jacqueline, d'A. H., Babajide, O. O. (eds.), *Proceedings of a Conference on Plant Virology in Sub Saharan Africa* (Organized by IITA, Ibadan, pp. 233–260). International Institute of Tropical Agriculture. Oyo State, Nigeria..

Nehra, N. S., & Kartha, K. K. (1994). Meristem and Shoot Tip Culture: Requirements and Applications. In Vasil, I. K., Thorpe, T. A. (eds.), *Plant Cell and Tissue Culture* (pp. 37–70). Springer Netherlands.

Nicaise, V. (2014). Crop immunity against viruses: Outcomes and future challenges. *Frontiers in Plant Science*, *5*, 660.

Okole, B. N., Odhav, B., & Bornman, C. H. (2004). Commercialisation of plants in Africa. *South African Journal of Botany*, *70*(1), 109–115.

Omiti, J. M., Chacha, R. N., & Andama, M. S. (2002). Biotechnology can improve food security in Africa. *African Journal of Food, Agriculture, Nutrition and Development*, *2*(2), 14–21.

Oves, E., & Gaytova, N. (2016). New elements in technology of improvement and obtaining of basic potato clones of promising cultivars and hybrids. *Achievements of Science and Technology in Agro-Industrial Complex*, *30*(11), 60–62.

Panattoni, A., Luvisi, A., & Triolo, E. (2013). Elimination of viruses in plants: Twenty years of progress. *Spanish Journal of Agricultural Research*, *11*(1), 173–188.

Pandey, S., & Pandey, V. C. (2018). Plant Tissue Culture Techniques and Its Achievements. In Bharadwaj, D.N. (ed.) *Advanced Molecular Plant Breeding: Meeting the Challenge of Food Security* (pp. 329–366). CRC Press.

Pant, B. (2014). Application of Plant Cell and Tissue Culture for the Production of Phytochemicals in Medicinal Plants. In Adhikari, R., Thapa, S. (eds.), *Infectious Diseases and Nanomedicine II: First International Conference (ICIDN–2012), Dec. 15-18, 2012, Kathmandu, Nepal* (pp. 25–39). Springer.

Pant, M. (2016). A minimal cost micropropagation protocol for *Dianthus caryophyllus* L.—a commercially significant venture. *Indian Journal of Experimental Biology*, 54(3), 203–211.

Pant, M., Bhandari, A., & Husen, A. (2022). Adventitious Root Formation and Clonal Propagation of Forest-Based Tree Species. In Husen, A. (ed.), *Environmental, Physiological and Chemical Controls of Adventitious Rooting in Cuttings* (pp. 471–490). Academic Press.

Pant, M., & Husen, A. (2022). Micropropagation in Mature Trees by Manipulation of Phase Change, Stress, and Culture Environment. In Husen, A. (ed.), *Environmental, Physiological and Chemical Controls of Adventitious Rooting in Cuttings* (pp. 421–437). Academic Press.

Parmessur, Y., Aljanabi, S., Saumtally, S., & Dookun-Saumtally, A. (2002). Sugarcane yellow leaf virus and sugarcane yellows phytoplasma: Elimination by tissue culture. *Plant Pathology*, *51*(5), 561–566.

Pathak, M. R., & Abido, M. S. (2014). The role of biotechnology in the conservation of biodiversity. *Journal of Experimental Biology*, *2*, 352–363.

Pe, P. P. W., Naing, A. H., Soe, M. T., Kang, H., Park, K. I., & Kim, C. K. (2020). Establishment of meristem culture for virus-free and genetically stable production of the endangered plant Hosta capitata. *Scientia Horticulturae*, *272*, 109591.

Phillips, G. C., & Garda, M. (2019). Plant tissue culture media and practices: An overview. *In Vitro Cellular & Developmental Biology-Plant*, *55*, 242–257.

Quiroz, K. A., Berríos, M., Carrasco, B., Retamales, J. B., Caligari, P. D., & García-Gonzáles, R. (2017). Meristem culture and subsequent micropropagation of Chilean strawberry (*Fragaria chiloensis* (L.) Duch.). *Biological Research*, *50*, 20.

Rahimi, S., Bayati, M., Kordrostami, M., & Ghasemi-Soloklui, A. A. (2023). Tissue Culture Techniques to Conserve Endangered Medicinal Plants with Antimicrobial and Antiviral Activities. In Jha, S., Halder, M. (eds.), *Medicinal Plants: Biodiversity, Biotechnology and Conservation* (pp. 675–710). Springer Nature Singapore.

Rani, V., Sengar, R. S., & Gupta, S. (2022). *In vitro* techniques for the development of tissue culture plantlets. *Progressive Agriculture*, *22*(2), 104–109.

Raza, A., Haris, M., Riaz, A., Shaheen, M., Sabir, S., Ali, M. Z., … Rehman, M. T. (2023). Overview of diverse universe of mycoviruses and their role in pathogenesis of fungi. *Science Letters*, *11*(1), 35–49.

Rohit, S. (2014). Micropropagation: An essential tool to flourish endangered medicinal plants. *Global Journal of Research on Medicinal Plants & Indigenous Medicine*, *3*(6), 252.

Roossinck, M. J. (2015). Plants, viruses and the environment: Ecology and mutualism. *Virology*, *479*, 271–277.

Rubio, L., Galipienso, L., & Ferriol, I. (2020) Detection of plant viruses and disease management: Relevance of genetic diversity and evolution. *Frontiers in Plant Science*, 11:1092. doi: 10.3389/fpls.2020.01092

Sabharwal, P. (2019). *Molecular Insights into the Structure and Function of Pepper Vein Banding Virus Encoded Proteins and Endocytic Uptake Pathway of Virus-like Particles into Mammalian Cells* (Doctoral dissertation).

Saenz, P., Salvador, B., Simón-Mateo, C., Kasschau, K. D., Carrington, J. C., & García, J. A. (2002). Host-specific involvement of the HC protein in the long-distance movement of potyviruses. *Journal of Virology*, *76*(4), 1922–1931.

Sahijram, L. (2015). Somaclonal Variation in Micropropagated Plants. In Bahadur, B., Venkat-Rajam, M., Sahijram, L., Krishnamurthy, K.V. (eds.), *Plant Biology and Biotechnology: Volume II: Plant Genomics and Biotechnology* (pp. 407–416). Springer.

Shan, M., Khan, Y. A., & Nazar, H. (2018). Application of biotechnology in medicinal field. *Single Cell Biology*, *7*, 174.

Shukla, S. K. (2023). Conservation of medicinal plants: Challenges and opportunities. *Journal of Medicinal Botany*, *7*, 5–10.

Shuler, M. L., Pyne, J. W., & Hallsby, G. A. (1984). Prospects and problems in the large scale production of metabolites from plant cell tissue cultures. *Journal of the American Oil Chemists' Society*, *61*(11), 1724–1728.

Singh, B., & Kaur, A. (2016). *In vitro* production of PLRV and PSTVd-free plants of potato using electrotherapy. *Journal of Crop Science and Biotechnology*, *19*, 285–294.

Souza, P. F., Garcia-Ruiz, H., & Carvalho, F. E. (2019). What proteomics can reveal about plant–virus interactions? Photosynthesis-related proteins on the spotlight. *Theoretical and Experimental Plant Physiology*, *31*, 227–248.

Souza, P. F. N., & Carvalho, F. E. L. (2019). Killing two birds with one stone: How do plant viruses break down plant defenses and manipulate cellular processes to replicate themselves? *Journal of Plant Biology*, *62*, 170–180.

Sudheer, W. N., Thiruvengadam, M., & Nagella, P. (2023). A comprehensive review on tissue culture studies and secondary metabolite production in *Bacopa monnieri* L. Pennell: A nootropic plant. *Critical Reviews in Biotechnology*, *43*, 956–970.

Tsay, H. S. (1982). Application of plant tissue culture in agriculture. *Journal-Agricultural Association of China*, *120*, 9–21.

Viswanath, M., Ravindra Kumar, K., Chetanchidambar, N. M., & Mahesh, S. S. N. M. (2023). Regeneration mechanisms in plant tissue culture: A. *Journal of Pharmaceutical Innovation*, *12*, 2948–2952.

Walkey, D. G. (1991). Production of Virus-Free Plants. In *Applied Plant Virology* (pp. 270–292). Springer Netherlands.

Wang, A. (2021). Cell-to-cell movement of plant viruses via plasmodesmata: A current perspective on potyviruses. *Current Opinion in Virology*, *48*, 10–16.

Wang, M. R., Chen, L., Zhang, Z., Blystad, D. R., & Wang, Q. C. (2018a). Cryotherapy: A Novel Method for Virus Eradication in Economically Important Plant Species. In Loyola-Vargas, V., Ochoa-Alejo, N. (eds.), *Plant Cell Culture Protocols* (pp. 257–268). Humana Press.

Wang, M. R., Cui, Z. H., Li, J. W., Hao, X. Y., Zhao, L., & Wang, Q. C. (2018b). *In vitro* thermotherapy-based methods for plant virus eradication. *Plant Methods*, *14*(1), 1–18.

Wang, M. R., Li, B. Q., Feng, C. H., & Wang, Q. C. (2016). Culture of shoot tips from adventitious shoots can eradicate Apple stem pitting virus but fails in Apple stem grooving virus. *Plant Cell, Tissue and Organ Culture (PCTOC)*, *125*, 283–291.

Wang, P., Liu, J., Lyu, Y., Huang, Z., Zhang, X., Sun, B., … Zhang, C. (2022a). A review of vector-borne rice viruses. *Viruses*, *14*(10), 2258. doi: 10.3390/v14102258

Wang, Y., Pruitt, R. N., Nürnberger, T., & Wang, Y. (2022b). Evasion of plant immunity by microbial pathogens. *Nature Reviews Microbiology*, *20*(8), 449–464.

Xu, T., Duan, H., Cai, X., Yang, R., Yao, F., Tan, B., & Liu, F. (2023). Cutting efficiency of a newly developed incubator-based laser cutting method and its effects on potato tissue culture plantlet growth. *Potato Research*, *66*(3), 889–903.

Yadav, N., & Khurana, S. P. (2016). Plant Virus Detection and Diagnosis: Progress and Challenges. In Shukla, P. (ed.), Frontier *Discoveries* and *Innovations* in *Interdisciplinary Microbiology* (pp. 97–132). Springer.

Yakhshiboeva, D. T., & Jumaniyozova, N. D. (2023). Prospects of microclonal propagation of medicinal plant species. *American Journal of Agriculture and Horticulture Innovations*, *3*(3), 11–14.

Yan, K., Du, X., & Mao, B. (2022). Production of Virus-Free Chrysanthemum (*Chrysanthemum morifolium* Ramat) by Tissue Culture Techniques. Wang, A., Li, Y. (eds.), *Plant Virology: Methods and Protocols* (pp. 171–186). Humana.

Yarra, R., Jin, L., Zhao, Z., & Cao, H. (2019). Progress in tissue culture and genetic transformation of oil palm: An overview. *International Journal of Molecular Sciences*, *20*(21), 5353.

Zaitlin, M., & Palukaitis, P. (2000). Advances in understanding plant viruses and virus diseases. *Annual Review of Phytopathology*, *38*(1), 117–143.

Zanardo, L. G., de Souza, G. B., & Alves, M. S. (2019). Transcriptomics of plant–virus interactions: A review. *Theoretical and Experimental Plant Physiology*, *31*, 103–125.

Zanotti, G., & Grinzato, A. (2021). Structure of filamentous viruses. *Current Opinion in Virology*, *51*, 25–33.

Zhao, S., & Li, Y. (2021). Current understanding of the interplays between host hormones and plant viral infections. *PLoS Pathogens*, *17*(2), e1009242.

9 Somatic Embryogenesis

A Pathway to Conservation and Medical Advancement

Arundhati Singh, Vedanshi Pal, and Manu Pant

9.1 INTRODUCTION

Nature has long been a reservoir of healing, with approximately 85% of traditional medicinal practices drawing from the rich tapestry of plant extracts. About 80% of the population still uses these herbal treatments in developed and developing countries (Vieira and Skorupa, 1993). There has been advancement in numerous primary therapies, but there is a fast-growing shift toward herbal medicine driven by increasing worries about the toxicities associated with conventional treatments (Akram et al., 2021). However, there is an urgent need to protect some plant species since they are in danger of extinction. Plant tissue culture is an ingenious strategy poised at the intersection of rapid mass propagation, widescale revegetation, and germplasm preservation (Fay, 1992). At the heart of this biotechnological marvel lies the enigmatic process known as somatic embryogenesis (SE), a remarkable journey that holds the promise of revolutionizing the conservation and production of medicinal plants (Simonovic et al., 2020).

SE, which results in the creation of somatic embryos, is the process through which somatic cells go through differentiation. Evolved from asexual embryogenesis, somatic embryogenesis occurs in response to varied environmental and genetic barriers to fertilization. SE only happens in a limited setting in nature; this is seen, for example, in *Paeonia* ovules, and numerous *Kalanchoe* species have tiny bipolar structures visible at the leaf margins, and a portion of these species create these structures on their own, whereas others do so in reaction to environmental stressors (Garcês & Sinha, 2009).

Under *in vitro* conditions, somatic embryos can develop either directly from the cultured explant or indirectly through an intermediate stage of embryogenic calli (Sharp et al., 1980; Zhang et al., 2021). The somatic embryos formed may be unicellular or multicellular in origin. The developmental path includes globular-shaped, heart-shaped, and torpedo-shaped embryo development before the cotyledonary phase appears, following the parallel developmental path of a zygotic embryo. The embryogenic response varies significantly with the choice of explants, plant growth regulators, and nitrogen source in the media. The technique has been implemented for mass production and conservation of a wide range of medicinal plant species. Large-scale production and conservation through the development of artificial seeds (encapsulated somatic embryos) and adoption of bioreactors for production, ensuring clonal propagation due to the single cell origin of somatic embryos, and rapid plant production from a single cell (eliminating the need for a separate rooting procedure) are some of the significant applications of SE that distinguishes it from the other methods of tissue culture–based propagation.

This chapter elucidates the intricate process and multifaceted applications of SE, revealing its pivotal role in advancing the propagation, preservation, and sustainable deployment of medicinal plant resources.

9.2 TYPES OF SOMATIC EMBRYOGENESIS

There are broadly two types of SE: Direct and indirect. It can occur in explants due to abiotic stress or using plant growth regulators like 2,4-dichlorophenoxyacetic acid (2,4-D), a synthetic auxin. A

DOI: 10.1201/b23374-9

crucial first step in the transition toward totipotent developmental pathways is the synthesis of internal auxins, which is promoted by administering both auxin and stress (Feher, 2015). At the histological level, the advancement of developmental processes following the start of SE within a cultured explant has been widely described with the word "undifferentiated" or "dedifferentiated" which has historically been assigned to the callus due to its initial lack of defined structure, and this nomenclature still holds today (Ikeuchi et al., 2013). Subsequently, there are two possible developmental trajectories, the direct or the indirect pathway, of embryo formation. However, practical application frequently makes it difficult to tell the two apart because they frequently appear in the same explanation. Creating a callus, a collection of cells initially characterized by vacuolation and exhibiting variable degrees of compactness, marks the beginning of the indirect SE pathway, which predominates. Proembryogenic masses (PEMs) emerge on the surface or within the collective callus after the formation of the embryogenic callus (Guzzo et al., 1995; Raghavan, 2004). Following the initiation of an embryogenic callus, individual cells or clusters of cells depart from these PEMs, commencing their trajectory toward embryonic development (Toonen et al., 1994; Halperin, 1966). Noteworthy is that both callus and PEMs frequently manifest after auxin treatment, especially in the presence of 2,4-D. Despite auxin's role in fostering the expansion and formation of callus and PEMs, its subsequent removal is often imperative to facilitate histogenesis, encompassing processes like apical-basal and bilateral patterning and the elongation of the embryo (Hazubska-Przybył et al., 2020).

When there is no callus formation, it reflects the key distinguishing feature of the alternative method, known as direct embryogenesis (Long et al., 2022). Within this paradigm, the explant displays a less prolific pattern of compact cell divisions but one that is more controlled and constrained. As stated earlier, while somatic embryos are generated through direct and indirect SE, they exhibit numerous physical similarities; the genetic milieu can change embryos originating from the indirect SE route, a phenomenon termed somaclonal variation (Miguel and Marum, 2011).

Furthermore, somatic embryos can initiate a subsequent SE cycle, termed secondary or cyclic embryogenesis (Raemakers et al., 1995). These secondary embryos can originate directly from primary embryos or indirectly through the progression of embryogenic calluses. In the past, it was thought that an explant's age would determine whether it could go through direct or indirect embryogenesis. According to a generalization made by Merkle et al. (1995), the differences in explants from their zygotic embryo stage lead to more callus formation.

9.3 STAGES OF SOMATIC EMBRYOGENESIS

The complexity of higher plant embryogenesis has been the subject of major scientific efforts throughout the past century (Armenta-Medina et al., 2021). However, comprehension of the complex mechanisms regulating the transition from somatic cells to the complex patterns of embryo morphogenesis shows a clear imbalance. Studies utilizing light and electron microscopy have fully illustrated the morphological and anatomical changes underlying embryonic development (Raghavan, 1986), but our knowledge of the underlying molecular pathways still needs to be improved. To bridge the gap between observable morphological changes and the underlying molecular reasons, researchers are now paying more attention to this area.

9.3.1 Embryogenic Initiation

Somatic cells that have undergone differentiation gain the ability to develop embryonic characteristics and undergo proliferation, refer to Figure 9.1, eventually transforming into cells with embryonic properties during the initial stages of SE. This shift is supported by the activation of genes, which drives an embryogenic gene expression program while changing the tissue's current gene profile (Quiroz-Figueroa et al., 2006). Determining the critical elements activating the embryogenic pathway is essential to this procedure. According to Dudits et al. (1991) and De Jong et al. (1993), plant

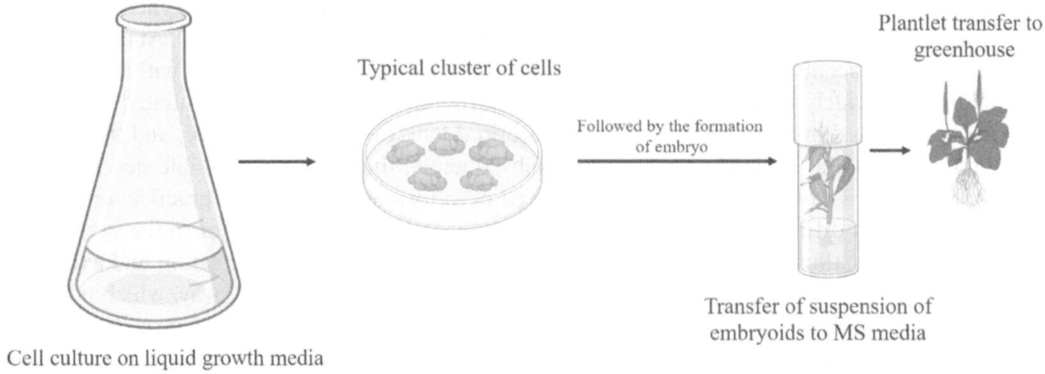

FIGURE 9.1 Stages of somatic embryogenesis.

growth regulators (PGRs) and stress responses operate as signal initiators, causing a cascade of gene reprogramming that causes cell division and, subsequently, SE. Notably, auxin acts as the principal inducer of auxin responses, which promote adaptability by causing genetic, metabolic, and physiological changes (Feher et al., 2003).

Growth regulators such as cytokinin or abscisic acid play a role in SE responses in addition to auxin (Choi et al., 1998; Sagare et al., 2000; Senger et al., 2001). Numerous plant species have shown the ability of contemporary growth regulators, such as oligosaccharides, jasmonates, polyamines, and brassino steroids, to initiate SE (Jiménez and Thomas, 2005). PEM development results from the subsequent embryogenic cell proliferation, which is triggered by auxin and controlled by species-specific embryo differentiation (Li et al., 2022). Reduced nitrogen is a new component that affects embryogenic induction and development in various ways (Stuart and Strickland, 1984; Trigiano et al., 1992).

Direct embryogenesis requires minor reprogramming for calli to develop embryogenic potential, but indirect embryogenesis necessitates significant modifications (Williams and Maheswaran, 1986; Yeung, 1995). Explant source, physiological condition, and growth regulators are among the factors. Microspores, ovules, somatic embryos, and seedlings favor direct embryogenesis, whereas indirect embryo formation is extensively studied and used in transformation and hybridization, for instance (Jin et al., 2005; Li et al., 2006; Yang et al., 2007).

9.3.2 Origin of Somatic Embryos

As reported by Raghavan (1976), and Haccius (1978), somatic embryos often have one of two origins: Either unicellular or multicellular. The coordinated behavior of nearby cells within a morphogenetic cluster directly affects whether somatic embryos are single celled or multicellular. Synchronized cell divisions occur in cases of unicellular origin, occasionally connecting the embryo to maternal tissue via a suspensor-like structure (Williams and Maheswaran, 1986). In contrast, protrusions lacking synchronized cell divisions characterize embryos with multicellular origins, frequently united with the maternal tissue in their basal region (Quiroz-Figueroa et al., 2006).

9.3.3 Development Stages of the Embryos

Conceptually, the morphogenetic phase that establishes the embryo's basic structure comes first, followed by the metabolic phase, which is characterized by biochemical processes that prepare the embryo for quiescence (Lopes and Larkins, 1993; Méndez-Hernández et al., 2019). Dicot zygotic embryo development occurs in four successive phases throughout the morphogenetic stage: Globular, heart-shaped, torpedo-shaped, and cotyledonal. Surprisingly, these morphological stages are mirrored in somatic embryos (Kawahara and Komamine, 1995; Rogo et al., 2023). PEMs transform into torpedo- and heart-shaped shapes before coalescing into a typical globular embryo (Calabuig-Serna et al., 2023). Somatic embryos depart from their zygotic counterparts at the cotyledon stage; instead of going into developmental stasis, they shoot meristems and grow seedlings. Somatic embryos differ from their zygotic counterparts in that they are more manageable, have controllable culture conditions, and provide a wealth of material for experiments (Kawahara and Komamine, 1995).

9.4 FACTORS INFLUENCING SOMATIC EMBRYOGENESIS

SE is a three-step process involving embryogenesis induction, embryo development, and embryo maturation, with the latter phase influenced by several key factors, as discussed next.

9.4.1 Choice of Explant

Various types of explants can be used to initiate SE, such as anthers, pollen, leaves, petioles, embryos, cotyledons, and tubers (Solangi et al., 2023). In many cases, immature zygotic embryos serve as optimal explants for SE. For example, conifers use immature zygotic embryo explants for this process, particularly cells in the suspensor region of the zygotic embryo (Zhang et al., 2021). However, in soybeans, cotyledons show a better response to embryogenesis than immature zygotic embryos. Meanwhile, in alfalfa, sections of petioles from young, fully expanded leaves are capable of inducing SE (Gupta and Durzan, 1986; Bhansali, 1990). The developmental stage of the explant is crucial for the formation of embryogenic callus. Young and juvenile explants tend to produce more somatic embryos than older ones, and even tissues from the same plant may respond differently (Lu and Vasil, 1982; Woodward and Puonti, 2001; Panaia et al., 2004).

9.4.2 Genetic Variation

Genotypic variation between plants can significantly influence the embryogenesis process, possibly due to differences in endogenous hormone levels (Zdravković-Korać et al., 2023). For example, in a study it was shown that while most alfalfa varieties have a regenerative capacity of 10%, the Rangelander variety has a higher growth frequency.

9.4.3 Growth Regulators

The most used protocol for initiating SE involves the initial development of a callus in a medium supplemented with auxins, followed by SE when the callus is transferred to a low-growth regulator medium (Raghavan, 1976; Smith and Krikorian, 1990). The establishment and maintenance of embryogenic cultures in most species has relied primarily on the manipulation of growth regulators. In particular, the presence of auxin promotes callus proliferation but inhibits differentiation. Removing or reducing auxin levels triggers somatic embryo development (Smith and Krikorian, 1990).

Auxin plays a crucial role in the first step of embryogenesis, especially in the induction phase (Hazubska-Przybył et al., 2020). In most well-studied plant species, synthetic auxin is typically required for the induction of SE (Hazubska-Przybył et al., 2020). However, some studies have shown that induction of embryogenesis in explants can also be achieved by adjusting factors such as higher sucrose concentrations or varying pH. For carrots, for example, increasing the pH to 5.6 promotes explant proliferation. Staple grains or grasses typically require auxin (2,4-D) in media, typically at concentrations in the range of 1–2.5 mg/L. Excessive auxin levels can inhibit embryogenesis in explants of citrus plants (Gupta and Durzan, 1986; Bhansali, 1990).

It is from the culture medium that serves as a signal for callus cells to initiate organized growth patterns. This suggests that the presence of auxin may prepare PEMs for the globular stage of embryogenesis and contain inhibitors to the progression of embryogenesis (Zimmerman, 1993). These are known to establish cell polarity, particularly in early spherical embryos, which is crucial for bilateral symmetry during plant embryogenesis. The correct distribution of auxin gradients is crucial for polarity determination (Liu et al., 1993).

Cytokinin, an important tissue culture growth regulator, induces cell division in explants. While some studies have reported that adding BAP (a type of cytokinin) to the medium can inhibit embryogenesis, others have found that cytokinin promotes embryogenesis, as seen in coffee, where cytokinin alone is sufficient to promote embryogenesis to induce explants (Gupta and Durzan, 1986; Bhansali, 1990). Cytokinins, like thidiazuron (TDZ), can stimulate SE and shoot regeneration, possibly by influencing endogenous auxin/cytokinin levels within cell-cultured tissue (Panaia et al., 2004).

Different plant species show different sensitivities to different growth regulators, leading to different responses in SE. This variation exists even within the same plant family or between cultivars (Panaia et al., 2004). Due to species-specific responses, it is essential in research on SE to first test a range of PGRs to determine the most effective combination (Vasil and Vasil, 1981).

9.4.4 Nitrogen Source

The form of nitrogen used in the culture media influences the embryogenesis process (Duarte-Aké et al., 2022). For example, reduced nitrogen sources such as KNO_3 are required to induce embryogenesis in carrots, while ammonia is required for alfalfa and orchard grass (Gupta and Durzan, 1986). In addition, studies have shown that the presence of specific amino acids, particularly proline and serine/threonine, in the media enhances embryogenic responses in many plant species (Bhansali, 1990; Duarte-Aké et al., 2022).

9.4.5 Polyamines

The concentration of polyamines (e.g., putrescine, spermidine, and spermine) in explants or media can influence SE. Typically, higher concentrations of polyamines are found in polyembryonic plants than in monoembryonic plants (Gupta and Durzan, 1986; Chen et al., 2019). However, the exact relationship between polyamine concentration and induction of SE requires further research (Bhansali, 1990).

9.5 OTHER BIOCHEMICAL FACTORS AFFECTING SOMATIC EMBRYOGENESIS

Certain bioactive compounds, including amino acids such as glutamine, proline, and tryptophan and polyamines such as putrescine, have been found to promote SE in certain plant species. These compounds contribute to cellular processes, enhance signal transduction pathways, serve as precursor molecules for growth regulators, and regulate DNA synthesis (Astarita et al., 2003). Some researchers have highlighted the importance of using complex organic extracts such as coconut water (CW), taro extract (TE), potato extract (PE), corn extract (CE), and papaya extract (PAE) in SE for certain species (Mose et al., 2020). Although organic extracts have been beneficial for SE, it is important to note that these extracts contain undefined components. This can make it difficult to identify the specific components of the extract that promote SE and can also lead to inconsistencies in the composition of the extract with each preparation (Nic-Can et al., 2015).

9.6 ADVANTAGES OF SOMATIC EMBRYOGENESIS

In terms of plant production, this method is commercially viable for mass-producing plants (Jiménez, 2001). It stands out from traditional micropropagation by enabling the growth of large quantities of somatic embryos in bioreactors, potentially up to 1.35 million somatic embryos per liter of culture medium, with simultaneous root and shoot development, thus obviating the need for separate rooting stages typical of conventional micropropagation (Giri et al., 2004; Egertsdotter et al., 2019). It also simplifies the process by eliminating the need for separate root induction, thus speeding up the entire cycle of plantlet growth and adjustment to new environments (Gupta and Durzan, 1986; Bhansali, 1990). The bipolar nature of the embryos facilitates direct development into plantlets, bypassing the rooting stage necessary in organogenesis (Burg et al., 2007). These make SE a method with a lot of potential for mass production of medicinal plant species.

Besides, SE is frequently used for large-scale production and the creation of synthetic seeds, which further accentuates the advantages of this technique. The single-cell origin of somatic embryos ensures clonal plant propagation and makes somatic embryos an excellent method for mutation induction and minimizing the occurrence of chimeras (Normah et al., 2013). While the germination rate of somatic embryos can be a limitation, particularly for large-scale propagation of desired mutations, the method streamlines the development and growth of superior and new plant varieties (Shareef, 2017).

Overall, this cultivation technique is easily scalable and requires less labor for subculturing (Mondal, 2020). It allows for the synchronization of embryo development and germination, leading to optimized plant yields and labor reduction (Mondal, 2020).

9.7 SOMATIC EMBRYOGENESIS IN MEDICINAL PLANTS

Biotechnological methods like *in vitro* propagation and genetic techniques are pivotal in improving medicinal herbs, shrubs, and trees (Merkle and Dean, 2000). In hard-to-propagate woody plants, SE is specifically a preferred method (Germanà and Lambardi, 2016). For instance, in eucalyptus, a tree species valued for pulp and eucalyptus oil, SE has been reported from leaf and shoot tip culture (Prakash and Gurumurthi, 2010; Corredoira et al., 2015). Sandalwood (*Santalum album*), a valuable source of sandalwood essential oil, has also been propagated via bioreactor-based cultivation of SE (Misra and Dey, 2013). SE in sandalwood has also been employed for enhanced production of medicinal raw materials such as santalols, phenols, and arabinogalactan proteins (Peeris and Senarath, 2015).

SE has also been studied other medicinal plants. In *Catahranthus roseus*, indirect embryogenesis was reported with semi-solid media supplemented with 2,4-D for callus development, GA_3 for embryo maturation, and BAP for germination (Mujib et al., 2014). The study also showed that while solid media fostered the highest number of normal somatic embryos, the transition to liquid media in bioreactors markedly enhanced growth and germination rates. Embryogenic callus from leaves of the endangered medicinal plant *Wedelia calendulacea* was initiated on MS medium containing 2,4-D, while somatic embryo germination was reported on medium supplemented with BAP and GA_3 (Sharmin et al., 2014). In *Citrullus colocynthis*, leaf explants were used for somatic embryo development on MS medium containing BAP and NAA under intermittent light conditions (Ramakrishna and Shasthree, 2016).

In *Portulaca oleracea*, Sedaghati et al. (2019) utilized stem and leaf explants for SE, emphasizing the significance of cytokinins and auxins in the process. In soybean, immature cotyledons have been shown to play a significant role in the development of somatic embryos besides inclusion of 2,4-D and BAP in the culture medium (Raza et al., 2019). The members of the Gentianaceae family (genus examples include *Swertia*, *Gentiana*, *Exacum*, *Centaurium*, *Eustoma*, and *Bupleurum*) have particularly responded well to SE treatments on a variety of media like MS (Murashige and Skoog's, 1962), woody plant medium (WPM, Lloyd and McCown, 1981), and B5 medium (Gamborg et al., 1968) supplemented with auxins and cytokinins (Tomiczak et al., 2019).

SE in *Ferula jaeschkeana*, an endangered medicinal herb, was reported from petiole explants on medium supplemented with 2,4-D and Kn (during callus induction) and auxins during embryo development (Sharma and Khajuria, 2020). The study showed that the presence of abscisic acid, proline in the medium, and the number of subcultures significantly affected embryo maturation. In another study on *Viola canescens*, MS medium supplemented with L-glutamine and casein hydrolysate (1–4%) proved to be beneficial for somatic embryo development, while the inclusion of abscisic acid in the medium was essential for embryo maturation (Khajuria et al., 2021).

9.8 CHALLENGES OF SOMATIC EMBRYOGENESIS

Despite the numerous advantages of the technique of SE, it is beset with some challenges.

Asynchronous development: A significant challenge in SE is the asynchronous development of somatic embryos, leading to a mixture of embryos at various developmental stages in the same culture (Egertsdotter et al., 2019). Although strategies like sorting cell aggregates by size and density have been employed to achieve synchrony in embryo development, these methods can diminish the overall formation rate of somatic embryos (Fujimura and Komamine, 1979). Besides, reducing the size of cell clusters may adversely affect the frequency of embryo development (Chée and Cantliffe, 1989).

Cell line stability and somaclonal variations: Another challenge involves the stability of cell lines over extended periods. The ability of cells to undergo or complete the embryogenic process can decline over time, potentially leading to unsuccessful regeneration (Burg et al., 2007). While this issue of stability is concerning, it can simultaneously serve as a mechanism to limit the accumulation of unwanted mutations (somaclonal variations) that arise from prolonged culture durations (Ammirato, 1983). Initiating new cultures regularly can help produce phenotypically normal somatic embryos and plants, thus reducing the incidence of somaclonal variations. While somaclonal variations pose a challenge in maintaining genetic fidelity, they also offer an opportunity to generate unique and beneficial plant varieties. Although considered a drawback, this variation can be exploited for the development of plants with novel traits (Duta-Cornescu et al., 2023).

9.9 CONCLUSION

The investigation and application of SE within the realm of medicinal plants mark a transformative chapter in the convergence of agriculture, conservation, and pharmaceutical sciences. SE stands as a beacon of innovation, heralding the potential for large-scale cultivation of medicinal plants, safeguarding of genetic diversity, and reduction of extractive pressures on wild plant populations. This advanced biotechnological approach not only ensures a consistent supply of essential bioactive compounds but also champions the cause of biodiversity conservation through *ex situ* strategies. Moreover, the integration of SE with genetic and genomic research offers unprecedented opportunities for enhancing the production of specific medicinal compounds. By employing genetic engineering techniques, scientists can tailor plant metabolic pathways to optimize the synthesis of desired therapeutic agents, thus revolutionizing the field of pharmaceuticals and herbal medicine. Additionally, the establishment of robust cryopreservation techniques for somatic embryos represents a vital stride toward the long-term conservation of rare and endangered medicinal plant species, ensuring their availability for future generations. The exploration of advanced analytical technologies such as metabolomics and proteomics can further refine our understanding of SE, enabling the optimization of culture conditions and enhancement of product yields.

As we venture into the future, the scalability and economic feasibility of SE for commercial purposes remain paramount. Addressing the technical and economic challenges associated with automation, bioreactor designs, and cost reduction will be crucial in transitioning SE from laboratory research to widespread agricultural and pharmaceutical applications. The pursuit of these advancements promises not only to fortify the supply chain of medicinal plants but also to forge new paths in the sustainable and ethical utilization of botanical resources for human health and well-being.

REFERENCES

Akram, M., Mohiuddin, E., Adetunji, C. O., Oladosun, T. O., Ozolua, P., Olisaka, F. N., Egbuna, C., Micheal, O. S., Adetunji, J. B., Hameed, L., Awuchi, C. G., Patrick-Iwuanyanwu, K. C., & Olaniyan, O. T. 2021. Prospects of phytochemicals for the treatment of helminthiasis. In: Egbuna, C., Akram, M., Ifemejee, J. C. (eds) *Neglected Tropical Diseases and Phytochemicals in Drug Discovery*: 199–223. Wiley.

Ammirato, P. V. 1983. Embryogenesis. In: Evans, D. A., Sharp, W. R., Ammirato, P. V., and Yamada, Y. (eds) *Handbook of Plant Cell Culture Volume 1*: 82–123, New York: MacMillan.

Armenta-Medina, A., Gillmor, C. S., Gao, P., Mora-Macias, J., Kochian, L. V., Xiang, D., & Datla, R. 2021. Developmental and genomic architecture of plant embryogenesis: From model plant to crops. *Plant Communications*, 2(1):100136.

Astarita, L. V., Handro, W., & Floh, E. I. 2003. Changes in polyamines content associated with zygotic embryogenesis in the Brazilian pine, *Araucaria angustifolia* (Bert.) O. Ktze. *Brazilian Journal of Botany*, 26:163–168.

Atta, R., Laurens, L., Boucheron-Dubuisson, E., Guivarc'h, A., Carnero, E., Giraudat-Pautot, V., Rech, P., & Chriqui, D. 2009. Pluripotency of *Arabidopsis xylem* pericycle underlies shoot regeneration from root and hypocotyl explants grown *in vitro*. *The Plant Journal*, 57(4):626–644.

Bhansali, R. R. 1990. Somatic embryogenesis and regeneration of in plantlets in pomegranate. *Annals of Botany*, 66(3):249–253.

Burg, K., Helmersson, A., Bozhkov, P., & Von Arnold, S. 2007. Developmental and genetic variation in nuclear microsatellite stability during somatic embryogenesis in pine. *Journal of Experimental Botany*, 58(3):687–698.

Calabuig-Serna, A., Mir, R., Arjona, P., & Seguí-Simarro, J. M. 2023. Calcium dynamics and modulation in carrot somatic embryogenesis. *Frontiers in Plant Science*, 14: 1150198.

Che, P., Lall, S., & Howell, S. H. 2007. Developmental steps in acquiring competence for shoot development in Arabidopsis tissue culture. *Planta*, 226:1183–1194.

Chée, R. P., & Cantliffe, D. J. 1989. Composition of embryogenic suspension cultures of *Ipomoea batatas* Poir. and production of individualized embryos. *Plant Cell, Tissue and Organ Culture*, 17:39–52.

Chen, D., Shao, Q., Yin, L., Younis, A., & Zheng, B. 2019. Polyamine function in plants: Metabolism, regulation on development, and roles in abiotic stress responses. *Frontiers in Plant Science*, 9:1945.

Choi, Y. E., Yang, D. C., Park, J. C., Soh, W. Y., & Choi, K. T. 1998. Regenerative ability of somatic single and multiple embryos from cotyledons of Korean ginseng on hormone-free medium. *Plant Cell Reports*, 17:544–551.

Corredoira, E., Ballester, A., Ibarra, M., & Vieitez, A. M. 2015. Induction of somatic embryogenesis in explants of shoot cultures established from adult *Eucalyptus globulus* and *E. saligna* × *E. maidenii* trees. *Tree Physiology*, 35(6):678–690.

De Jong, A. J., Schmidt, E. D. L., & de Vries, S. C. 1993. Early events in higher plant embryogenesis. *Plant Molecular Biology*, 22:367–377.

Duarte-Aké, F., Márquez-López, R. E., Monroy-González, Z., Borbolla-Pérez, V., & Loyola-Vargas, V. M. 2022. The source, level, and balance of nitrogen during the somatic embryogenesis process drive cellular differentiation. *Planta*, 256(6):113.

Dudits, D., Bogre, L., & Gyorgyey, J. 1991. Molecular and cellular approaches to the analysis of plant embryo development from somatic cells *in vitro*. *Journal of Cell Science*, 99:475–484.

Duta-Cornescu, G., Constantin, N., Pojoga, D. M., Nicuta, D., & Simon-Gruita, A. 2023. Somaclonal variation—Advantage or disadvantage in micropropagation of the medicinal plants. *International Journal of Molecular Sciences*, 24(1):838.

Egertsdotter, U., Ahmad, I., & Clapham, D. 2019. Automation and scale up of somatic embryogenesis for commercial plant production, with emphasis on conifers. *Frontiers in Plant Science*, 10:436563.

Fay, M. F. 1992. Conservation of rare and endangered plants using *in vitro* methods. *In vitro Cellular & Developmental Biology- Plant*, 28:1–4.

Feher, A. 2015. Somatic embryogenesis—Stress-induced remodeling of plant cell fate. *BBA Gene Regulation Mechanisms*, 1849:385–402.

Feher, A., Pasternak, T. P., & Dudits, D. 2003. Transition of somatic plant cells to an embryogenic state. *Plant Cell Tissue Organ Culture*, 74:201–228.

Fujimura, T., & Komamine, A. 1979. Synchronization of somatic embryogenesis in a carrot cell suspension culture. *Plant Physiology*, 64(1):162–164.

Gamborg, O. L., Miller, R., & Ojima, K. 1968. Nutrient requirements of suspension cultures of soybean root cells. *Experimental cell research*, 50(1):151–158.

Garcês, H., & Sinha, N. 2009. The 'mother of thousands' (*Kalanchoë daigremontiana*): A plant model for asexual reproduction and CAM studies. *Cold Spring Harbor Protocols*, 2009(10): pdb-emo133.

Germanà, M. A., & Lambardi, M. (Eds.). 2016. *In Vitro Embryogenesis in Higher Plants*. New York/Heidelberg: Humana Press.

Giri, C. C., Shyamkumar, B., & Anjaneyulu, C. 2004. Progress in tissue culture, genetic transformation and applications of biotechnology to trees: An overview. *Trees*, 18:115–135.

Gupta, P. K., & Durzan, D. J. 1986. Somatic polyembryogenesis from callus of mature sugar pine embryos. *Biotechnology*, 4(7):643–645.

Guzzo, F., Baldan, B., Levi, M., Sparvoli, E., Loschiavo, F., Terzi, M., & Mariani, P. 1995. Early cellular events during induction of carrot explants with 2,4-D. *Protoplasma*, 185:28–36.

Haccius, B. 1978. Question of unicellular origin of non-zygotic embryos in callus cultures. *Phytomorphology*, 28:74–81.

Halperin, W. 1966. Alternative morphogenetic events in cell suspensions. *American Journal of Botany*, 53:443–453.

Hazubska-Przybył, T., Ratajczak, E., Obarska, A., & Pers-Kamczyc, E. 2020. Different roles of auxins in somatic embryogenesis efficiency in two *Picea* species. *International Journal of Molecular Sciences*, 21(9):3394.

Ikeuchi, M., Sugimoto, K., & Iwase, A. 2013. Plant callus: Mechanisms of induction and repression. *The Plant Cell*, 25(9):3159–3173.

Jiménez, V. M. 2001. Regulation of *in vitro* somatic embryogenesis with emphasis on to the role of endogenous hormones. *Revista Brasileira de Fisiologia Vegetal*, 13:196–223.

Jiménez, V. M., & Thomas, C. 2005. Participation of plant hormones in determination and progression of somatic embryogenesis. In: *Somatic Embryogenesis*: 103–118. Berlin, Heidelberg: Springer Berlin Heidelberg.

Jin, S., Zhang, X., Liang, S., Nie, Y., Guo, X., & Huang, C. 2005. Factors affecting transformation efficiency of embryogenic callus of upland cotton (*Gossypium hirsutum*) with *Agrobacterium tumefaciens. Plant Cell Tissue Organ Culture*, 81:229–237.

Kawahara, R., & Komamine, A. 1995. Molecular basis of somatic embryogenesis. In: Bajaj, Y. P. S. (ed.) *Biotechnology in Agriculture and Forestry, Vol. 30. Somatic Embryogenesis and Synthetic Seed*: 30–40, Berlin: Springer-Verlag.

Khajuria, A. K., Hano, C., & Bisht, N. S. 2021. Somatic embryogenesis and plant regeneration in *Viola canescens* Wall. Ex. Roxb.: An endangered Himalayan herb. *Plants*, *10*(4), 761.

Li, M., Wrobel-Marek, J., Heidmann, I., Horstman, A., Chen, B., Reis, R., Angenent, G. C., & Boutilier, K. 2022. Auxin biosynthesis maintains embryo identity and growth during BABY BOOM-induced somatic embryogenesis. *Plant Physiology*, 188(2):1095–1110.

Li, Z. T., Dhekeny, S., Dutt, M., van Anan, M., Tattersll, M. J., Kelley, K. T., & Gray, J. 2006. Optimizing *Agrobacterium-mediated* transformation of grapevine. *In Vitro Cellular & Developmental Biology Plant*, 42:220–227.

Liu, C. M., Xu, Z. H., & Chua, N. H. 1993. Auxin polar transport is essential for the establishment of bilateral symmetry during early plant embryogenesis. *The Plant Cell*, 5(6):621–630.

Lloyd G and McCown B. 1981. Commercially feasible micropropagation of mountain laurel, *Kalmia latifolia*, by use of shoot tip culture. *Proceedings of International Plant Propagation Society*, 30:421–427.

Long, Y., Yang, Y., Pan, G., & Shen, Y. 2022. New insights into tissue culture plant-regeneration mechanisms. *Frontiers in Plant Science*, 13:926752.

Lopes, M. A., & Larkins, B. A. 1993. Endosperm origin, development, and function. *Plant Cell*, 5:1383–1399.

Lu, C. Y., & Vasil, I. K. 1982. Somatic embryogenesis and plant regeneration in tissue cultures of Panicum maximum Jacq. *American Journal of Botany*, 69(1): 77–81.

Méndez-Hernández, H. A., Ledezma-Rodríguez, M., Avilez-Montalvo, R. N., Skeete, A., De-la-Peña, C., & Loyola-Vargas, V. M. 2019. Signaling overview of plant somatic embryogenesis. *Frontiers in Plant Science*, 10:414470.

Merkle, S. A., & Dean, J. F. 2000. Forest tree biotechnology. *Current Opinion in Biotechnology*, 11(3):298–302.

Merkle, S. A., Parrott, W. A., & Flinn, B. S. 1995. Morphogenic aspects of somatic embryogenesis. In: *In Vitro Embryogenesis in Plants*: 155–203. Dordrecht: Springer Netherlands.

Miguel, C., & Marum, L. 2011. An epigenetic view of plant cells cultured *in vitro*: Somaclonal variation and beyond. *Journal of Experimental Botany*, 62:3713–3725.

Misra, B. B., & Dey, S. 2013. Culture of East Indian sandalwood tree somatic embryos in air-lift bioreactors for production of santalols, phenolics and arabinogalactan proteins. *AoB Plants*, 5:plt025.

Mondal, T. K. 2020. Somatic embryogenesis and alternative *in vitro* techniques. In: *Tea: Genome and Genetics*: 85–126. Springer: Singapore.

Mose, W., Daryono, B. S., Indrianto, A., Purwantoro, A., & Semiarti, E. 2020. Direct somatic embryogenesis and regeneration of an Indonesian orchid *Phalaenopsis amabilis* (L.) Blume under a variety of plant growth regulators, light regime, and organic substances. *Jordan Journal of Biological Sciences*, 13(4):509–518.

Mujib, A., Ali, M., & Isah, T. 2014. Somatic embryo mediated mass production of *Catharanthus roseus* in culture vessel (bioreactor)–A comparative study. *Saudi Journal of Biological Science*s, 21(5):442–449.

Murashige, T., & Skoog, F. 1962. A revised medium for rapid growth and bio assays with tobacco tissue cultures. *Physiologia Plantarum*, 15(3).

Nic-Can, G. I., Galaz-Ávalos, R. M., De-la-Peña, C., Alcazar-Magaña, A., Wrobel, K., & Loyola-Vargas, V. M. 2015. Somatic embryogenesis: Identified factors that lead to embryogenic repression. A case of species of the same genus. *PLoS One*, 10(6):e0126414.

Normah, M. N., Rohani, E. R., & Mohamed-Hussein, Z. A. 2013. Somatic embryogenesis in higher plants. *Malaysian Applied Biology*, 42(2):1–12.

Panaia, M., Senaratna, T., Dixon, K. W., & Sivasithamparam, K. 2004. High-frequency somatic embryogenesis of koala fern (*Baloskion tetraphyllum*, Restionaceae). *In Vitro Cellular & Developmental Biology-Plant*, 40:303–310.

Peeris, M. K., & Senarath, W. T. 2015. *In vitro* propagation of *Santalum album* L. *Journal of the National Science Foundation of Sri Lanka*, 43(3):265–272.

Prakash, M. G., & Gurumurthi, K. 2010. Effects of type of explant and age, plant growth regulators and medium strength on somatic embryogenesis and plant regeneration in *Eucalyptus camaldulens. Plant Cell Tissue Organ Culture*, 100:13–20.

Quiroz-Figueroa, F. R., Rojas-Herrera, R., Galaz-Avalos, R. M., & Loyola-Vargas, V. M. 2006. Embryo production through somatic embryogenesis can be used to study cell differentiation in plants. *Plant Cell Tissue Organ Culture*, 86:285–301.

Raemakers, C. J. J. M., Jacobsen, E., & Visser, R. G. F. 1995. Secondary somatic embryogenesis and applications in plant-breeding. *Euphytica*, 81:93–107.

Raghavan, V. 1976. Adventive embryogenesis: Induction of diploid embryoids. In: Sutchliffe, J. F. (eds) *Experimental Embryogenesis in Vascular Plants*: 349–381, London: Academic Press.

Raghavan, V. 1986. *Embryogenesis in Angiosperms: A Developmental and Experimental Study* (Vol. 17). CUP Archive.

Raghavan, V. 2004. Role of 2,4-dichlorophenoxyacetic acid (2,4-D) in somatic embryogenesis on cultured zygotic embryos of *Arabidopsis*: Cell expansion, cell cycling, and morphogenesis during continuous exposure of embryos to 2,4-D. *American Journal of Botany*, 91:1743–1756.

Ramakrishna, D., & Shasthree, T. 2016. High efficient somatic embryogenesis development from leaf cultures of *Citrullus colocynthis* (L.) Schrad for generating true type clones. *Physiology and Molecular Biology of Plants: An International Journal of Functional Plant Biology*, 22(2):279–285.

Raza, G., Singh, M. B., & Bhalla, P. L. 2019. Somatic embryogenesis and plant regeneration from commercial soybean cultivars. *Plants*, 9(1):38.

Rogo, U., Fambrini, M., & Pugliesi, C. 2023. Embryo rescue in plant breeding. *Plants*, 12(17):3106.

Sagare, A. P., Lee, Y. L., Lin, T. C., Chen, C. C., & Tsay, H. S. 2000. Cytokinin induced somatic embryogenesis and plant regeneration in *Corydalis yanhusuo* (Fumariaceae) - A medicinal plant. *Plant Science*, 160:139–147.

Sedaghati, B., Haddad, R., & Bandehpour, M. 2019. Efficient plant regeneration and Agrobacterium-mediated transformation via somatic embryogenesis in purslane (*Portulaca oleracea* L.): An important medicinal plant. *Plant Cell, Tissue and Organ Culture (PCTOC)*, 136:231–245.

Senger, S., Mock, H. P., Conrad, U., & Manteuffel, R. 2001. Immunomodulation of ABA function affects early events in somatic embryo development. *Plant Cell Reports*, 20:112–120.

Shareef, H. J. 2017. Desiccation and cold hardening of date palm somatic embryos improve germination. In: *Date Palm Biotechnology Protocols Volume I: Tissue Culture Applications*: 119–128.

Sharma, R. K., & Khajuria, A. K. 2020. Somatic embryogenesis and plant regeneration in *Ferula jaeschkeana* Vatke: A threatened medicinal herb. *Vegetos*, 33:658–664.

Sharmin, S. A., Alam, M. J., Sheikh, M. M., Sarker, K. K., Khalekuzzaman, M., Haque, M. A., Alam, M. F., & Alam, I. 2014. Somatic embryogenesis and plant regeneration in *Wedelia calendulacea* Less, an endangered medicinal plant. *Brazilian Archives of Biology and Technology*, 57(3):394–401. 10.1590/S1516-8913201401840.

Sharp, W. R., Sondahl, M. R., Caldas, L. S., & Maraffa, S. B. 1980. The physiology of *in vitro* asexual embryogenesis. *Horticultural Reviews*, 2:268–310.

Simonović, A. D., M. Trifunović-Momčilov, M., Filipović, B. K., Marković, M. P., Bogdanović, M. D., & Subotić, A. R. 2020. Somatic embryogenesis in *Centaurium erythraea* Rafn—Current status and perspectives: A review. *Plants*, 10(1):70.

Smith, D. L., & Krikorian, A. D. 1990. Somatic embryogenesis of carrot in hormone-free medium: External pH control over morphogenesis. *American Journal of Botany*, 77(12):1634–1647.

Solangi, N., Jatoi, M. A., Abul-Soad, A. A., Mirani, A. A., Solangi, M. A., & Markhand, G. S. 2023. Factors influencing somatic embryogenesis and plantlet regeneration of date palm using immature floral buds. *Sarhad Journal of Agriculture*, 39(2):323–331.

Stuart, D. A., & Strickland, S. G. 1984. Somatic embryogenesis from cell cultures of *Medicago sativa* L. II. The interaction of amino acids with ammonium. *Plant Science Letters*.

Tomiczak, K., Mikuła, A., Niedziela, A., Wójcik-Lewandowska, A., Domżalska, L., & Rybczyński, J. J. 2019. Somatic embryogenesis in the family *Gentianaceae* and its biotechnological application. *Frontiers In Plant Science*, 10:762.

Toonen, M. A. J., Hendriks, T., Schmidt, E. D. L., Verhoeven, H. A., Vankammen, A., & Devries, S. C. 1994. Description of somatic-embryo-forming single cells in carrot suspension-cultures employing video cell tracking. *Planta*, 194:565–572.

Trigiano, R. N., May, R. A., & Conger, B. V. 1992. Reduced nitrogen influences somatic embryo quality and plant regeneration from suspension cultures of orchard grass. *In Vitro Cellular & Developmental Biology*, 28:87–191.

Vasil, V., & Vasil, I. K. 1981. Somatic embryogenesis in cereals. *BioScience*, 31(3):246–248.

Vieira, R. F., & Skorupa, L. A. 1993. Brazilian medicinal plants gene bank. *Acta Horticulture*, 330:51–58.

Williams, E. G., & Maheswaran, G. 1986. Somatic embryogenesis: Factors influencing coordinated behaviour of cells as an embryogenic group. *Annals of Botany*, 57:443–462.

Woodward, B., & Puonti-Kaerlas, J. 2001. Somatic embryogenesis from floral tissue of cassava (*Manihot esculenta* Crantz). *Euphytica*, 120:1–6.

Yang, X. Y., Zhang, X. L., Jin, S. X., Fu, L. L., & Wang, L. G. 2007. Production and characterization of asymmetric hybrids between upland cotton Coker 201 (*Gossypium hirsutum*) and wild cotton (G. *klozschianum* Anderss). *Plant Cell Tissue Organ Culture*, 89:225–235.

Yeung, E. C. 1995. Structural and development patterns in somatic embryogenesis. In: Thorpe, T. A. (eds) *In Vitro Embryogenesis in Plants*: 205–247, Netherlands: Kluwer.

Zdravković-Korać, S., Belić, M., Ćalić, D., & Milojević, J. 2023. Somatic embryogenesis in spinach—A review. *Horticulturae*, 9(9):1048.

Zhang, M., Wang, A., Qin, M., Sun, Y., & Zhang, L. 2021. Direct and indirect somatic embryogenesis induction in *Camellia oleifera* Abel. *Frontiers in Plant Science*, 12:644389.

Zhang, S., Yan, S., An, P., Cao, Q., Wang, C., Wang, J., Zhang, H., & Zhang, L. 2021. Embryogenic callus induction from immature zygotic embryos and genetic transformation of *Larix kaempferi* 3x *Larix gmelinii* 9. *PLoS One*, 16(10):e0258654.

Zimmerman, J. L. 1993. Somatic embryogenesis: A model for early development in higher plants. *The Plant Cell*, 5(10):1411.

10 Haploid Plant Production and Its Applications for Medicinal Plants

Alison M. R. Ferrie

10.1 INTRODUCTION

Medicinal plants have been described as plants that have therapeutic properties for humans or animals. Different parts of the plants can be used in traditional medicine or remedies, including roots, leaves, bark, flowers, fruits, or seeds. The compound of interest may also be extracted from the plant and used as a pharmaceutical (Namdeo, 2018). According to the Medicinal Plant Names Service, there is a list of over 35,000 plants with reference to medicinal properties. For many of the species, there is little commercial production and there is a reliance on collecting from the natural populations which has resulted in overharvesting (Medicinal Plants Names Service Portal, Royal Botanical Graden, Kew, 2023). Of those 35,000 species, about 1280 are on the endangered list (The Convention on International Trade in Endangered Species of Wild Fauna and Flora; CITES). The CITES database indicates that between 2006 and 2015, there were 54 million kg of medicinal plant products exported, mainly from 43 species of which 47% of these exports came from "wild" plants (CITES). Other estimates have indicated 60–90% of the plant species are sourced from the wild (Jenkins et al., 2018). Since many of the species are being harvested from natural areas, there has been very little genetic improvement for those plants and therefore there can be variability in the phenotypes, agronomics, and active ingredients and compounds of therapeutic value. This variation in the active ingredients can cause discrepancies in clinical trial data as well as in the industry that wants to sell a uniform product.

Doubled haploidy (DH) methodology is the fastest way to fix beneficial traits and generate true breeding plants and therefore uniformity in the growth of the plant and its constituents (Ferrie, 2017). This is not new technology; it was first exploited in the 1970s when the DH cultivar "Maris Haplona" rapeseed (*Brassica napus* L.) was released (Thompson, 1972). Now, DH methodology is routinely used in breeding programs and for basic research in a wide range of species (Figure 10.1). A plant that has the gametic chromosome number is classified as a haploid plant. As these haploid plants are sterile, chromosome doubling, either spontaneously or induced by chemicals, is necessary for the utilization of the technology. The resulting plants are termed DH and have a 2n chromosome number.

This chapter reviews the published literature on DH plant production in medicinal plants, with emphasis placed on advances reported in the last 10 years. The response of medicinal plants to DH protocols was reviewed previously (Ferrie et al., 2005; Ferrie, 2007; Ferrie, 2009; Ferrie, 2012). Compared to large acreage crop species, there have been very few reports for the production or utilization of DH protocols in medicinal plants. Seguí-Simarro et al. (2021) published a comprehensive review of DH methodology. This three-volume manual provided DH protocols for 44 different species as well as a listing of the plant species where haploids and doubled haploids have been produced (383 species). The manual included protocols for cereals (13), Apiaceae (4), Brassicaceae (6), Solanaceae (8), Cucurbits (6), woody species (5), and Alliaceae (2), as well as 7 protocols for other species (chickpea [*Cicer arietinum* L.], red beet [*Beta vulgaris* L. subsp. *vulgaris*], sugar beet [*Beta vulgaris* L.], cow cockle [*Saponaria vaccaria* L.], borage [*Borago officinalis* L.], African violet [*Saintpaulia ionantha*], and marigold [*Tagetes* spp.]).

DOI: 10.1201/b23374-10

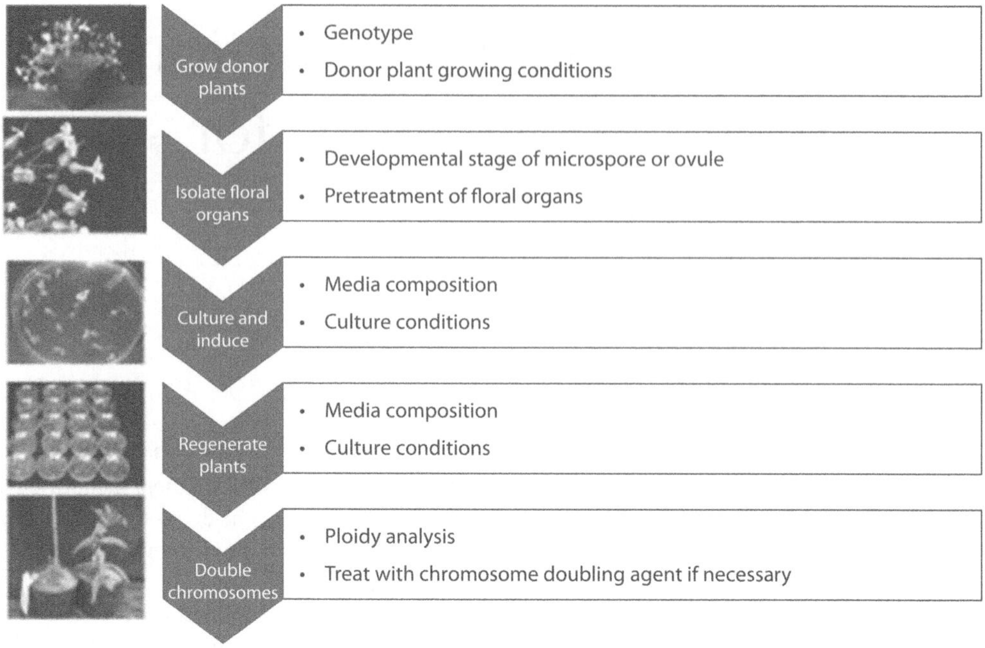

FIGURE 10.1 Steps involved in the double haploid process and factors evaluated to enhance the protocol.

10.2 METHODS OF PRODUCING DOUBLED HAPLOIDS

Haploid plants can occur naturally, and this has been observed in over 100 species; however, the mechanism of haploidy is often unknown and the frequency of resulting haploid plants is too low for practical application (Dunwell, 2010). There are *in vitro* methods and *in vivo* methods for producing haploids. *In vitro* methods include the culture of the male gametophyte (anthers and isolated microspores), termed androgenesis, or the culture of the female gametophyte (unfertilized ovules), termed gynogenesis. *In vivo* methods include pollination within the same species (intraspecific hybridization), pollination between wild relatives or unrelated species (interspecific hybridization), pollination with irradiated pollen or chemical treatment, induction of parthenogenesis through hormones or genetic manipulation, and newer techniques such as manipulation of the CENH3 protein. There have been several reviews of DH technology via *in vitro* or *in vivo* methods (Chen et al., 2011; Dwivedi et al., 2015; Kalinowska et al., 2019; Seguí-Simarro et al., 2021; Hale et al., 2022). This chapter will focus on *in vitro* methods, androgenesis and gynogenesis, as there has been very few publications on *in vivo* methods for DH production in medicinal plants.

Androgenesis is the most commonly used DH method, and there are many published reports of haploid embryogenesis from anthers or isolated microspore culture. Guha and Maheshwari (1964) were the first to describe the culture of anthers in *Datura*, and much of the early work focused solely on anther culture. Ten years later, Nitsch (1974) observed development from microspores that had shed from the anther. This observation led to mechanically isolating the microspores from the anthers to achieve embryo development (Lichter, 1982), which has advantages over culturing the whole anther (Seguí-Simarro et al., 2021.) The anther wall, which is somatic tissue, may also respond to the culture treatments and produce a callus and subsequent embryos, which are not desired in a haploidy program. For some species, anther culture can be very time-consuming and may require a microscope to extract the anthers from the buds. Isolated microspore culture techniques are well established in some species and are being routinely used in breeding organizations as well as for basic research (reviewed in Weyen, 2021). Tracking and studying microspore and embryo development are more efficient using isolated microspore culture (Indrianto et al., 2001; Rivas-Sendra et al., 2019; Corral-Martínez et al., 2020).

In comparison to the female gametophytic cells (unfertilized ovules), microspores are more synchronized in their development, uniform in size, easier to manipulate physically, and more abundant. Anthers contain thousands of microspores, each having the potential to regenerate into an embryo or callus, depending on the species. Although there are fewer female gametes available in a plant, gynogenesis is an alternative haploid production system when the male gametophytic pathway is recalcitrant or defective. Gynogenesis has been shown to have a lower rate of albinism and higher genetic stability when compared to androgenesis. Gynogenic haploids have been routinely produced in several species including sugar beet, gerbera, and onion but have also been produced in barley, sunflower, cucumber, and wheat (Chen et al., 2011).

Despite the methodology used, i.e., androgenesis or gynogenesis, a number of factors influence the response of the gametophyte, and these could be classified as pre-isolation (e.g., genotype, donor plant conditions, pretreatments, developmental stage of the pollen grain or ovule) and post-isolation (e.g., media composition and culture conditions). These factors are manipulated in order to develop an efficient protocol, and there are countless publications describing the modifications to achieve an efficient DH protocol.

10.3 ANDROGENESIS

10.3.1 Genotype

There are differences in androgenic response among different genotypes of the same species, and in some cases, there are differences between the plants within the genotype, especially for wild germplasm (Phippen and Ockendon, 1990; Hiramatsu et al., 1995). Genotype dependency is more predominant in androgenesis than gynogenesis. This has been shown in many species including medicinal plants. For initial protocol development, work usually focuses on one genotype and then expanding the protocol to other genotypes. Ari et al. (2022) evaluated 66 wild Turkish *Vaccaria hispanica* genotypes for microspore culture response. The authors based their screening study on previously published protocols (Ferrie et al., 2005; Kernan and Ferrie, 2006) with modifications. Significant differences were observed among the lines with embryo yields of 2.6–300.6 embryos/100 buds. Three genotypes did not respond. From the 63 responding genotypes, there were 7068 embryos produced. Because of low germination, development of hairy roots, poor root establishment, and precocious flowering, only three plants grew to maturity. Ploidy analysis indicated that two plants were diploid or doubled haploid and one was a haploid.

An in-depth study evaluated 46 lemon balm (*Melissa officinalis* L.) accessions to anther culture, microspore culture, or ovule culture (Kästner et al., 2016). Multicellular structures and microcalli were observed from anther cultures of 10 genotypes; however, the response was sporadic and no conclusions could be drawn. Similarly for microspore culture, sporadic production of microcalli was observed from six genotypes, but again there was no regeneration. In *Althaea officinalis* (marshmallow), comparison was made between two genotypes, white flowers and pink flowers. It was determined that callus production was slightly better from the pink flowers than from the white flowers (Ewais et al., 2019). Genotypic studies have also been conducted in the *Asteraceae* species, marigold (*Tagetes* spp.), and purple coneflower (*Echinacea purpurea* L. Moench). Differences in callus production and shoots per anther were observed among five genotypes of African marigold (*Tagetes erecta* L.) as well as the five genotypes of French marigold (*Tagetes patula* L.). When comparing the two marigold species, the higher anther culture response was from the French marigold genotypes rather than with the African marigold genotypes. Cultivars "Pusa Arpita" and "Dainty Marietta" yielded 4.6 and 4.5 shoots/anther, respectively. The most responsive African marigold variety produced 2.3 shoots/anther (Kumar et al., 2019). Evaluation of four echinacea genotypes resulted in callus induction rates ranging from 13.28% to 44.38% (Chen et al., 2016).

10.3.2 Donor Plant Conditions

The environmental conditions in which the donor plants are grown can influence the embryogenic response of the microspores. Floral organs can be harvested from naturally growing areas or from plants grown under field conditions, in a greenhouse, or in environmentally controlled units (Table 10.1). There are benefits in growing plants in environmentally controlled chambers as temperature, photoperiod, light quality, watering and fertilizer regimens, and pests are controlled. This means that treatments resulting in positive embryogenic response have a greater likelihood of being repeated. However, with some species, indoor controlled environments may not be possible (i.e., trees).

Most of the anther/microspore experiments for medicinal plants have utilized field-grown material (Table 10.1). For lemon balm, floral organs were collected from field, greenhouse, and environmentally controlled chambers; however there were very little response and no conclusions could be drawn (Kästner et al., 2016).

Other studies have utilized greenhouses, but temperatures cannot be controlled as in growth chambers. Hoveida et al. (2017) grew borage in the greenhouse with a temperature of 28–32 °C during the day and 18–24 °C at night. Similarly, for stevia (*Stevia rebaudiana*, Bertoni), a temperature range of 20–35 °C was observed (Uskutoğlu et al., 2019). For some crop species (e.g., *Brassica* spp.), a precise temperature of 10/5 °C is used (Keller et al., 1987; Takahata and Keller, 1991; Baillie et al., 1992).

10.3.3 Pretreatments

Pretreatment of the whole plants or isolated floral organs by a chemical or physical stress can enhance microspore embryogenesis. The most common pretreatment is a temperature stress, either a cold or elevated temperature. The length of the temperature stress is also critical, and this can vary depending on the genotype.

A cold treatment (4 °C for 1–2 days) of *Panax ginseng* C.A. Mey buds resulted in more callus growth than the control (no cold) treatment; however, a longer pretreatment (8–14 days) resulted in a decrease in callus induction as well as a decrease in fresh weight of the resulting callus (Lee et al., 2013). Similarly, a cold pretreatment (4 °C) was beneficial in marigold (Kumar et al., 2019) and in bitter melon (*Momordica charantia* L.; Usman et al., 2015). For marigold, callus production and shoot production were observed from all cold pretreatments (2–16 days) and the control (no cold treatment), with the best response at 10 days, which resulted in 3.7 shoots per anther (Kumar et al., 2019). Uskutoğlu et al. (2019) evaluated a cold pretreatment on stevia anthers for 2, 4, 6, and 8 days but observed no beneficial effect of the cold treatment. Further experiments with a high-temperature pretreatment of 35 °C resulted in embryogenic structures developing from the calli. For lemon balm, a cold (6 °C) treatment and an elevated temperature (30 °C) were evaluated on anther or microspore culture, microcalli was observed in both treatments but not further development (Kästner et al., 2016). *Camellia assamica* ssp. *Lasiocalyx* (Planch MS; cambod tea) anthers were subjected to a cold treatment (5 °C) for 5 days, an elevated temperature (33 °C) for 5 days, or no pretreatment (25 °C). The most callus development was in the control; however, the best callus production from inside the anther (i.e., microspores) was with the cold pretreatment. The elevated temperature was not beneficial, and a cold temperature stress longer than 5 days was not advantageous (Mishra et al., 2022).

Other stress treatments used in androgenesis include centrifugation and electroporation of anthers to enhance callus production in borage. The best response in terms of callus production and embryo-like structures was 100 V electric discharge or a 6-minute centrifugation of anthers at 300× g (Hoveida et al., 2017).

10.3.4 Developmental Stage of the Pollen Grain

The one factor that seems to be fairly consistent between species is the responsiveness of the mid-to-late uninucleate microspore for embryogenic culture, although a wider range has also been utilized

TABLE 10.1
List of Medicinal Plant Species in Which Androgenic Protocols (Anther Culture or Microspore Culture) Have Been Published Since 2013

Species	Geno Type	Donor Plant Conditions	Pretreatment	Developmental Stage of the Pollen Grain	Media Constituents	Culture Conditions	Culture Response	Reference
Abelmoschus esculentus	Breeding lines	NA	None	Mid-to-late uninucleate	MSNB (Mangat and Roy modifications)	25 °C in dark	Callus	Varandani et al. 2021
Althaea officinalis	White, Pink	Field	None	Uninucleate	MS, 5 mg/L 2,4-D, 1.5 mg/L BA, 3% sucrose, 4 mg/L amino18	26 °C 16-hour photoperiod	Callus	Ewais et al. 2019
Borago officinalis	1 local Iranian variety	Field	4 °C for 4 days	Mid-to-late uninucleate, 5–7 mm buds	B5 media salts with NLN vitamin, 3% maltose, 2 mg/L 2,4-D, 1 mg/L BA,	32 °C for 3 days	Haploid embryos	Eshaghi et al. 2015
Borago officinalis	1 local Iranian variety	Greenhouse, 28–32 °C/18–24 °C	One of the following: 200 mg/L colchicine for 4 days, 0.2% n-butanol for 5 hours, 6 minutes centrifugation at 300× g, or 100 V electrical shock	Mid-to-late uninucleate, 5–7 mm buds	B5 media salts with NLN vitamins	4 °C for 4 days dark, then 24 °C, 16/8-hour photoperiod	Plants	Hoveida et al. 2017
Borago officinalis	1 local Iranian variety	Greenhouse, 28–32 °C/18–24 °C	1 of the following: electroporation, centrifugation, colchicine, n-Butanol	Mid-to-late uninucleate, 5–7 mm buds	B5 media salts with NLN vitamin, 3% maltose, 2 mg/L 2,4-D, 1 mg/L BA, 5 mg/L $AgNO_3$	4 °C for 4 days dark, then 25 °C, 16/8-hour photoperiod	Plants	Abdollahi et al. 2021
Camellia assamica	1 - TV19	Field	5 °C dark 5 days	Early-to-late uninucleate	MS, 6% sucrose, 5 uM BA, 1 µM 2,4D, 1 µM NAA	25 °C dark	Plants	Mishra et al. 2022
Echinacea purpurea	4	NA	NA	Early-to- late uninucleate	N6, 4% sucrose, 0.5 mg/L BA, 0.1 mg/L NAA	25–27 °C, 12-hour photoperiod	Plants	Chen et al. 2016
Melissa officinalis	46 lines	Field, greenhouse, climate-controlled chambers	35 °C for 6 days	Late uninucleate	Not conclusive	27 °C dark for 14 days, then 16/8-hour photoperiod	Calli	Kästner et al. 2016
Momordica charantia	1 - Faisalabad long	NA	4 °C for 48 hours, dark	Uninucleate-to-early binucleate	MS, 3% sucrose, 3.22 µM/L NAA, 0.88 µM/L BA	25 °C, dark	Embryos	Usman et al. 2015

(*Continued*)

TABLE 10.1 (*Continued*)
List of Medicinal Plant Species in Which Androgenic Protocols (Anther Culture or Microspore Culture) Have Been Published Since 2013

Species	Geno Type	Donor Plant Conditions	Pretreatment	Developmental Stage of the Pollen Grain	Media Constituents	Culture Conditions	Culture Response	Reference
Panax ginseng	NA	NA	4 °C for 1–2 days	NA	MS, 3% sucrose, 4.53 µM 2,4-D, 4.44 µM BA	25 °C, dark	Plants	Lee et al. 2013
Physalis peruviana	1	NA	4 °C 24 hours	Late uninucleate	B5	24 °C, 16 h ours	Plants	Mállap-Detquizán et al. 2023
Stevia rebaudiana	NA	Greenhouse, 20–35 °C	35 °C for 2 days	Uninucleate	MS, 3% sucrose, 2 mg/L NAA, 0.5 mg/L kinetin, 2 mg/L $AgNO_3$	24 °C, 16-hour photoperiod	Embryos	Uskutoğlu et al. 2019
Tagetes erecta	5	Greenhouse, 20–35 °C/5–18 °C	4 °C for 10 days	Early uninucleate-to-early binucleate	EMS1, 4.5% sucrose, 5% coconut water, PVP, $AgNO_3$, 2 mg/L BAP, 0.5 mg/L NAA	20 days dark	Plants	Kumar et al. 2019
Tagetes spp.	1	Greenhouse, 20–25 °/15–18 °C	4 °C, 10 days cold	Late uninucleate-to-early binucleate	EMS1, 4.5% sucrose, 5% coconut water, PVP, $AgNO_3$, 2 mg/L BAP, 0.5 mg/L NAA	25 °C for 20 days dark	Plants	Kurimella et al. 2021
Tagetes patula	5	Greenhouse, 20–35 °C/5–18 °C	4 °C for 10 days	Early uninucleate-to-early binucleate	EMS1, 4.5% sucrose, 5% coconut water, PVP, $AgNO_3$, 0.5 BAP, 0.25 mg/L NAA	20 days dark	Plants	Kumar et al. 2019
Vaccaria hispanica	1	Climate-controlled chamber	None	Mid-to-late uninucleate	NLN, 15% sucrose	32 °C 3 days then 24 °C, dark	Plants	Ferrie and Ehlert, 2021
Vaccaria hispanica	66 lines from different regions of Turkey	NA	NA	Late uninucleate-to-early binucleate	NLN, 15% sucrose	32 °C 3 days then 24 °C, dark	Plants	Ari et al. 2022

(early to late uninucleate) (Table 10.1). Studies have correlated the developmental stage of the pollen grain with the bud size for ease of selection. The bud size can change depending on growing conditions and genotypes; therefore, buds should be restaged when changing growth conditions or cultivars. For borage (*Borago officinalis* L.) under the donor plant conditions evaluated, floral buds 5.0–7.0 mm long which had anthers 2.5–3.5 mm in length produced microspores at the mid-to-late uninucleate stage (Eshaghi et al., 2015; Hoveida et al., 2017). Uninucleate microspores were used in stevia (Uskutoğlu et al., 2019) and cambod tea (Mishra et al., 2022) as well as echinacea (Chen et al., 2016). Late uninucleate to early binucleate microspores were beneficial for marigold species (Kurimella et al., 2021). In an earlier study on marigold, the authors found that the early uninucleate to early binucleate microspores were found in 3.5–4.0 mm buds for French marigold and 3.0–3.5 mm buds of African marigold; however, for some marigold genotypes, the buds were 2.5 mm (Kumar et al., 2019).

10.3.5 Isolation Method

Most studies on DH of medicinal plants have focused on anther culture, although there are a few publications utilizing isolated microspore culture. As mentioned previously, there are advantages and disadvantages on using anther culture or microspore culture.

In either androgenic method, sterilization of the explant is very important to avoid contamination and at the same time not decrease the viability of the microspore. The sterilization method depends on the species and the laboratory. Usually, this is a quick wash with ethanol (30 seconds to 3 minutes) followed by a wash with sodium hypochlorite (0.8–2.5% NaClO) for 7–15 minutes. After the sodium hypochlorite treatment, the explants are usually rinsed with sterile water a few times. This method of ethanol wash followed by 2% NaClO for 15 minutes was effective for *Physalis peruviana* (Mállap-Detquizán et al., 2023). Others have used 0.8% NaClO for 7 minutes with 93% sterile conditions in *Camellia assamica* ssp. *lasiocalyx* (Mishra et al., 2022). Mercuric chloride (HgCl) has also been used for sterilization (Kumar et al., 2019; Varandani et al., 2021).

10.3.6 Media Composition

The media composition is a major factor influencing the androgenic response. The basal media, carbohydrate source and concentration, macro- and micro-nutrients, growth regulators, and other additives are the major components that contribute to the response. There are countless papers evaluating media modifications to enhance androgenic response.

Some of the more commonly used basal media in androgenesis is Gamborg's B5 (Gamborg et al., 1968), MS (Murashige and Skoog, 1962), N6 (Chu, 1978), or NLN (Lichter, 1982). MS media has been used to generate calli or microspore-derived embryos in *P. ginseng*, *A. officinalis*, stevia, *Abelmoschus esculentis*, and *Camellia assamica* (Lee et al., 2013; Ewais et al., 2019; Uskutoğlu et al., 2019; Varandani et al., 2021). B5 has been used in borage and *Physalis peruviana* (Eshaghi et al., 2015; Hoveida et al., 2017; Mállap-Detquizán et al., 2023). Four basal media were evaluated (N6, C17, W14, potato II) for echinacea microspore culture (Chen et al., 2016). Callus induction was significantly higher with N6 media than with the other basal media formulations (Chen et al., 2016). Five basal media (AT-3, KA, B5-NLN, EDM, and NN) compositions were evaluated in borage, AT-3 provided the most calli, but the most embryo-like structures were obtained from the B5-NLN media composition (Hoveida et al., 2017). Nine basal media formulations were evaluated in marigold, with B5, MS, and three modified MS media (differences in vitamins and amino acids) resulting in callus formation and shoot production. There was no response from N6, Nitsch (Nitsch and Nitsch, 1969), or NLN, as the anthers dried up and died (Kumar et al., 2019).

Different carbohydrates have been used in microspore culture or anther culture studies, although sucrose is the most commonly used. For the recent studies on androgenesis in medicinal plants, only one species, borage, responded to maltose (Eshaghi et al., 2015; Hoveida et al., 2017). Calli were produced in both sucrose and maltose, but embryogenic calli were only found in the 3% maltose cultures. Studies in *P. ginseng* indicated that sucrose was better than maltose, and 3% or 6% sucrose resulted in greater fresh weight of calli as well as the size of the calli (Lee et al., 2013). Seven sucrose concentrations ranging from 3% to 12% were evaluated in marigold (Kumar et al., 2019) with the best response (3.8 shoots/anther) at 4.5% sucrose. An increase in the sucrose concentration resulted in a decrease in callus induction and the number of shoots. At 12% sucrose, callus production was inhibited. Four sucrose concentrations were evaluated in *Echinacea* with the best response at 4%, although there was no significant difference between the sucrose concentrations (4–7%) evaluated. Sucrose was also more responsive than glucose in *Camellia assamica* ssp. *lasiocalyx* at all concentrations evaluated (3%, 6%, 9%, 12% carbohydrate) (Mishra et al., 2022). It was also observed that 6% sucrose was significantly better than lower sucrose concentrations (3%) or higher concentrations (9%, 12%) and at 12% sucrose, callus production was inhibited.

Growth regulators such as auxins (e.g., 2,4-D, NAA) and cytokinins (e.g., BA, kinetin) are used extensively to promote callus or embryo development from anthers or microspores. In stevia, callus was obtained from three different media formulations: MS media with NAA and BA or 2,4-D and BA; however, globular and torpedo embryos were obtained from 1.0 mg/L NAA and 0.5 mg/L, and kinetin and "bar shaped" embryos were obtained with 2.0 mg/L NAA and 0.5 mg/L kinetin (Uskutoğlu et al., 2019). There was no regeneration to plants from any of the structures. In *A. officinalis*, 12 combinations of 2,4-D and BA or NAA and BA were evaluated (Ewais et al., 2019). There was no callus development from hormone-free media, and differences were observed in the other growth regulator combinations in the percentage of anthers forming callus, callus quantity, and callus quality (i.e., texture and color). With the media formulation of 5.0 mg/L 2,4-D and 1.5 mg/L BA, 100% of the anthers formed a callus, which was compact and cream to white in color. Results were similar for both genotypes (pink flowers, white flowers) (Ewais et al., 2019). For the maximum number of shoots per anther in marigolds, 12 different growth regulator combinations were evaluated. For *T. patula* the best response was with a media containing 2.0 mg/L BA and 0.5 mg/L NAA, whereas in *T. erecta*, the best direct differentiation to shoots was in a media with lower BA (0.5 mg/L) and lower NAA (0.25 mg/L) concentrations (Kumar et al., 2019). A media with only TDZ resulted in no callus production for both of the *Tagetes* species. Ten growth regulators were evaluated in *Abelmoschus esculentis* anther cultures, with callus being induced only on media with NAA (Varandani et al., 2021). The addition of 2,4-D (1 μM) and 1 μM NAA and 5 μM of BA resulted in the highest rate of callus production from *Camellia assamica* ssp. *Lasiocalyx* anther cultures (Mishra et al., 2022), and for *Momordica charantia*, both NAA and BA were necessary for callus induction, as other plant growth regulators tested were not beneficial (Usman et al., 2015).

Other additives have also been used to enhance microspore embryogenesis. In *A. officinalis*, the addition of 18 amino acids (4.0 mg/L amino 18, a pharmaceutical mix of essential and non-essential amino acids) resulted in maximum embryogenic callus induction for both genotypes (white and pink); however, there was no plant regeneration (Ewais et al., 2019). Biotin, which is a water-soluble B vitamin, also enhanced callus production when compared to the control but was not as beneficial as the amino 18 product (Ewais et al., 2019). A treatment of 0.2% or 0.3% n-butanol on borage buds enhanced callus induction from the anthers (41.25%) compared to the control (21.25%) (Hoveida et al., 2017). The number of embryos per anther was also significantly higher in the n-butanol treatment than in the control, and plants were regenerated from these embryos. Kumar et al. (2019) recommended the addition of coconut water (5%), silver nitrate ($AgNO_3$), and polyvinylpyrrolidone (PVP) to a modified MS media to improve callus induction and shoot production of *T. erecta* and *T. patula*.

10.3.7 Culture Conditions

Most anthers or microspores, once isolated, are grown in the dark; however, light has shown to play a role in enhancing the androgenic response in marigold (Kumar et al., 2019). The best response was

20 days dark followed by light (3.1 shoots per anther); although shoots developed from the 2 days or 7 days dark followed by light, they were slower to develop. Dark was beneficial for *Camellia assamica* ssp. *Lasiocalyx* (Mishra et al., 2022). For *Momordica charantia*, comparisons were made between completely dark and a 16-hour photoperiod. Callus induction was greater and earlier under the completely dark conditions (Usman et al., 2015).

The culture temperature and duration can influence the embryogenic response of the microspores. This usually ranges from 24 to 27 °C, although a higher temperature (32 °C) has been used in *V. hispanica* (Ari et al., 2022) or borage (Eshaghi et al., 2015). This 3-day heat shock treatment is similar to the conditions used for *Brassica* (Ferrie and Keller, 1997). A cold temperature (4 °C) for 4 days has also been beneficial (Hoveida et al., 2017).

10.4 GYNOGENESIS

In 1964, Tulecke (1964) observed haploid callus from the female gametophyte of *Ginkgo biloba*, but it was another 12 years before the first plants from the female gametophyte were reported (San Noeum, 1976). Compared to androgenesis, there are fewer publications on gynogenesis of medicinal plants; however, the factors influencing the gynogenic response are similar to those of androgenesis.

10.4.1 Genotype

Differences in gynogenic response were observed among four ecotypes of Persian shallot (*Allium hirtifolium* Boiss.) (Panahandeh and Farhadi, 2019). Embryogenic structures occurred in all genotypes, although one ecotype (Yasoj) was slower than the others. The ecotype Lorestan gave the highest number of embryos per flower (1.64). Doubled haploids have been produced in the cultivated gentian species via androgenesis (Doi et al., 2010; 2013; 2011) or gynogenesis (Pathirana et al., 2011); however, there were no publications on the response of wild gentian species to doubled haploid methodologies. A study by Takamura et al. (2021) evaluated 19 gentian species in which embryogenic structures were observed in 15 of them. The number of embryo-like structures per flower bud ranged from 0 to 1.99. Of the 15 species that produced embryos, regenerated plants were observed in 11 species. The number of regenerated plants per responding genotype ranged from 1 to 78 for a total of 212 plants. Androgenic doubled haploids have been produced in *Tagetes* spp. (Kumar et al., 2019); however, the response was low, and a higher frequency of haploids are required for a breeding program. Another issue with marigold is that some of the genotypes in the breeding program were male sterile, making androgenic methods not feasible. Gynogenic studies evaluated two genotypes, and differences were observed between the two lines (Kumar et al., 2020). Gynogenic embryos were observed in only one genotype: "Local Orange".

10.4.2 Donor Plant Conditions

Most of the recent gynogenesis studies have utilized plants that were grown under field conditions, under natural conditions, or in greenhouses (Table 10.2). As previously mentioned, there are advantages in utilizing environmentally controlled units, as plants can be grown year-round and conditions like temperature, photoperiod, and light quality can be accurately measured and controlled.

10.4.3 Pretreatments

Temperature is a common pretreatment that can be modified to enhance embryogenesis; however, in recent gynogenesis publications for medicinal plants, no pretreatment appears to be the best. A cold pretreatment (4 °C for 5 days) of the ovules was beneficial for embryogenic response in black cumin (El-Mahrouk et al., 2018). This is opposite to the response for *T. erecta*, where a cold treatment of 2 days or longer was detrimental. The best response for callus induction and embryo production was

TABLE 10.2
List of Medicinal Plant Species in Which Gynogenic Protocols Have Been Published Over the Last 10 Years

Species	Genotype	Donor Plant Conditions	Pretreatments	Developmental Stage	Media Composition	Post Isolation Culture Conditions	Culture Response	Reference
Allium hirtifolium	4 ecotypes: Arak, Lorestan, Sanandaj, Yasoj	Field	None	3–5 days before anthesis	B5, 7.5% sucrose, 2 mg/L BA + 2 mg/L NAA	25 °C, 16-hour photoperiod	Haploid plants	Panahandeh and Farhadi 2019
Gentiana spp.	19 wild species	Field, greenhouse	4 °C until culture	Anther dehiscent stage	1/2 NLN, 10% sucrose	25 °C, dark	Haploid plants	Takamura et al. 2021
Nigella sativa	1 pure line	Field	4 °C for 5 days	1–2 days before anthesis	MS, 3% sucrose, 2 mg/L 2,4-D	25 °C, dark	Haploid plants	El-Mahrouk et al. 2018
Tagetes erecta	2 male sterile lines	Net house	None	Ovules from half-open buds	EMS, coconut water 5%, $AgNO_3$, PVP, 2 mg/L BA, 0.5 mg/L NAA or 2 mg/L BA and 2 mg/L 2,4-D, 8% sucrose	25 °C, dark	Haploid plants	Kumar et al. 2020

the control treatment of 25 °C (Kumar et al., 2020). No pretreatment was required for the *Gentiana* species (Takamura et al., 2021) or for *A. hirtifolium* (Panahandeh and Farhadi, 2019).

10.4.4 Developmental Stage

Like the microspore, the developmental stage of the ovule can influence the embryogenic response, which also varies among species. Chen et al. (2011) noted that gynogenic response from ovules can occur from a number of developmental stages. This is compared to the microspore, which has a narrow range of stages, i.e., uninucleate microspore. Studies have shown that young uninuclear, nearly mature embryos sac, and mature embryo sacs have responded in culture (reviewed by Chen et al., 2011). These stages are sometimes correlated to days before anthesis. In *A. hirtifolium*, flower buds were collected 3–5 days before anthesis (Panahandeh and Farhadi, 2019), whereas with *Nigella*, buds were collected 1–2 days before anthesis (El-Mahrouk et al., 2018). Takamura et al. (2021) evaluated two stages of ovule development in wild gentian species and correlated this with anther development: Anther indehiscent and anther dehiscent. The flower buds at the anther-dehiscent stage had more than twice as many embryos than the anther-indehiscent stage. In marigold, the flower bud was used to correlate with the developmental stage (Kumar et al., 2020). Three stages were evaluated: Fully developed unopen flower buds, half-open flower buds, and fully open flower buds. Callus developed from all developmental stages, although significantly more callus and embryo production were observed from the half-open flower buds.

10.4.5 Media Composition

Similar to androgenesis, the media composition plays a major role in gynogenic response of the female gametophyte. Several basal media formulations have been used including B5 (Persian shallot; Panahandeh and Farhadi, 2019), NLN (*Gentian* spp.; Takamura et al., 2021) and MS (marigold; Kumar et al., 2020). A comparison on several basal media formulations was conducted with *T. erecta* ovules (Kumar et al., 2020). Embryos were observed with EMS (modified MS), MS, and B5 but not with N6 or NLN-13. This was similar to their results utilizing anther culture methods (Kumar et al., 2019).

Sucrose is the most widely used carbohydrate in tissue culture and in gynogenic protocols. Sucrose concentrations range from 3% (*Nigella*; El-Mahrouk et al., 2018) to 7.5% (*A. hirtifolium*; Panahandeh and Farhadi, 2019) to 10% (*Gentian* spp.; Takamura et al., 2021). For marigold, six sucrose concentrations were evaluated (3–12%), and embryo formation was the highest at 8%, although embryos were produced at all carbohydrate concentrations (Kumar et al., 2020).

Growth regulators are necessary for callus induction and embryogenic response via gynogenesis as they are for androgenesis. The type of growth regulators as well as the concentration depends on the species. In black cumin, callus and embryogenic calli were only observed on media with auxins (1.0–2.0 mg/L 2,4-D or 2.0 mg/L NAA), as there was no response when BA or kinetin was used in the media (El-Mahrouk et al., 2018). Two concentrations of the auxins, 2,4-D and NAA, were evaluated in Persian shallot (Panahandeh and Farhadi, 2019). The highest number of gynogenic embryos per flower and the highest number of gynogenic plants were observed to be with 2.0 mg/L NAA + 2.0 mg/L BA, which was observed for all four ecotypes. For the marigold gynogenic experiments, Kumar et al. (2020) evaluated BA (0.25–2.0 mg/L) in combination with 2,4-D (0.5–2.0 mg/L) or NAA (0.25–0.5 mg/L). The best combination in terms of callus induction in the shortest time span as well as the most embryo-like structures were from 2.0 mg/L BA and 2.0 mg/L 2,4-D. The highest shoot production was observed on the media with BA (2.0 mg/L) and NAA (0.5 mg/L).

Besides growth regulators, coconut water, $AgNO_3$, and PVP have also been used to enhance callus induction and embryo-like structures (Kumar et al., 2020). In most cases, the media is solidified with agar (El-Mahrouk et al., 2018; Panahandeh and Farhadi, 2019; Takamura et al., 2021) or Gelrite (Kumar et al., 2020).

10.4.6 Culture Conditions

There have been very few studies looking at the culture conditions for continued development after the isolation of the ovules. In most cases, the temperature is 25–27 °C (Table 10.2), and cultures are in the dark for callus induction and the light for plant regeneration. A study with *T. erecta* found that more embryo development was observed with cultures that were under continuous dark for 30 days rather than a 16/8 photoperiod (Kumar et al., 2020). A continuous dark treatment was also used in wild gentian species (Takamura et al., 2021) and *Nigella sativa* (El-Mahrouk et al., 2018); however, a 16-hour photoperiod was used for *A. hirtifolium* (Panahandeh and Farhadi, 2019).

10.5 DOUBLED HAPLOID PLANT PRODUCTION

Plants can be regenerated from the male or female gametophyte via direct embryogenesis or indirectly through a callus phase. The direct regeneration route is the preferred method. Again, media composition and culture conditions are modified to enhance further development of the embryos or calli and subsequent root and shoot production. Various media compositions are evaluated; in many cases, there is a reduction in carbohydrate concentrations and modification of the growth regulators. Light conditions also become a requirement. For medicinal plants, MS medium supplemented with gibberellic acid (GA3) was beneficial for shoot induction followed by B5 medium with indole-3-butyric acid (IBA) to induce roots in *P. ginseng* (Lee et al., 2013). Ari et al. (2022) obtained over 7000 embryos from the *V. hispanica* experiments; from this 124 germinated. In most cases, these embryos produced hairlike roots rather than shoots or a normal root system. Only three plantlets survived: One was haploid and the other two were diploid.

It has been shown in cereals (i.e., wheat, barley) that albinism is a major problem in the resulting haploid/doubled haploid plants derived from microspore or anther culture (Żur et al., 2021). This phenomenon has also been observed from microspore or anther culture of *Aesculus hippocastanum* L. (horse chestnut) (Ćalić et al., 2013). Abscisic acid (ABA) at a concentration of 0.01 mg/L was used to reduce the incidence of albinism to 6% from 11%. The number of chimeric embryos was also reduced.

The ploidy of the regenerated plants can be determined by flow cytometry, chromosome counts, or the number of chloroplasts in the stomata guard cells. For those studies that did produce plants and ploidy levels were determined, haploid, diploid, and polyploid plants were usually detected. In *Echinacea* (Chen et al., 2016), the majority (94.5%) were dihaploid, whereas the parental line was a tetraploid plant, and for *A. hirtifolium*, the frequency of gynogenic plantlets was 70–77%, depending on the genotype (Panahandeh and Farhadi, 2019). In marigold, 424 plants were regenerated via the androgenic process and 56 were analyzed for ploidy. Of those 56 plants, 14.1% were dihaploid, 66.1% were tetraploid, and 19.6% were polyploid (Kumar et al., 2019). The phenotype of the dihaploid plants was short and had poor vigor. Regenerated marigold plantlets were also developed through the gynogenesis route (Kumar et al., 2020). Of the 60 plants that developed via gynogenesis, 17 were evaluated for ploidy. Seven plants were determined to be haploids, eight were diploid, and two plants were polyploids. The haploid plants had pale green leaves and were less vigorous than the parental donor line, similar to the androgenic-derived plantlets. In both cases, the polyploid plants were highly vigorous, as discussed in Tables 10.1 and 10.2).

For haploid plants to be utilized in a breeding program, a chromosome doubling agent is required, such as colchicine or antimitotic agents like trifluralin. The diploid plants could be a spontaneous doubled haploid, or it could be generated from somatic tissue. Marker analysis is required to confirm the doubled haploid status.

10.5.1 Utilization of DH

Haploid and doubled haploid plants have been used in practical applications as well as basic research. For those species where there are well-established DH protocols, breeding, mutagenesis, transformation, genomic selection and genomic prediction, genome editing, and reverse breeding studies have been carried out utilizing either the haploid material (i.e., microspore-derived embryos) or the haploid/doubled haploid plants (Weyen, 2021). There have been very few publications utilizing the DH plants generated from medicinal plants. It was mentioned that the marigold DH lines generated through gynogenesis were incorporated into the breeding program of the Indian Agricultural Research Institute (Kumar et al., 2020). A study evaluated the metabolites found in *Camellia assamica* ssp. *Lasiocalyx* haploid callus and embryos derived through anther culture. A similar metabolite profile was observed in the androgenic cultures and the donor plants (Mishra et al., 2022).

10.6 CONCLUSION

DH methodology has proven to be an invaluable breeding tool for the improvement of crop species. These methods are routinely used by the industry to accelerate the development of new cultivars with improved agronomics, enhanced quality, and adaptability to changing environments. Developing DH protocols for the medicinal plants would be beneficial in yielding uniform true-breeding lines with consistent components of interest. This may alleviate some of the pressures of overharvesting from the wild. Progress has been made over the last few years in the area of DH for some of the medicinal plants, although more research is required.

REFERENCES

Abdollahi, M. R., Eshaghi, Z. C., & Seguí-Simarro, J. M. (2021). Haploid plant production in Borage (*Borago officinalis* L.) by Anther Culture. *Methods in Molecular Biology* 2289:237–248.

Ari, E., Bedir, H., Deniz, I.G., Genc, I., & Seguí-Simarro, J.M. 2022. Evaluation of the androgenic competence of 66 wild Turkish *Vaccaria hispanica* (Mill.) Rauschert genotypes through microspore culture. *Plant Cell Tissue Organ Culture* 148:209–214.

Baillie, A.M.R., Epp, D.J., Hutcheson, D., & Keller, W.A. 1992. *In vitro* culture of isolated microspores and regeneration of plants in *Brassica campestris*. *Plant Cell Reports* 11:234–237.

Ćalić, D., Bohanec, B., Devrnja, N., Milojević, J., Tubić, L., Kostić, I., & Zdravković-Korać, S. 2013. Impact of abscisic acid in overcoming the problem of albinism in horse chestnut androgenic embryos. *Trees* 27:755–762.

Chen, J.-F., Cui, L., Malik, A.A., & Mbira, K.G. 2011. *In vitro* haploid and dihaploid production via unfertilized ovule culture. *Plant Cell Tissue Organ Culture* 104:311–319.

Chen, X., Nilanthi, D., Yang, Y., & Wu, H. 2016. Anther culture and plant regeneration of tetraploid purple coneflower (*Echinacea purpurea* L.). *Journal of Biosciences and Medicines* 4:89–96.

Chu, C.C. 1978. The N6 medium and its applications to anther culture of cereal crops. In: *Proceedings of Symposium on Plant Tissue Culture* Beijing, China: Science Press, pp. 43–50.

Corral-Martínez, P., Siemons, C., Horstman, A., Angenent, G.C., de Ruijter, N., & Boutilier, K. 2020. Live imaging of embryogenic structures in *Brassica napus* microspore embryo cultures highlights the developmental plasticity of induced totipotent cells. *Plant Reproduction* 33:143–158.

Doi, H., Hoshi, N., Yamada, E., Yokoi, S., Nishihara, M., Hikage, T., & Takahata, Y. 2013. Efficient haploid and doubled haploid production from unfertilized ovule culture of gentians (*Gentiana* spp.). *Breeding Science* 63:400–406.

Doi, H., Takahashi, R., Hikage, T., & Takahata, Y. 2010. Embryogenesis and doubled haploid production from anther culture in gentian (*Gentiana triflora*). *Plant Cell Tissue Organ Culture* 102:27–33.

Doi, H., Yokoi, S., Hikage, T., Nishihara, M., Tsutsumi, K., & Takahata, Y. 2011. Gynogenesis in gentians (*Gentiana triflora, G. scabra*): Production of haploids and doubled haploids. *Plant Cell Reports* 30:1099–1106.

Dunwell, J.M. 2010. Haploids in flowering plants: Origins and exploitation. *Plant Biotechnology Journal* 8:377–424.

Dwivedi, S.L., Britt, A.B., Tripathi, L., Sharma, S., Upadhyaya, H.D., & Ortiz, R. 2015. Haploids: Constraints and opportunities in plant breeding. *Biotechnology Advances* 33:812–829.

El-Mahrouk, M.E., Maamoun, M.K., EL-Banna, A.N., Omran, S.A., Dewir, Y.H., & El-Hendawy, S. 2018. *In vitro* gynogenesis and flow cytometry analysis of the regenerated haploids of black cumin (*Nigella sativa*). *HortScience* 53:681–686.

Eshaghi, Z.C., Abdollahi, M.R., Moosavi, S.S., Deljou, A., & Seguí-Simarro, J.M. 2015. Induction of androgenesis and production of haploid embryos in anther cultures of borage (*Borago officinalis* L.). *Plant Cell Tissue Organ Culture* 122:321–329.

Ewais, E.A., Ismail, M.A., Amin, M.A., & Abd-El-moety, E.S. 2019. Phytochemical contests of white and pink flowers of marshmallow (*Althaea officinalis* L.) plants and their androgenesis potential on anther culture in response to chemical elicitors. *Bioscience Research* 16:1276–1289.

Ferrie, A.M.R. 2007. Doubled haploid production in nutraceutical species: A review. *Euphytica* 158:347–357.

Ferrie, A.M.R. 2009. Current status of doubled haploids in medicinal plants. In: *Advances in Haploid Production in Higher Plants*, Touraev A., Forster B.P., and Jain S.M. (eds) Springer Science + Business Media, B.V, pp. 209–218.

Ferrie, A.M.R. 2012. Advances in microspore culture technology: A biotechnological tool for the improvement of medicinal plants. Chapter 8 In: *Biotechnology for Medicinal Plants*, Chandra, S. (eds) Berlin: Springer-Verlag, pp. 191–206. DOI 10.1007/978-3-642-29974-2_8

Ferrie, A.M.R. 2017. Doubled haploid production in higher plants. In: *Encyclopedia of Applied Plant Sciences*, Vol 2, Thomas, B., Murray, B.G, and Murphy, D.J. (eds) Waltham: MA:Academic Press, pp. 147–151.

Ferrie, A.M.R., Bethune, T., & Kernan, Z. 2005. An overview of preliminary studies on the development of doubled haploid protocols for nutraceutical species. *Acta Physiol Plant* 27:735–741.

Ferrie, A.M.R., & Ehlert, Z. 2021. Doubled haploidy for cow cockle (*Saponaria vaccaria* L.). In: *Doubled Haploidy Technology: Methods and Protocols*, Vol 3, Segui-Simarro, J.M. (ed). Springer Science+ Business Media, LLC, Part of Springer Nature, pp. 263–270.

Ferrie, A.M.R., & Keller, W.A. 1997. Production of haploids in *Brassica* spp via microspore culture. In: *Plant Tissue Culture Manual, Supplement 7*, Lindsey, K. (ed). Dordrecht: Section E. Kluwer Academic Publishers, pp. 155–164.

Gamborg, O.L., Miller, R.A., & Ojima, K. 1968. Nutrient requirements of suspension cultures of soybean root cells. *Experimental Cell Research* 50:151–158.

Guha, S., & Maheshwari, S.C. 1964. *In vitro* production of embryos from anthers of *Datura*. *Nature* 204:497.

Hale, B., Ferrie, A.M.R., Chellamma, S., Samuel, J.P., & Phillips, G.C. 2022. Androgenesis-based doubled haploidy: Past, present, and future perspectives. *Frontiers in Plant Science* 12. DOI 10.3389/fpls.2021.751230

Hiramatsu, M., Odahara, K., & Matsue, Y. 1995. A survey of microspore embryogenesis in leaf mustard (*Brassica juncea*). *Acta Horticulturae* 392:139–145.

Hoveida, Z.S., Abdollahi, M.R., Mirzaie-Asl, A., Moosavi, S.S., & Seguí-Simarro, J.M. 2017. Production of doubled haploid plants from anther cultures of borage (*Borago officinalis* L.) by the application of chemical and physical stress. *Plant Cell Tissue and Organ Culture* 130:369–378.

Indrianto, A., Barinova, I., Touraev, A., & Heberle-Bors, E. 2001. Tracking individual wheat microspores *in vitro*: Identification of embryogenic microspores and body axis formation in the embryo. *Planta* 212:163–174.

Jenkins, M., Timoshyna, A., & Cornthwaite, M. 2018. *Wild at Home. Exploring the Global Harvest, Trade and Use of Wild Plant Ingredients*. Cambridge, UK: TRAFFIC.

Kalinowska, K., Chamas, S., Unkel, K., Demidov, D., Lermontova, I., Dresselhaus, T., Kumlehn, J., Dunemann, F., & Houben, A. 2019. State-of-the-art and novel developments of *in vivo* haploid technologies. *Theoretical and Applied Genetics* 132:593–605.

Kästner, U., Kittler, J., & Marthe, F. 2016. Comparison of *in vitro* haploid induction in balm (*Melissa officinalis*). *Plant Cell, Tissue and Organ Culture* 126:561–566.

Keller, W.A., Arnison, P.G., & Cardy, B.K. 1987. Haploid from gametophytic cells: Recent developments and future prospects. In: *Plant Tissue and Cell Culture*, Green, C.E., Somers, D.A., Hackett, W.P. & Biesboer, D.D. (eds) New York: Allan R. Liss, pp. 233–241.

Kernan, Z., & Ferrie, A.M.R. 2006. Microspore embryogenesis and the development of a double haploidy protocol for Cow Cockle (*Saponaria vaccaria*). *Plant Cell Reports* 25:274–280.

Kumar, K.R., Singh, K.P., Bhatia, R., Raju, D.V.S., & Panwar, S. 2019. Optimising protocol for successful development of haploids in marigold (*Tagetes* spp.) through *in vitro* androgenesis. *Plant Cell, Tissue and Organ Culture* 138:11–28.

Kumar, K.R., Singh, K.P., Raju, D.V.S., Bhatia, R., & Panwar, S. 2020. Maternal haploid induction in African marigold (*Tagetes erecta* L.) through *in vitro* culture of un-fertilized ovules. *Plant Cell, Tissue and Organ Culture* 143:549–564.

Kurimella, R.K., Bhatia, R., Singh, K.P., & Panwar, S. 2021. Production of haploids and doubled haploids in marigold (*Tagetes* spp.) using anther culture. In: D*oubled Haploid Technology: Vol. 3: Emerging Tools, Cucurbits, Trees, Other Species*, Seguí-Simarro, J. (ed). Methods in Molecular Biology, pp. 271–287.

Lee, H.-Y., Khorolragchaa, A., Sun, M.-S., Kim, Y.-J., Kim, Y.-J., Kwon, W.-S., & Yang, D.-C. 2013. Plant regeneration from anther culture of *Panax ginseng*. *Korean J. Plant Research* 26:383–388.

Lichter, R. 1982. Induction of haploid plants from isolated pollen of *Brassica napus*. *Zeitschrift Fur Pflanzenzuchtung* 105:427–434.

Mállap-Detquizán, G., Meléndez-Mori, J.B., Huaman-Huaman, E., Vilca-Valqui, N.C., & Oliva, Y.M. 2023. Cultivo de anteras en *Physalis peruviana* L. estadios de microspores, métodos de esterilización y medios de cultivo. *Bioagro* 35:33–42.

Medicinal Plant Names Service Portal, Royal Botanic Gardens, Kew, accessed on 31/05/2023 Medicinal Plant Names Services | Kew

Mishra, V.K., Bajpai, R., & Chaturvedi, R. 2022. Androgenic haploid plant development via embryogenesis with simultaneous determination of bioactive metabolites in Cambod tea (*Camellia assamica* ssp. *lasiocalyx*). *Plant Cell, Tissue and Organ Culture* 148:515–531.

Murashige, T., & Skoog, F. 1962. A revised medium for rapid growth and bioassays with tobacco tissue cultures. *Physiologia Plantarum* 15:473–497.

Namdeo, A.G. 2018. Cultivation of medicinal and aromatic plants. In: *Natural Products and Durg Discovery*, Mandal, S.C., Mandal, V., Konishi, T. (ed) Chapter 20 Elsevier, pp. 525–553.

Nitsch, C. 1974. La culture de pollen isolé sur mileu synthétique. Comptes Rendus de l'Académie des *Sciences Paris* 278:1031–1034.

Nitsch, J.P., & Nitsch, C. 1969. Haploid plants from pollen grains. *Science* 163:85–87.

Panahandeh, J., & Farhadi, N. 2019. Haploid induction via *in vitro* gynogenesis in Persian shallot (*Allium hirtifolium*). *Journal of Horticultural Research* 27:91–98.

Pathirana, R., Frew, T., Hedderley, D., Timmerman-Vaughan, G., & Morgan, E.R. 2011. Haploid and doubled haploid plants from developing male and female gametes of *Gentiana triflora*. *Plant Cell Reports* 30:1055–1065.

Phippen, C., & Ockendon, D.J. 1990. Genotype, plant, bud size and media factors affecting anther culture of cauliflowers (*Brassica oleracea* var *botrytis*). *Theoretical and Applied Genetics* 79:33–38.

Rivas-Sendra, A., Corral-Martínez, P., Porcel, R., Camacho-Fernández, C., Calabuig-Serna, A., & Seguí-Simarro, J. 2019. Embryogenic competence of microspores is associated with their ability to form a callosic, osmoprotective subintimal layer. *Journal of Experimental Botany* 70:1267–1281.

San Noeum, L.H. 1976. Haploïdes d'*Hordeum vulgare* L. par culture *in vitro* d'ovaires no fecondés. *Ann Amelior Plantes* 26:751–754.

Seguí-Simarro, J. M., Moreno, J. B., Fernández, M. G., & Mir, R. 2021. Species with Haploid or Doubled Haploid Protocols. *Methods in Molecular Biology* 2287:41–103.

Takahata, Y., & Keller, W.A. 1991. High frequency embryogenesis and plant regeneration in isolated microspore culture of *Brassica oleracea* L. *Plant Science* 74:235–242.

Takamura, Y., Takahashi, R., Hikage, T., Hatakeyama, K., & Takahata, Y. 2021. Production of haploids and doubled haploids from unfertilized ovule culture of various wild species of gentians (*Gentiana* spp.). *Plant Cell, Tissue, Organ Culture* 146:505–514.

Thompson, K.F. 1972. *Oilseed Rape. Reports of the Plant Breeding Institute*. Cambridge: Cambridge University Press, pp. 94–96.

Tulecke, W. 1964. A haploid tissue culture from the female gametophyte of *Ginkgo biloba*. *Nature* 203:94–95.

Uskutoğlu, T., Uskutoğlu, D., & Turgut, K. 2019. Effects on pre-treatment and different tissue culture media for androgenesis in *Stevia rebaudiana* Bertoni. *Sugar Tech* 21:1016–1023.

Usman, M., Bakhsh, K., Fatima, B., Zaman, Q., & Shah, M.H. 2015. Exploring embryogenic competence in anthers of bitter gourd (*Momordica charantia* L.) cultivar Faisalabad long. *The Journal of Animal & Plant Sciences* 25:181–188.

Varandani, S., Bhattacharya, A., & Char, B.R. 2021. An approach towards induction of double haploids in okra (*Abelmoschus esculentus* L. Moench). *Journal of Applied Horticulture Lucknow* 23:89–92.

Weyen, J. 2021. Applications of doubled haploids in plant breeding and applied research. In: *Doubled Haploid Technology: Volume 1: General Topics, Alliaceae, Cereals*, Vol 2287, Seguí-Simarro, J.M. (ed) Methods in Molecular Biology, pp. 23–39.

Żur, I., Gajecka, M., Dubas, E., Krzewska, M., & Szarejko, I. 2021. Albino plant formation in androgenic cultures: An old problem and new facts. In: *Doubled Haploid Technology: Volume 2: Hot Topics, Apiaceae, Brassicaceae, Solanaceae*, Vol 2288, Seguí-Simarro, J.M. (ed) Methods in Molecular Biology, pp. 3–23.

11 Synthetic Seed Technology for the Conservation of Medicinal Plants

Hashitha N. Aluthgamage, H.I.G Kokila Anuruddi, and D.L.C Kumari Fonseka

11.1 INTRODUCTION

Synthetic seeds can be identified as artificially encapsulated plant parts such as somatic embryos, shoot tips, nodal segments, cell aggregates, or any other living tissues that can be sown as seeds and possess the ability to develop into a plant under suitable *in vitro* or *ex vitro* conditions even after storage (Magray et al., 2017). The concept of synthetic seeds was initially introduced by Murashige in the year 1977. Murashige's concept elucidated the possibility of generating artificial seeds encapsulating somatic embryos capable of sprouting and giving rise to a plant. Following Murashige's concept, the first successful batch of synthetic seeds was developed for carrots (*Daucus carota*) in the 1980s (Kitto and Janick, 1982). Similarly, synthetic seeds for alfalfa (*Medicago sativa*) were successfully developed in 1986 (Redenbaugh et al., 1986). Due to the difficulty in producing somatic embryos for many plant species, the production of synthetic seeds using *in vitro*–generated plant parts was introduced by Bapat et al. (1987) through encapsulated shoot buds of mulberry (*Morus indica*).

There is a notable surge in interest regarding the creation of synthetic seeds, particularly for plants that face challenges in propagation due to various factors (Qahtan et al., 2019). Some prominent advantages of synthetic seeds such as genetic uniformity, easiness of handling due to the small size, low cost of production compared to the hybrid seeds, and the possibility of direct sowing made it a rapidly growing area of biotechnology (Rai et al., 2009). On the other hand, reduced usage of materials, space, and time for synthetic seed production technology made it the economically viable and ideal tool for the conservation of rare and endangered plants. In recent years the encapsulating technique has been increasingly applied as a successful tool for cryopreservation of *in vivo*-derived plant parts, which is a promising manner of long-term conservation (Sharma et al., 2019). Synthetic seeds have become an excellent procedure for propagating hybrid varieties, genetically modified plants, elite genotypes, and endangered plant species (Mandal et al., 2000).

Despite their considerable economic worth, medicinal plants are rarely grown commercially. The majority of medicinal plants are harvested from their natural environments, which often leads to overexploitation and the threat of extinction. Hence, there must be a viable technique for a continual supply of medicinal plants in mass quantities while conserving them in both *in situ* and *ex situ* approaches. Though mass propagation is a practicable technique for supplying common medicinal plant species, a more promising technique should be used for rare and endangered medicinal plants. Because of their compatibility with both mass propagation and conservation methods, synthetic seeds present a compelling choice for the rapid production and preservation of medicinal plants.

DOI: 10.1201/b23374-11

11.2 SYNTHETIC SEED PRODUCTION TECHNOLOGY

11.2.1 Explants

Synthetic seeds can be produced with various explants which are depicted in the Figure 11.1. Since the most productive explant type varies with the plant species, preliminary studies should be conducted to identify the ideal species-specific explant. Somatic embryos are the first form of the explant type used in synthetic seed production (Murashige, 1977). The bipolar nature of somatic embryos containing shoot and root poles resembling zygotic embryos provides the capacity for rapid development of a plant, which is ideal for synthetic seed production (Mariani et al., 2014). However, some other explant types are preferred over somatic embryos due to the complexity of somatic embryogenesis. Presently, nodal segments containing axillary buds are recognized as the prevailing choice of explant for synthetic seed production, as noted by Qahtan et al. (2019). Likewise, apical shoot buds are also frequently employed for the same purpose, as highlighted in the work of Rai et al. (2009). Due to their undifferentiated nature, which requires several components for a successful differentiation, callus is not preferable for the production of synthetic seeds (Gantait et al., 2015).

11.2.2 Encapsulation Matrix

When producing synthetic seeds, the material used to encase them plays a crucial role in determining the overall success of the process. The encapsulation matrix possesses the capacity to function as a storage unit for nutrients, thereby facilitating the survival and expeditious growth of plants. It is essential to consider the encapsulation material carefully to ensure optimal outcomes. In the context of artificial seeds, the encapsulation matrix serves a similar function as the seed coat in natural seeds.

The successful growth and development of the synthetic seed rely significantly on the essential contribution of the seed coat. It must fulfill several functions, such as safeguarding the explants, providing essential nutrients and growth factors, ensuring the safety of the artificial seed during storage and handling, and having a mechanism to initiate germination. Additionally, the seed coat must be compatible with biological and chemical systems and be environmentally friendly by being biodegradable (Khor and Loh, 2005). These requirements are crucial in creating an optimal environment for the artificial seed to flourish and achieve its full potential.

A range of gels, such as agar, alginate, carboxy methyl cellulose, carrageenan, Gelrite, and sodium pectate, have been employed for producing synthetic seeds. Following a series of experiments, it was concluded that the alginate encapsulation technique proved to be the most optimal and efficient approach for achieving this objective (Gantait et al., 2015). Among the variety of gel types utilized for encapsulation purposes, sodium alginate has emerged as the most widely employed matrix due to its affordability, effective gelling characteristics, and non-hazardous composition (Cheruvathur et al., 2013). The firmness of hydrogels is influenced by the concentration of Na^+ ions found in the solution of sodium alginate. These ions are then substituted by Ca^{2+} ions in the

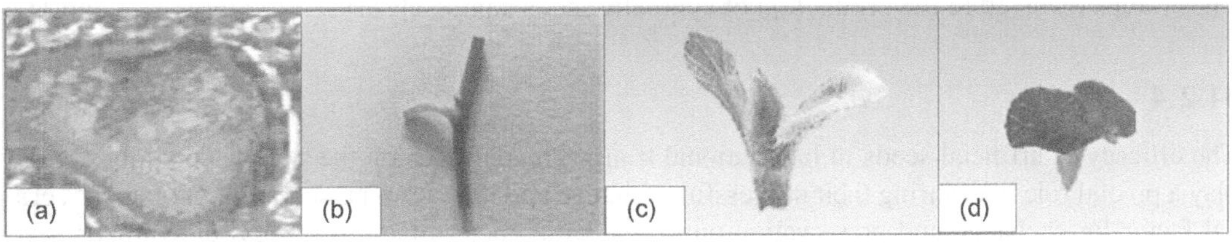

FIGURE 11.1 Explant types used for the production of synthetic seeds: (a) Somatic embryos, (b) nodal segments, (c) shoot tips, and (d) microshoots.

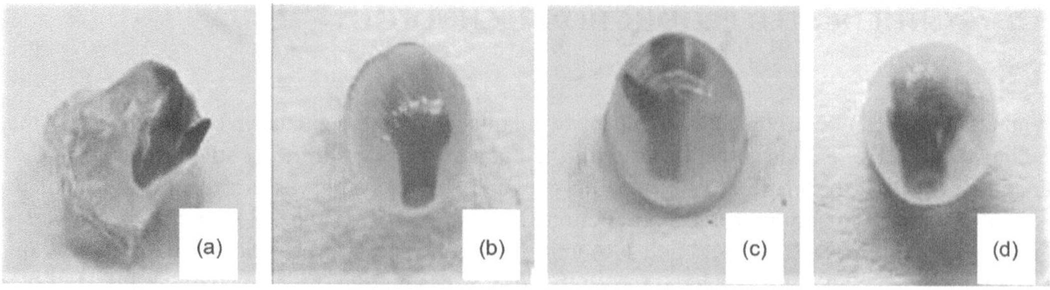

FIGURE 11.2 Influence of sodium alginate concentration on the shape and consistency of synthetic seeds: (a) 2%, (b) 3%, (c) 4%, and (d) 5%.

Source: Fonseka et al., 2019.

$CaCl_2H_2O$ solution, which leads to the creation of an impermeable calcium alginate structure (Daud et al., 2008). The sodium alginate concentration required for a successful encapsulation is species-specific (Redenbaugh et al., 1986). However, studies have shown that the best synthetic seeds were made using a 3% solution of sodium alginate and a 100 mM $CaCl_2H_2O$ solution (Gantait et al., 2015). The encapsulation matrix was prepared using several solutions such as distilled water, MS solution, distilled water + plant growth hormones, and MS solution + plant growth hormones (Daud et al., 2008) as well. The effects of sodium alginate concentration on the shape and consistency of synthetic seeds can be observed in Figure 11.2.

11.2.3 Encapsulation Procedure

The steps in synthetic seed production (Magray et al., 2017; Qahtan et al., 2019) are as follows (Figures 11.3 and 11.4):

1. Introduce explants to the sodium alginate solution.
2. Carefully drop the sodium alginate solution containing explants into a 100 mM calcium chloride solution in a 4:1 ratio.
3. Keep the newly formed synthetic seeds in the calcium chloride solution for 30 minutes for hardening.
4. Wash the synthetic seeds with sterilized distilled water three times.
5. Transfer the synthetic seeds to a sterilized filter paper and let dry for 5 minutes under the laminar airflow.
6. Store the seeds at the optimum temperature according to the requirement.

These steps involved in the production of synthetic seeds can be observed in Figures 11.3 and 11.4.

11.2.4 Storage

The efficacy of artificial seeds in international transit greatly relies on the storage conditions, which play a pivotal role in ensuring their successful exchange and long-term preservation. Storage is a critical factor for exchange and preservation purposes, which determine the success of synthetic seeds after their transportation abroad. To ensure successful germplasm exchange and utilization of synthetic seed technology in commercial applications, it is essential to implement appropriate storage conditions and adhere to specific storage periods. These prerequisites are crucial for the maintenance of viability and overall quality of the genetic material involved. By carefully regulating storage parameters, the potential benefits of synthetic seed technology can be fully utilized (Ikhlaq et al., 2010).

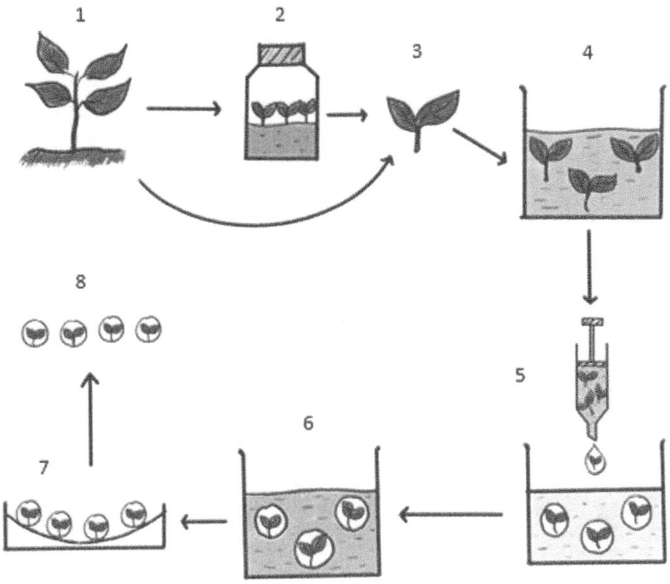

1 – Mother plant

2 – *In vitro* culture

3 – Explant

4 – Transferring explants into the sodium alginate solution

5 – Dropping the sodium alginate solution containing explants into the calcium chloride solution and letting the synthetic seeds harden for 30 minutes

6 – Washing synthetic seeds with sterilized double distilled water (three times)

7 – Transferring the synthetic seeds to a sterile filter paper and let dry for 5 minutes

8 – Synthetic seeds

FIGURE 11.3 Synthetic seed production procedure.

It is essential to establish effective storage methods to ensure the successful transformation of synthetic seeds containing clonal germplasm belonging to high-quality and threatened cultivars within a short period (Maruyama et al., 1997).

11.2.5 Germination and Regrowth

For synthetic seeds to be an effective conservation technique, they must be capable of functioning like regular seeds even after being stored. Hence, the success of synthetic seeds depends on their ability to germinate and convert into complete plants after their respective storage period when introduced to a germination media (Micheli et al., 2007). When considering the germination of synthetic seeds, the concentration of sodium alginate in the encapsulation matrix and germination media are the most crucial factors (Fonseka et al., 2019). Figure 11.5 shows the germinating synthetic seeds of *Celastrus paniculatus* in a test tube. According to Mandal et al. (2000), a nutrient-rich medium is more effective in the germination of synthetic seeds. Though synthetic seeds can be directly introduced to the field, the *in vitro* establishment is advised for a higher germination and

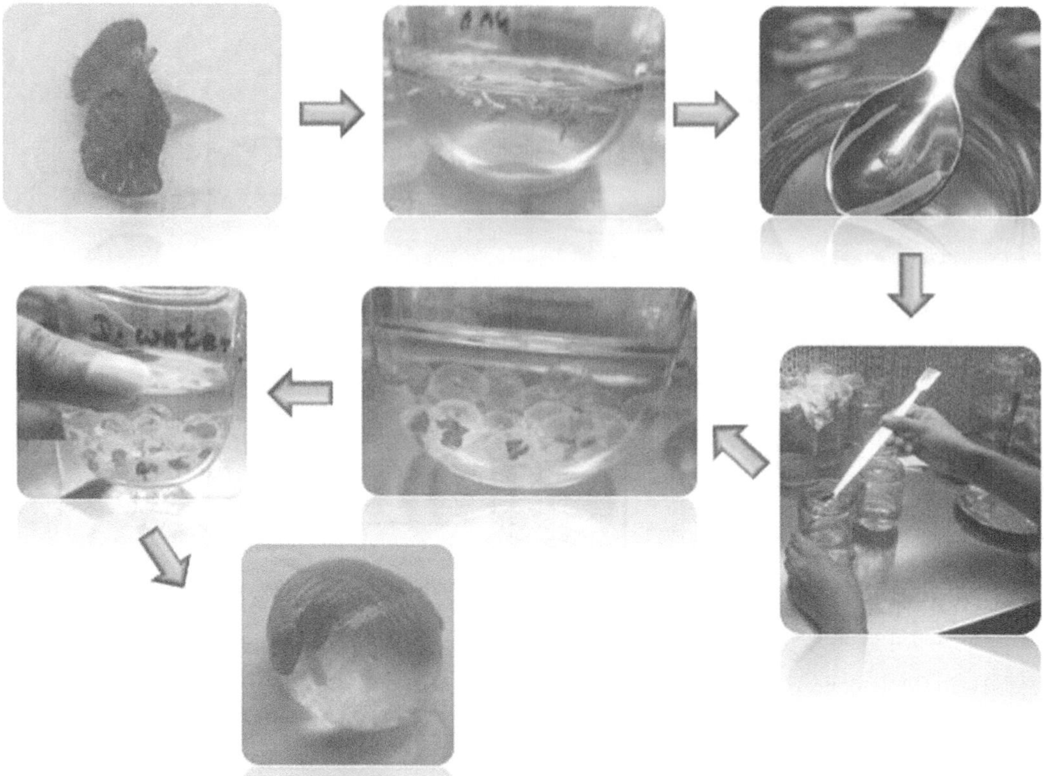

FIGURE 11.4 Production of synthetic seeds with a sterilized spoon.

Source: Aluthgamage et al., 2016.

survival percentage. Both tissue culture media and sterilized substrates like perlite, vermiculite, or sand can be used for the *in vitro* establishment of synthetic seeds (Sharma et al., 2013). Similar to the nutrient content, the type and concentration of hormones in the media also play a crucial role in the re-growth process (Cheruvathur et al., 2013).

FIGURE 11.5 Germinating synthetic seeds of *Celastrus paniculatus*.

Source: Fonseka et al., 2019.

11.3 APPLICATION OF SYNTHETIC SEED TECHNOLOGY FOR THE CONSERVATION OF MEDICINAL PLANTS

It is widely acknowledged that human actions are imposing a significant danger to natural ecosystems, resulting in the gradual extinction of numerous plant species. Among the leading causes of this perilous situation is agriculture, which often involves clearing the indigenous vegetation, causing erosion, salinization, and the invasion of alien species. The situation has only worsened with climate change becoming a significant danger to biodiversity. Despite the efforts of *in situ* conservation, it is often not sufficient to protect endangered species. Seed banking is also not always an option given the various challenges that come with it. Therefore, the need to safeguard genetic diversity has led to the development of micropropagation, which can produce numerous propagules from a region's genetic diversity. This tool is proving to be crucial in conservation efforts, as highlighted by Rogers (2003).

The literature indicates that certain investigations have been carried out concerning the production of synthetic seeds for widely recognized medicinal plants, as well as for some plants possessing medicinal attributes. As gathering from their natural habitats is the primary form of collection for most medicinal plants, producing synthetic seeds is a promising measure of conservation. Due to the simplicity of *in vitro* propagation and encapsulation, a higher number of studies were undertaken for medicinal plants categorized as herbs and shrubs (Table 11.1). Additional investigations were also conducted for plants in the vines and woody climber category (Table 11.2), along with medicinal trees (Table 11.3).

Nonetheless, instances of producing synthetic seeds using aquatic plants remain uncommon. Among the limited examples, synthetic seeds have been successfully produced from two ornamental aquatic plants, namely *Cryptocoryne lutea* and *Rotala rotundifolia* (Micheli and Benelli, 2019). Notably, *R. rotundifolia* holds medicinal and antimicrobial attributes, contributing to its utilization in disease treatment.

11.4 THE FUTURE POTENTIAL OF APPLYING SYNTHETIC SEED TECHNOLOGY FOR THE CONSERVATION OF MEDICINAL PLANTS

Reintroduction to their natural habitats is of the highest importance in the case of conserving rare and endangered plants. With an increasing number of plants facing eradication over time, researchers have to move toward conservation measures to facilitate the easy reintroduction process. Several plant conservation facilities have identified and used *in vitro* culture and micropropagation as a successful *ex situ* conservation technique for decades (Saxena et al., 2019). Synthetic seeds have gained attention as a promising measure of plant conservation in recent years as a new branch in plant tissue culture technology.

Due to its capability of acting as regular seeds, synthetic seeds have a vast scope in the future of plant conservation. Synthetic seeds of several plant varieties have been successfully tested for direct sowing in some substrates as well as soil (Mandal et al., 2000). Hence synthetic seeds have a comparative advantage over micropropagated plants due to the absence of an acclimatization period (Gantait and Kundu, 2017). Similarly, the storage and transportation of synthetic seeds are much more convenient due to their compact nature (Gantait et al., 2017). Short- and long-term conservation is possible for synthetic seeds with low-temperature storage and cryopreservation (Saxena et al., 2019).

11.5 CHALLENGES IN APPLYING SYNTHETIC SEED TECHNOLOGY FOR THE CONSERVATION OF MEDICINAL PLANTS

Even though synthetic seeds emerge as a promising measure of plant conservation, several factors challenge the effectiveness of this procedure. Viability loss with time, difficulties in germination,

TABLE 11.1
Herbs and Shrubs with Medicinal Properties Conserved Using Synthetic Seed Technology

Species	Explant	Sodium Alginate %	Storage Condition	Storage Period	Germination %	References
Adhatoda vasica Nees	SB	3%	–	–	66.28%	Anand and Bansal, 2002
Angelica glauca	SE	3%	–	–	75%	Bisht et al., 2015
Arnebia euchroma	SE	3%	–	–	60.6%	Manjkhola et al., 2005
Bacopa chamaedryoides	ST	3%	15°C	8 weeks	90%	Haque et al., 2015
Camellia sinensis	SE	4%	–	–	–	Muslihatin et al., 2018
Cannabis sativa	NS	5%	15°C	24 weeks	60%	Lata et al., 2012
Cassia angustifolia	MS	3%	4°C	8 weeks	78%	Gantait et al., 2015
Catharanthus roseus	SE	2.5%	4°C	30 days	81%	Maqsood et al., 2012
Centella asiatica (L.)	AB/NS	4%	25°C ± 3°C	200 days	85%	Prasad et al., 2014
Coelogyne breviscapa	PLBs	3%	4°C	15 days	–	Mohanraj et al., 2009
Crinum malabaricum	SE	2%	In darkness at 4°C	1 year	92.5%	Priyadharshini et al., 2020
Curcuma amada	MS	3%	In Luffa storage matrix	4 weeks	75%	Banerjee et al., 2012
Dendrobium densiflorum	PLBs	3%	8°C	60 days	90%	Mohanty and Das, 2013
Eclipta alba	ST, NS	3%	4°C	8 weeks	82.6–85.4%	Ray and Bhattacharya, 2010
Gerbera jamesonii	MS and SE	3%	4°C	1–3 months	75–95%	Taha et al., 2008
Hedychium coronarium J.	MS	3%	4°C	4 weeks	73%	Behera et al., 2020
Mentha arvensis	ST, NS	–	4°C	60 days	44–46%	Islam and Bari, 2012
Picrrhiza kurrooa	MS	3%	In a moist environment at 25°C	3 months	89.33%	Mishra et al., 2011
Plumbago zeylanica L.	Embryoid callus	3%	25°C ± 2°C	12 days	36.8%	Jain et al., 2018
Rauvolfia serpentina	MS	3%	4°C	4 weeks	80%	Faisal et al., 2012
Rauvolfia tetraphylla	MS	3%	4°C	4 weeks	72%	Alatar and Faisal, 2012
Rhinacanthus nasutus (L.)	SE	3%	25°C ± 2°C	45 days	92%	Cheruvathur et al., 2013
Ruta graveolens	NS	3%	4°C	4 weeks	86.7%	Ahmad et al., 2012
Simmondsia chinensis	ST	3%	–	–	66.6%	Kumar et al., 2010
Stevia rebaudiana	ST, NS	3%	4°C	45 days	50%	Ali et al., 2012
Swertia chirayita	SE	4%	–	–	84%	Kumar and Chandra, 2014
Urginea altissima	SE	3%	In darkness at 25°C ± 2°C	15 days	91%	Baskaran et al., 2018
Vitex negundo	NS	3%	4°C	8 weeks	50%	Ahmad and Anis, 2010
Zingiber officinale	MS	4%	In cryovials at 25°C	8 weeks	53%	Sundararaj et al., 2010

Abbreviations: AB, Axillary buds; MS, Microshoots; NS, Nodal segments; PLBs, Protocorm-like bodies; SE, Somatic embryos; ST, Shoot tips.

TABLE 11.2
Vines and Woody Climbers with Medicinal Properties Conserved Using Synthetic Seed Technology

Species	Explant	Sodium Alginate %	Storage Condition	Storage Period	Germination %	References
Ceropegia bulbosa	NS	3%	4°C in dark	60 days	50.7%	Dhir and Shekhawat, 2013
Clitoria ternatea	SE	4%	4°C	5 months	86%	Kumar and Thomas, 2012
Decalepis hamiltonii	NS	4%	4°C	4 weeks	47%	Sharma and Shahzad, 2012
Gymnema sylvestre	NS	3%	4°C	8 weeks	41.8%	Saeed et al., 2018
Hemidesmus indicus	SE	3%	4°C	120 days	86%	Cheruvathur et al., 2013
Mondia whitei	SE	3%	4°C	50 days	51.6%	Baskaran et al., 2015
Tylophora indica	NS	3%	4°C	8 weeks	50.3%	Faisal and Anis, 2007
Vitis vinifera	SE	2%	25°C	3 months	–	Nirala et al., 2007

Abbreviations:˜ MS, Microshoots; NS, Nodal segments; SE, Somatic embryos; ST, Shoot tips.

and malformation of beads are some of the major drawbacks (Qahtan et al., 2019). Similarly, the synthetic seed production procedure is a delicate practice that needs skilled laboratory workers. High-technology laboratories are required with higher biosecurity levels. Hence, the current synthetic seed production procedure involves higher costs. Therefore, a cost-effective procedure should be optimized for commercialization purposes. Further studies should be done in the area of atomization as a successful solution.

TABLE 11.3
Trees with Medicinal Properties Conserved Using Synthetic Seed Technology

Species	Explant	Sodium Alginate %	Storage Condition	Storage Period	Germination %	References
Acca sellowiana	SE	1%	–	–	51.4%	Inocente et al., 2007
Dalbergia sissoo	SE	2.5%	4°C	30 days	21.4%	Singh and Chand, 2010
Erythrina variegate	NS	3%	4°C	8 weeks	47%	Javed et al., 2017
Juglans regia	ST	3%	8°C	3 months	60–80%	Sota et al., 2023
Khaya senegalensis	ST	3%	25°C in dark	8 weeks	73–88%	Hung and Trueman, 2011
Kinnow mandarin (*Citrus nobilis* × *Citrus deliciosa*)	SE	4%	Liquid nitrogen (–20°C)	7 months	58.33	Singh et al., 2007
Moringa oleifera	SE	4%	–	–	–	Muslihatin et al., 2018
Paulownia elongate	SE	3%	4°C	30 days	43.2%	Ipekci and Gozukirmizi, 2003
Pinus radiata	SE	3%	–	–	66%	Aquea et al., 2008
Quercus suber L.	SE	5%	4°C	2 months	26%	Pintos et al., 2008
Salix tetrasperma	NS	3%	4°C	8 weeks	49.67%	Khan et al., 2018
Terminalia arjuna	ST	3%	–	–	91.6%	Gupta et al., 2014

Abbreviations: MS, Microshoots; NS, Nodal segments; SE, Somatic embryos; ST, Shoot tips.

11.6 CONCLUSION

With the increasing population, the vegetative cover of the earth is rapidly declining due to the increase in farmlands, expansion of urban areas, industrialization, and so many other factors. The current rapid deforestation results in the depletion of natural environments for numerous rare and indigenous plant species. The predicament becomes graver for endangered medicinal plants, as they are extensively harvested from their native habitats. Hence, conservation plays a vital role in the sustainable usage of medicinal plants. Various branches of plant tissue culture have been tested for both long-term and short-term *ex situ* conservation of medicinal plants over the decades. Synthetic seed production has gained attention in recent years due to some favorable features compared to other plant tissue culture techniques. The ability to act as regular seeds, absence of an acclimatization period, ease in storage and transportation, and capability in cryopreservation can be identified as the major advantages of synthetic seeds. Similarly, synthetic seeds have some drawbacks that should be optimized through further studies. The high cost of production, the requirement for skilled laborers, viability loss, difficulties in germination and regrowth, and deformations can be identified as major drawbacks. Once these factors are optimized, synthetic seeds can be used more effectively for the conservation and commercialization of synthetic seeds.

ABBREVIATIONS

°C	Celsius
$CaCl_2H_2O$	Calcium chloride
mM	Millimolar
MS	Murashige and Skoog

REFERENCES

Ahmad, N. & Anis, M.2010. Direct plant regeneration from encapsulated nodal segments of *Vitex negundo*, *Biologia Plantarum*, 54 (4): 748–752. doi:10.1007/s10535-010-0134-8

Ahmad, N., Faisal, M., Fatima, N. & Anis, M. 2012. Encapsulation of microcuttings for propagation and short-term preservation in *Ruta graveolens* L.: A plant with high medicinal value. *Acta Physiologiae Plantarum*, 34: 2303–2310. https://doi.org/10.1007/s11738-012-1031-x

Alatar, A. & Faisal, M. 2012. Encapsulation of *Rauvolfia tetraphylla* microshoots as artificial seeds and evaluation of genetic fidelity using RAPD and ISSR markers, *Journal of Medicinal Plants Research*, 6 (7): 1367–1374.

Ali, A., Gull, I., Majid, A., Saleem, A., Naz, S. & Naveed, N. H. 2012. *In vitro* conservation and production of vigorous and desiccate tolerant synthetic seeds in *Stevia rebaudiana*, *Journal of Medicinal Plants Research*, 6 (7): 1327–1333.

Aluthgamage, H. N., Herath, H. M. I. & Fonseka, D. L. C. K. 2016. Synthetic seed production in *Chirita zeylanica* as a conservation method. In: *Proceedings of International Symposium on Agriculture and Environment (ISAE) 2016*, Faculty of Agriculture, University of Ruhuna, Matara, Sri Lanka. 46–52.

Anand, Y. & Bansal, Y. K. 2002. Synthetic seeds: A novel approach of *in vitro* plantlet formation in Vasaka (*Adhatoda vasica* Nees.), *Plant Biotechnology*, 19 (3): 159–162.

Aquea, F., Poupin, M. J., Matus, J. T., Gebauer, M., Medina, C. & Arce-Johnson, P. 2008. Synthetic seed production from somatic embryos of *Pinus radiata*, *Biotechnology Letters*, 30: 1847–1852.

Banerjee, S., Singh, S., Pandey, H., Pandey, P. & ur Rahman, L. 2012. Conservation and storage of *Curcuma amada* Roxb. synseeds on Luffa sponge matrix and RAPD analysis of the converted plantlets, *Industrial Crops and Products*, 36 (1): 383–388.

Bapat, V. A., Mhatre, M. & Rao, P. S. 1987. Propagation of *Morus indica* L. (Mulberry) by encapsulated shoot buds, *Plant Cell Reports*, 6: 393–395.

Baskaran, P., Kumari, A. & Van Staden, J. 2015. Embryogenesis and synthetic seed production in *Mondia whitei*, *Plant Cell, Tissue and Organ Culture (PCTOC)*, 121: 205–214.

Baskaran, P., Kumari, A. & Van Staden, J. 2018. *In vitro* propagation via organogenesis and synthetic seeds of *Urginea altissima* (Lf) Baker: A threatened medicinal plant, *3 Biotech*, 8: 1–8.

Behera, S., Rout, K. K., Panda, P. C. & Naik, S. K. 2020. Production of non-embryogenic synthetic seeds for propagation and germplasm transfer of *Hedychium coronarium* J. Koenig, *Journal of Applied Research on Medicinal and Aromatic Plants*, 19: 100271.

Bisht, A. K., Bhatt, A., Dhar, U., Bhatt, A. & Dhar, U. 2015. Note on somatic embryogenesis and synthetic seed production in *Angelica glauca*: A valuable medicinal plant of Himalaya, *Journal of Medicinal Plants Research*, 9 (12): 419–425.

Cheruvathur, M. K., Kumar, G. K. & Thomas, T. D. 2013. Somatic embryogenesis and synthetic seed production in *Rhinacanthus nasutus* (L.) Kurz, *Plant Cell, Tissue and Organ Culture (PCTOC)*, 113: 63–71.

Cheruvathur, M. K., Najeeb, N. & Thomas, T. D. 2013. *In vitro* propagation and conservation of Indian sarsaparilla, *Hemidesmus indicus* L. R. Br. Through somatic embryogenesis and synthetic seed production. *Acta Physiologiae Plantarum*, 35: 771–779. https://doi.org/10.1007/s11738-012-1117-5

Daud, N., Taha, R. M. & Hasbullah, N. A. 2008. Artificial seed production from encapsulated micro shoots of *Saintpaulia ionantha* Wendl. (African Violet), *Journal of Applied Sciences*, 8: 4662–4667.

Dhir, R. & Shekhawat, G. S. 2013. Production, storability and morphogenic response of alginate encapsulated axillary meristems and genetic fidelity evaluation of *in vitro* regenerated *Ceropegia bulbosa*: A pharmaceutically important threatened plant species, *Industrial Crops and Products*, 47: 139–144.

Faisal, M. & Anis, M. 2007. Regeneration of plants from alginate-encapsulated shoots of *Tylophora indica* (Burm. f.) Merrill, an endangered medicinal plant, *The Journal of Horticultural Science and Biotechnology*, 82 (3): 351–354.

Faisal, M., Alatar, A. A., Ahmad, N., Anis, M.& Hegazy, A. K. 2012. Assessment of genetic fidelity in *Rauvolfia serpentina* plantlets grown from synthetic (encapsulated) seeds following *in vitro* storage at 4°C, *Molecules*, 17: 5050–5061.

Fonseka, D. L. C. K., Wickramaarachchi, W. W. U. I. & Madushani, R. P. S. 2019. Synthetic seed production as a tool for the conservation and domestication of *Celastrus paniculatus*: A rare medicinal plant, *Annual Research & Review in Biology*, 32 (4): 1–8.

Gantait, S. & Kundu, S. 2017. Artificial seed technology for storage and exchange of plant genetic resources. In: *Advanced Technologies for Crop Improvement and Agricultural Productivity*. Agrobios (International), 135–159, Jodhpur.

Gantait, S., Kundu, S., Ali, N. & Sahu, N. C. 2015. Synthetic seed production of medicinal plants: A review on influence of explants, encapsulation agent and matrix. *Acta Physiologiae Plantarum*, 37(5): 1–12. doi:10.1007/s11738-015-1847-2

Gantait, S., Vijayan, J. & Majee, A. 2017. Artificial seed production of *Tylophora indica* for interim storing and swapping of germplasm, *Horticultural Plant Journal*, 3 (1): 41–46.

Gupta, A. K., Harish, Rai, M. K., Phulwaria, M., Agarwal, T. & Shekhawat, N. S. 2014. *In vitro* propagation, encapsulation, and genetic fidelity analysis of *Terminalia arjuna*: A cardioprotective medicinal tree, *Applied Biochemistry and Biotechnology*, 173: 1481–1494.

Haque, S. M., Kundu, S., Das, A. & Ghosh, B. 2015. *In vitro* mass propagation and synthetic seed production combined with phytochemical and antioxidant analysis of *Bacopa chamaedryoides*: An ethnomedicinally important plant, *In Vitro*, 8 (2: 377–383.

Hung, C. D. & Trueman, S. J. 2011. Encapsulation technology for short-term preservation and germplasm distribution of the African mahogany *Khaya senegalensis*, *Plant Cell, Tissue and Organ Culture (PCTOC)*, 107 (3): 397–405. doi:10.1007/s11240-011-9990-y

Ikhlaq, M., Hafiz, I. A., Micheli, M., Ahmad, T., Abbasi, N. A. & Standardi, A. 2010. *In vitro* storage of synthetic seeds: Effect of different storage conditions and intervals on their conversion ability, *African Journal of Biotechnology*, 9 (35: 5712–5721.

Inocente, G. C. C., Dal Vesco, L. L., Steinmacher, D., Torres, A. C. & Guerra, M. P. 2007. Improvements in somatic embryogenesis protocol in feijoa (*Acca sellowiana* (Berg) Burret): Induction, conversion and synthetic seeds, *Scientia Horticulturae*, 111 (3): 228–234. doi:10.1016/j.scienta.2006.10.030

Ipekci, Z. & Gozukirmizi, N. 2003. Direct somatic embryogenesis and synthetic seed production from *Paulownia elongata*, *Plant Cell Reports*, 22: 16–24.

Islam, M. S. & Bari, M. A. 2012. *In vitro* regeneration protocol for artificial seed production in an important medicinal plant *Mentha arvensis* L., *Journal of Bio-Science*, 20: 99–108.

Jain, P., Danwra, K., Sharma, H. P. & Mahato, D. 2018. *In vitro* tissue culture studies and synthetic seed formation from *Plumbago zeylanica* L., *Indian Journal of Experimental Biology*, 56: 769–773.

Javed, S. B., Alatar, A. A., Anis, M. & Faisal, M. 2017. Synthetic seeds production and germination studies, for short term storage and long distance transport of *Erythrina variegata* L.: A multipurpose tree legume. *Industrial Crops and Products*, 105: 41–46. doi:10.1016/j.indcrop.2017.04.053

Khan, M. I., Ahmad, N., Anis, M., Alatar, A. A. & Faisal, M. 2018. *In vitro* conservation strategies for the Indian willow (*Salix tetrasperma* Roxb.), A vulnerable tree species via propagation through synthetic seeds. *Biocatalysis and Agricultural Biotechnology*, 16: 17–21. doi:10.1016/j.bcab.2018.07.002

Khor, E. & Loh, C. S. 2005. Artificial seeds. In: Nedovic, V. & Willaert, R. (eds) *Applications of Cell Immobilization Biotechnology*, 527–537, Springer, Berlin.

Kitto, S. L. & Janick, J. 1982. Polyox as an artificial seed coat for asexual embryos, *Horticultural Science*, 17: 448.

Kumar, G. K. & Thomas, T. D. 2012. High frequency somatic embryogenesis and synthetic seed production in *Clitoria ternatea* Linn. *Plant Cell, Tissue and Organ Culture*, 110: 141–151. https://doi.org/10.1007/s11240-012-0138-5

Kumar, S., Rai, M. K., Singh, N. & Mangal, M. 2010. Alginate-encapsulation of shoot tips of jojoba [*Simmondsia chinensis* (Link) Schneider] for germplasm exchange and distribution, *Physiology and Molecular Biology of Plants*, 16 (4): 379–382. doi:10.1007/s12298-010-0043-5

Kumar, V. & Chandra, S. 2014. High frequency somatic embryogenesis and synthetic seed production of the endangered species *Swertia chirayita*, *Biologia*, 69: 186–192.

Lata, H., Chandra, S., Mehmedic, Z., Khan, I. A. & El-Sohly, M. A. 2012. *In vitro* germplasm conservation of high D9-tetrahydrocannabinol yielding elite clones of *Cannabis sativa* L. under slow growth conditions, *Acta Physiologiae Plantarum*, 34: 743–750.

Magray, M. M., Wani, K. P., Chatto, M. A. & Ummyiah, H. M. 2017. Synthetic seed technology, *International Journal of Current Microbiology and Applied Sciences*, 6 (11): 662–674.

Mandal, J., Pattnaik, S. & Chand, P. K. 2000. Alginate encapsulation of axillary buds of *Ocimum americanum* L. (hoary basil), *O. basilicum* L. (sweet basil), *O. gratissimum* L. (shrubby basil), and *O. sanctum* (sacred basil), *In Vitro Cellular and Developmental Biology – Plant*, 36: 287–92.

Manjkhola, S., Dhar, U. & Joshi, M. 2005. Organogenesis, embryogenesis, and synthetic seed production in *Arnebia euchroma*—A critically endangered medicinal plant of the Himalaya, *In Vitro Cellular & Developmental Biology-Plant*, 41: 244–248.

Maqsood, M., Mujib, A. & Siddiqui, Z. H. 2012. Synthetic seed development and conversion to plantlet in *Catharanthus roseus* (L.) G. Don, *Biotechnology*, 11 (1): 37–43.

Mariani, T. S., Sasmitamiharja, D., Mienanti, D., Latif, S., Ginting, G. & Miyake, H. 2014. Somatic embryogenesis of oil palm (*Elaeis guineensis* Jacq.) for synthetic seed production, *Asian Journal of Applied Sciences*, 2 (3: 358–367..

Maruyama, E., Kinoshita, I., Ishii, K., Shigenaga, H., Ohba, K. & Saito, A. 1997. Alginate-encapsulation technology for the propagation of the tropical forest trees: *Cedrela odorata* L., *Guazama crinite* Mart., *Jacaranda mimosaefolia* D. Don, *Silvae Genetica*, 46: 17–23.

Micheli, M. & Benelli, C. 2019. Synthetic seeds of two aquatic plants. In: Faisal, M., & Alatar, A. A. (eds) *Synthetic Seeds: Germplasm Regeneration, Preservation and Prospects*, 233–239, Springer Nature, Cham.

Micheli, M., Hafiz, I. A. & Standardi, A. 2007. Encapsulation of *in vitro*-derived explants of olive (*Oleaeuropaea* L. Cv. Moraiolo): II. Effects of storage on capsule and derived shoots performance, *Scientia Horticulturae*, 113: 286–292.

Mishra, J., Singh, M., Palni, L. M. S. & Nandi, S. K. 2011. Assessment of genetic fidelity of encapsulated microshoots of *Picrorhiza kurrooa*, *Plant Cell, Tissue and Organ Culture (PCTOC)*, 104 (2): 181–186. doi:10.1007/s11240-010-9816-3

Mohanraj, R., Ananthan, R. & Bai, V. N. 2009. Production and storage of synthetic seeds in *Coelogyne breviscapa* Lindl, *Asian Journal of Biotechnology*, 1 (3): 124–128.

Mohanty, P. & Das, J. 2013. RETRACTED ARTICLE: Synthetic seed technology for short term conservation of medicinal orchid *Dendrobium densiflorum* Lindl. Ex Wall and assessment of genetic fidelity of regenerants, *Plant Growth Regulation*, 70 (3): 297–303.

Murashige, T. 1977. Plant cell and organ cultures as horticultural practices, *Acta Horticulturae*, 78: 17–30.

Muslihatin, W., Jadid, N., Saputro, T. B., Purwani, K. I., Himayani, C. E. S. & Calandry, A. W. 2018. Characteristic of synthetic seeds from two medicinal plants (*Moringa oleifera* and *Camellia sinensis*), *Journal of Physics: Conference Series*, 1040: 012005.

Nirala, N. K., Das, D. K., Reddy, M. K., Srivastava, P. S., Sopory, S. K. & Upadhyaya, K. C. 2007. Encapsulated somatic embryos of grape (*Vitis vinifera* L.): An efficient way for storage, transport and multiplication of pathogen free plant material. In: *Proceedings Asia Pacific Conference on Plant Tissue and Agribiotechnology (APaCPA)*, 17: 21.

Pintos, B., Bueno, M. A., Cuenca, B. & Manzanera, J. A. 2008. Synthetic seed production from encapsulated somatic embryos of cork oak (*Quercus suber* L.) and automated growth monitoring, *Plant Cell, Tissue and Organ Culture*, 95: 217–225.

Prasad, A., Singh, M., Yadav, N. P., Mathur, A. K. & Mathur, A. 2014. Molecular, chemical and biological stability of plants derived from artificial seeds of *Centella asiatica* (L.) Urban—An industrially important medicinal herb, *Industrial Crops and Products*, 60: 205–211.

Priyadharshini, S., Manokari, M. & Shekhawat, M. S. 2020. *In vitro* conservation strategies for the critically endangered Malabar river lily (*Crinum malabaricum* Lekhak & Yadav) using somatic embryogenesis and synthetic seed production, *South African Journal of Botany*, 135: 172–180.

Qahtan, A. A., Abdel-Salam, E. M., Alatar, A. A., Wang, Q. C. & Faisal, M. 2019. An introduction to synthetic seeds: Production, techniques, and applications. In: Faisal, M., & Alatar, A. A. (eds) *Synthetic Seeds: Germplasm Regeneration, Preservation and Prospects*, 1–20, Springer Nature, Cham.

Rai, M. K., Asthana, P., Singh, S. K., Jaiswal, V. S. & Jaiswal, U. 2009. The encapsulation technology in fruit plants—A review, *Biotechnology Advances*, 27 (6): 671–679. doi:10.1016/j.biotechadv.2009.04.025

Ray, A. & Bhattacharya, S. 2010. Storage and conversion of *Eclipta alba* synseeds and RAPD analysis of the converted plantlets, *Biologia Plantarum*, 54 (3): 547–550. doi:10.1007/s10535-010-0097-9

Redenbaugh, K., Paason, B. D., Nichol, J. W., Kossler, M. E., Viss, P. R. & Walker, K. A. 1986. Somatic seeds: Encapsulation of asexual plant embryos, *Bio Technology*, 4: 797–801.

Rogers, S. 2003. Tissue culture and wetland establishment of the freshwater monocots *Carex*, *Juncus*, *Scirpus*, and *Typha*, *In Vitro Cellular & Developmental Biology-Plant*, 39: 1–5.

Saeed, T., Shahzad, A., Ahmad, N. & Parveen, S. 2018. High frequency conversion of non-embryogenic synseeds and assessment of genetic stability through ISSR markers in *Gymnema sylvestre*, *Plant Cell, Tissue and Organ Culture (PCTOC)*, 134 (1): 163–168. doi:10.1007/s11240-018-1409-6

Saxena, A., Shukla, M. & Saxena, P. 2019. Synthetic seeds: Relevance to endangered germplasm conservation *in vitro*. In: Faisal, M., & Alatar, A. A. (eds) *Synthetic Seeds: Germplasm Regeneration, Preservation and Prospects*, 21–60, Springer Nature, Cham.

Sharma, N., Gowthami, R. & Pandey, R. 2019. Synthetic seeds: A valuable adjunct for conservation of medicinal plants. In: Faisal, M., & Alatar, A. A. (eds) *Synthetic Seeds: Germplasm Regeneration, Preservation and Prospects*, 181–216, Springer Nature, Cham.

Sharma, S. & Shahzad, A. 2012. Encapsulation technology for short-term storage and conservation of a woody climber, *Decalepis hamiltonii* Wight and Arn. *Plant Cell, Tissue and Organ Culture*, 111: 191–198. https://doi.org/10.1007/s11240-012-0183-0

Sharma, S., Shahzad, A. & Teixeira da Silva, J. A. 2013. Synseed technology—A complete synthesis, *Biotechnology Advances*, 31: 186–207.

Singh, A. K. & Chand, S. 2010. Plant regeneration from alginate encapsulated somatic embryos of *Dalbergia sissoo* Roxb, *Indian Journal of Biotechnology*, 9: 319–324.

Singh, B., Sharma, S., Rani, G., Virk, G. S., Zaidi, A. A. & Nagpal, A. 2007. *In vitro* response of encapsulated and non-encapsulated somatic embryos of Kinnow mandarin (*Citrus nobilis* Lour × *C. deliciosa* Tenora), *Plant Biotechnology Reports*, 1 (2): 101–107. doi:10.1007/s11816-007-0015-6

Sota, V., Benelli, C., Myrselaj, M., Kongjika, E. & Gruda, N. S. 2023. Short-term conservation of *Juglans regia* L. Via synthetic seed technology, *Horticulturae*, 9 (5): 559.

Sundararaj, S. G., Agrawal, A. & Tyagi, R. K. 2010. Encapsulation for *in vitro* short-term storage and exchange of ginger (*Zingiber officinale* Rosc.) germplasm, *Scientia Horticulturae*, 125 (4): 761–766. doi:10.1016/j.scienta.2010.06.001

Taha, R. M., Hasbullah, N. A. & Awal, A. 2008, August. Production of synthetic seeds from micro shoots and somatic embryos of *Gerbera jamesonii* Bolus ex. Hook f. In: *Proceedings of VI International Symposium on In Vitro Culture and Horticultural Breeding*, 829: 91–98.

12 Silicon Nanoparticles (SiNPs)–Mediated Morpho-histochemical Improvements in the Micropropagation of Medicinal Plants

Mahipal S. Shekhawat and Nirmal Joshee

12.1 INTRODUCTION

Silicon (Si) stands out as a nearly indispensable plant nutrient, exerting significant influence on physiological processes and substantially enhancing growth, development, and stress tolerance across various plant species. As the Earth's second most abundant elements, silicon has garnered attention for its beneficial role in advancing crop improvement (Parveen and Ashraf, 2010). Recently, a growing trend has been observed on the production of nanoparticles (NPs) using biological sources and their utilization in plant tissue culture technology, as documented by Tomar et al. (2018) and Tymoszuk and Miler (2019). Notably, the effect of Si on the production of high-quality plants during *in vitro* growth has emerged as a fascinating area of exploration, showcasing enhanced survival rates and a notable reduction in morpho-physiological and structural disorders. This review delves into the multifaceted role of silicon, exploring its implications for plant development and the promising avenues it opens for the optimization of *in vitro* growth conditions.

Silicon nanoparticles (SiNPs) have been recognized for their distinctive physiological attributes, positioning them as effective carriers for nutrients, minerals, proteins, and nucleotides within plant systems (Bapat et al., 2020). Notably, SiNPs serve a dual role by aiding in water absorption and acting as carriers for larger molecules, as highlighted in the study by Rastogi et al. (2021). The manifold positive impacts of SiNPs on plant growth constitute a compelling narrative, encompassing enhancements in photopigments, enzymatic activities, biomass production, oil deposition, enhancement of vegetative growth, augmentation of flower traits, improvements in seed germination, and an increase in the growth of seedlings (Siddiqui and Al-Whaibi, 2014; Attia and Elhawat, 2021; Mukarram et al., 2021; Cokul Raj et al., 2023a).

The hormetic effects induced by SiNPs play a crucial role in fostering the formation of robust roots, expediting regeneration and shoot proliferation, facilitating somatic embryo formation, and even inducing flowering *in vitro* (Cokul Raj et al., 2023a). Moreover, when incorporated into the nutrient medium at low concentrations (100 mg L^{-1}), SiNPs have been found to enhance seed setting, adding another layer to their versatile impact on plant physiology (Cokul Raj et al., 2023a). This review synthesizes the diverse contribution of SiNPs to plant growth and development, shedding light on their potential applications and future explorations in harnessing their benefits for optimized agricultural practices (Figure 12.1).

DOI: 10.1201/b23374-12

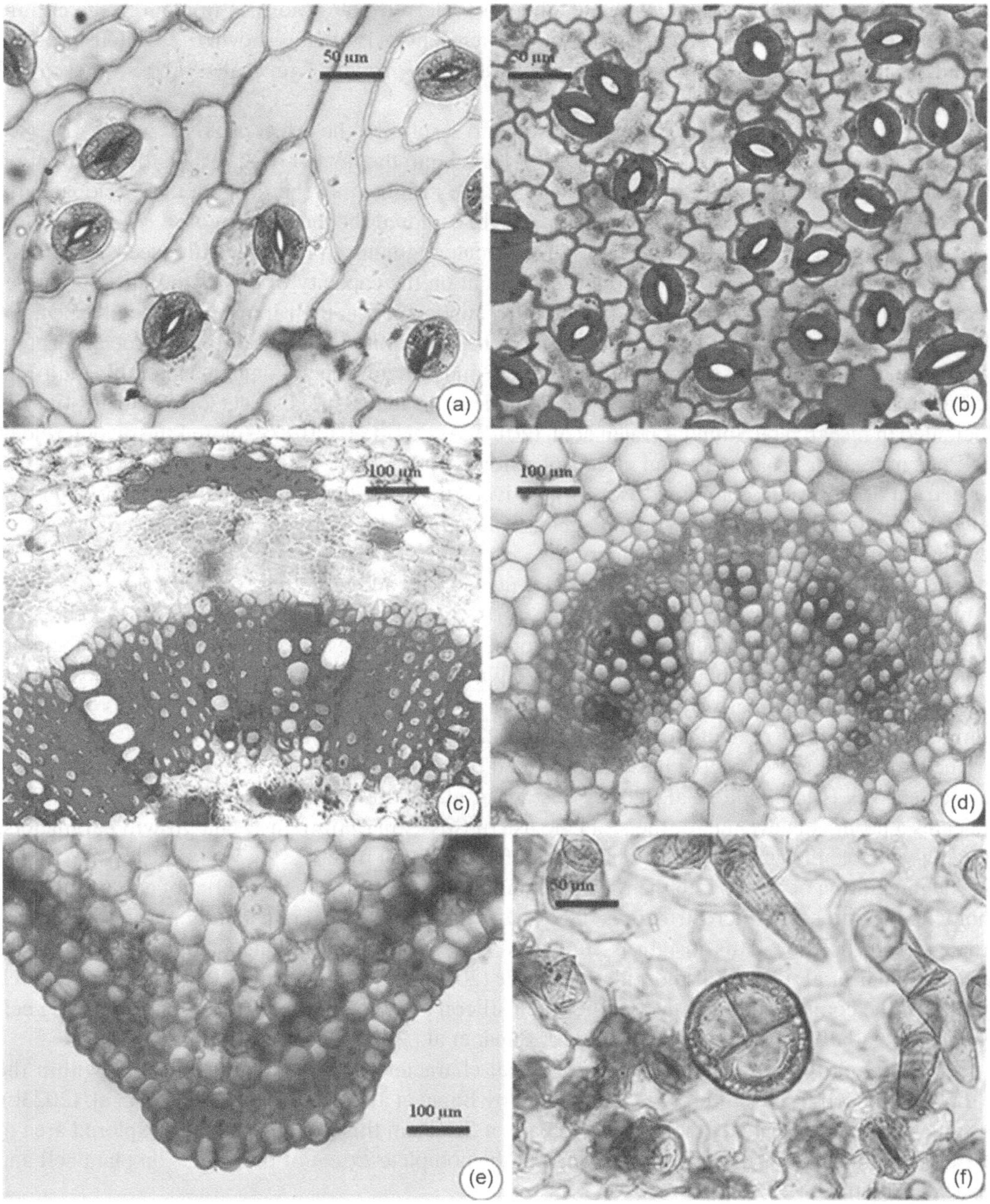

FIGURE 12.1 (a) Well-developed stomata of *Scaevola* spp. Treated with silicon nanoparticles. (b) Developed stomatal apparatuses in *Hemidesmus indicus* due to the presence of SiNPs in the medium. (c) Micro-structural developments in *Vitex* stem in the presence of SiNPs. (d) Well-developed vascular bundles in the leaves of *Thunbergia erecta* derived from SiNP-fortified medium. (e) Well-differentiated epidermal and mesophylls in SiNP-treated leaves of *Thunbergia* spp. (f) Development of trichomes on the SiNP-containing medium in *Vitex* spp.

The positive impact of silicon and SiNPs on plant growth and stress tolerance has been well documented (Rajput et al., 2021). SiNPs have demonstrated the ability to boost cell division and differentiation while improving photosynthesis and biomass accumulation. Additionally, they serve as a protective shield for plants, guarding against various biotic and abiotic stressors like drought, high salinity, and fungal infections. This makes them a preferred option for use in plant tissue culture (Goswami et al., 2022). Introducing SiNPs into the nutrient medium has shown tremendous promise in plant tissue culture, significantly enhancing overall plant growth (Kim et al., 2002; Sivanesan and Park, 2014).

Numerous investigations have been undertaken to scrutinize the effect of SiNPs across different facets of plant physiology. Siddiqui et al. (2020) delved into the impact of SiNPs on the germination of seeds and growth of seedlings, uncovering valuable insights into their role in these critical early stages of plant development. Cokul Raj et al. (2023b) explored the mechanism of *in vitro* flower formation, adding a level of understanding to the intricate regulatory processes influenced by SiNPs.

Studies by Mukarram et al. (2021) have shed light on the capacity of SiNPs to induce improvements in chl *a*, chal *b*, carotenoids, and metabolites, underscoring their impact on crucial aspects of plant biochemical processes. Furthermore, SiNPs have exhibited the ability to enhance absorption and distribution of nutrients, providing mechanical and structural help to plant cells. This multifaceted approach extends to decrease the *in vitro* made cells and tissue disorders, like vitrification, as elucidated in research by Soundararajan et al. (2017). Asgari et al. (2018) explored the pivotal role of SiNPs in mitigating the detrimental effects of oxidative stress, as well as offering resilience against a spectrum of biotic and abiotic stresses. Collectively, these studies contributed to a comprehensive understanding of the diverse and advantageous roles played by SiNPs in shaping different ways of development and stress response in plants.

Silicon and SiNPs serve as nano-carriers for nucleotides and proteins, exerting a positive impact on biochemical processes, as underscored in reports of Bapat et al. (2020) and Rastogi et al. (2020). The involvement of SiNPs spans various physiochemical activities, imparting an effect on the morphology and physiology of plantlets and fortifying mechanisms against disease-causing organisms. These contributions collectively play a pivotal role in increasing overall growth of plants and yield, as highlighted in the works of Goswami et al. (2022) and Kovács et al. (2022).

The exogenous use of SiNPs has demonstrated their efficacy in improving plant growth across various species (Mahmoud et al., 2020; Rastogi et al., 2020; Mustafa and Toaiema, 2022). This indicates a promising avenue for the strategic use of SiNPs in optimizing plant growth parameters, offering valuable insights for the advancement of agricultural practices.

The integration of SiNPs into the MS medium has demonstrated the potential of enhancing morphological features and microstructures of cells and tissues, thereby contributing to improve micropropagation efficiency. This insight is supported by various studies conducted by Siddiqui et al. (2020), Tomar et al. (2018), Tymoszuk and Miler (2019), and Bapat et al. (2020). Recent research highlights the complex interconnection between silicon and metabolites crucial for reinforcing cell walls, especially lignin (Siddiqui et al., 2020; Bapat et al., 2020; Cokul Raj et al., 2023b).

SiNPs, owing to their distinctive physiological characteristics, find significant uses within the realm of plant cell tissue culture, as documented by Bapat et al. (2020) and Cokul Raj et al. (2023b). Despite these advancements, the detailed impact of SiNPs in this remains an underexplored area of research, urging further investigation to unravel the complete extent of the effects on plant cell and tissue development.

12.2 SILICON NANOPARTICLES: PREPARAION AND PROPERTIES

SiNPs, characterized by a girth of 12 nm and an area of 275–340 m^2 g^{-1}, function as versatile metal-free quantum dots exhibiting the capacity to confer resistance to stresses in plants, improving growth and activating defense mechanisms. Plants tend to develop storage tissues by suppressing catabolism under stress environments, as observed in studies by Khan et al. (2016), Tina et al. (2023), and Manokari et al. (2023a). *In vitro* application of SiNPs is advantageous for normal plant development and the enhancement of protective tissue systems (Manokari et al., 2023b). SiNPs act as stress signals, triggering a series of self-protective responses, including the production of active biochemicals which facilitate stress adaptation, as highlighted by Wang et al. (2022). The incorporation of SiNPs also contributes to the formation of self-protective features against stresses. This is achieved through the modulation of the cell wall and an increase in the thickness of the leaf, as demonstrated in studies by Chen et al. (2018) and Frew et al. (2018).

To prepare SiNP suspensions, various silica derivatives such as K_2SiO_3, Na_2SiO_3, and SiO_2 have been tested, as documented in reports by Baoshan et al. (2004), Siddiqui et al. (2020), and Soares et al. (2011). In plant tissue culture experiments, different grades of Aerosil, such as Aerosil^{300} Aerosil^{200}, etc., have been predominantly utilized. Aerosil^{300}, hydrophilic silica with a 10-nm diameter and a surface area of 270–330 $m^2 g^{-1}$ and a pH range of 3.7–4.5, sourced from Evonik Industries in Germany, is composed entirely of amorphous silica according to literature provided by Evonik. We followed the silica suspension protocols outlined by Attia and Elhawat (2021), with required changes in various studies (Cokul Raj et al., 2023a, b).

12.3 SILICON IMPROVED SHOOT REGENERATION AND MICRO-MORPHOLOGY

SiNPs have been widely recognized for their positive impact on the formation of plants. In the context of *in vitro* plant tissue culture, the balance of essential minerals in the growth medium is crucial for the growth and development of shoots. However, challenges related to morphology and structural issues and associated morpho-physiochemical anomalies have been significant impediments in cell and tissue culture (Manokari et al., 2021a, b). Various studies have demonstrated that silicon plays important function in enhancing the morpho-structural growth of micropropagated plants (Manokari et al., 2021c, 2023c, 2022). Silicon contributes to the stabilization of cell walls during cell divisions and elongation, maintaining cells and biochemical processes and promoting the survival of plants (Cokul Raj et al., 2023b). The unique physiochemical properties of SiNPs have a positive effect on plant biochemical processes. This includes benefits such as improved photosynthetic rate, exchange of gases, increased leaf N_2 content, boosting enzyme functions, and increasing the concentration of essential oils when utilized as a spray on leaves (Mukarram et al., 2021; 2022). Consequently, the use of SiNPs has been associated with increased productivity in various industrially useful plants (Ali et al., 2019; Sánchez-Navarro et al., 2021). Furthermore, plants grown on SiNPs have demonstrated a thickened cell wall and increased tolerance to stresses (Lux et al., 2020; Shen et al., 2022). Overall, the application of SiNPs appears to be a very useful way to address challenges in plant tissue culture and enhance the health of plants.

In the cultivation of micropropagated shoots of *Hemidesmus indicus*, the incorporation of SiNPs into Murashige and Skoog (MS) growth medium resulted in a notable improvement in quality, as reported by Manokari et al. (2023a). The shoots derived from explants displayed enhanced growth compared to those devoid of SiNPs. Structurally, the shoots cultivated with SiNPs exhibited the development of functional stomata, a normal cuticle, mesophyll with many layers, cortex, and many numbers of vascular elements in both leaves and stems. This was in stark contrast to the control shoots, which showed stomata and mesophylls with poor development and a limited number of epidermal, cortical, and food and water transporting tissues (Figure 12.1a–e). Manokari et al. (2023b) also investigated the impact of SiNPs on micropropagation and morphological and anatomical traits in tissue-cultured plants of *Santalum album*.

Shoots generated *in vitro* were cultivated on MS nutrient medium augmented with SiNPs and a preoptimized combination of growth regulators. Micro-morphological analysis of the shoots demonstrated several SiNP-derived changes, including formation of a thick stem, the multiple-layered ground tissues and sclerenchymatous bundle-sheath caps, and development of cambium, facilitating the development of phloem and xylem tissues. This improvement of a robust shoot system *in vitro* contributed to the overall enhancements in *S. album*.

The introduction of SiNPs into the preoptimized medium for regeneration not only led to an increase in photosynthetic tissues but also influenced the formation of mesophyll tissues. Additionally, the use of SiNPs resulted in a good thickness of leaf tissues, deposition of a wax layer on the cuticle, and increased leaf biomass across different *in vitro* plantlets (Asmar et al., 2013a, b). Furthermore, the supplementation of SiNPs in the nutrient solution led to a progressive enhancement in ground tissue density due to better transport of nutrients and plant growth regulators (PGRs) from the medium to the explants. This could be attributed to the steady progress in spongy and palisade tissues and the organized formation of foliar tissues (Manokari et al., 2023c). In leaves derived from SiNP treatment, well-differentiated vascular tissues were observed; contrasting with this, the tissues formed under controlled conditions displayed a smaller number of vascular elements in this experiment. The treated leaves exhibited good wall thickening in the wall of xylem cells and collenchymas tissues. Extensive studies have highlighted that Si promotes lignin formation and deposition in cell walls, particularly in the xylem and phloem elements (Głazowska et al., 2018).

The anatomical traits of leaves play a crucial role in plant survival within a specific environment, given that leaves are essential photosynthetic organs (Mollick et al., 2021). The development of phloem elements regulates the transport of nutrients and PGRs, while the density of xylem vessels influences the hydraulic transport of water (Gorsuch and Oberbauer, 2002). Indicators of photosynthetic capacity, such as well-differentiated mesophyll tissues and developed stomata in the foliage, are pivotal for adapting to abiotic pressures (Grassi and Magnani, 2005).

NPs are absorbed by xylem tissues through microchannels in the cell walls, as observed in studies by (Krupa-Małkiewicz and Calomme, 2021). Silicon plays a crucial role in enhancing the growth and development of plants by facilitating efficient water and mineral transportation and boosting metabolic processes (Li et al., 2023). Previous research has indicated that the addition of silicon to a medium yields improvements in the morpho-anatomical and physiochemical traits of plants. It contributes to the development of callus and formation of stems and roots and aids in the acclimatization of plants to biotic and abiotic stresses (Soares et al., 2012; Mustafa and Toaiema, 2022).

Current research on SiNPs is directed to enhance shoot multiplication, the formation of structures related to defense, and the tolerance of stress in *in vitro* systems, with the aim of alleviating all types of stresses (Kim et al., 2002; Sivanesan and Park, 2014). Determining the optimum amount of SiNPs is crucial to achieving complete and healthy cells and tissue growth. To mitigate the potential toxicity of SiNPs, it is necessary to identify the appropriate source and optimal concentration of nano silica for promoting healthy cell growth (Mantovani et al., 2020).

12.4 EFFECT OF SiNPs ON INDUCTION AND MATURATION OF ADVENTITIOUS ROOTS AND ACCLIMATIZATION OF PLANTLETS

A crucial phase in the practical application of *in vitro* technology involves the successful establishment of roots and acclimatization of micropropagated plants, followed by their transfer to the field (Manokari et al., 2023c). Adventitious root formation is the consequence of complex interactions between internal and external elements, including phytohormones, light, temperature, and nutrient availability (Cokul Raj et al., 2023b). The inadequate rooting of shoots raised *in vitro* poses a significant obstacle, profoundly impacting the rate of survival of plants *in vivo*. Manokari et al. (2023c) utilized *Hedyotis biflora* as a model to examine the influence of SiNPs on the formation of roots and development of structures. The shoots proliferated and cultivated on ½ MS nutrient medium having IBA and SiNPs demonstrated enhanced morpho-anatomy of the roots. Morphologically, the roots treated with SiNPs exhibited notable superiority: They were thicker, sturdier, had higher biomass, and were more numerous compared to the fragile roots in the control group. Microscopic observations of the developed roots revealed that SiNP supplementation resulted in better development of dermal tissues, secondary cortex, and vascular tissues. In contrast, roots developed on the medium devoid of SiNPs displayed primary structures only and lacked the complexity observed in SiNP-treated roots. An interesting finding was the significantly higher amount of secondary xylem elements in the SiNP-derived roots compared to the roots developed in the control medium. This observation connected completely to the enhanced adaptability of roots *in vivo*.

Half-strength MS medium has proved efficient in stimulating root formation across various plant species, as highlighted in studies by Manokari et al. (2021a, 2022). The deliberate dietary shortage in the nutrient medium has been observed to expedite the procedure of formation of roots, particularly in the control of root-inducing hormones. While previous research utilized full-strength MS medium in the mixture with IBA for the formation of roots from the shoot of *Vitex trifolia*, the combination of ½ MS + NAA was found to yield robust and premature roots from the base of shoots (Hiregoudar et al., 2006; Ahmed and Anis, 2012; Ahmad et al., 2013).

The development of structures *in vitro* was found to support root induction in the shoots when NAA was employed (Manokari et al., 2023a). Healthy shoots originating from the SiNP medium demonstrated superior performance in inducing the major root and promoting the formation of side roots. This, in turn, improved the assimilation of water and food from the nutrient solution and subsequently from the potting combination in the greenhouse conditions. The result was an enhanced survival rate of plants *in vivo*. Notably, the silicon-incorporated *in vitro*–raised plants exhibited comparatively higher or nearly equivalent survival in the field to that of many plant species under both greenhouse and natural environments (Samantaray et al., 2013; Ahmed and Anis, 2014).

As per Avila et al. (2020), the addition of Si in the solution has been shown to mitigate the adverse effects of a water deficit on the synthesis of food, alleviate the impact of drought on foliage potential, and foster the formation of roots. Furthermore, the addition of silicon enhances the presence of mitochondria in cells, thereby promoting oxidative phosphorylation. This elevation in respiration rate leads to an increased production of adenosine triphosphate (ATP) in roots, enhancing the roots' capacity to take food and water, as noted by Ke et al. (1997).

Manokari et al. (2023d) conducted a study on *Thunbergia erecta* mediated by SiNPs of 2.0 mg L^{-1} and demonstrated effective acclimatization to the greenhouse system and exhibited a notably higher survival rate *in vivo* too. SiNP-exposed plantlets of *H. biflora* exhibited superior acclimatization efficiency and survival rates compared to control-grown plants. Such findings underscore the function of SiNP-mediated anatomical traits in facilitating the good acclimatization of tissue-cultured *H. biflora* plants to greenhouse and field environments, ensuring their robust survival post-transplantation (Manokari et al., 2023c).

In the case of *V. trifolia* shoots treated with SiNPs, effective rooting was observed with 1/2 MS nutrient medium containing NAA, with shoots from the SiNPs displaying the maximum rooting competence. These findings emphasize that the application of SiNPs considerably enhanced the tissue formation and morpho-physiology of *V. trifolia*, thereby promoting the endurance of plants *in vivo* (Cokul Raj et al., 2023b).

A review of the literature indicates that the use of SiNPs contributes to the structural development of roots by influencing mineral nutrition (Tripathi et al., 2012). It was noted that the optimal concentration of SiNPs resulted in more pronounced improvements in both qualitative and quantitative rooting parameters compared to the control group. This examination is corroborated by the findings of Soares et al. (2011), who demonstrated that the use of Si in the medium as K_2SiO_3 and Na_2SiO_3 improved rooting effectiveness, increased the root numbers, and promoted the development of stem and leaves in the orchid *Cattleya loddigesii*.

Nano silicon has been reported to enhance morphogenesis and rhizogenesis in *Thymus serpyllum* (Mustafa and Toaiema, 2022). Additionally, in *Musa acuminata* 'Grand Nain', nano silicon has demonstrated the ability to alleviate oxidative stress, induce the formation of photopigments, enhance photosynthesis, and influence the potassium-sodium percentage (Mahmoud et al., 2020). The use of silicon dioxide NPs externally has been shown to improve morphological features of both shoot and root systems in the seedlings of *Larix olgensis*, as indicated by studies conducted by Baoshan et al. (2004) and Soares et al. (2011).

Microscopic examination of *in vitro* roots of *H. biflora* revealed that control roots exhibited poorly developed tissue systems, whereas optimal concentrations (3.0 mg L^{-1}) of SiNPs significantly supported considerable tissue development (Manokari et al., 2023c). The underdeveloped tissue systems and microstructures of *in vitro*-induced roots must strengthen during greenhouse adaptation, allowing them to withstand *in vivo* situations (Shekhawat et al., 2021). The positive contribution of SiNPs in promoting the formation of dermal, cortical, and other secondary tissues is evident in this study.

The inclusion of silicon in the rooting medium was found to enhance foliar defense traits, like the leaf's internal structures, thickness of tissues, and deposition of wax on the cuticle, as indicated by studies conducted by Braga et al. (2009), Luz et al. (2012), and Asmar et al. (2013b). Silicon plays a pivotal role in bolstering the structural integrity of cell walls during the processes of elongation and division. This function is essential for preserving cell shape, a critical factor in ensuring the survival and optimal function of cells, particularly in roots (He et al., 2013). The presence of silicon influences the formation of cell wall components, contributing to alterations in their composition. This effect can lead to a heightened thickness of the root epidermis (Hossain et al., 2002).

Within the secondary cortex of SiNP-derived roots, the presence of air spaces serves a crucial purpose: Retaining oxygen even in hypoxic or anoxic conditions. This retention significantly contributes to bolstering survival rates by ensuring an adequate oxygen supply (Seago et al., 2005). In SiNP-derived roots, the Casparian bands exhibited a notably heightened prominence in comparison to the control roots. Specifically, the Casparian bands within the exodermis function as a formidable protective shield, impeding the entry of soil solution into the middle cortex through the apoplast. Additionally, the suberin lamellae play a critical role by inhibiting absorption from the apoplast to the symplast, further fortifying the root's defenses (Namyslov et al., 2020). The development of the exodermis serves to shield tissues from unnecessary water loss, with its differentiation being strongly influenced by deficiencies in nitrogen, phosphorus, and potassium (Namyslov et al., 2020).

In SiNP-derived roots, distinct reductions were observed in the sizes of both endodermis and secondary phloem regions compared to the control. This shrinkage primarily stemmed from nutrient diversion mechanisms. The endodermis, acting as a crucial apoplastic barrier, plays a pivotal role in regulating plant water retention and the transportation of solutes within the system (Xiang et al., 2019). SiNP uptake in roots happens through two pathways: Intracellularly or extracellularly via xylem and plasmodesmata. Silicon deposition specifically in the endodermis of root cell walls serves a crucial role. It aids in the selective absorption of nutrients while acting as a barrier that prevents harmful ions from reaching and affecting aerial organs (Yeo et al., 1999).

Silicon facilitates root elongation and mitigates NaCl stress by diminishing salt gathering and promoting lignification in tissues, as demonstrated by Siddiqui et al. (2020). Earlier investigations have indicated that root lignifications may stop food translocation from lower parts to areal parts, including foliage and reproductive parts (Chen et al., 2019; Xie et al., 2023).

Roots cultivated from the proper SiNPs concentrations in *H. biflora* showcased heightened tissue density along with an observable increase in secondary vascular tissues. Silicon plays a role in water retention in plants, potentially linked to the thick xylem elements found at the middle for the roots derived from SiNPs (Luyckx et al., 2017). Silicon/SiNPs have been shown to increase lignification of xylem walls and thickening of cell walls (Liu et al., 2015). The formation and maturation of periderm alongside the development of secondary vascular tissues within the roots of *H. biflora* represent pivotal structural attributes that facilitate adaptation to various abiotic stress conditions (Manokari et al., 2023c). Therefore, improving rhizogenesis during *in vitro* propagation becomes significant, as it may contribute to successful adaptation in greenhouse and natural conditions, ultimately enhancing the post-transplantation response of *in vitro* plants.

Transitioning and relocating plantlets from laboratory to field conditions marks a notably challenging phase within *in vitro* propagation technology (Lebedev et al., 2019). A successfully micropropagation protocol necessitates the provision of optimal conditions for *in vitro* structural development, thereby enhancing root induction and acclimatization and maximizing the survival of regenerated plantlets. The superiority of the roots of lab-grown plants significantly affects the success of greenhouse adaptation (Mežaka et al., 2023).

The impact of SiNPs on *in vitro* formation of plants ensures the production of good-quality plants to increased survival rates and a decrease in the occurrence of morphological, physiochemical, and microscopic disorders (Ziv, 2010; Krupa-Małkiewicz and Calomme, 2021). The study affirms that SiNPs effectively enhance the root systems of lab-grown *H. biflora* plants, resulting in successful acclimatization and achieving 100% survival rate.

Diverse concentrations of SiNPs used during root induction experiments distinctly influenced the acclimatization process and subsequent survival capacity of *H. biflora* (Manokari et al., 2023c). Plantlets that were rooted in SiNPs exhibited notably enhanced responsiveness during greenhouse hardening and acclimatization experiments, surpassing the control counterparts. The collected data strongly indicated that *H. biflora* plantlets originating from the optimized SiNP medium displayed remarkable survival rates (100%) along with enhanced morphological growth and development, surpassing the performance of the control and other SiNP concentrations (Manokari et al., 2023c).

As per assertions, the *in vitro* use of silicon has been found to increase cell wall stiffness by escalating the deposition of hemicelluloses and lignin. The reinforcement of structural stability within cell walls during processes like cell elongation and divisions is known to significantly contribute to the successful acclimatization and overall survival of plants (Camargo et al., 2007; He et al., 2013; Głazowska et al., 2018).

12.5 SiNPs INCREASE THE BIOMASS AND RATE OF PHOTOSYNTHESIS

The use of Si has demonstrated a positive influence on the absorption and distribution of essential minerals. Silicon supplementation has been shown to promote the efficient uptake of N and P, resulting in enhanced overall biomass (Greger et al., 2018; Frazão et al., 2020). Numerous experiments have demonstrated that Si supplementation enhances the formation of chl *a* and chl *b*, improves food production, and considerably enhances foliar and shoot thickness in many important plants (Pei et al., 2015; Rastogi et al., 2021; Xu et al., 2023). The increased number of shoots and foliage facilitates light interception and boosts CO_2 absorption, leading to elevated photosynthesis rates and ultimately enhancing crop yields, as observed in *Oryza sativa* (Detmann et al., 2012). The active transportation pathway for silicon is facilitated by Nod26-like intrinsic proteins (NIPs) and Lsi2 homologs (Ratcliffe et al., 2017). The proper amount of SiNPs (1.0 mg L^{-1}) acts as a mild stressor, regulating morphometric growth and leaf formation *in vitro* by promoting effective mineral absorption from the nutrient medium. Silicon's role extends to fostering the production of crucial photosynthetic pigments. Its contribution lies in bolstering plant tolerance to UV-B stress, thereby aiding in the enhancement of these pigments (Braga et al., 2009; Chen et al., 2018).

Cokul Raj et al. (2023b) examined the effect of SiNPs on shoot multiplication and biomass production in *V. trifolia* during laboratory cultivation. Among the various concentrations of Aerosil^{300} tested, the use of 6.0 mg L^{-1} SiNPs with optimized PGRs in regeneration solution positively influenced both shoot multiplication and the formation of fresh biomass. The microscopic examination revealed that foliage formed in the absence of SiNPs exhibited less developed chlorophylls and conducting tissues. Conversely, the use of SiNPs led to the formation of well-differentiated tissue systems, resulting in an increased lamina width, development of mesophyll tissues, cell wall thickening, higher trichome numbers, well-ordered gaud cells of stomata, and improved conduction elements (Figure 12.1f). These morphological changes contributed to enhanced photosynthetic efficiency in the plants (Cokul Raj et al., 2023b).

Exposure to SiO_2 nanoparticles to the plants demonstrated a favorable impact on the development of adventitious roots, leading to improved biomass and enhanced morphological traits in banana plants (El-Kady et al., 2017). The application of silicon was found to enhance the plant's water status, induce alterations in the microstructure of foliage, activate plant defense mechanisms, and reduce free radicals, thereby promoting overall growth and biomass (Parveen and Ashraf, 2010). SiNPs were observed to enhance the biomass of roots by facilitating the absorbance of silicon by cells. In contrast to the present findings, Hossain et al. (2002) reported that the biomass of roots increased proportionally with SiO_2 concentration. This contrasts with the other studies, where reduction in dry weight was reported with the optimized SiNP concentration. The underdeveloped root system *in vitro*, attributed to readily available carbon and other nutrients and PGR disturbances, poses challenges to the success of plant production via micropropagation (Bairu and Kane, 2011).

12.6 FORMATION OF CRYSTALS, PHYTOLITHS, RAPHIDES, AND IDIOBLASTS

Crystals and raphides demonstrated superior structural support capabilities compared to carbon-based compounds (Ma and Yamaji, 2008), contributing to the development of robust and sturdy shoots via intricate physiochemical pathways (Quigley and Anderson, 2014). Silicon provides mechanical protection, influences plant gene expression, and acts in defense pathways in plants (Debona et al., 2017; Souri et al., 2021). The presence of silicon leads to a notable rise in phytolith deposition on stem surfaces, resulting in a significant reinforcement of the sclerenchyma cell walls of stem tissues due to increased thickness (Bokor et al., 2019). Once absorbed through the roots, silicon can manifest in diverse forms within plant tissues. Silica cells or bodies are typically widespread in the leaf epidermis or often create a concentrated stratum beneath the cuticle (Coskun et al., 2019).

Stegmata, specialized cells where silicon deposition commonly occurs, encase sclerenchyma fibers linked to vascular bundles, notably bolstering the plant's resistance capabilities (Reynolds et al., 2009; Bokor et al., 2019). The optimal SiNPs had a substantial impact on structural maturation, leading to the formation of healthy plants with improved tolerance attributes under a controlled lab environment. These advancements have been thoroughly documented through microscopic observations (Manokari et al., 2023b). It has been shown that the polymerization of Si happens just beneath the cuticle and in the cell walls. Plants fortified with silicon display heightened resilience against pest assaults and greater adaptability to diverse environmental stresses like drought, frost, and metal toxicity when grown in open fields (Debona et al., 2017; Goswami et al., 2022).

Within the parenchymatous tissues originating from SiNP-derived roots, the detection of idioblasts housing needle-shaped raphides was noted. The formation of these raphide crystals serves as an additional defense mechanism, imparting increased mechanical protection to *H. biflora*, potentially aiding in overcoming abiotic stress conditions under *ex vitro* circumstances (Manokari et al., 2023c). Additionally, the type of crystal is a vital diagnostic for the Rubiaceae family. The accumulation of raphides serves multiple purposes, including controlling sunlight at the time of the food making process, storing calcium, and eliminating oxalate from the biochemical processes (Franceschi and Nakata, 2005).

Silicon enhances a plant's tolerance to stresses by influencing the antioxidant enzymatic process, the cation-binding ability of cell walls, levels of internal plant hormones, and the formation of crystals (Soares et al., 2012; Asmar et al., 2013; Adrees et al., 2015). Moreover, Si decreases the absorption of heavy metals by enhancing the formation of cell wall products (Adrees et al., 2015; Krupa-Małkiewicz and Calomme, 2021). The increased presence of crystals in the foliage of *Dioscorea pentaphylla* of SiNP-treated stems aims to considerably increase the constancy of shoots formed in the laboratory (Manokari et al., 2023f).

12.7 INDUCTION OF CALLUS AND SOMATIC EMBRYOGENESIS

Introducing SiNPs into the growth medium provides a practical remedy to address disruptions in morphological, structural, and physiochemical traits induced during *in vitro* cultivation (Mahmoud et al., 2020; Manokari et al., 2023e). Previous studies have highlighted the constructive effect of Si on the initiation of callus and somatic embryogenesis in monocots (Máthé et al., 2012). The observed enhancement in callus induction attributed to silicon supplementation may be linked to heightened nutrient absorption and increased photosynthetic activity (Cokul Raj et al., 2023a). Notably, the addition of SiNPs to the nutrient medium has been found to promote embryogenesis in root-induced calli of *Phragmites australis* (Máthé et al., 2012).

12.8 *IN VITRO* FLOWERING AND FRUIT SETTING INDUCED BY SiNPs

Remarkably, the integration of SiNPs into the medium has been reported to increase both qualitative and quantitative characteristics, mitigating morphological, physiochemical, and microstructural anomalies and consequently bolstering the adaptation of plants (Asmar et al., 2013b; Sivanesan and Jeong, 2014; Adrees et al., 2015; Krupa-Małkiewicz and Calomme, 2021). SiNPs are considered to activate key pathway genes involved in the synthesis and mobilization of soluble sugars, as well as regulate hormone metabolism (Asmar et al., 2013; Adress et al., 2015; Krupa-Małkiewicz and Calomme, 2021). This activation, in turn, improves photosynthetic efficiency, stress resistance, and overall formation of plantlets (Li et al., 2023). The heightened levels of proline, total phenol, monoaldehyde, and catalase activities resulting from SiNP supplementation contribute to enhanced responses and flower formation in the greenhouse in *Petunia* × *atkinsiana* (Krupa-Małkiewicz and Calomme, 2021). Similarly, the introduction of Si into the medium has been shown to promote *in vitro* flowering and fruit setting in *Vitex negundo* (Cokul Raj et al., 2023a).

12.9 REDUCTION IN HYPERHYDRICITY

The exogenous addition of SiNPs to nutrient media finds diverse uses in plant cell, tissue, and organ cultures, particularly in the formation of structurally resilient plant materials. This supplementation serves to counteract laboratory-driven anomalies such as vitrification and hyperhydricity, ensuring the natural adaptation of plants (Havas and Ghaderi, 2018; Rahmawati et al., 2022; Tymoszuk and Kulus, 2022; Kulus et al., 2022). The incorporation of Si into the medium emerges as a crucial strategy for enhancing morphology and anatomy (Luz et al., 2012; Lux et al., 2020) and alleviating morphological and physiochemical anomalies in plants, like vitrification, and shoot tip necrosis (Krupa-Małkiewicz and Calomme, 2021). The noted rise in root weight implies that SiNPs aid in enhancing water absorption and movement within the plant. These beneficial impacts of silicon oxide might be credited to its hydrophobic characteristics (Mathur and Roy, 2020).

12.10 CHANGES IN FOLIAR PARAMETERS: STOMATA AND TRICHOMES

The anatomical characteristics of leaves play a very important role in deciding a plant's adaptability to diverse conditions (Padilha et al., 2018; Shekhawat et al., 2021). The leaf microstructural developments in foliar tissues are openly impacted by sunlight and water content. In the specific conditions of *in vitro* environments characterized by highly specialized light and humidity aspects, poor tissue systems are induced in *Stachys natalensis*, *V. negundo*, and *Mertensia maritima* (Kalicharan et al., 2018; Manokari et al., 2020; Copetta et al., 2021). Unfortunately, these plants exhibit difficulties during acclimatization and subsequent survival when transitioning to field conditions marked by enhanced light and decreased humidity (Shekhawat et al., 2021). Examinations using microscopy in certain studies have shown that applying nano silica externally results in heightened tissue formation, growth, and deposits in structural tissues like the cuticle, epidermis, cortical tissues, and xylem vessels (Manokari et al., 2023b). In contrast, control leaves showed a less formed cuticle and underdeveloped tissues (Manokari et al., 2023f).

The use of SiNPs in the nutrient solution proved effective in alleviating abnormal stomatal development induced by *in vitro* culture, promoting increased epidermal, cortical, and conductive tissue thickening in both stem and foliage. Abnormal and nonfunctional stomata developed *in vitro* can impede the net food formation process and gas conductance, leading to the formation of heterotrophic plantlets. Manokari et al. (2023d, f) demonstrated that control leaves exhibited abnormal and underdeveloped stomatal cells, hindering proper stomatal closure. On the contrary, leaves treated with SiNPs exhibited structured guard cells and reinforcement in the guard cell walls, hinting at a possible link to the mechanism regulating stomatal movement.

The stomatal shape and aperture size are determined by the turgor pressure within stomatal guard cells, with the thickening of the cell wall playing a crucial role (Woolfenden et al., 2017). This thickening not only influences stomatal morphology but also contributes to regulating closure of the stomata and providing potency to abiotic stress (Zargar et al., 2019). Silicon has been shown to enhance resistance against diseases and pathogens by reinforcing plant structures, making them stronger and more resistant (Ye et al., 2013; Marxen et al., 2016). The study reported SiNP-mediated defense responses against herbivores and insects in *V. trifolia*, attributed to the increased trichome density, potentially aiding in the plant's survival in field conditions (Cokul Raj et al., 2023b). This study demonstrated that applying SiNPs externally, coupled with finely tuned PGRs, encouraged the development of numerous shoots exhibiting enhanced morphometric and microstructural characteristics in *V. trifolia.*

12.11 CONCLUSION AND FUTURE PROSPECTS

SiNPs present an opportunity to boost the overall health of plantlets cultivated through tissue culture. When paired with carefully adjusted growth regulators, SiNPs showcased both qualitative and quantitative enhancements in shoot structure, elevating node and leaf counts while increasing epicuticular wax content. Additionally, SiNPs played a role in improving tissue system differentiation and density within the stems of micropropagated medicinal plant shoots. These SiNP-treated shoots, grown in an SiNP-enriched medium, displayed strong performance during *ex vitro* rooting, hardening, and acclimatization trials. Incorporating optimal SiNPs into the nutrient medium emerged as a promising strategy to address morpho-anatomical and physiological issues encountered *in vitro*, thereby boosting the chances of survival for micropropagated plantlets. Short-term application of mild SiNPs demonstrated potential in enhancing the creation of desired root inoculums for adventitious root cultures aimed at augmenting secondary metabolites. Notably, leaves from SiNP-treated plants exhibited well-defined photosynthetic and vascular tissue systems, with fewer anatomical irregularities compared to control leaves. This positive impact on acclimatization and survival in *ex vitro* and *in vivo* conditions underscored the improved quality of shoots and rooting, ensuring successful acclimatization and plantlet survival. However, further ultra-histological research is required to document changes and improvements in various tissues affected by SiNP exposure during different stages of *in vitro* organogenesis. Nonetheless, additional research is crucial to uncover the specific mechanisms by which SiNPs exert their influence on plant cells and tissues.

ACKNOWLEDGMENTS

Author Mahipal S. Shekhawat gratefully acknowledges the National Medicinal Plants Board, Ministry of AYUSH, Government of India, for providing financial support for research and development work on *Hemidesmus indicus.*

AUTHOR CONTRIBUTIONS

Both authors equally contributed to the preparation of this manuscript and approved the final version of it.

CONFLICT OF INTEREST

The authors declare no conflicts of interest.

HUMAN AND ANIMAL RIGHTS

This research did not involve experiments with human or animal participants.

INFORMED CONSENT

Informed consent was obtained from all individual participants included in the study.

REFERENCES

Adrees, M., Ali, S., Rizwan, M., Zia-ur-Rehman, M., Ibrahim, M., Abbas, F., Farid, M., Qayyum, M. F. & Irshad, M. K. 2015. Mechanisms of silicon-mediated alleviation of heavy metal toxicity in plants: A review. *Ecotoxicology and Environmental Safety*. 119: 186–197. https://doi.org/10.1016/j.ecoenv.2015.05.011.

Ahmad, N., Javed, S. B., Khan, M. I. & Anis, M. 2013. Rapid plant regeneration and analysis of genetic fidelity in micropropagated plants of *Vitex trifolia*: An important medicinal plant. *Acta Physiologiae Plantarum*. 35: 2493–2500. https://doi.org/10.1007/s11738-013-1285-y.

Ahmed, M. R. & Anis, M. 2012. Role of TDZ in the quick regeneration of multiple shoots from nodal explant of *Vitex trifolia* L. – An important medicinal plant. *Applied Biochemistry and Biotechnology*. 168: 957–966. https://doi.org/10.1007/s12010-012-9799-0.

Ahmed, R. & Anis, M. 2014. Rapid *in vitro* propagation system through shoot tip cultures of *Vitex trifolia* L. – an important multipurpose plant of the pacific traditional medicine. *Physiology and Molecular Biology of Plants*. 20(3): 385–392. https://doi.org/10.1007/s12298-014-0233-7.

Ali, S., Rizwan, M., Hussain, A., Zia Ur Rehman, M., Ali, B., Yousaf, B., Wijaya, L., Alyemeni, M. N. & Ahmad, P. 2019. Silicon nanoparticles enhanced the growth and reduced the cadmium accumulation in grains of wheat (*Triticum aestivum* L.). *Plant Physiology and Biochemistry*. 140: 1–8. https://doi.org/10.1016/j.plaphy.2019.04.041.

Asgari, F., Majd, A., Jonoubi, P. & Najafi, F. 2018. Effects of silicon nanoparticles on molecular, chemical, structural and ultrastructural characteristics of oat (*Avena sativa* L.). *Plant Physiology and Biochemistry*. 127: 152–160. https://doi.org/10.1016/j.plaphy.2018.03.021

Asmar, S., Castro, E., Pasqual, M., Pereira, F. & Soares, J. 2013a. Changes in leaf anatomy and photosynthesis of micropropagated banana plantlets under different silicon sources. *Scientia Horticulturae*. 161: 328–332. https://doi.org/10.1016/j.scienta.2013.07.021.

Asmar, S. A., Pasqual, M., de Araujo, A. G., Silva, R. A. L., Rodrigues, F. A. & Pio, L. A. S. 2013b. Morphophysiological characteristics of acclimatized "Grande Naine" banana plants in response to *in vitro* use of silicon, *Semina-Ciencias Agrarias*. 34: 73–82.

Attia, E. A. & Elhawat, N. 2021. Combined foliar and soil application of silica nanoparticles enhances the growth, flowering period and flower characteristics of marigold (*Tagetes erecta* L.). *Scientia Horticulturae*. 282: 110015. https://doi.org/10.1016/j.scienta.2021.110015.

Avila, R. G., Magalhães, P. C., da Silva, E. M., Júnior, C. C. G., de Paula Lana, U. G., de Alvarenga, A. A. & de Souza, T. C. 2020. Silicon supplementation improves tolerance to water deficiency in sorghum plants by increasing root system growth and improving photosynthesis. *Silicon*. 12(11): 2545–2554. https://doi.org/10.1007/s12633-019-00349-5.

Bairu, M. W. & Kane, M. E. 2011. Physiological and developmental problems encountered by *in vitro* cultured plants. *Plant Growth Regulation*. 63: 101–103. https://doi.org/10.1007/s10725-011-9565-2.

Baoshan, L., Chunhui, L., Lijun, F., Shuchun, Q. & Min, Y. 2004. Effect of TMS (nanostructured silicon dioxide) on growth of Changbai Larch seedlings. *Journal of Forestry Research*. 15: 138–140. https://doi.org/10.1007/BF02856749.

Bapat, G., Zinjarde, S. & Tamhane, V. 2020. Evaluation of silica nanoparticle mediated delivery of protease inhibitor in tomato plants and its effect on insect pest *Helicoverpa armigera*. *Colloids and Surfaces B: Biointerfaces*. 193: 111079. https://doi.org/10.1016/j.colsurfb.2020.111079.

Bokor, B., Soukup, M., Vaculík, M., Weidinger, M., Lichtscheidl, I., Vávrová, S., Šoltys, K., Sonah, H., Deshmukh, R., Bélanger, R. R., White, P. J. A. H. & Lux, A. 2019. Silicon uptake and localisation in date palm (*Phoenix dactylifera*) – A unique association with sclerenchyma. *Frontiers in Plant Sciences*. 10. https://doi.org/10.3389/fpls.2019.00988.

Braga, F. T., Nunes, C. F., Favero, A. C., Pasqual, M., Carvalho, J. G. & Castro, E. M. 2009. Anatomical characteristics of the strawberry seedlings micropropagated using different sources of silicon. *Pesquisa Agropecuária Brasileira*. 44: 128–132. https://doi.org/10.1590/S0100-204X2009000200003.

Camargo, M. S., Korndorfer, G. H. & Pereira, H. S. 2007. Solubility of silicon in soils: Effect of lime and silicic acid applied. *Bragantia*. 66: 637–647. https://doi.org/10.1590/S0006-87052007000400014.

Chen, D., Wang, S., Yin, L. & Deng, X. 2018. How does silicon mediate plant water uptake and loss under water deficiency? *Frontiers in Plant Sciences*. 9: 281. https://doi.org/10.3389/fpls.2018.00281.

Chen, L. L., Chu, S. S., Zhang, L., Xie, J., Dai, M., Wu, X. & Peng, H. S. 2019. Tissue-specific metabolite profiling on the different parts of bolting and unbolting *Peucedanum praeruptorum* Dunn (Qianhu) by laser microdissection combined with UPLC-Q/TOF-MS and HPLC-DAD. *Molecules*. 24(7): 1439. https://doi.org/10.3390/molecules24071439.

Cokul Raj, M., Manokari, M., Arumugam, N., Dey, A., Faisal, M., Alatar, A. A., Alok, A. & Shekhawat, M. S. 2023a. Silicon nanoparticles mediated *in vitro* flowering and study of pollen viability in *Vitex negundo* L. *Silicon*. https://doi.org/10.1007/s12633-023-02397-4.

Cokul Raj, M., Manokari, M., Dey, A., Faisal, M., Alatar, A. A., Joshee, N. & Shekhawat, M. S. 2023b. Improvements in morphometric and structural traits of *Vitex trifolia* L. to exogenous application of nano-silicon *in vitro*. *South African Journal of Botany*. https://doi.org/10.1016/j.sajb.2023.07.049.

Copetta, A., Bazzicalupo, M., Cassetti, A., Marchioni, I., Mascarello, C., Cornara, L., Pistelli, L. & Ruffoni, B. 2021. Plant production and leaf anatomy of *Mertensia maritima* (L.) Gray: Comparison of *in vitro* culture methods to improve acclimatization. *Horticulturae*. 7(5): 111. https://doi.org/10.3390/horticulturae7050111.

Coskun, D., Deshmukh, R., Sonah, H., Menzies, J. G., Reynolds, O., Ma, J. F., Kronzucker, H. J. & Bélanger, R. R. 2019. In defence of the selective transport and role of silicon in plants. *New Phytologist*. 223(2): 514–516. https://doi.org/10.1111/nph.15764.

Debona, D., Rodrigues, F. A. & Datnoff, L. E. 2017. Silicon's role in abiotic and biotic plant stresses. *Annual Reviews*. https://doi.org/10.1146/annurev-phyto-080516-035312.

Detmann, K. C., Araújo, W. L., Martins, S. C. V., Sanglard, L. M. V. P., Reis, J. V., Detmann, E., Rodrigues, F., Nunes-Nesi, A., Fernie, A. R. & DaMatta, F. M. 2012. Silicon nutrition increases grain yield, which, in turn, exerts a feed-forward stimulation of photosynthetic rates via enhanced mesophyll conductance and alters primary metabolism in rice. *New Phytologist*. 196(3): 752–762. https://doi.org/10.1111/j.1469-8137.2012.04299.x.

El-Kady, M. E., El-Boray, M. S., Shalan, A. M. & Mohamed, L. M. 2017. Effect of silicon dioxide nanoparticles on growth improvement of banana shoots *in vitro* within rooting stage. *Journal of Plant Production*. 8: 913–916. https://doi.org/10.21608/JPP.2017.40900.

Franceschi, V. R. & Nakata, P. A. 2005. Calcium oxalate in plants: Formation and function. *Annual Reviews in Plant Biology*. 56: 41–75. https://doi.org/10.1146/annurev.arplant.56.032604.144106.

Frazão, J. J., Prado, R. D., Pereira, J. & Rossatto, D. R. 2020. Silicon changes c:N:P stoichiometry of sugarcane and its consequences for photosynthesis, biomass partitioning and plant growth. *Scientific Reports*. 10: 1–10. https://doi.org/10.1038/s41598-020-69310-6.

Frew, A., Weston, L. A., Reynolds, O. L. & Gurr, G. M. 2018. The role of silicon in plant biology: A paradigm shift in research approach. *Annals of Botany*. 121(7): 1265–1273, https://doi.org/10.1093/aob/mcy009.

Głazowska, S., Baldwin, L., Mravec, J., Bukh, C., Hansen, T. H., Jensen, M. M., Fangel, J. U., Willats, W. G. T., Glasius, M., Felby, C. & Schjoerring, J. K. 2018. The impact of silicon on cell wall composition and enzymatic saccharification of *Brachypodium distachyon*. Biotechnol. *Biofuels*. 11: 171. https://doi.org/10.1186/s13068-018-1166-0.

Gorsuch, D. M. & Oberbauer, S. F. 2002. Effects of mid-season frost and elevated growing season temperature on stomatal conductance and specific xylem conductivity of the arctic shrub. Salix pulchra, *Tree Physiol*. 22 (14): 1027–1034.

Goswami, P., Mathur, J. & Srivastava, N. 2022. Silica nanoparticles as novel sustainable approach for plant growth and crop protection. *Heliyon*. 8(7): e09908. https://doi.org/10.1016/j.heliyon.2022.e09908.

Grassi, G. & Magnani, F. 2005. Stomatal, mesophyll conductance and biochemical limitations to photosynthesis as affected by drought and leaf ontogeny in ash and oak trees. *Plant Cell and Environment*. 28(7): 834–849. https://doi.org/10.1111/j.1365-3040.2005.01333.x.

Greger, M., Landberg, T. & Vaculík, M. 2018. Silicon influences soil availability and accumulation of mineral nutrients in various plant species. *Plants*. 7(2). https://doi.org/10.3390/plants7020041.

Havas, F. & Ghaderi, N. 2018. Application of iron nanoparticles and salicylic acid in *in vitro* culture of strawberries (*Fragaria*× *ananassa* Duch.) To cope with drought stress. *Plant Cell Tissue and Organ Culture*. 132: 511–523. https://doi.org/10.1007/s11240-017-1347-8.

He, C., Wang, L., Liu, J., Liu, X., Li, X., Ma, J., Lin, Y. & Xu, F. 2013. Evidence for 'silicon' within the cell walls of suspension-cultured rice cells. *New Phytologist*. 200(3): 700–709. https://doi.org/10.1111/nph.12401.

Hiregoudar, L. V., Murthy, H. N., Bhat, J. G., Nayeem, A., Hema, B. P., Hahn, E. J. & Paek, K. Y. 2006. Rapid clonal propagation of *Vitex trifolia*. *Biologia Plantarum*. 50: 291–294. https://doi.org/10.1007/s10535-006-0023-3.

Hossain, M. T., Mori, R., Soga, K., Wakabayashi, K., Kamisaka, S., Fujii, S., Yamamoto, R. & Hoson, T. 2002. Growth promotion and an increase in cell wall extensibility by silicon in rice and some other Poaceae seedlings. *Journal of Plant Research*. 115: 23–27. https://doi.org/10.1007/s102650200004.

Kalicharan, B., Naidoo, Y., Nakhooda, M. & Dewir, Y. 2018. Micromorphological evaluation of the foliar trichomes of field grown and micropropagated *Stachys natalensis* Hochst. (Lamiaceae). *South African Journal of Botany*. 119: 369–376. https://doi.org/10.1016/j.sajb.2018.08.025.

Ke, Y. S., Huang, X. H., Zhang, Z. T., Xiao, C. M., Wu, L. F., Li, Y. Z., Jian, H. Q. & Wu, Y. Q. 1997. Effect of silicon fertilizer on nitrogen, phosphorus and potassium nutrition of rice and analysis of reasons for increasing production. Guangdong *Agriculture Science*. 5: 25–27. https://doi.org/10.16768/j.issn.1004-874x.1997.05.011.

Khan, M. A., Gemenet, D. C. & Villordon, A. 2016. Root system architecture and abiotic stress tolerance: Current knowledge in root and tuber crops. *Frontiers in Plant Science*. 7: 209303. https://doi.org/10.3389/fpls.2016.01584.

Kim, S. G., Kim, K. W., Park, E. W. & Choi, D. 2002. Silicon-induced cell wall fortification of rice leaves: A possible cellular mechanism of enhanced host resistance to blast. *Phytopathology*. 92: 1095–1103. https://doi.org/10.1094/PHYTO.2002.92.10.1095.

Kovács, S., Kutasy, E. & Csajbók, J. 2022. The multiple role of silicon nutrition in alleviating environmental stresses in sustainable crop production. *Plants*. 11(9): 1223. https://doi.org/10.3390/plants11091223.

Krupa-Małkiewicz, M. & Calomme, M. 2021. Actisil application affects growth, flowering, and biochemical parameters in petunia *in vitro* and greenhouse. *Plant Cell Tissue and Organ Culture*. 146: 449–459. https://doi.org/10.1007/s11240-021-02078-3.

Kulus, D., Tymoszuk, A., Jedrzejczyk, I. & Winiecki, J. 2022. Gold nanoparticles and electromagnetic irradiation in tissue culture systems of bleeding heart: Biochemical, physiological, and (cyto) genetic effects. *Plant Cell Tissue and Organ Culture*. 149: 715–734. https://doi.org/10.1007/s11240-022-02236-1.

Lebedev, V., Arkaev, M., Dremova, M., Pozdniakov, I. & Shestibratov, K. 2019. Effects of growth regulators and gelling agents on ex vitro rooting of raspberry. *Plants*. 8(1). https://doi.org/10.3390/plants8010003.

Li, Y., Xi, K., Liu, X., Han, S., Han, X., Li, G., Yang, L., Ma, D., Fang, Z., Gong, S., Yin, J. & Zhu, Y. 2023. Silica nanoparticles promote wheat growth by mediating hormones and sugar metabolism. *Journal of Nanobiotechnology*. 21: 2. https://doi.org/10.1186/s12951-022-01753-7.

Liu, P., Yin, L., Wang, S., Zhang, M., Deng, X., Zhang, S. & Tanaka, K. 2015. Enhanced root hydraulic conductance by aquaporin regulation accounts for silicon alleviated salt-induced osmotic stress in *Sorghum bicolor* L. *Environmental and Experimental Botany*. 111: 42–51. https://doi.org/10.1016/j.envexpbot.2014.10.006.

Lux, A., Lukačová, Z., Vaculík, M., Švubová, R., Kohanová, J., Soukup, M., Martinka, M. & Bokor, B. 2020. Silicification of root tissues. *Plants*. 9(1): 111. https://doi.org/10.3390/plants9010111.

Luyckx, M., Hausman, J. F., Lutts, S. & Guerriero, G. 2017. Silicon and plants: Current knowledge and technological perspectives. *Frontiers in Plant Sciences*. 8: 411. https://doi.org/10.3389/fpls.2017.00411.

Luz, J. M. Q., Asmar, S. A., Pasqual, M., de Araujo, A. G., Pio, L. A. S. & Resende, R. F. 2012. Modification in leaf anatomy of banana plants cultivar "Maca" subjected to different silicon sources *in vitro*. *Acta Horticulturae*. 961: 239–243. https://doi.org/10.17660/ActaHortic.2012.961.30.

Ma, J. F. & Yamaji, N. 2008. Functions and transport of silicon in plants. *Cell and Molecular Life Science*. 65(19): 3049–3057. https://doi.org/10.1007/s00018-008-7580-x.

Mahmoud, L. M., Dutt, M., Shalan, A. M., El-Kady, M. E., El-Boray, M. S., Shabana, Y. M. & Grosser, J. W. 2020. Silicon nanoparticles mitigate oxidative stress of *in vitro*-derived banana (*Musa acuminata* 'Grand Nain') under simulated water deficit or salinity stress. *South African Journal of Botany*. 132: 155–163. https://doi.org/10.1016/j.sajb.2020.04.027.

Manokari, M., Badhepuri, M. K., Sandhya, D., Dey, A., Kumar, V., Faisal, M., Alatar, A. A., Singh, R. K. & Shekhawat, M. S. 2021b. Validation of *meta*-topolin in organogenesis, improved morpho-physio-chemical responses, and clonal fidelity analysis in *Dioscorea pentaphylla* L. – An underutilized yam species. *South African Journal of Botany*. 145: 284–292. https://doi.org/10.1016/j.sajb.2021.12.039.

Manokari, M., Cokul Raj, M., Dey, A., Faisal, M., Alatar, A. A., Alok, A. & Shekhawat, M. S. 2022. Micro-morpho-anatomical transitions at various stages of *in vitro* development of *Crinum malabaricum* Lekhak and Yadav: A critically endangered medicinal plant. *Plant Biology*. https://doi.org/10.1111/plb.13464.

Manokari, M., Cokul Raj, M., Dey, A., Faisal, M., Alatar, A. A., Joshee, N. & Shekhawat, M. S. 2023c. Improvements in morpho-anatomical traits of adventitious roots of *Hedyotis biflora* (L.) Lam. Using silicon nanoparticles. *Silicon*. https://doi.org/10.1007/s12633-023-02484-6.

Manokari, M., Cokul Raj, M., Dey, A., Faisal, M., Alatar, A. A., Joshee, N. & Shekhawat, M. S. 2023e. Structural alterations of *Cymbopogon citratus* (DC.) Stapf leaves and roots caused by silicon nanoparticles during *in vitro* propagation. *Industrial Crops and Products.* https://doi.org/10.1016/j.indcrop.2023.116648.

Manokari, M., Cokul Raj, M., Dey, A., Faisal, M., Alatar, A. A., Singh, R. K. & Shekhawat, M. S. 2023d. Silicon nanoparticles moderated morphometric deficiencies by improving micro-morpho-structural traits in *Thunbergia erecta* (Benth.) T. Anderson. *Silicon.* https://doi.org/10.1007/s12633-023-02451-1.

Manokari, M., Cokul Raj, M., Priyadharshini, S., Phulwaria, M. & Shekhawat, M. S. 2020. Foliar micromorphology: A promising tool to improve survival percentage of tissue culture raised plantlets with special reference to *in vitro* propagation of *Vitex negundo* L. *Vegetos* 33: 504–515. https://doi.org/10.1007/s42535-020-00134-5.

Manokari, M., Dey, A., Faisal, M., Alatar, A. A., Singh, R. K. & Shekhawat, M. S. 2023b. Silicon nanoparticles (SiNPs) positively affect morpho-structural differentiation in micropropagated plantlets of *Santalum album* L. *Silicon.* https://doi.org/10.1007/s12633-023-02558-5.

Manokari, M., Dey, A., Faisal, M., Alatar, A. A., Singh, R. K. & Shekhawat, M. S. 2023f. *In vitro* tuberization using silicon nanoparticles and short-term cold storage of mini-tubers of *Dioscorea pentaphylla* L. *BioNanoScience.* https://doi.org/10.1007/s12668-023-01185-z.

Manokari, M., Faisal, M., Alatar, A. A., Barboni, D. & Shekhawat, M. S. 2023a. Silicon nanopraticles on the enhanced micropropagation of *Hemidesmus indicus* (L.) R. Br. Ex Schult. *Silicon.* https://doi.org/10.1007/s12633-023-02714-x.

Manokari, M., Mehta, S. R., Priyadharshini, S., Badhepuri, M. K., Sandhya, D., Jayaprakash, K., Cokul Raj, M., Dey, A., Rajput, B. S. & Shekhawat, M. S. 2021a. Meta-topolin mediated improved micropropagation, foliar micro-morphological traits, biochemical profiling, and assessment of genetic fidelity in *Santalum album* L. *Industrial Crops and Products.* 171: 113931. https://doi.org/10.1016/j.indcrop.2021.113931.

Manokari, M., Priyadharshini, S. & Shekhawat, M. S. 2021c. Micro-structural stability of micropropagated plants of *Vitex negundo* L. *Microscopy and Microanalysis.* 27: 626–634. https://doi.org/10.1017/S1431927621000283.

Mantovani, C., Pivetta, K. F. L., de Mello Prado, R., de Souza, J. P., Nascimento, C. S., Nascimento, C. S. & Gratão, P. L. 2020. Silicon toxicity induced by different concentrations and sources added to *in vitro* culture of epiphytic orchids. *Scientia Horticulturae*, 265: 109272. https://doi.org/10.1016/j.scienta.2020.109272.

Marxen, A., Klotzbucher, T., Jahn, R., Kaiser, K., Nguyen, V. S., Schmidt, A., Schadler, M. & Vetterlein, D. 2016. Interaction between silicon cycling and straw decomposition in a silicon deficient rice production system. *Plant Soil.* 398: 153–163. https://doi.org/10.1007/s11104-015-2645-8.

Máthé, C., Mosolygó, Á, Surányi, G., Beke, A., Demeter, Z., Tóth, V. R., Beyer, D., Mészáros, I. & M-Hamvas, M. 2012. Genotype and explant-type dependent morphogenesis and silicon response of common reed (*Phragmites australis*) tissue cultures. *Aquatic Botany.* 97(1):57–63. https://doi.org/10.1016/j.aquabot.2011.11.005.

Mathur, P. & Roy, S. 2020. Nanosilica facilitates silica uptake, growth and stress tolerance in plants. *Plant Physiology and Biochemistry.* 157: 114–127. https://doi.org/10.1016/j.plaphy.2020.10.011.

Mežaka, I., Kļaviņa, D., Kaļāne, L. & Kronberga, A. 2023. Large-scale *in vitro* propagation and ex vitro adaptation of the endangered medicinal plant *Eryngium maritimum* L. *Horticulturae*, 9(2): 271. https://doi.org/10.3390/horticulturae9020271.

Mollick, A. S., Sultana, R. & Azad, M. S. et al. 2021. Leaf morphological plasticity in three dominant tree species in the Sundarbans mangrove forest of Bangladesh in different salinity zones. *Wetlands Ecology and Management.* 29(2): 265–279.

Mukarram, M., Khan, M. M. A. & Corpas, F. J. 2021. Silicon nanoparticles elicit an increase in lemongrass (*Cymbopogon flexuosus* (Steud.) Wats) agronomic parameters with a higher essential oil yield. *Journal of Hazardous Materials.* 412: 125254. https://doi.org/10.1016/j.jhazmat.2021.125254.

Mukarram, M., Petrik, P., Mushtaq, Z., Khan, M. M. A., Gulfishan, M. & Lux, A. 2022. Silicon nanoparticles in higher plants: Uptake, action, stress tolerance, and crosstalk with phytohormones, antioxidants, and other signalling molecules. *Environmental Pollution.* 310: 119855. https://doi.org/10.1016/j.envpol.2022.119855.

Mustafa, S. S. S. & Toaiema, W. E. 2022. *In vitro* propagation and influence of silicon nanoparticles on growth of *Thymus serpyllum. Eastern Journal of Agriculture and Biological Sciences.* 2: 17–24. https://www.qabasjournals.com/index.php/ejabs/article/view/117.

Nagaveni, C. & Rajanna, L. 2013. *In-vitro* flowering in *Vitex trifolia* L. *Botanical Research International.* 6: 13–16.

Namyslov, J., Bauriedlová, Z., Janoušková, J., Soukup, A. & Tylová, E. 2020. Exodermis and endodermis respond to nutrient deficiency in nutrient-specific and localized manner. *Plants*. 9(2): 201. https://doi.org/10.3390/plants9020201.

Padilha, J. H. D., Ribeiro, A. Z., Steinmacher, D., Amano, É & Quoirin, M. 2018. Leaf histology of greenhouse plants, *in vitro* cultured and somatic embryo-derived plants of *Bactris gasipaes* Kunth. *Acta Scientiarum Biological Sciences*. 40: 1–7. https://doi.org/10.4025/actascibiolsci.v40i1.37340.

Parveen, N. & Ashraf, M. 2010. Role of silicon in mitigating the adverse effects of salt stress on growth and photosynthetic attributes of two maize (*Zea mays* L.) Cultivars grown hydroponically, *Pakistan Journal of Botany*. 42 (3): 1675–1684.

Pei, F. Y., Dong, C. W., Chen, W. Z., Yang, Y., Fang, Q. F., Duan, J. X., Huang, P. Z. & Wang, H. D. 2015. Preparation and the effect of nano-silicon fertilizer on the growth of amaranth. *Horticulture and Seeds*. 6: 12–17. https://doi.org/10.16530/j.cnki.cn21-1574/s.2015.06.005.

Quigley, K. M. & Anderson, T. M. 2014. Leaf silica concentration in Serengeti grasses increases with watering but not clipping: Insights from a common garden study and literature review. *Frontiers in Plant Sciences*. 5: 1–10. https://doi.org/10.3389/fpls.2014.00568.

Rahmawati, M., Mahfud, C., Risuleo, G. & Jadid, N. 2022. Nanotechnology in plant metabolite improvement and in animal welfare. *Applied Science*. 12(2): 838. https://doi.org/10.3390/app12020838.

Rajput, V. D., Minkina, T., Feizi, M., Kumari, A., Khan, M., Mandzhieva, S., Sushkova, S., El-Ramady, H., Verma, K. K., Singh, A., Hullebusch, E., Dv., Singh, R. K., Jatav, H. S. & Choudhary, R. 2021. Effects of silicon and silicon-based nanoparticles on rhizosphere microbiome. *Plant Stress Growth Biology*. 10(8): 791. https://doi.org/10.3390/biology10080791.

Rastogi, A., Tripathi, D. K., Yadav, S., Chauhan, D. K., Živčák, M., Ghorbanpour, M., El-Sheery, N. I. & Brestic, M. 2020. Application of silicon nanoparticles in agriculture. *3 Biotech*. 9: 90. https://doi.org/10.1007/s13205-019-1626-7.

Rastogi, A., Yadav, S., Hussain, S., Kataria, S., Hajihashemi, S., Kumari, P., Yang, X. & Brestic, M. 2021. Does silicon really matter for the photosynthetic machinery in plants…? *Plant Physiology and Biochemistry*. 169: 40–48. https://doi.org/10.1016/j.plaphy.2021.11.004.

Ratcliffe, S., Jugdaohsingh, R., Vivancos, J., Marron, A., Deshmukh, R., Ma, J. F., Mitani-Ueno, N., Robertson, J., Wills, J., Boekschoten, M. V., Müller, M., Mawhinney, R. C., Kinrade, S. D., Isenring, P., Bélanger, R. R. & Powell, J. J. 2017. Identification of a mammalian silicon transporter. *American Journal of Physiology: Cell Physiology*. 312(5): C550–C561. https://doi.org/10.1152/ajpcell.00219.2015.

Reed, B. M., Wada, S., DeNoma, J. & Niedz, R. P. 2013. Mineral nutrition influences physiological responses of pear *in vitro*. *In Vitro Cell and Developmental Biology –Plant*. 49: 699–709. https://doi.org/10.1007/s11627-013-9556-2.

Reynolds, O. L., Keeping, M. G. & Meyer, J. H. 2009. Silicon-augmented resistance of plants to herbivorous insects: A review. *Annals of Applied Biology*. 155(2): 171–186. https://doi.org/10.1111/j.1744-7348.2009.00348.x.

Samantaray, S., Bishoyi, A. K. & Maiti, S. 2013. Plant regeneration from callus cultures of *Vitex trifolia* (Lamiales: Lamiaceae): A potential medicinal plant, *Revista De Biología* Tropical. 61 (3): 1083–94.

Sánchez-Navarro, J. F., González-García, Y., Benavides-Mendoza, A., Morales-Díaz, A. B., González-Morales, S., Cadenas-Pliego, G., García-Guillermo, M. D. S. & Juárez-Maldonado, A. 2021. Silicon nanoparticles improve the shelf life and antioxidant status of *Lilium*. *Plants*. (Basel). 29: 2338. https://doi.org/10.3390/plants10112338.

Seago, J. L. Jr, Marsh, L. C., Stevens, K. J., Soukup, A., Votrubová, O. & Enstone, D. E. 2005. A re-examination of the root cortex in wetland flowering plants with respect to aerenchyma. *Annals of Botany*. 96: 565–579. https://doi.org/10.1093/aob/mci211.

Shekhawat, M. S., Mehta, S. R., Manokari, M., Priyadharshini, S., Badhepuri, M. K., Jogam, P., Dey, A. & Rajput, B. S. 2021. Morpho-anatomical and physiological changes of Indian sandalwood (*Santalum album* L.) Plantlets in *ex vitro* conditions to support successful acclimatization for plant mass production. *Plant Cell Tissue and Organ Culture*. 147: 423–435 https://doi.org/10.1007/s11240-021-02136-w.

Shen, Z., Cheng, X., Li, X., Deng, X., Dong, X., Wang, S. & Pu, X. 2022. Effects of silicon application on leaf structure and physiological characteristics of *Glycyrrhiza uralensis* Fisch. and *Glycyrrhiza inflata* Bat. Under salt treatment. *BMC Plant Biology*. 22: 390. https://doi.org/10.1186/s12870-022-03783-7.

Siddiqui, H., Ahmed, K. B. M., Sami, F. & Hayat, S. 2020. Silicon nanoparticles and plants: Current knowledge and future perspectives. In: Hayat, S., Pichtel, J., Faizan, M., Fariduddin, Q. (eds.). *Sustainable Agriculture Reviews 41. Nanotechnology for Plant Growth and Development*. Springer International Publishing, Cham, Switzerland. pp. 129–142.

Siddiqui, M. H. & Al-Whaibi, M. H. 2014. Role of nano-SiO_2 in germination of tomato (*Lycopersicum esculentum* seeds Mill.). *Saudi Journal of Biological Science*. 21: 13–17. https://doi.org/10.1016/j.sjbs.2013.04.005.

Sivanesan, I. & Jeong, B. R. 2014. Silicon promotes adventitious shoot regeneration and enhances salinity tolerance of *Ajuga multiflora* Bunge by altering activity of antioxidant enzyme. *Scientific World Journal*. https://doi.org/10.1155/2014/521703.

Sivanesan, I. & Park, S. W. 2014. The role of silicon in plant tissue culture. *Frontiers Plant Sciences*. 21(5): 571. https://doi.org/10.3389/fpls.2014.00571.

Soares, J. D. R., Pasqual, M., de Araujo, A. G., De Castro, E. M., Pereira, F. J. & Braga, F. T. 2012. Leaf anatomy of orchids micropropagated with different silicon concentrations. *Acta Scientiarum Agron*. 34: 413–421. https://doi.org/10.4025/actasciagron.v34i4.15062.

Soares, J. D. R., Pasqual, M., Rodrigues, F. A., Villa, F. & de Araujo, A. G. 2011. Silicon sources in the micropropagation of the *Cattleya* group orchid. *Acta Scientiarum Agron* 33: 503–507. https://doi.org/10.4025/actasciagron.v33i3.6281.

Soundararajan, P., Manivannan, A., Cho, Y. S. & Jeong, B. R. 2017. Exogenous supplementation of silicon improved the recovery of hyperhydric shoots in *Dianthus caryophyllus* L. by stabilizing the physiology and protein expression. *Frontiers Plant Sciences*. 8. https://doi.org/10.3389/fpls.2017.00738.

Souri, Z., Khanna, K., Karimi, N. & Ahmad, P. 2021. Silicon and plants: Current knowledge and future prospects. *Journal of Plant Growth Regulation*. 40: 906–925. https://doi.org/10.1007/s00344-020-10172-7.

Tina, P. V., Chauhan, K., Pant, K., Pant, G. & Pant, M. 2023. Plant response to silicon nanoparticles: Growth performance and defense mechanisms. In: Husen, A. (ed.). *Nanomaterials and Nanocomposites Exposures to Plants. Smart Nanomaterials Technology*. Springer, Singapore. https://doi.org/10.1007/978-981-99-2419-6_9.

Tomar, P. C., Kalrs, T., Ahuja, G. & Taneja, S. 2018. Plants emerging with promising nanoworld. *BioTechnologia*. 99 (2): 165–179. https://doi.org/10.5114/bta.2018.75660.

Trejo-Téllez, L. I., García-Jiménez, A., Escobar-Sepúlveda, H. F., Ramírez-Olvera, S. M., Bello-Bello, J. J. & Gómez-Merino, F. C. 2020. Silicon induces hormetic dose-response effects on growth and concentrations of chlorophylls, amino acids and sugars in pepper plants during the early developmental stage. *Peer J*. 8: e9224. https://doi.org/10.7717/peerj.9224.

Tripathi, D. K., Singh, V. P., Kumar, D. & Chauhan, D. K. 2012. Impact of exogenous silicon addition on chromium uptake, growth, mineral elements, oxidative stress, antioxidant capacity, and leaf and root structures in rice seedlings exposed to hexavalent chromium, Acta *Physiologiae Plantarum*. 34 (1): 279–289.

Tymoszuk, A. & Kulus, D. 2022. Effect of silver nanoparticles on the *in vitro* regeneration, biochemical, genetic, and phenotype variation in adventitious shoots produced from leaf explants in *Chrysanthemum*. *International Journal of Molecular Sciences*. 23(13): 7406. https://doi.org/10.3390/ijms23137406.

Tymoszuk, A. & Miler, N. 2019. Silver and gold nanoparticles impact on *in vitro* adventitious organogenesis in chrysanthemum, gerbera and cape primrose. *Scientia Horticulturae*. 257: 108766. https://doi.org/10.1016/j.scienta.2019.108766.

Wang, L., Ning, C., Pan, T. & Cai, K. 2022. Role of silica nanoparticles in abiotic and biotic stress tolerance in plants: A review. *International Journal of Molecular Sciences*. 23(4). https://doi.org/10.3390/ijms23041947.

Woolfenden, H. C., Bourdais, G., Kopischke, M., Miedes, E., Molina, A., Robatzek, S. & Morris, R. J. 2017. A computational approach for inferring the cell wall properties that govern guard cell dynamics. *The Plant Journal*. 92(1): 5–18. https://doi.org/10.1111/tpj.13640.

Xiang, J., Ming, J., Yin, H., Zhu, Y., Li, Y., Long, L., Ye, Z., Wang, H., Wang, X., Zhang, F., Yang, Y. & Yang, C. 2019. Anatomy and histochemistry of the roots and shoots in the aquatic selenium hyperaccumulator *Cardamine hupingshanensis* (Brassicaceae)". *Open Life Sciences*. 14(1): 318–326. https://doi.org/10.1515/biol-2019-0035.

Xie, J., Tang, X., Xie, C., Wang, Y., Huang, J., Jin, J., Liu, H., Zhong, C., Zhou, R., Ren, G. & Zhang, S. 2023. Comparative analysis of root anatomical structure, chemical components and differentially expressed genes between early bolting and unbolting in *Peucedanum praeruptorum* Dunn. *Genomics*. 115(2): 110557. https://doi.org/10.1016/j.ygeno.2023.110557.

Xu, R., Huang, J., Guo, H., Wang, C. & Zhan, H. 2023. Functions of silicon and phytolith in higher plants. *Plant Signal and Behavior*. 18(1). https://doi.org/10.1080/15592324.2023.2198848.

Ye, M., Song, Y., Long, J., Wang, R., Baerson, S. R., Pan, Z., Zhu-Salzman, K., Xie, J., Cai, K., Luo, S. & Zeng, R. 2013. Priming of jasmonate-mediated antiherbivore defense responses in rice by silicon. *Proceedings of National Academy of Science-USA*. 110: E3631–E3639. https://doi.org/10.1073/pnas.1305848110.

Yeo, A. R., Flwoers, S. A., Rao, G., Welfare, K., Senanayake, N. & Flowers, T. J. 1999. Silicon reduces sodium uptake in rice (*Oryza sativa* L.) in saline conditions and this is accounted for by a reduction In the transpirational bypass flow. *Plant Cell and Environment.* 22: 559–565. https://doi.org/10.1046/j.1365-3040.1999.00418.x.

Zargar, S. M., Mahajan, R., Bhat, J. A., Nazir, M. & Deshmukh, R. 2019. Role of silicon in plant stress tolerance: Opportunities to achieve a sustainable cropping system. *3 Biotech.* 9(3). https://doi.org/10.1007/s13205-019-1613-z.

Ziv, M. 2010. Silicon effects on growth acclimatization and stress tolerance of bioreactor cultured *Ornithogalum dubium* plants, *Acta Horticulturae.* 865: 29–36.

13 Cryogenic Conservation of Medicinal Plants

Importance and Techniques

Ravi Gowthami, Subhash Chander, Sandhya Gupta, Vartika Srivastava, G.M. Puneeth, D.A. Deepak, Muthusamy Shankar, and Kartar Singh

13.1 INTRODUCTION

Plants possess specific secondary metabolites that have a significant physiological impact on the mammalian system, and these plants are called "medicinal plants". These substances have curative, additive, or synergistic effects on human health. Medicinal plants as whole plants or its parts such as the stem, root, bark, seeds, leaves, rhizomes, etc., have been used in the treatment of numerous ailments. Many of the secondary metabolites found in plants are being employed as medications or as a raw material for their production. Many of the world's earliest recognized medical systems, including the Chinese and Tibetan traditional medicines of China, Arabian medicine, the Indus civilization's Ayurveda, and the Japanese Kempo system, are mostly based on plants. Medicinal herbs have long been used to protect human health, both in ancient and modern cultures.

Medicinal plants play a pivotal role in the treatment of a number of diseases for a large population in the developing world. According to reports, two-thirds of plant species worldwide have therapeutic qualities (Krishnaiah et al., 2011), though there is no reliable figure for the total number of medicinal plants on Earth. However, a few reports document the total number of medicinal plant species is 35,000–70,000 or 53,000 worldwide; 10,000–11,250 in China; 7500 in India; and 2237 in Mexico (Chouhan, 2022). The most significant source of therapeutic substances is nature itself. Many of the plant species used in allopathy, traditional medicine, tribal and folk medicine, and other medical systems are harvested from naturally occurring wild populations. Around 80% of individuals in developing nations depend on traditional medicine for their basic healthcare needs, with plant extracts being used in roughly 85% of cases. People are switching more and more to herbal drugs due to the high costs and negative side effects of synthetic drugs; this trend is expanding not only in developing but also in developed countries. The herbal market has witnessed an astounding increase in widespread acceptance. According to estimates from the World Health Organization (WHO), the current market for medicinal plants is worth $14 billion USD annually, and the demand for medical plants is expected to rise to more than $5 trillion USD by 2050. The demand for raw materials derived from medicinal plants is expanding at a pace of 15–25% annually.

Lack of access to therapeutic plants, alien species encroaching on land, and habitat destruction have led to the uncontrolled collection of medicinal plants from the wild. The natural population of medicinal plants is under pressure from this, and the rate of extinction has accelerated, especially in developing nations like China, India, Nepal, Kenya, Tanzania, and Uganda (Chen et al., 2016). Destructive harvesting methods and overharvesting for medical purposes are putting the genetic diversity of medicinal plants in the globe in severe jeopardy. Other reasons are the widespread devastation of the plant-rich habitat caused by urbanization, agricultural encroachment, and forest degradation (Gowthami et al., 2021). The natural habitat for many herbs is disappearing due to the world's population growth, increasing anthropogenic activities, and rapidly eroding natural

DOI: 10.1201/b23374-13

ecosystems. Thus, many of the highly valuable medicinal plants are at different levels of threat, and some are on the verge of extinction. Furthermore, climate change has a significant impact on medicinal plants; some of these effects include higher carbon dioxide concentrations, which favor C_3 plants over C_4 plants, an increase in pests and diseases, higher rainfall, higher soil salinity, etc. Globally, many plant species are threatened to varying degrees by about 13.49% (40,468 species), which includes many medicinal plants (Gowthami et al., 2021).

13.2 CONSERVATION OF MEDICINAL PLANTS

The preservation of medicinal plants is achieved both by conservation and cultivation. Under the conservation approach, medicinal plant species are safely conserved either *in situ* and/or *ex situ*. The majority of medicinal plants are endemic species, and the primary source of their therapeutic qualities is the presence of secondary metabolites, which are expressed into stimuli in their natural habitats (Figueiredo and Grelle, 2009). *In situ* conservation of medicinal plants is achieved through different approaches including biosphere reserves, national parks, gene sanctuaries, home gardens, and on-farm fields (Figure 13.1). *Ex situ* conservation of medicinal plants can be achieved outside of their natural habitat in genebanks. Seed conservation in the seed genebank is the most effective and easy method for the conservation of medicinal plants. However, the majority of medicinal plants either do not produce seeds or produce very small seeds; many species are also not amenable to seed propagation, and some plants propagated through seeds are highly heterozygous and exhibit significant variations in growth, habit, and yield. Maintenance of such species in a field genebank often comes with challenges like high maintenance costs and susceptibility to natural calamities,

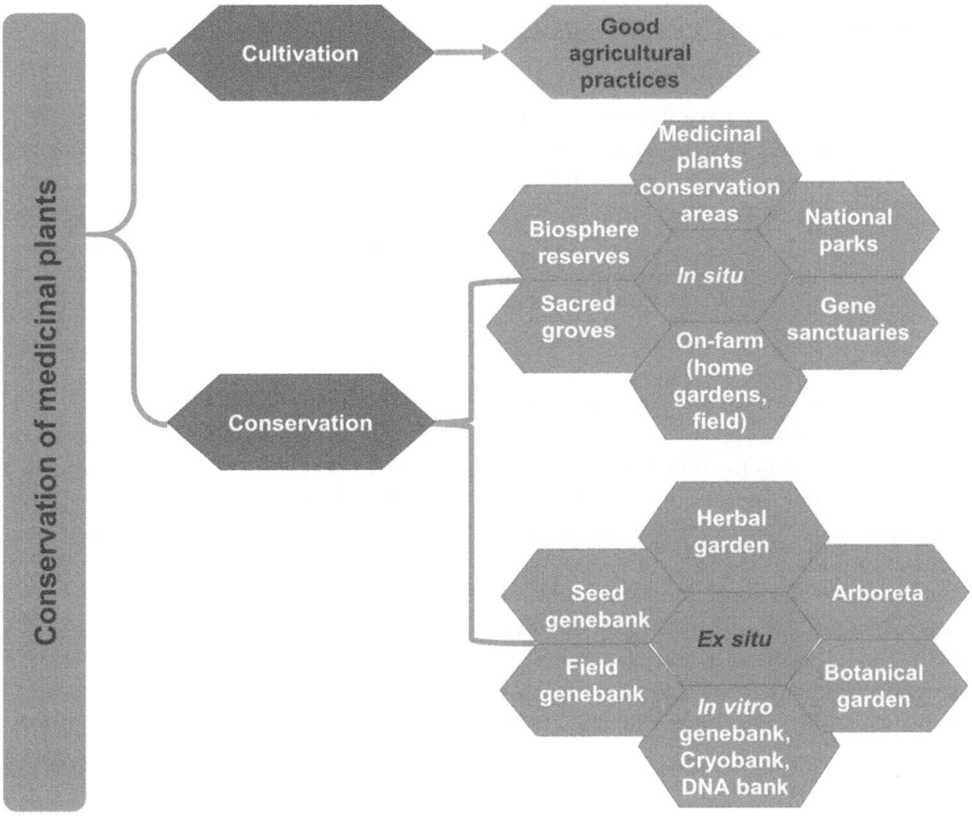

FIGURE 13.1 Conservation approaches for medicinal plants.

pests, and diseases. Due to its advantages over other approaches, *in vitro* techniques have become more and more popular for germplasm conservation. In particular, any medicinal plant species are safely conserved in the *in vitro* genebanks. The fact that *in vitro* methods require a lot less storage space than field storage is one of the more widespread advantages associated with employing them. However, *in vitro* conservation through an *in vitro* genebank allows conservation for only short- to medium-term periods, and frequent subculturing may lead to the induction of genetic variability through somaclonal variations (Sharma et al., 2020). Long-term conservation of medicinal plants is accomplished through cryopreservation, which is the freezing of biological materials at extremely low temperatures: –196°C in liquid nitrogen or –150°C in the liquid nitrogen vapor phase. Cryopreservation is the best and only option for the long-term storage of clonal germplasm, genetic variants with unique medical or industrial value, non-orthodox (recalcitrant, dehydration-sensitive) seed species, vegetatively propagated plants, rare and endangered species, disease-free plants, pollen, embryogenic cultures, and genomic resources (Agrawal et al. 2022a,b, 2024; Gowthami et al. 2024a,b). The theory relies on the concept that freezing the water inside plant cells will stop the biochemical reactions that take place within the cells. Cell viability remains intact while metabolic activity and cell division are inhibited as a result of freezing. There are several cryopreservation techniques, and the choice of technique depends on the kind and family of the plant as well as the tissue or explant. Different explants can be cryopreserved, such as whole seeds, apical or axillary shoot tips, dormant buds, pollen, zygotic embryos, embryonic axes, bulbils, and explants obtained *in vitro* including apical or axillary shoot tips, nodal segments, embryogenic cultures, somatic embryos, hairy roots, protocorms, protocorm-like bodies (PLBs), cell suspensions, callus, protoplasts, spores, etc. (Agrawal et al., 2022a, b, 2024; Sharma et al., 2020; Gowthami et al. 2024a,b).

13.3 CRYOPRESERVATION TECHNIQUES FOR CONSERVATION OF MEDICINAL PLANTS

Significant advancements have been made in plant cryopreservation since its inception in 1956. Many novel approaches have emerged, moving from traditional controlled freezing methods to vitrification-based techniques including vitrification, droplet-freezing, encapsulation-dehydration, and encapsulation-vitrification protocols and the more modern cryo-plate, cryo-mesh, and vacuum infiltration–based methods. Many of these techniques have been successfully employed for cryopreservation of medicinal plants using different explants (Table 13.1).

13.3.1 Desiccation

Desiccation is the process of directly immersing dehydrated explants in liquid nitrogen. Dehydration of explants can be achieved by (1) air desiccation, (2) silica gel desiccation, and (3) laminar flow desiccation for 1–5 hours to attain the critical moisture content. This is the simplest method of cryopreservation, as it does not require expensive equipment like a programmed freezer or chemicals (cryoprotectants). Many species have been successfully cryopreserved using this method, and it is also suitable for various explants including whole seeds, zygotic embryos, embryonic axes, pollen, dormant buds, and the shoot tips of desiccation-resistant plants. The degree of success obtained with these explants differs depending on the species (Engelmann, 2000; Kulus and Zalewska, 2014). This technique has been utilized in medicinal crops using mature seeds (Decruse and Seeni, 2003; Varghese and Naithani, 2008 and Kushwaha et al., 2010), somatic embryos (Popova et al., 2010; Krishnan et al., 2011), and callus (Lu et al., 2009).

13.3.2 Slow/programmable Freezing

The basic principle of slow or programmable freezing, also known as classical thawing or controlled rate cooling, is the dehydration of samples caused by freezing in the presence of extracellular ice.

TABLE 13.1
Selected Examples of Cryopreservation of Medicinal Plants

Species	Explant Used	Cryo-technique	Survival/Regrowth/ Germination (%)	References
Aconitum heterophyllum	Mature seeds	Desiccation	88	Kushwaha et al. (2010)
Agave peacockii	Shoot tips	Droplet-vitrification	96	Delgado-Aceves et al. (2022)
Allium ampeloprasum	*In vitro* shoot bases	Vitrification	31.67	Gowthami et al. (2023a)
	In vitro shoot bases	Droplet-vitrification	58.33	Gowthami et al. (2023a)
Allium cepa	Shoot tips	Droplet-vitrification	56–60	Wang et al. (2020a)
Allium chinense	Shoot tips	V cryo-plate	100	Tanaka et al. (2019)
Allium sativum	Shoot apices from bulbs	Vitrification	90	Baek et al. (2003)
	Apices of bulbs	Vitrification	75	Volk et al. (2004)
	Shoot tips	Vitrification	70	Keller (2005)
	Apices of bulbs (1–2 mm)	Droplet-vitrification	100	Ellis et al. (2006)
	Immature involucres (bulbil primordial, floral buds)	Droplet-vitrification	83	Kim et al. (2006)
	Unripe inflorescences	Droplet-vitrification	78.2	Ellis et al. (2006)
	Stem discs	Encapsulation-dehydration	55	Kim et al. (2007)
	Shoot apices	Droplet-vitrification	75.9	Lynch et al. (2012)
	Shoot tips	D-cryo-plate	94	Liu et al. (2017)
Alpinia nigra	Shoot tips	Vitrification	27	Rungjindamai et al. (2010)
Artemisia herba-alba	Shoot tips	Encapsulation-dehydration	70	Shatnawi (2011)
	Shoot tips	Encapsulation-dehydration	6	Sharaf et al. (2012)
	Shoot tips	Encapsulation-vitrification	12	
Arundina graminifolia	Seeds	D cryo-plate	82	Thammasiri et al. (2019)
		Encapsulation-dehydration	74	

(*Continued*)

TABLE 13.1 (*Continued*)
Selected Examples of Cryopreservation of Medicinal Plants

Species	Explant Used	Cryo-technique	Survival/Regrowth/ Germination (%)	References
Azadirachta indica	Seed	Desiccation followed by fast freezing	45	Varghese and Naithani 2008
Bacopa monnieri	Shoot tips	Vitrification	20	Sharma et al. (2011, 2017)
	Shoot tips	Improved vitrification	>50	Gowthami et al. (2023b)
	Shoot tips	Droplet-vitrification	80	Gowthami et al. (2023b)
Baliospermum montanum	Shoot meristems	Vitrification	63.33	Parvathi et al. (2023)
Bletilla striata	Immature seeds	Direct plunge in LN vitrification	81–92	Hirano et al. (2005)
	Mature seeds	Droplet-vitrification	93	Jitsopakul et al. (2008)
	Germinating seeds	Droplet-vitrification	91	
Brassidium Shooting Star	Protocorm-like bodies	Droplet-vitrification	30	Rahmah et al. (2015)
Buxus hyrcana Pojark	Shoot tips	Encapsulation-dehydration	60	Kaviani and Negahdar (2017)
Buxus sempervirens	Shoot tips	Encapsulation-dehydration	66.30	Negahdar et al. (2021)
	Shoot tips	Encapsulation-vitrification	60	
Castanea sativa	Embryonic axes	Encapsulation vitrification	64	Gaidamashvili et al. (2021)
Catharanthus roseus	ECa	Vitrification	88.9	Fatima et al. (2009)
	Embryogenic tissue	Two-step cryopreservation	72	Mujib et al. (2022)
Celisostoma areitinum	Protocorms	Encapsulation-dehydration	49	Maneerattanarungroj et al. (2007)
Chlorophytum borivilianum	Meristems	Vitrification	33	Chauhan et al. (2021)
Cleome rosea	Shoot tips	V cryo-plate	100	da Silva Cordeiro et al. (2015)
Clinopodium odorum	Shoot tips	V cryo-plate	71	Engelmann-Sylvestre and Engelmann (2015)
Clitoria ternatea	Somatic embryos	Encapsulation-dehydration	55	Nair and Reghunath (2008)
Cocos nucifera	Shoot tips	Droplet-vitrification	>50	Wilms et al. (2019)

Coriandrum sativum	Somatic embryos	Desiccation	98.2	Popova et al. (2010)
Dahlia cv	*In vitro* shoot tips	Droplet-vitrification	44.44	Gowthami et al. (2023c)
Dioscorea alata	Shoot tips	Encapsulation-dehydration	21.8	Mandal et al. (1996)
	Shoot tips	Droplet-vitrification	20	Leunufna and Keller (2005)
	Shoot tips	Vitrification	39	Mukherjee et al. (2009)
	Shoot tips	Encapsulation-dehydration	5–15	Uchendu and Keller (2016)
D. batatas	Shoot tips	Encapsulation-dehydration	40–45	Shin et al. (2006)
D. bulbifera	Shoot tips	Encapsulation-dehydration	26.1	Mandal et al. (1996)
	Embryogenic calli	Vitrification	60–70	Hong et al. (2009)
	Somatic embryos	Encapsulation-dehydration	67.8	Mandal et al. (2009)
	Shoot tips	Vitrification	58	Mukherjee et al. (2009)
	Embryogenic calli	Encapsulation-vitrification	70	Ming-Hua and Sen-Rong (2010)
	Shoot tips	Droplet-vitrification	30–50	Leunufna and Keller (2005)
D. deltoidea	Shoot tips	Vitrification	83	Mandal and Dixit-Sharma (2007)
	Shoot tips	Encapsulation-dehydration	57	Mukherjee et al. (2009)
	Shoot tips	Encapsulation-dehydration	76	Mandal and Dixit-Sharma (2007)
	Shoot tips	Vitrification	30–54	Sharma et al. (2022)
		Droplet-vitrification	21–51	Sharma et al. (2022)
D. cayenensis	Shoot tips	Droplet-vitrification	30–50	Leunufna and Keller (2005)
	Shoot tips	Encapsulation-dehydration	15–35	Uchendu and Keller (2016)

(Continued)

TABLE 13.1 (*Continued*)
Selected Examples of Cryopreservation of Medicinal Plants

Species	Explant Used	Cryo-technique	Survival/Regrowth/ Germination (%)	References
D. floribunda	Shoot tips	Encapsulation-dehydration	58	Mandal et al. (1996)
	Shoot tips	Vitrification	30	Ahuja et al. (2002)
	Shoot tips	Vitrification	30	Mandal and Ahuja-Ghosh (2007)
	Shoot tips	Encapsulation-dehydration	29	
D. prazeri	Nodal segments, shoot tips	Vitrification	92	Thankappan and Morawala-Patell (2021)
	Nodal segments, shoot tips	Encapsulation-vitrification	38	
	Nodal segments, shoot tips	Encapsulation-dehydration	75	
D. polystachya	Shoot tips	Droplet-vitrification	30–50	Leunufna and Keller (2005)
D. rotundata	Shoot tips	Vitrification	71	Mandal et al. (2008)
	Shoot tips	Encapsulation-dehydration	67	
D. wallichii	Shoot tips	Encapsulation-dehydration	37	Mandal et al. (1996)
Eucalyptus grandis	Shoot tips	V cryo-plate	31.1	Oliveira et al. (2022)
Fritillaria anhuiensis	Shoot tips	Vitrification	52.3	Zhu et al. (2011)
F. cirrhosa	Callus	Vitrification	87.4	Yue-hua (2011)
	Callus	Vitrification	80	Wang et al. (2012)
F. imperialis	Bulb scale	Encapsulation-dehydration	74.30	Seydi et al. (2020)
	Bulb scale	Encapsulation-vitrification	81.6	
Fritillaria thunbergii	Pollen	Controlled freezing	56.4	Su (1991)
Gentiana cruciata	Proembryonic callus	Slow cooling	2.5–2.7	Mikuła et al. (2005)
	Proembryonic callus	Vitrification	85.7–91	
	Proembryonic callus	Droplet-vitrification	82	Mikuła et al. (2011a)
	Proembryonic callus	Encapsulation-dehydration	83	Mikuła et al. (2008)

G. kurroo	Cell suspension	Vitrification	100	Mikuła et al. (2006)
	Cell suspension	Encapsulation-dehydration	95	Mikuła et al. (2011b)
	Shoot tips	Vitrification	35	Sharma et al. (2021)
	Shoot tips	Droplet-vitrification	60	
Gentiana macrophylla	Hairy roots	Encapsulation-dehydration	75.7	Xue et al. (2008)
G. scabra	Axillary buds	Encapsulation-dehydration	87	Suzuki et al. (2005)
	Shoot tips	Encapsulation-vitrification	80–93.3	Tanaka et al. (2004)
	Shoot tips	Vitrification	60–76.7	
G. tibetica	CSC	Vitrification	80	Mikuła (2006)
	Proembryonic callus	Encapsulation-dehydration	68	Mikuła et al. (2008)
G. triflora	Shoot tips	Encapsulation-vitrification	43.3–86.7	Tanaka et al. (2004)
	Shoot tips	Vitrification	16.7–46.7	
G. pneumonanthe	Shoot tips	Encapsulation-vitrification	80	Tanaka et al. (2004)
	Shoot tips	Vitrification	40	
Ginkgo biloba	Callus	Desiccation	54	Lu et al. (2009)
Grammatophyllum speciosum	Protocorms	Droplet-vitrification	38	Sopalun et al. (2010)
	Protocorms	Encapsulation-dehydration	24	
	Protocorms	Encapsulation-vitrification	14	
Helianthus tuberosus	Shoot tips	Vitrification	34	Volk and Richards (2006)
	Shoot tips	Droplet-vitrification	83	Zhang et al. (2017)
Hladnikia pastinacifolia	Shoot tips	Encapsulation-vitrification	64	Ciringer et al. (2018)
H. pastinacifolia	Shoot tips	Encapsulation-dehydration	53	

(Continued)

TABLE 13.1 (*Continued*)
Selected Examples of Cryopreservation of Medicinal Plants

Species	Explant Used	Cryo-technique	Survival/Regrowth/ Germination (%)	References
Holostemma annulare	Shoot tips	Pregrowth, encapsulation-dehydration	54.2	Decruse et al. (1999, 2004); Decruse and Seeni (2002)
Hypericum perforatum	Shoot tips	Vitrification	71.3	Bruňáková et al. (2011)
Hypericum richer spp. *transsilvanicum*	Axillary buds	Droplet-vitrification	68	Coste et al. (2012)
H. umbellatum	Axillary buds	Droplet-vitrification	71	
Indigofera tinctoria	Axillary shoot	Encapsulation-dehydration	62.2	Nair and Reghunath (2008)
Kaempferia galanga	Shoot tips	Pregrowth; vitrification	50	Krishnan et al. (2011)
	Somatic embryos	Pregrowth; desiccation	42.8	
Lilium japonicum	Apical meristems	Vitrification	68	Matsumoto et al. (1995)
Lilium spp.	Shoot tips	Droplet-vitrification	42–87	Yi et al. (2014)
Lilium × siberia	Apical meristems	Vitrification	35–45	Chen et al. (2011)
Lilium × siberia	Apical meristems	Droplet-vitrification	35–84	
Maesa lanceolata	Hairy roots	Encapsulation-dehydration	90	Lambert et al. (2009)
M. truncatula	Hairy roots	Encapsulation-dehydration	53	
Magnolia macrophylla	Shoot tips	Droplet-vitrification	30	Folgado and Panis (2019)
Magnolia sirindhorniae	Shoot tips	Encapsulation-vitrification	33	Kesa et al. (2005)
Mentha spp.	Shoot tips	Droplet-vitrification	89	Senula et al. (2007)
		Slow cooling	95	Uchendu and Reed (2008)
		V cryo-plate	100	Yamamoto et al. (2012)
Nelumbo nucifera	Seeds	Vitrification	92.1	Yong et al. (2023)
Panax ginseng	Adventitious roots	Desiccation	67	Oh et al. (2009)
		Vitrification	60	

Penthorum chinense	nodal explants	Droplet-vitrification	64	Zilani et al. (2022)
Petiveria alliacea	Shoot tips	D cryo-plate	85	Pettinelli et al. (2017)
	Somatic embryos	Vitrification	85	
	Somatic embryos	Encapsulation-dehydration	70	
Physalis angulata	Apical meristem	Droplet-vitrification	90	de Jesús Romo-Paz et al. (2021)
Picrorhiza kurroa	Shoot tips	Vitrification	35	Sharma and Sharma (2003)
Piper nigrum	Seeds	Desiccation	45	Chaudhury and Chandel (1994)
	Seeds	Desiccation	85	Decruse and Seeni (2003)
	Somatic embryo	Encapsulation-dehydration	62	Yamuna (2007)
		Vitrification	71	
Podophyllum hexandrum	Mature seeds	Desiccation	89	Kushwaha et al. (2010)
	Zygotic embryos	Vitrification	73.3	Parasher et al. (2023)
	Zygotic embryos	V cryo-plate	90	
Postemon yatabeanus	Shoot tips	Droplet-vitrification	91	Lee et al. (2023)
Rauvolfia serpentina	Nodal segments	Pregrowth; vitrification	66	Ray and Bhattacharya (2008)
Sideritis leucantha Cav. subsp. leucantha	Shoot tips	Vitrification	70	Juan-Vicedo et al. (2021)
Stevia rebaudiana	Shoot tips	Vitrification	65.6	Shatnawi et al. (2011)
	Shoot tips	Droplet-vitrification	80	Benelli et al. (2021)
	Shoot tips	V cryo-plate	93	
Zingiber officinale	Shoot buds	Vitrification	80	Yamuna et al. (2007)
Ziziphora tenuior	Shoot tips	Encapsulation-dehydration	40	Al-Baba et al. (2015)

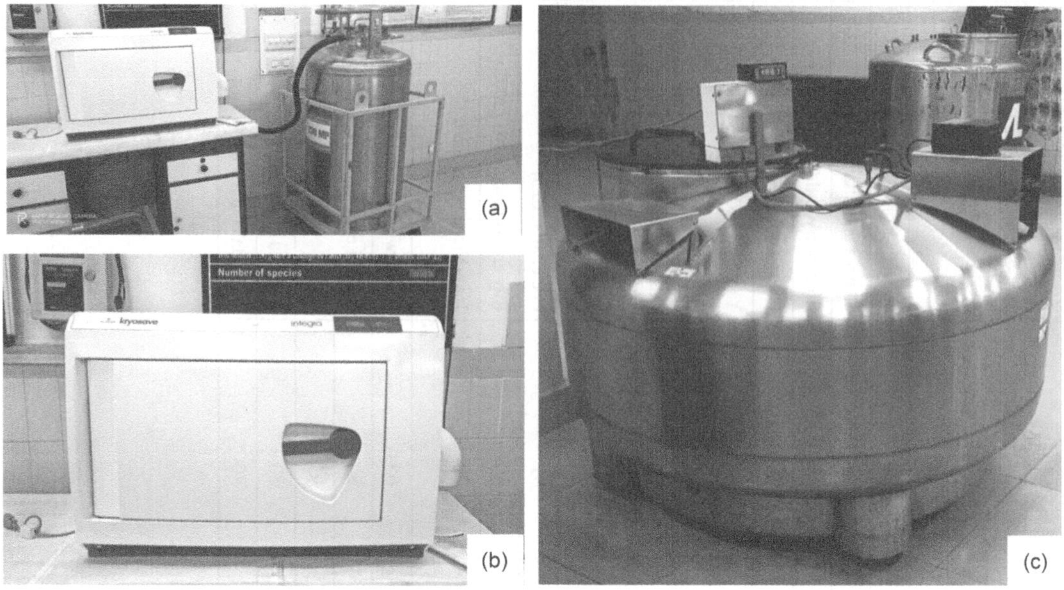

FIGURE 13.2 (a and b) Programmable freezers. (c) Cryotank.

Plant tissues are cooled gradually at a rate of 0.5–2°C minute^{-1} to approximately – 40°C in automatically programmed freezers, and then they are submerged in liquid nitrogen (Nausch and Buyel, 2021; Mosa et al., 2023) (Figure 13.2). Ice crystals develop during the extended freezing process, either naturally or through artificial nucleation in an extracellular solution. As a result, there is a water vapor deficit between the inside and outside of cells, allowing water to escape the cells and freeze extracellularly. Because of the ensuing osmotic imbalance, slow freezing results in the development of extracellular ice prior to the formation of intracellular ice crystals, which causes water to flow outside of the cell. To enhance dehydration, a cryoprotectant solution (such as dimethyl sulfoxide [DMSO], glycerol, ethylene glycol, and sucrose) could be added, albeit high quantities of these agents may be harmful to plant cells (Kaczmarczyk et al., 2012). This technique was the first to successfully cryopreserve plant cells (Kartha, 1985). The main drawback of this approach is that it requires a costly programmed freezer and exposes samples to subzero temperatures for extended periods, which can be harmful to species that are sensitive to cold. This has been used in *Mentha* spp. (Uchendu and Reed, 2008).

13.3.3 Vitrification

The definition of vitrification according to Fahy et al. (1984) is "the solidification of a liquid caused not by crystallization but by an extreme elevation in viscosity during cooling".

Plant cells were the first to be reported to be cryopreserved using vitrification solutions (Uragami et al., 1989; Sakai et al., 1990). Vitrification-based techniques employing vitrification solutions are now thought to be the most commonly used. In many plant species, vitrification is rapidly being used as the equipment of choice for the cryopreservation of plant germplasm. Vitrification is the process of treating explants with concentrated cryoprotective chemicals (penetrating and non-penetrating) or vitrification solutions at non-freezing temperatures for 15 minutes to 2 hours. The application of cryoprotective agents causes dehydration, which inhibits the intracellular and extracellular formation of ice crystals. An accumulation of these elements causes the cells' water content to decrease, and the major vacuole splits into multiple smaller ones (Zamecnik et al., 2021). The most commonly used vitrification solution is Plant Vitrification Solution 2 (PVS2), which is composed (*w/v*) of 30% glycerol, 15% ethylene glycol, 15% DMSO, and 0.4 M sucrose in MS (Murashige and Skoog, 1962)

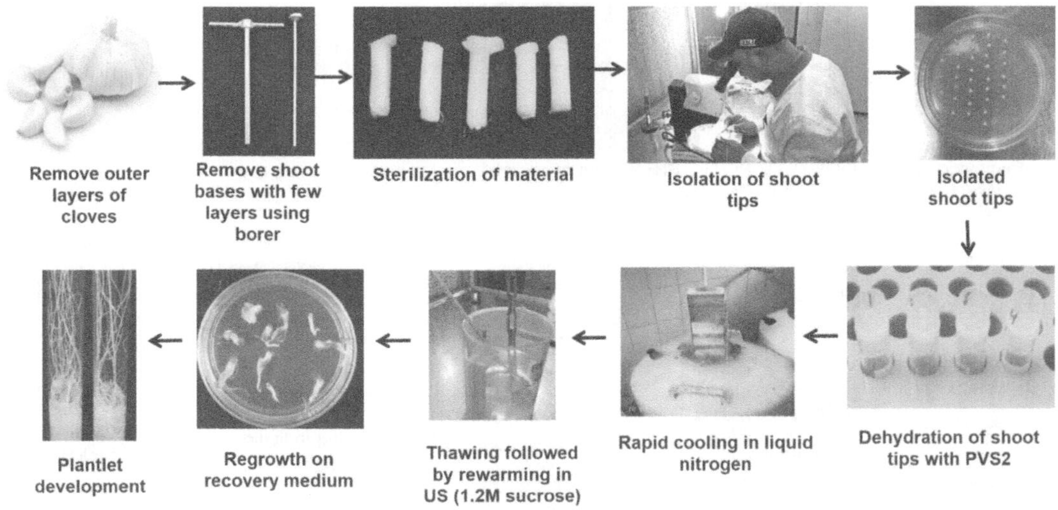

FIGURE 13.3 Generic protocol for vitrification technique of cryopreservation in *Allium sativum*.

(Sakai et al., 1990). The second most commonly used solution is Plant Vitrification Solution 3 (PVS3), which is composed of 40% (*w/v*) sucrose and 40% (*w/v*) glycerol in MS medium (Nishizawa et al., 1993). While exposure to the vitrification solution is necessary to achieve successful dehydration, the toxicity of the cryoprotective chemicals might cause cell damage if the solution is exposed excessively. Therefore, it is crucial to optimize the exposure duration and incubation temperature to these solutions (Sakai et al., 2008; Sharma et al., 2021; Gowthami et al., 2023b) to obtain significant regrowth following cryopreservation. A pre-cryoprotection step known as the loading solution stage is used to improve the penetration of the cryoprotectant through the plant cell membrane and to induce dehydration tolerance to prevent the potential damage to shoot tips caused by direct exposure to vitrification solutions (Sakai and Engelmann, 2007; Sakai et al., 2008; Gowthami et al., 2023c). The procedures involved in the vitrification process vary depending on the species; generic steps include (1) preconditioning of mother cultures; (2) preculturing explants on the medium containing cryoprotective chemicals/high sucrose concentration; (3) osmotic dehydration using a loading solution; (4) cryoprotection using vitrification solutions; (5) freezing quickly in liquid nitrogen (LN); (6) rapid thawing; (7) in high sucrose solution (1.2 M solution); and (8) Recovery (Figure 13.3). This procedure has effectively been used for an array of medicinal plant species using various explants *viz. in vitro* shoot bases (Sharma and Sharma, 2003; Gowthami et al., 2023a), seeds (Yong et al., 2023), somatic embryos (Pettinelli et al., 2017), shoot apices from bulbs (Baek et al., 2003; Volk et al., 2004), shoot buds (Yamuna et al., 2007), shoot tips (Ahuja et al., 2002; Tanaka et al., 2004; Keller, 2005; Mikuła, 2006; Volk and Richards, 2006; Mandal and Dixit-Sharma, 2007; Mandal and Ahuja-Ghosh, 2007; Mandal et al., 2008; Mukherjee et al., 2009; Rungjindamai et al., 2010; Bruňáková et al., 2011; Zhu et al., 2011; Yue-hua, 2011; Shatnawi et al., 2011; Sharma et al., 2017; Thankappan and Morawala-Patell, 2021; Sharma et al., 2021; Juan-Vicedo et al., 2021; Sharma et al., 2022), adventitious roots (Oh et al., 2009), shoot meristem (Chen et al., 2011; Chauhan et al., 2021; Parvathi et al., 2023), embryogenic callus (Fatima et al., 2009; Hong et al., 2009), callus (Wang et al., 2012), proembryonic callus (Mikuła et al., 2005), cell suspension (Mikuła et al., 2006), and zygotic embryos (Parasher et al., 2023).

13.3.4 Droplet Vitrification

The droplet freezing approach was developed by Kartha et al. (1982), wherein cassava shoot tips were placed in droplets of the cryoprotective medium on aluminum foil in petri dishes using a two-step cooling procedure. Subsequently, Pennycooke and Towill (2000) employed PVS2-based droplet

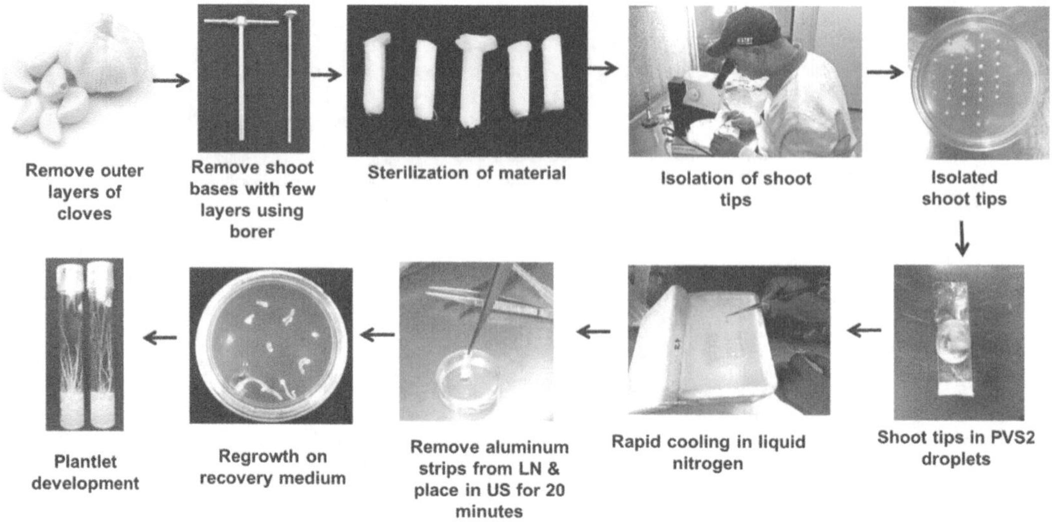

FIGURE 13.4 Generic protocol for droplet-vitrification technique of cryopreservation in *Allium sativum*.

freezing to cryopreserve sweet potato shoot tips. Panis et al. (2005) refined and improved this technique in Musa, resulting in the development of droplet vitrification (DV), one of the most effective cryopreservation methods. The major difference between vitrification and DV is the freezing rate/cooling rate. Because DV cools at a pace of around 130°C second^{-1} quicker than vitrification, which cools at a rate of about 6°C second^{-1}, intracellular water can transition from a liquid to a glassy state more quickly while water crystallization is reduced (Towill and Bonnart, 2003). Due to its high post-thaw recovery, the DV technique has been successfully applied to a wide range of plant species (Wang et al., 2020b). The generic steps involved in the DV process include (1) preconditioning of mother cultures; (2) preculturing explants on medium containing cryoprotective chemicals/high sucrose concentration; (3) osmotic dehydration using a loading solution; (4) cryoprotection using vitrification solutions and placement of shoot tips in cryoprotectant droplets contained in aluminum foil strips; (5) quick freezing of strips in LN; (6) thawing; (7) 7ehydration in high sucrose solution (1.2 M solution); and (8) recovery (Figure 13.4). This method has been employed in various medicinal plants by using explants, ranging from shoot tips (Leunufna and Keller, 2005; Senula et al., 2007; Wilms et al., 2019; Yi et al., 2014; Zhang et al., 2017; Folgado and Panis, 2019; Benelli et al., 2021; Sharma et al., 2021; Delgado-Aceves et al., 2022; Sharma et al., 2022; Gowthami et al., 2023a, 2023b, 2023c; Lee et al., 2023), *in vitro* shoot bases (Gowthami et al., 2023a), apices of bulbs (Ellis et al., 2006), unripe inflorescence (Ellis et al., 2006), shoot apices (Lynch et al., 2012), mature seeds (Jitsopakul et al., 2008), germinating seeds (Jitsopakul et al.,2008), protcorm like bodies (Rahmah et al., 2015), proembryonic callus (Mikuła et al., 2011a), protocorms (Sopalun et al., 2010), axillary buds (Coste et al., 2012), apical meristem (Chen et al., 2011; de Jesús Romo-Paz et al., 2021), and nodal explants (Zilani et al., 2022).

13.3.5 Encapsulation Dehydration

The encapsulation-dehydration technique of cryopreservation is mostly based on synthetic seed technology developed by Fabre and Dereuddre (1990) for cryopreservation of *Solanum* shoot apices. In encapsulation dehydration, explants are encapsulated in alginate beads formed through 3–4% sodium and 50–100 mM calcium chloride and held in liquid growth medium with 0.75 M sucrose for several hours or days to load the cells with sucrose. The beads are then physically dehydrated under laminar airflow or in silica gel desiccators to attain optimum moisture content (20–25%).

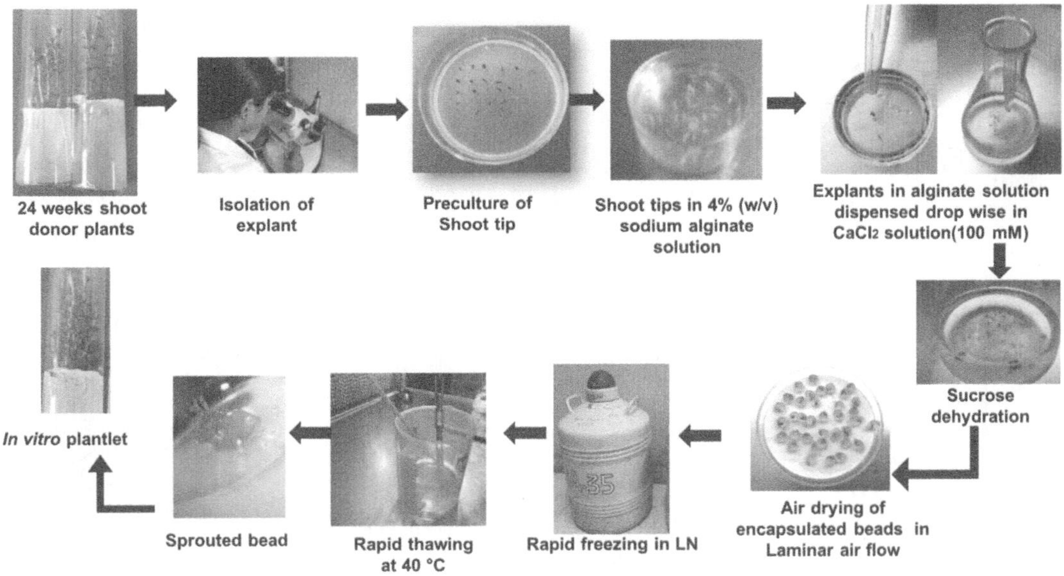

FIGURE 13.5 Generic protocol for encapsulation-dehydration technique of cryopreservation in *Bacopa monnieri*.

After dehydration, the beads are placed in cryotubes and directly immersed in LN (Figure 13.5). Encapsulation dehydration avoids potential chemical toxicity by using sucrose as the only cryoprotectant, as opposed to conventional vitrification techniques that include glycerol or DMSO as cryoprotectants. Since its establishment, this technique has been employed for conserving medicinal species using stem discs (Kim et al., 2007), shoot tips (Mandal et al., 1996; Decruse et al., 2004; Shin et al., 2006; Mandal and Ahuja-Ghosh, 2007; Mukherjee et al., 2009; Shatnawi, 2011; Sharaf et al., 2012; Al-Baba et al., 2015; Uchendu and Keller, 2016; Kaviani and Negahdar, 2017; Ciringer et al., 2018; Negahdar et al., 2021), seeds (Thammasiri et al., 2019), protocorms (Maneerattanarungroj et al., 2007), somatic embryos (Yamuna, 2007; Nair and Reghunath, 2008; Mandal et al., 2009; Pettinelli et al., 2017), bulb scale (Seydi et al., 2020), proembryonic callus (Mikuła et al., 2008), cell suspension (Mikuła et al., 2011b), hairy roots (Xue et al., 2008; Lambert et al., 2009), axillary buds (Suzuki et al., 2005), protocorms (Sopalun et al., 2010), and axillary shoots (Nair and Reghunath, 2008).

13.3.6 Encapsulation Vitrification

Encapsulation vitrification is a combination of the two methods, namely vitrification and encapsulation-dehydration. This technique was first reported by Matsumoto et al. (1995) for shoot apices of *Wasabia japonica*. Encapsulation vitrification is similar to encapsulation dehydration, except dehydration is achieved by vitrification solutions. The generic steps involved in the encapsulation-vitrification technique include (1) preconditioning of mother cultures; (2) preculturing explants on a medium containing cryoprotective chemicals/high sucrose concentration medium; (3) encapsulation of explants in alginate solution; (4) osmotic dehydration using a loading solution; (5) cryoprotection using vitrification solutions; (6) quick freezing in LN; (7) thawing; (8) rehydration in high sucrose solution (1.2 M solution); and (9) recovery (Figure 13.6). Encapsulation vitrification has been reported for cryopreservation of shoot tips (Tanaka et al., 2004; Kesa et al., 2005; Sharaf et al., 2012; Ciringer et al., 2018; Negahdar et al., 2021; Thankappan and Morawala-Patell 2021),

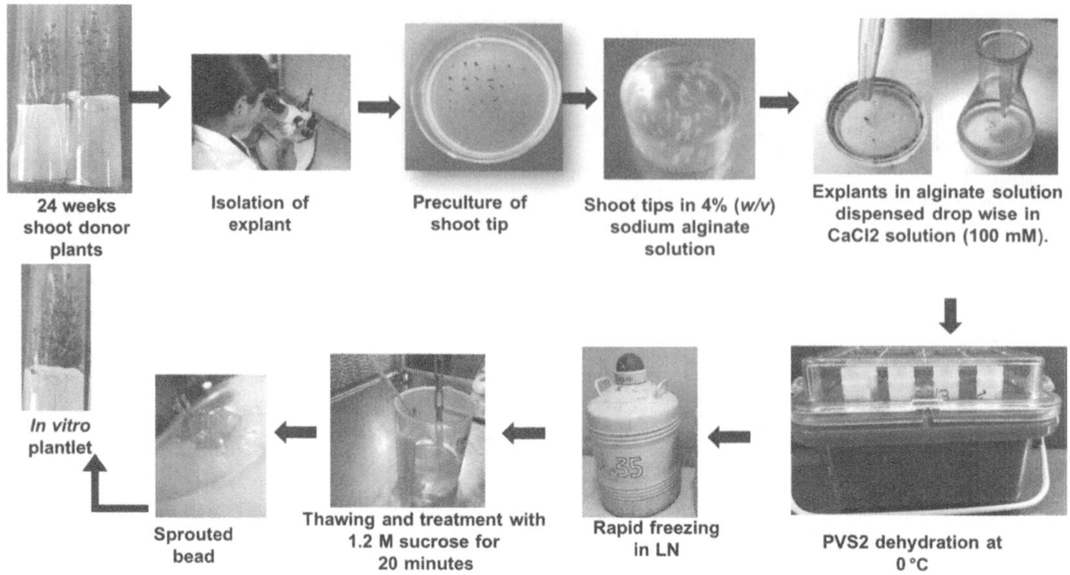

FIGURE 13.6 Generic protocol for encapsulation-vitrification technique of cryopreservation in *Bacopa monnieri.*

embryonic axes (Gaidamashvili et al., 2021), embryogenic calli (Ming-Hua and Sen-Rong, 2010), nodal segments (Thankappan and Morawala-Patell, 2021), bulb scale (Seydi et al., 2020), and protocorms (Sopalun et al., 2010).

13.3.7 V Cryo-Plate

Aluminum cryo-plates have recently been reported as a method for cryopreservation of several plant species. The two new cryogenic procedures utilizing cryo-plates are the D cryo-plate, also known as the dehydration cryo-plate, and the V cryo-plate, also known as the vitrification cryo-plate. The V cryo-plate technique was developed by Yamamoto et al. (2011) by combining the encapsulation and droplet-vitrification techniques for cryopreservation of *Tanacetum cinerariifolium* shoot tips, where dehydration was achieved by a vitrification solution. In the V-cryo-plate method, precultured shoot tips are placed into small wells of a cryo-plate (one shoot tip per one well), and alginate solution is poured on the cryo-plate. Next, a calcium solution (0.1 M calcium chloride and 0.4 M sucrose in MS medium) is added dropwise until the explant (such as shoot tips) is covered. Following polymerization, the shoot tip–containing cryo-plate is moved to a Petri plate that has a loading solution (LS) and then treated with vitrification solution, PVS2, or another vitrification solution. The cryo-plates are placed in cryovials and then submerged in LN. Rewarming involves placing the cryo-plate with shoot tips in an unloading solution (MS media containing 1.0 M sucrose) at room temperature and then plating in the recovery medium (Figure 13.7). Shoot tips (Yamamoto et al., 2012; da Silva Cordeiro et al., 2015; Engelmann-Sylvestre and Engelmann, 2015; Tanaka et al., 2019; Benelli et al., 2021; Oliveira et al., 2022) are being used in this technique.

13.3.8 D Cryo-Plate

Droplet vitrification and encapsulation dehydration are combined in the D cryo-plate method. The D cryo-plate method was first developed by Niino et al. (2013) for the cryopreservation of mat rush (*Juncus decipiens*) shoot tips. The D cryo-plate method involves placing precultured shoot tips (one

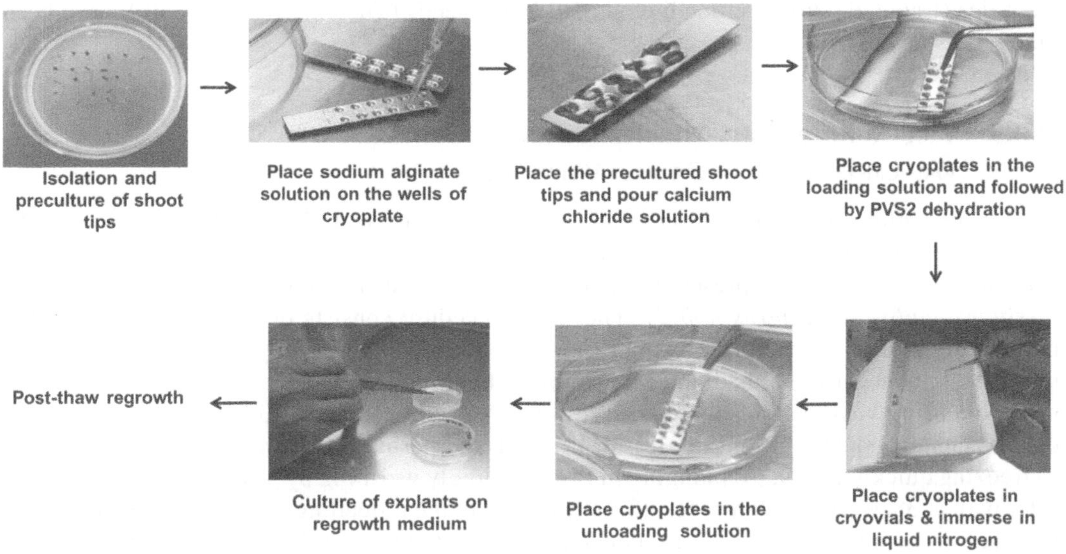

FIGURE 13.7 Generic protocol for V cryo-plate technique of cryopreservation in *Bacopa monnieri*.

shoot tip per well) into a small well of a cryo-plate and covering them with an alginate solution. Next, a calcium solution (0.1 M calcium chloride and 0.4 M sucrose in MS medium) is added drop-wise until the shoot tips are covered. Following polymerization, the shoot tip–containing cryo-plate is placed in an LS, and either silica gel or air desiccation (in a laminar air flow cabinet) is performed. After desiccation, cryo-plates with shoot tips are then placed in cryovials and submerged directly in LN. Rewarming involves placing the cryo-plate with shoot tips in an unloading solution (MS media containing 1.0 M sucrose) at room temperature and then plating the recovery medium (Figure 13.8).

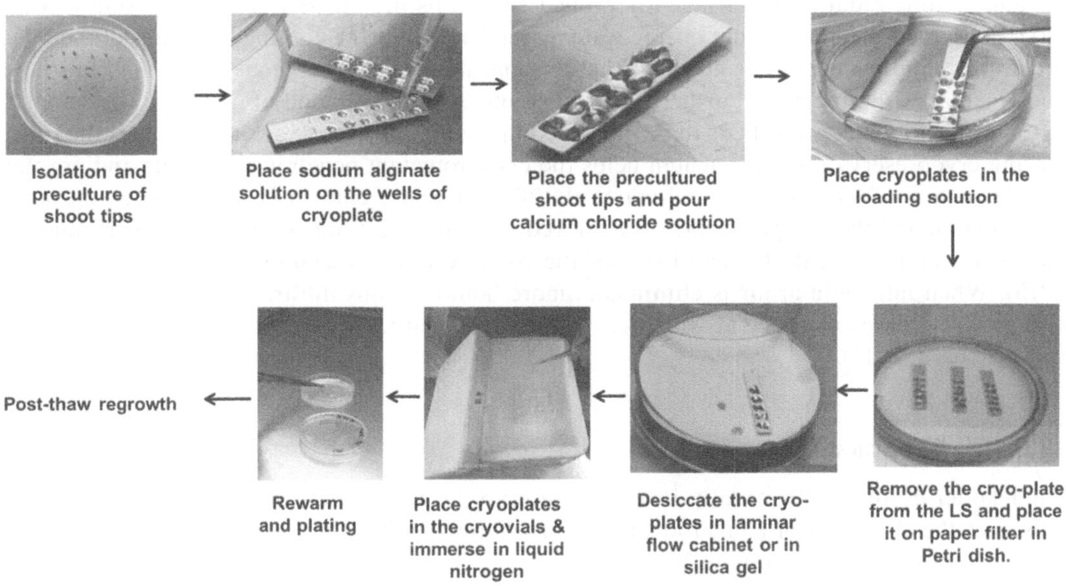

FIGURE 13.8 Generic protocol for D cryo-plate technique of cryopreservation in *Bacopa monnieri*.

Shoot tips (Liu et al., 2017; Pettinelli et al., 2017) and seeds (Thammasiri et al., 2019) have been used for this process.

13.3.9 Vacuum-Infiltration-Vitrification and Vacuum-Infiltration-Droplet-Vitrification

The vacuum-infiltration-vitrification (VIV) cryopreservation technique is the same as the DV technique except for exposing the shoot tips to a vacuum. VIV was first reported by Nadarajan and Pritchard (2014) for cryopreservation of zygotic embryos/embryonic axes of *Carica papaya*, *Passiflora edulis*, and *Laurus nobilis*. The VIV procedure consists of (1) preconditioning the mother cultures; (2) preculturing explants on a medium containing cryoprotective chemicals/high sucrose concentration medium; (3) osmotic dehydration using a loading solution; (4) cryoprotection using vitrification solutions (PVS2) in a vacuum chamber (10 L volume) and treated under vacuum (381 mm [15 in] of Hg) for the required period and then releasing the vacuum; (5) freezing quickly in LN; (6) plunging in LN; (7) rapidly warming by placing the cryovials in a water bath at 38 ± 2°C for 2 minutes; (8) rehydration in the unloading solution at room temperature; and (9) recovery. According to their findings, VIV significantly shortened the amount of time needed to incubate PVS2 (from 30 to 60 minutes to 1–5 minutes) in order to achieve comparable post-cryopreservation development. The advantages of VIV over vitrification include shorter cryoprotectant treatment times (few minutes), less temperature dependence in cryoprotectant application, and enhanced post-cryopreservation regrowth. An additional advantage of this technology is that, in contrast to ultra-rapid cooling methods, where the tissues are in direct contact with the cryogen, there is less possibility of cross-contamination during LN storage because the specimens are enclosed in cryovials. Since VIV cryopreservation also solves the challenges of specimen heterogeneity (mass, volume, oil composition, and thermal properties), it has significant potential as the basis for a generic technique for the preservation of tissues from different plant species. Funnekotter et al. (2015) demonstrated vacuum-infiltration-droplet-vitrification (VIDV) in *Androcalva perlaria*, *Anigozanthos viridis* sp. *terrapectans*, *Lomandra sonderi*, and *Loxocarya cinerea* and observed that the VIDV technique led to a much shorter PVS2 incubation period for cryogenic survival and regeneration, which cut the amount of time required to sufficiently protect shoot tips in half to a quarter. Prystalov et al. (2021) observed that the amount of free water crystallizing during cooling by means of the VIV process is significantly lower and that the concentration of cryoprotective solutions in grape buds increases significantly. Furthermore, this reduces the likelihood of exposing the samples to toxins by allowing grape buds to be incubated in the cryoprotective solution for half the time. During cryopreservation, the vacuum helps the cryoprotectant penetrate the tissue and prevent the production of ice crystals (Soejima et al., 2012). Greater contact between the cryoprotectant solution and the sample surface is achieved by removing trapped air bubbles surrounding the micro-topographical characteristics of the tissue through vacuuming (Funnekotter et al., 2015). When intracellular air is eliminated, more homogeneous infiltration results, which significantly increases the pace at which cryoprotective chemicals can enter the cell (Nadarajan and Pritchard, 2014).

13.3.10 Cryo-mesh

Funnekotter et al. (2017) recently invented the cryo-mesh approach, which successfully cryopreserves the shoot tips of kangaroo paw (*A. viridis*) with 83% of post-thaw regrowth. The primary distinction between this technique and the VC method is the use of stainless steel wire mesh strips, or cryo-mesh, in place of aluminum cryo-plates. Cryo-mesh is a stainless steel mesh strip with a wire diameter of 0.224 mm and dimensions of 25 × 7 mm (length × breadth) with 0.4-mm apertures. For the cryopreservation of different explants, the cryo-mesh has been suggested as an easy substitute for cryo-plates and aluminum foil (Wang et al., 2021). The generic steps involved in the cryo-mesh technique include (1) preconditioning of mother cultures; (2) preculturing explants on a medium containing cryoprotective chemicals/high sucrose concentration; (3) placement of precultured shoot tips on cryo-mesh (one shoot tip per one well) and pouring the alginate solution on the cryo-mesh; (4) adding the calcium solution dropwise until the shoot tips are covered to attain polymerization; (5) placing the cryo-mesh in the loading solution; (6) Cryoprotection using vitrification solutions; (7) quick freezing in LN; (8) rehydration in high sucrose solution (1.2 M solution); and (9) recovery.

13.4 GENETIC STABILITY OF CRYOPRESERVED MEDICINAL PLANTS

Apart from mere conservation, the objective of cryoconservation is to sustain the genetic fidelity of cryo-regenerated plants (Tahtamouni et al., 2015). Cryopreservation ensures true to typeness of the germplasm conserved, as metabolic activity at ultra-low temperatures is theoretically ceased (Panis et al., 2002; Agrawal et al., 2021). Thus, the credibility of the cryopreservation protocol remains worthwhile if the genetic composition of the cryopreserved plant is unaffected (Zarghami et al., 2008). To ensure genetic conformity, the meristem/shoot tip (differentiated) are targeted for cryoconservation instead of the callus (undifferentiated) to evade genetic changes (Benson et al., 1996; Harding, 2004). During cryopreservation, the explant undergoes several stresses viz., chemical sterilization to *in vitro* establishment, explant isolation to treatment with different cryoprotective solutions, and freezing at ultra-low temperature. Consequently, the chain of events may cause genetic and epigenetic variations in the conserved germplasm (Wang et al., 2021). Therefore, assessment of the genetic identity of the cryopreserved medicinal germplasm is paramount to ensure true to typeness. The genetic fidelity of the cryopreserved germplasm can be assessed at various levels including morphological, chromosomal, biochemical, and molecular (Dixit et al., 2003; Agrawal et al., 2019) (Table 13.2). Among these molecular markers have been found to be more reliable and reproducible, as they are not altered with environmental variations (Kamle et al., 2014). Numerous molecular markers have been used for the assessment of the genetic fidelity of medicinal plants (Harding, 2004). The most prevalent markers used in the assessment of the genetic stability of medicinal plants are randomly amplified polymorphic DNA (RAPD) and intersimple sequence repeat (IISR) (Table 13.2). However, occasionally a combination of molecular markers is used such as amplified fragment length polymorphism (AFLP) and RAPD in *Chrysanthemum morifolium* (Martin et al., 2011) and ISSR and expressed sequence tag- simple sequence repeat (EST-SSR) in *Dioscorea deltoidei* (Sharma et al., 2022). In most of the cases no genetic variation has been detected in cryopreserved medical plants; however, in some cases variation has been detected at the molecular level viz., *C. morifolium* (Martin et al., 2011), *Mentha* × *piperita* (Martin et al., 2013; Gonzalez-Benito et al., 2016). Martin et al., 2015, in *Mentha* × *piperita*, reported that plantlets developed using DV (99% stable in MEN 198 genotype) were found to be more stable than the plantlets recovered from encapsulation dehydration (24% in MEN 186). Apart from this, a genotypic response to genetic variation was also detected viz., in MEN 186 genotype (58% variation), whereas in MEN 198 only 3% variation was observed. Methylation-sensitive amplified polymorphism (MSAP) is another set of markers used for analysis of epigenetic alteration after cryopreservation, which works on the principle of the differential sensitivity of restriction nucleases to cytosine methylation to know the methylation level of the DNA sequences (Reyna-López et al., 1997). Ibáñez et al. (2019) reported that altered methylation (increased) was detected as analyzed just after each step of cryopreservation; however, after 1 day of cryo-recovery, the methylation level was comparable to the control shoot tips.

TABLE 13.2
Selected Examples on Genetic Stability Assessment of Cryopreserved Medicinal Plants

Crop	Explant Used for Conservation (Conservation Technique)	Genetic Stability	Variation Detected (Yes/No)	References
Bacopa monnieri	Shoot tips (V)	RAPD, HPLC	No	Sharma et al. (2017)
Chrysanthamum morifolium	Shoot tips (ED)	AFLP, RAPD	Yes	Martin et al. (2011)
	Shoot tips (DV)	ISSR, RAPD and FCM	No	Bi et al. (2016)
Cleome rosea	Roots (V)	RAPD	No	Cordeiro et al. (2016)
Dioscorea bulbifera	Embryogenic tissues (ED)	RAPD, biochemical and morphological	No	Dixit et al. (2003)
Dioscorea deltoidea	Shoot tips (V & DV)	Biochemical analysis (HPLC) diosgenin content	No	Dixit-Sharma et al. (2005)
	Shoot tips (V, DV)	ISSR, EST-SSR	No	Sharma et al. (2022)
D. floribunda	Root segments (V)	Morphological, biochemical and RAPD	No	Ahuja et al. (2002)
Gentiana kurroo	Shoot tips (V, DV)	ISSR, EST-SSR	No	Sharma et al. (2021)
Mentha × piperita	Shoot tips (ED & DV)	RAPD	Yes (MEN 186: 20–76%); Yes (MEN 198: 1–13%)	Martin et al. (2015)
	Shoot tips (DV)	RAPD	Yes	Martin et al. (2013)
	Shoot tips (ED)	RAPD, AFLP	Yes	Gonzalez-Benito et al. (2016)
	Shoot tips (DV)	ISSR	No	Ogur et al. (2023)
	Shoot tips (ED)	RAPD, AFLP, MSAP	No	Ibáñez et al. (2019)
Rabdosia rubescens	Shoot tips (ED)	FCM & SRAP	No	Ai et al. (2012)
Valeriana jatamansi	Shoot tips (V)	Biochemical (HPLC)	No	Sharma et al. (2021)
Ziziphora tenuior	Shoot tips (EV & DV)	AFLP	No	Al-Baba et al. (2015)

Abbreviations: AFLP: Amplified fragment length polymorphism; DV: Droplet vitrification; ED: Encapsulation dehydration; EST-SSR: Expressed sequence tag-derived simple sequence repeat markers; FCM: Flow cytometry; HPLC: High-performance liquid chromatography; ISSR: Inter-simple sequence repeat; MSAP: Methylation sensitive amplified polymorphism; RAPD: Random amplified polymorphic DNA; SRAP: sequence-related amplified polymorphism; V: Vitrification.

13.5 CONCLUSION

There is an urgent need to conserve the vast diversity of medicinal plants at the ecological, species, and gene pool levels due to the alarming rate at which medicinal plants are in jeopardy of being threatened with various dangers. Cryopreservation is an effective and economical method for the long-term storage of medicinal plants. Different cryopreservation techniques have been established for plant cells, organs, and tissues. This chapter examined several cryopreservation techniques used in various species and the effects of cryopreservation on genetic integrity and survival rates. These techniques can be effectively used in conservation. Integration of biotechnological and cryogenic tools and the development of easier and more effective protocols for the cryopreservation of medicinal plant tissues ensure the safe conservation of medicinal plants.

REFERENCES

Agrawal, A., Gowthami, R., Chander, S., Srivastava, V., Shankar, M., & Deepak, D. A. 2024. Biotechnological tools for conservation of plant genetic diversity. In: Al-Khayri J. M., Jain S. M., & Penna S. (eds.). *Sustainable Utilization and Conservation of Plant Genetic Diversity* (pp. 711–752). Springer, Singapore. https://doi.org/10.1007/978-981-99-5245-8_21

Agrawal, A., Gowthami, R., Chander, S. & Srivastava, V. 2022a. Sustainability of *in vitro* genebanks and cryogenebanks, *Indian Journal of Plant Genetic Resources* 35: 180–184.

Agrawal, A., Sharma, N., Gupta, S., Bansal, S., Srivastava, V., Malhotra, E. V., Chander, S., Gowthami, R. & Singh, K. 2022b. Biotechnological applications for plant germplasm conservation at ICAR-national bureau of plant genetic resources, India- recent achievements. *Acta Hortic.* 1339: 29–42.

Agrawal, A., Sharma, N., Gupta, S., Bansal, S., Srivastava, V., Malhotra, E. V., Chander, S., Gowthami, R. & Singh, K. 2021. Biotechnological applications for plant germplasm conservation at ICAR-National Bureau of Plant Genetic Resources, India–recent achievements. In IX International Scientific and Practical Conference on Biotechnology as an Instrument for Plant Biodiversity Conservation 1339, pp. 29–42.

Agrawal, A., Singh, S. & Malhotra, E. V., et al. 2019. *In vitro* conservation and cryopreservation of clonally propagated horticultural species. In: Rajasekharan PE, Rao VR (eds). *Conservation and Utilization of Horticultural Genetic Resources.* Springer Singapore, Singapore, pp 529–578. https://doi.org/10.1007/978-981-13-3669-0_18

Ahuja, S., Mandal, B. B., Dixit, S. & Srivastava, P. S. 2002. Molecular, phenotypic and biosynthetic stability in *Dioscorea floribunda* plants derived from cryopreserved shoot tips, *Plant Science* 163 (5): 971–977.

Ai, P. F., Lu, L. P. & Song, J. J. 2012. Cryopreservation of *in vitro*-grown shoot-tips of *Rabdosia rubescens* by encapsulation-dehydration and evaluation of their genetic stability, *Plant Cell, Tissue and Organ Culture* (PCTOC) 108: 381–387.

Al-Baba, H., Shibli, R. A., Akash, M., Al-Qudah, T. S., Tahtamouni, R. W. & Al-Ruwaiei, H. 2015. Cryopreservation and genetic stability assessment of threatened medicinal plant (*Ziziphora tenuior* L.) Grown wild in Jordan, *Jordan Journal of Biological Sciences* 8 (4): 247–256.

Baek, H. J., Kim, H. H., Cho, E. G., Chae, Y. A. & Engelmann, F. 2003. Importance of explant size and origin and of preconditioning treatments for cryopreservation of garlic shoot apices by vitrification, *CryoLetters* 24 (6): 381–388.

Benelli, C., Carvalho, L. S. O. & El Merzougui, S., et al. 2021. Two advanced cryogenic procedures for improving *Stevia rebaudiana* (Bertoni) cryopreservation. *Plants* 10: 277. https://doi.org/10.3390/plants10020277

Benson, E. E., Reed, B. M., Brennan, R. M., Clacher, K. A. & Ross, D. A. 1996. Use of thermal analysis in the evaluation of cyropreservation protocols for *Ribes nigrum* L. Germplasm, *CryoLetters* 17: 347–362.

Bi, W. L., Pan, C., Liu, J. & Wang, Q. C. 2016. Greenhouse performance, genetic stability and biochemical compounds in *Chrysanthemum morifolium* 'Hangju' plants regenerated from cryopreserved shoot tips. *Acta Physiologiae Plantarum* 38: 1–10. https://doi.org/10.1007/s11738-016-2288-2

Bruňáková, K., Zámečník, J., Urbanová, M. & Čellárová, E. 2011. Dehydration status of ABA-treated and cold-acclimated *Hypericum perforatum* L. Shoot tips subjected to cryopreservation, *Thermochimica Acta* 525: 62–70.

Chaudhury, R. & Chandel, K. P. S. 1994. Germination studies and cryopreservation of seeds of black pepper (*Piper nigrum* L.): A recalcitrant species, *Cryoletters* 15: 145–150.

Chauhan, R., Singh, V., Keshavkant, S. & Quraishi, A. 2021. Vitrification-based cryopreservation of *in vitro*-grown apical meristems of *Chlorophytum borivilianum* Sant et Fernand: A critically endangered species, *Proceedings of the National Academy of Sciences, India Section B: Biological Sciences* 91 (2): 471–476.

Chen, S. L., Yu, H., Luo, H. M., Wu, Q., Li, C. F. & Steinmetz, A. 2016. Conservation and sustainable use of medicinal plants: Problems, progress, and prospects, *Chinese Medicine* 11: 1–10.

Chen, X. L., Li, J. H., Xin, X., Zhang, Z. E., Xin, P. P. & Lu, X. X. 2011. Cryopreservation of *in vitro*-grown apical meristems of *Lilium* by droplet-vitrification, *South African Journal of Botany* 77: 397–403.

Chouhan, S. S. 2022. Endangered plants used as medicine: A review, *International Journal of Pharmaceutical Research and Applications* 7 (6): 131–140.

Ciringer, T., Martín, C., Šajna, N., Kaligarič, M. & Ambrožič-Dolinšek, J. 2018. Cryopreservation of an endangered *Hladnikia pastinacifolia* Rchb. By shoot tip encapsulation-dehydration and encapsulation-vitrification, *In Vitro Cellular & Developmental Biology-Plant* 54: 565–575.

Cordeiro, L. D. S., Simões-Gurgel, C. & Albarello, N. 2016. Cryopreservation of adventitious roots of *Cleome rosea* Vahl (Cleomaceae) using a vitrification technique and assessment of genetic stability, *CryoLetters* 37 (4): 231–242.

Coste, A., Halmagyi, A., Butiuc-Keul, A. L., Deliu, C., Coldea, G. & Hurdu, B. 2012. *In vitro* propagation and cryopreservation of Romanian endemic and rare *Hypericum* species, *Plant Cell, Tissue and Organ Culture* 110: 213–226.

da Silva Cordeiro, L., Simões-Gurgel, C. & Albarello, N., et al. 2015. Cryopreservation of *in vitro*-grown shoot tips of *Cleome rosea* Vahl (Cleomaceae) using the v cryo-plate technique. *In Vitro Cell Dev Biol Plant* 51: 688–695. https://doi.org/10.1007/s11627-015-9714-9

de Jesús Romo-Paz, F., Folgado, R. & Delgado-Aceves, L., et al. 2021. Tissue culture of *Physalis angulata* L. (Solanaceae): Techniques for micropropagation and germplasm long-term preservation. *Plant Cell, Tissue and Organ Culture* 144: 73–78. https://doi.org/10.1007/s11240-020-01970-8

Decruse, S. W. & Seeni, S. 2002. Ammonium nitrate in the culture medium influences regeneration potential of cryopreserved shoot tips of *Holostemma annulare, Cryoletters* 23: 55–60.

Decruse, S. W. & Seeni, S. 2003. Seed cryopreservation is a suitable storage procedure for a range of *Piper* species, *Seed Science and Technology* 31: 213–217.

Decruse, S. W., Seeni, S. & Nair, G. M. 2004. Preparative procedures and culture media effect on the success of cryostorage of *Holostemma annulare* shoot tips, *Plant Cell, Tissue and Organ Culture* 76: 179–182.

Decruse, S. W., Seeni, S. & Pushpangadan, P. 1999. Cryopreservation of alginate coated shoot tips of *in vitro* grown *Holostemma annulare* (Roxb.) K. Schum, an endangered medicinal plant: Influence of preculture and DMSO treatment on survival and regeneration, *Cryoletters* 20: 243–250.

Delgado-Aceves, L., Portillo, L., Folgado, R., de Jesús Romo-Paz, F. & Gonzalez-Arnao, M. T. 2022. New approaches for micropropagation and cryopreservation of *Agave peacockii*, an endangered species, *Plant Cell, Tissue and Organ Culture (PCTOC)* 150: 85–95.

Dixit, S., Mandal, B. B., Ahuja, S. & Srivastava, P. S. 2003. Genetic stability assessment of plants regenerated from cryopreserved embryogenic tissues of *Dioscorea bulbifera* L. Using RAPD, biochemical and morphological analysis, *CryoLetters* 24 (2): 77–84.

Dixit-Sharma, S., Ahuja-Ghosh, S., Mandal, B. B. & Srivastava, P. S. 2005. Metabolic stability of plants regenerated from cryopreserved shoot tips of *Dioscorea deltoidea*—An endangered medicinal plant, *Scientia Horticulturae* 105 (4): 513–517.

Ellis, D., Skogerboe, D., Andre, C., Hellier, B. & Volk, G. 2006. Implementation of garlic cryopreservation techniques in the national plant germplasm system, *CryoLetters* 27: 99–106.

Engelmann, F. 2000. Importance of cryopreservation for the conservation of plant genetic resources. In: Engelmann, F., Takagi, H. (eds.). *Cryopreservation of Tropical Plant Germplasm: Current Research Progress and Application*. IPGRI, Tsukuba, pp. 8–20.

Engelmann-Sylvestre, I. & Engelmann, F. 2015. Cryopreservation of *in vitro*-grown shoot tips of *Clinopodium odorum* using aluminium cryo-plates, *In Vitro Cellular & Developmental Biology - Plant* 51: 185–191.

Fabre, J. & Dereuddre, J. 1990. Encapsulation-dehydration: A new approach to cryopreservation of *Solanum* shoot-tips, *CryoLetters* 11: 413–426.

Fahy, G. M., MacFarlane, D. R., Angell, C. A., & Meryman, H. T. 1984. Vitrification as an approach to cryopreservation, *Cryobiology* 21(4): 407–426.

Fatima, S., Mujib, A., Nasim, S. A. & Siddiqui, Z. H. 2009. Cryopreservation of embryogenic cell suspensions of *Catharanthus roseus* L. (G) Don, *Plant Cell, Tissue and Organ Culture* 98: 1–9.

Figueiredo, M. S. L. & Grelle, C. E. V. 2009. Predicting global abundance of a threatened species from its occurrence: Implications for conservation planning. *Diversity and Distributions* 15: 117–121. https://doi.org/10.1111/j.1472-4642.2008.00525.x.

Folgado, R. & Panis, B. 2019. Cryopreservation of Ashe magnolia shoot-tips by droplet-vitrification, *Acta Hortic* 1234: 233–240.

Funnekotter, B., Bunn, E. & Mancera, R. L. 2017. Cryo-mesh: A simple alternative cryopreservation protocol, *CryoLetters* 38: 155–159.

Funnekotter, B., Whiteley, S. E., Turner, S. R., Bunn, E. & Mancera, R. L. 2015. Evaluation of the new vacuum infiltration vitrification (VIV) cryopreservation technique for native Australian plant shoot tips, *CryoLetters* 36 (2): 104–113.

Gaidamashvili, M., Khurtsidze, E., Kutchava, T., Lambardi, M. & Benelli, C. 2021. Efficient protocol for improving the development of cryopreserved embryonic axes of chestnut (*Castanea sativa* Mill.) By encapsulation-vitrification, *Plants* 10 (2): 231.

Gonzalez-Benito, M. E., Kremer, C. & Ibáñez, M. A., et al. 2016. Effect of antioxidants on the genetic stability of cryopreserved mint shoot tips by encapsulation–dehydration. *Plant Cell, Tissue and Organ Culture* 127: 359–368. https://doi.org/10.1007/s11240-016-1056-8

Gowthami, R., Rajashekaran, P. E., Martinez-Montero, M. E. 2024a. Cryoconservation of antiallergic medicinal plants: Techniques and applications. In: Gantait, S., Majumder, J., Sharangi, A. B. (eds.), *Biotechnology of Medicinal Plants with Antiallergy Properties* (pp. 611–634). Springer, Singapore. https://doi.org/10.1007/978-981-97-1467-4_21

Gowthami, R, Rajashekaran, P. E., Martinez-Montero, M. E., Chander, S., Singh, A. M., Puneeth, G. M,, Shankar, M., & Deepak, D. A. 2024b. In vitro techniques for safeguarding medicinal plants with antiallergic properties. In: Gantait, S., Majumder, J., & Sharangi, A. B. (eds.) *Biotechnology of Medicinal Plants with Antiallergy Properties* (pp. 635–672). Springer, Singapore. https://doi.org/10.1007/978-981-97-1467-4_22

Gowthami, R., Sharma, N., Chandra, R., Kurian, J. S., Malhotra, E. V. & Agrawal, A. 2023b. Development of an improved and simple shoot tip cryoconservation protocol for cryobanking of *Bacopa monnieri* (L.) Wettst. Germplasm, *In Vitro Cellular & Developmental Biology-Plant* 59: 744–756.

Gowthami, R., Sharma, N., Pandey, R. & Agrawal, A. 2021. Status and consolidated list of threatened medicinal plants of India. *Genetic Resources and Crop Evolution* 68: 2235–2263. https://doi.org/10.1007/s10722-021-01199-0

Gowthami, R., Chander, S., Pandey, R., Shankar, M. & Agrawal, A. 2023a. Development of cryoconservation protocol for in-vitro shoot bases of *Allium ampeloprasum* L. for long-term conservation. *Indian Journal of Plant Genetic Resources.*

Gowthami, R., Chander, S., Pandey, R., Shankar, M. & Agrawal, A. 2023c. Development of efficient and sustainable droplet-vitrification cryoconservation protocol for shoot tips for long-term conservation of *Dahlia* germplasm, *Scientia Horticulturae* 321: 112329.

Harding, K. 2004. Genetic integrity of cryopreserved plant cells: A review, *CryoLetters* 25: 3–22.

Hirano, T., Godo, T., Mii, M. & Ishikawa, K. 2005. Cryopreservation of immature seeds of *Bletilla striata* by vitrification, *Plant Cell Reports* 23: 534–539.

Hong, S., Yin, M., Shao, X., Wang, A. & Xu, W. 2009. Cryopreservation of embryogenic callus of *Dioscorea bulbifera* by vitrification, *CryoLetters* 30: 64–75.

Ibáñez, M. A., Alvarez-Mari, A., Rodríguez-Sanz, H., Kremer, C., González-Benito, M. E. & Martín, C. 2019. Genetic and epigenetic stability of recovered mint apices after several steps of a cryopreservation protocol by encapsulation-dehydration. A new approach for epigenetic analysis, *Plant Physiology and Biochemistry* 143: 299–307.

Jitsopakul, N., Thammasiri, K. & Ishikawa, K. 2008. Cryopreservation of *Bletilla striata* mature seeds, 3-day germinating seeds and protocorms by droplet-vitrification, *CryoLetters* 29 (6): 517–26.

Juan-Vicedo, J., Ramírez-Luna, J. E., Piqueras, A. & Casas, J. L. 2021. Micropropagation and cryopreservation by vitrification of the Spanish endemic medicinal plant *Sideritis leucantha* cav. subsp. *leucantha* (Lamiaceae). *In Vitro Cellular & Developmental Biology Plant* 57: 1057–1065. https://doi.org/10.1007/s11627-021-10173-5

Kaczmarczyk, A., Funnekotter, B., Menon, A., Phang, P. Y., Al-Hanbali, A., Bunn, E. & Mancera, R. 2012. Current issues in plant cryopreservation, in: Current Frontiers in Cryobiology, *InTech*, pp 417–438.

Kamle, M., Kumar, P. & Bajpai, A., et al. 2014. Assessment of genetic fidelity of somatic embryogenesis regenerated guava (*Psidium guajava* l.) Plants using DNA-based markers. *New Zealand Journal of Crop and Horticultural Science* 42: 1–9.

Kartha, K. K. 1985. *Cryopreservation of Plant Cells and Organs.* CRC Press, Boca Raton, FL.

Kartha, K. K., Leung, N. L. & Mroginski, L. A. 1982. *In vitro* growth responses and plant regeneration from cryopreserved meristems of cassava (*Manihot esculenta* Crantz). *Zeitschrift Für Pflanzenphysiologie* 107: 133–140. https://doi.org/10.1016/S0044-328X(82)80099-8

Kaviani, B. & Negahdar, N. 2017. Propagation, micropropagation and cryopreservation of *Buxus sempervirens* Pojark., an endangered ornamental shrub, *South African Journal of Botany* 111: 326–335.

Keller, E. 2005. Improvement of cryopreservation results in garlic using low temperature preculture and high-quality *in vitro* plantlets, *CryoLetters* 26: 357–366.

Kesa, S., Limpanavech, P., Kongchuensin, S., Chareonsap, P. & Juthamas, P. 2005. *In vitro* germplasm conservation of *Magnolia sirindhorniae* Noot. & Chalermglin in minimal growth condition and by cryopreservation. The 18th Ann. Meeting Thai Soc. Biotechnol. "Biotechnol: Benefits & Bioethics". IX-P-02.

Kim, H., Cho, E. & Park, S. 2006. Analysis of factors affecting the cryopreservation of garlic shoot tips, *Journal of Biomedical Nanotechnology* 2: 129–132.

Kim, H. H., Lee, J. K., Hwang, H. S. & Engelmann, F. 2007. Cryopreservation of garlic germplasm collections using the droplet vitrification technique, *CryoLetters* 28: 471–482.

Krishnaiah, D., Sarbatly, R. & Nithyanandam, R. 2011. A review of the antioxidant potential of medicinal plant species, *Food Bio Products Process* 89: 217–233.

Krishnan, P. N., Decruse, S. W. & Radha, R. K. 2011. Conservation of medicinal plants of Western ghats, India and its sustainable utilization through *in vitro* technology, *In Vitro Cell Dev Biol Plant* 47: 110–122.

Kulus, D. & Zalewska, M. 2014. Cryopreservation as a tool used in long-term storage of ornamental species—A review, *Scientia Horticulturae* 168: 88–107.

Kushwaha, R., Chanda, S., Ogra, R. & Bhattacharya, A. 2010. Cryopreservation criteria of *Podophyllum hexandrum* and *Aconitum heterophyllum* seeds based on storage behavior, *Seed Technol* 32: 117–127.

Lambert, E., Goossens, A., Panis, B., Van Labeke, M. C. & Geelen, D. 2009. Cryopreservation of hairy root cultures of *Maesa lanceolata* and *Medicago truncatula*, *Plant Cell, Tissue and Organ Culture* 96: 289–296.

Lee, H., Choi, B., Oh, S., Park, H., Popova, E., Paik, M. J. & Kim, H. 2023. Dynamics of organic acids during the droplet-vitrification cryopreservation procedure can be a signature of oxidative stress in *Pogostemon yatabeanus*, *Plants* 12 (19): 3489.

Leunufna, S. & Keller, E. R. J. 2005. Cryopreservation of yams using vitrification modified by including droplet method: Effects of cold acclimation and sucrose, *CryoLetters* 26 (2): 93–102.

Liu, X. X., Wen, Y. B., Cheng, Z. H. & Mou, S. W. 2017. Establishment of a garlic cryopreservation protocol for shoot apices from adventitious buds *in vitro*, *Sci. Hortic* 226: 10–18.

Lu, Z. W., Popova, E. V., Wu, C. H., Lee, E. J., Hahn, E. J., & Paek, K. Y. 2009. Cryopreservation of *Ginkgo biloba* cell culture: Effect of pretreatment with sucrose and ABA, *CryoLetters* 30(3): 232–243.

Lynch, P. T., Souch, G. R. & Harding, K. 2012. Effects of post-harvest storage of *Allium sativum* bulbs on the cryopreservation of stem-discs by encapsulation/dehydration, *J. Hortic. Sci. Biotechnol* 87: 588–592.

Mandal, B. B. & Ahuja-Ghosh, S. 2007. Regeneration of *Dioscorea floribunda* plants from cryopreserved encapsulated shoot tips: Effect of plant growth regulators, *CryoLetters* 28: 329–336.

Mandal, B. B., Ahuja-Ghosh, S. & Srivastava, P. S. 2008. Cryopreservation of *Dioscorea rotundata* Poir.: A comparative study with two cryogenic procedures and assessment of true-to-type of regenerants by RAPD analysis, *CryoLetters* 29 (5): 399–408.

Mandal, B. B., Chandel, K. P. & Dwivedi, S. 1996. Cryopreservation of yam (*Dioscorea* spp.) Shoot apices by encapsulation-dehydration, *CryoLetters* 17: 165–174.

Mandal, B. B. & Dixit-Sharma, S. 2007. Cryopreservation of *in vitro* shoot tips of *Dioscorea deltoidea* wall., An endangered medicinal plant: Effect of cryogenic procedure And storage duration, *CryoLetters* 28: 461–470.

Mandal, B. B., Dixit-Sharma, S. & Srivastava, P. S. 2009. Cryopreservation of embryogenic cultures of *Dioscorea bulbifera* L. By encapsulation-dehydration, *CryoLetters* 30: 440–448.

Maneerattanarungroj, P., Bunnag, S. & Monthatong, M. 2007. *In vitro* conservation of *Cleisostoma areitinum* (Rchb.f.) Garay, rare Thai orchid species by an encapsulation-dehydration method, *Asian J. Plant Sci* 6 (8): 1235–1240.

Martin, C., Cervera, M. T. & González-Benito, M. E. 2011. Genetic stability analysis of chrysanthemum (*Chrysanthemum* x *morifolium* Ramat) after different stages of an encapsulation–dehydration cryopreservation protocol, *J Plant Physiol* 168: 158–166.

Martin, C., Kremer, C., González, I. & González-Benito, M. E. 2015. Influence of the cryopreservation technique, recovery medium and genotype on genetic stability of mint cryopreserved shoot tips, *Plant Cell, Tissue and Organ Culture* (PCTOC) 122: 185–195.

Martin, C., Senula, A. & González, I., et al. 2013. Genetic identity of three mint accessions stored by different conservation procedures: Field collection, *in vitro* and cryopreservation. *Genet Resour Crop Evol* 60: 243–249.

Matsumoto, T., Sakai, A. & Yamada, K. 1995. Cryopreservation of *in vitro*-grown apical meristems of lily by vitrification, *Plant Cell, Tissue and Organ Culture* 41: 237–241.

Mikuła, A. 2006. Comparison of three techniques for cryopreservation and re-establishment of long-term *Gentiana tibetica* suspension culture, *CryoLetters* 27: 268–282.

Mikuła, A., Fiuk, A. & Rybczyński, J. J. 2005. Induction, maintenance and preservation of embryogenic competence of *Gentiana cruciata* L. cultures, *Acta Biol Crac Series Botanica* 47: 227–236.

Mikuła, A., Niedzielski, M. & Rybczyński, J. J. 2006. The use of TTC reduction assay for assessment of *Gentiana* spp. Cell suspension viability after cryopreservation. *Acta Physiol Plant* 28: 315–324. https://doi.org/10.1007/s11738-006-0027-9

Mikuła, A., Olas, M., Śliwińska, E. & Rybczyński, J. J. 2008. Cryopreservation by encapsulation of *Gentiana* spp. cell suspensions maintains re-growth, embryogenic competence and DNA content, *CryoLetters* 29: 409–418.

Mikuła, A., Tomiczak, K. & Rybczyński, J. J. 2011a. Cryopreservation enhances embryogenic capacity of *Gentiana cruciata* (L.) suspension culture and maintains (epi)genetic uniformity of regenerants. *Plant Cell Rep* 30: 565–574. https://doi.org/10.1007/s00299-010-0970-1

Mikuła, A., Tomiczak, K., Wójcik, A. & Rybczyński, J. J. 2011b. Encapsulation-dehydration method of cryopreservation elevates embryogenesity of *Gentiana kurroo* cell suspension and carrying on genetic stability of its regenerants after cryopreservation. *Acta Hortic*, 908: 143–154. https://doi.org/10.17660/ActaHortic.2011.908.16

Ming-Hua, Y. & Sen-Rong, H. 2010. A simple cryopreservation protocol of *Dioscorea bulbifera* l. Embryogenic calli by encapsulation-vitrification. *Plant Cell, Tissue and Organ Culture* 101: 349–358. https://doi.org/10.1007/s11240-010-9695-7

Mosa, K. A., Ahmed, A. E., Hazem, Y., Kanawati, I. S., Abdullah, A., Hernandez-Sori, L., Ali, M. A. & Vendrame, W. 2023. Insights into cryopreservation, recovery and genetic stability of medicinal plant tissues, *Fitoterapia* 169: 105555.

Mujib, A., Fatima, S. & Malik, M. Q. 2022. Cryo-derived plants through embryogenesis showed same levels of vinblastine and vincristine (anticancer) in *Catharanthus roseus* and had normal genome size, *Scientific Reports* 12 (1): 16635.

Mukherjee, P., Mandal, B. B., Bhat, K. V. & Biswas, A. K. 2009. Cryopreservation of Asian *Dioscorea bulbifera* L. and *D. alata* L. By vitrification: Importance of plant growth regulators, *CryoLetters* 30: 100–111.

Murashige, T. & Skoog, F. 1962. A revised medium for rapid growth and bioassays with tobacco tissue cultures. *Physiol Plant* 15: 473–497. https://doi.org/10.1111/j.1399-3054.1962.tb08052.x

Nadarajan, J. & Pritchard, H. W. 2014. Biophysical characteristics of successful oilseed embryo cryoprotection and cryopreservation using vacuum infiltration vitrification: An innovation in plant cell preservation, *PLoS One* 9 (5): e96169.

Nair, D. S. & Reghunath, B. R. 2008. Effective use of encapsulation-dehydration technique in cryopreserving somatic embryos of butterfly pea (*Clitoria ternatea* L.), *Journal of Herbs Spices & Medicinal Plants 13*: 83–95.

Nausch, H. & Buyel, J. F. 2021. Cryopreservation of plant cell cultures–Diverse practices and protocols, *New Biotechnology* 62: 86–95.

Negahdar, N., Hashemabadi, D. & Kaviani, B. 2021. *In vitro* conservation and cryopreservation of *Buxus sempervirens* L., a critically endangered ornamental shrub. *Russ J Plant Physiol* 68: 661–668. https://doi.org/10.1134/S1021443721040117

Niino, T., Yamamoto, S. & Fukui, K. et al. 2013. Dehydration improves cryopreservation of mat rush (*Juncus decipiens* Nakai) basal stem buds on cryo-plates. *CryoLetters* 34: 549–560

Nishizawa, S., Sakai, A. & Amano, Y., et al. 1993. Cryopreservation of asparagus (*Asparagus officinalis* L.) embryogenic suspension cells and subsequent plant regeneration by vitrification. *Plant Science* 91:67–73.

Ogur, E., Adanacioglu, N., Galatali, S., Ceylan, M. & Kaya, E. 2023. Cryopreservation of M*entha piperita* L. germplasm and confirmation of genetic stability after cryo-storage. *Journal of Animal & Plant Sciences* 33 (2).

Oh, S. Y., Wu, C. H., Popova, E., Hahn, E. J. & Paek, K. Y. 2009. Cryopreservation of *Panax ginseng* adventitious roots. *Journal of Plant Biology* 52: 348–354.

Oliveira, K. E. S. D., Souza, R. A. V. D., Carvalho, L. S. O. & Paiva, L. V. 2022. Influence of ethylene glycol on eucalyptus grandis cryopreservation using the v cryo-plate technique. *Crop Breeding and Applied Biotechnology* 22: e378422210.

Panis, B., Piette, B. & Swennen, R. 2005. Droplet vitrification of apical meristems: A cryopreservation protocol applicable to all Musaceae. *Plant Sci* 168: 45–55. https://doi.org/10.1016/j.plantsci.2004.07.022

Panis, B., Strosse, H., Van den Hende, S. & Swennen, R. 2002. Sucrose preculture to simplify cryopreservation of banana meristem cultures, *Cryo Letters* 23: 375–384.

Parasher, K., Sharma, S., Mukherjee, P. & Qazi, P. H. 2023. Cryopreservation of zygotic embryos of *Podophyllum hexandrum* Royle, an endangered medicinal plant, by vitrification and v cryo-plate techniques, *Cryoletters* 44 (4): 219–228.

Parvathi, K. P., Preetha, T. S. & Hemanthakumar, A. S. 2023. Cryopreservation of shoot meristems of *Baliospermum montanum* (Willd.) Muell. Arg. for long-term conservation and sustainable utilization, *Vegetos* 1–10. https://doi.org/10.1007/s42535-023-00720-3

Pennycooke, J. C. & Towill, L. E. 2000. Cryopreservation of shoot tips from *in vitro* plants of sweet potato (*Ipomoea batatas* (L.) Lam.) by vitrification. *Plant Cell Rep* 19: 733–737. https://doi.org/10.1007/s002999900171

Pettinelli, J. A., Soares, B. O., Cantelmo, L., Garcia, R. O., Mansur, E., Engelmann, F. & Gagliardi, R. F. 2017. Cryopreservation of somatic embryos from *Petiveria alliacea* L. by different techniques based on vitrification. *Vitro Cell Dev Biol Plant* 53: 339–345. https://doi.org/10.1007/s11627-017-9820-y

Popova, E., Kim, H. H. & Paek, K. Y. 2010. Cryopreservation of coriander (*Coriandrum sativum* L.) somatic embryos using sucrose preculture and air desiccation, *Scientia Horticulturae* 124: 522–528.

Prystalov, A., Bobrova, O. & Kuleshova, L. 2021. Application of vacuum infiltration in cryopreservation of isolated grape buds, *Problems of Cryobiology and Cryomedicine* 31 (1): 51–57.

Rahmah, S., Ahmad Mubbarakh, S., Soo Ping, K. & Subramaniam, S. 2015. Effects of droplet-vitrification cryopreservation based on physiological and antioxidant enzyme activities of *Brassidium* shooting star orchid. *The Scientific World Journal*. 2015(1): 961793.

Ray, A. & Bhattacharya, S. 2008. Cryopreservation of *in vitro* grown nodal segments of *Rauvolfia serpentina* by PVS2 vitrification, *Cryoletters* 29: 321–328.

Reyna-López, G. E., Simpson, J. & Ruiz-Herrera, J. 1997. Differences in DNA methylation patterns are detectable during the dimorphic transition of fungi by amplification of restriction polymorphisms, *Molecular and General Genetics* 253: 703–710.

Rungjindamai, C., Thammasiri, K., Chuenboonngarm, N. & Jenjittikul, T. 2010. Cryopreservation of *Alpinia nigra* (Gaertn.) Burtt by vitrification-based technique, *Thai Journal of Botany* 2: 151–154.

Sakai, A. & Engelmann, F. 2007. Vitrification, encapsulation-vitrification and droplet-vitrification: A review, *CryoLetters* 28: 151–172.

Sakai, A., Hirai, D., Niino, T. & Reed, B. 2008. *Plant Cryopreservation: A Practical Guide*. Springer, New York, NY, pp. 33–57.

Sakai, A., Kobayashi, S. & Oiyama, I. 1990. Cryopreservation of nucellar cells of navel orange (*Citrus sinensis* Osb. var. *brasiliensis* Tanaka) by vitrification. *Plant Cell* 9: 30–33. https://doi.org/10.1007/BF00232130

Senula, A., Keller, E. R., Sanduijav, T. & Yohannes, T. 2007. Cryopreservation of cold-acclimated mint (*Mentha* spp.) shoot tips using a simple vitrification protocol, *CryoLetters* 28 (1): 1–12.

Seydi, S., Sedaghathoor, S. & Kaviani, B. 2020. *In vitro* storage techniques of *Fritillaria imperialis* lubra maxima, a wild rare and critically endangered ornamental species, *Chiang Mai University Journal of Natural Sciences* 19: 612–628.

Sharaf, S. A., Shibli, R. A., Kasrawi, M. A. & Baghdadi, S. H. 2012. Cryopreservation of wild shih (*Artemisia herba-alba* Asso.) shoot-tips by encapsulation-dehydration and encapsulation vitrification, *Plant Cell, Tissue and Organ Culture* 108: 437–444.

Sharma, N., Gowthami, R., Devi, S. V., Malhotra, E. V., Pandey, R. & Agrawal, A. 2021. Cryopreservation of shoot tips of *Gentiana kurroo* Royle—a critically endangered medicinal plant of India, *Plant Cell, Tissue and Organ Culture* 144 (1): 67–72.

Sharma, N., Malhotra, E. V., Chandra, R., Gowthami, R., Sultan, S. M., Bansal, S. & Agrawal, A. 2022. Cryopreservation and genetic stability assessment of regenerants of the critically endangered medicinal plant *Dioscorea deltoidea* wall. Ex Griseb. for cryobanking of germplasm, *In Vitro Cellular & Developmental Biology-Plant* 58(4): 521–529.

Sharma, N., Pandey, R. & Gowthami, R. 2020. *In vitro* conservation and cryopreservation of threatened medicinal plants of India. In: Rajasekharan, P. E., Wani, S. H. (eds). *Conservation and Utilization of Threatened Medicinal Plants*. Springer, India, pp. 181–228.

Sharma, N., Satsangi, R. & Pandey, R. 2011. Cryopreservation of shoot tips of *Bacopa monnieri* (L.) Wettst by vitrification technique, *Acta Horticulturae* 908: 283–288.

Sharma, N. & Sharma, B. 2003. Cryopreservation of shoot tips of *Picrorhiza kurroa* Royle ex Benth an indigenous endangered medicinal plant through vitrification, *Cryoletters* 24: 181–190.

Sharma, N., Singh, R., Pandey, R. & Kaushik, N. 2017. Genetic and biochemical stability assessment of plants regenerated from cryopreserved shoot tips of a commercially valuable medicinal herb *Bacopa monnieri* (L.) Wettst, *In Vitro Cell Dev Biol Plant* 53: 346–351.

Shatnawi, M. A. 2011. Multiplication and cryogenic storage of *Artemisia herba-alba*: A medicinal plant, *Food Agriculture and Environment* 9: 340–344.

Shatnawi, M. A., Shibli, R. A., Abu-Romman, S. M., Al-Mazra'awi, M. S., Al Ajlouni, Z. I., Shatanawi, W. A. & Odeh, W. H. 2011. Clonal propagation and cryogenic storage of the medicinal plant *Stevia rebaudiana*, *Spanish Journal of Agricultural Research* 9: 213–220.

Shin, J. H., Kang, D. K., Bae, J. S., Lee, B. H. & Sohn, J. K. 2006. Development of cryopreservation system using shoot-apex in yam (*Dioscorea batatas*), *Journal of Plant Biotechnology* 8 (1): 43–50.

Soejima, K., Shimoda, K., Takeuchi, M., Yamaki, T., Kono, T., Sakurai, H. & Nakazawa, H. 2012. Preliminary report of novel technique for cryopreservation—vacuum- assisted cryoprotectant infiltration, *Cryobiology* 65: 21–26.

Sopalun, K., Kanchit, K. & Ishikawa, K. 2010. Vitrification-based cryopreservation of *Grammatophyllum speciosum* protocorm, *CryoLetters* 31: 347–357.

Su, X. 1991. Methods of cryopreservation of pollen in *Fritillaria thunbergii* Miq, *China Journal of Chinese Materia Medica* 16: 399–401.

Suzuki, M., Akihama, T. & Ishikawa, M. 2005. Cryopreservation of encapsulation gentian axillary buds following 2 step-preculture with sucrose and desiccation. *Plant Cell, Tissue and Organ Culture* 83: 115–121. https://doi.org/10.1007/s11240-005-4854-y

Tahtamouni, R., Shibli, R., Al-Abdallat, A., Al-Qudah, T., Younis, L., Al-Baba, H. & Al-Ruwaiei, H. 2015. *In vitro* conservation and cryopreservation of medicinal and aromatic plants: A review, *Jordan Journal of Agricultural Sciences* 11 (1): 147–167.

Tanaka, D., Niino, T., Isuzugawa, K., Hikage, T. & Uemura, M. 2004. Cryopreservation of shoot apices of *in-vitro* grown gentian plants: Comparison of vitrification and encapsulation-vitrification protocols, *CryoLetters* 25: 167–176.

Tanaka, D., Sakuma, Y., Yamamoto, S., Matsumoto, T. & Niino, T. 2019. Development of the v cryo-plate method for cryopreservation of *in vitro* rakkyo (*Allium chinense* G. Don), *Acta Horticulturae* 1234: 287–292.

Thammasiri, K., Prasongsom, S., Kongsawadworakul, P., Chuenboonngarm, N., Jenjittikul, T., Soonthornchainaksaeng, P., Viboonjun, U. & Muangkroot, A. 2019. Cryopreservation of *Arundina graminifolia* (D. Don) Hochr. seeds using d cryo-plate method. *Acta Hortic.* 1234: 301–308. https://doi.org/10.17660/ActaHortic.2019.1234.39

Thankappan, S. S. & Morawala-Patell, V. 2021. Germplasm conservation techniques and assessment of physiological, biochemical and molecular integrity of indigenous, endangered *Dioscorea prazeri*, *Cryoletters* 42 (3): 168–77.

Towill, L. E. & Bonnart, R. 2003. Cracking in a vitrification solution during cooling or warming does not affect growth of cryopreserved mint shoot tips, *CryoLetters* 24 (6): 341–346.

Uchendu, E. E. & Keller, E. R. J. 2016. Melatonin-loaded alginate beads improve cryopreservation of yam (*Dioscorea alata* and *D. cayenensis*), *CryoLetters* 37 (2): 77–87.

Uchendu, E. E. & Reed, B. M. 2008. A comparative study of three cryopreservation protocols for effective storage of *in vitro*-grown mint (*Mentha* spp), *CryoLetters 29* (3): 181–188.

Uragami, A., Sakai, A., Nagai, M. & Takahashi, T. 1989. Survival of cultured cells and somatic embryos of *Asparagus officinalis* cryopreserved by vitrification, *Plant Cell Rep* 8: 418–421.

Varghese, B. & Naithani, S. C. 2008. Oxidative metabolism-related changes in cryogenically stored neem (*Azadirachta indica* A. Juss) seeds, *Journal of Plant Physiology* 165 (7): 755–65.

Volk, G. M., Maness, N. & Rotindo, K. 2004. Cryopreservation of garlic (*Allium sativum* L.) Using plant vitrification solution 2, *CryoLetters* 25: 219–226.

Volk, G. & Richards, K. 2006. Preservation methods for Jerusalem artichoke cultivars, *HortScience* 41: 80–83.

Wang, M. R., Bi, W. & Shukla, M. R., et al. 2021. Epigenetic and genetic integrity, metabolic stability, and field performance of cryopreserved plants. *Plants* 10: 1889.

Wang, M. R., Lambardi, M., Engelmann, F., Pathirana, R., Panis, B., Volk, G. M. & Wang, Q. C. 2020b. Advances in cryopreservation of in vitro-derived propagules: Technologies and explant sources. *Plant Cell, Tissue and Organ Culture* (2020) 144: 7–20. https://doi.org/10.1007/s11240-020-01770-0

Wang, M. R., Zhang, Z., Zámečník, J., Bilavčík, A., Blystad, D. R., Haugslien, S. & Wang, Q. C. 2020a. Droplet-vitrification for shoot tip cryopreservation of shallot (*Allium cepa* var. *aggregatum*): Effects of PVS3 and PVS2 on shoot regrowth. *Plant Cell, Tissue and Organ Culture* 140: 185–195.

Wang, Y., Lin, K., Liu, Y., Ma, L., Dai, Y., Xu, S. & Wang, X. 2012. Research on cryopreservation of cultured *Fritillaria cirrhosa* D. Don tissues by vitrification, *Journal of Chengdu University* 31: 11–13.

Wilms, H., Rhee, J. H., Rivera, R. L., Longin, K. & Panis, B. 2019. Developing coconut cryopreservation protocols and establishing cryo-genebank at RDA; a collaborative project between RDA and bioversity international, *Acta Hortic* 1234: 343–348.

Xue, S. H., Luo, X. J., Wu, Z. H., Zhang, H. L. & Wang, X. Y. 2008. Cold storage and cryopreservation of hairy root cultures of medicinal plant *Eruca sativa* mill., *Astragalus membranaceus* and *Gentiana macrophylla* pall, *Plant Cell, Tissue and Organ Culture* 92: 251–260.

Yamamoto, S., Fukui, K., & Niino, T. 2011. A new cryopreservation method for vegetatively propagated plant genetic resources using aluminum cryo-plates, *Development Technology* 10, 10–11.

Yamamoto, S. I., Rafique, T., Fukui, K., Sekizawa, K. & Niino, T. 2012. V-cryo-plate procedure as an effective protocol for cryobanks: Case study of mint cryopreservation, *CryoLetters 33* (1): 12–23.

Yamuna, G. 2007. *Studies on Cryopreservation of Spices Genetic Resources.* Ph.D. Thesis, University of Calicut, Kerala, India.

Yamuna, G., Sumathi, V., Geetha, S., Praveen, K., Swapna, N. & Nirmal Babu, K. 2007. Cryopreservation of *in vitro* grown shoots of ginger (*Zingiber officinale* Rosc), *CryoLetters* 28: 241–252.

Yi, J. Y., Lee, G. A., Lee, S. Y. & Chung, J. W. 2014. Development of cryo-banking system of *Lilium* species, *Flower Research Journal* 22: 185–189.

Yong, S. H., Park, K. B., Seol, Y. W., Kim, D. H., Jeong, M. J. & Choi, M. S. 2023. Optimization conditions for cryopreservation of *Nelumbo nucifera* Gaertn. aquatic plant, *Plant Biotechnology Reports* 1–11.

Yue-hua, W. 2011. The key cryopreservation techniques of *Fritillaria cirrhosa* D. Don callus, *Journal of Anhui Agricultural Sciences* 23: 14023–14026.

Zamecnik, J., Faltus, M. & Bilavcik, A. 2021. Vitrification solutions for plant cryopreservation: Modification and properties, *Plants* 10 (12): 2623.

Zarghami, R., Pirseyedi, S. M., Hasrak, S. H. & Pakdaman, B. S. 2008. Evaluation of genetic stability in cryopreserved *Solanum tuberosum*, *African Journal of Biotechnology* 7: 2798–2802.

Zhang, J. M., Han, L., Lu, X. X., Volk, G. M., Xin, X., Yin, G. K., He, J. J., Wang, L. & Chen, X. L. 2017. Cryopreservation of Jerusalem artichoke cultivars using an improved droplet-vitrification method, *Plant Cell, Tissue and Organ Culture* 128: 577–587.

Zhu, Y., Chen, F., Xue, J., Zhang, A., Sheng, W. & Song, Y. 2011. Study on preservation of *Fritillaria anhuiensis* by vitrification *in vitro*, *China Journal of Chinese Materia Medica* 36 (18): 2462–2464.

Zilani, R. A., Lee, H., Popova, E. & Kim, H. 2022. *In vitro* multiplication and cryopreservation of *Penthorum chinense* shoot tips, *Life* 12 (11): 1759.

14 Optimization of Secondary Metabolites Production in Medicinal Plants

Aditi Sharma, Priti Krishna, Manu Pant, and Kumud Pant

14.1 INTRODUCTION

The term metabolite, typically constrained to small molecules, refers to intermediates and products of metabolism. In addition to performing catalytic activity on their own, metabolites play many other roles in signal transduction, enzyme modulation, molecular interactions, and other crucial processes (Baker and Rutter, 2023). Both primary and secondary metabolites make up this group of metabolites. As a result of primary metabolism, which aids in the maintenance of vital physiological processes, primary metabolites are recognized as the principal products produced by plants (Pagare et al., 2015). Plants can also biosynthesize more than a million other compounds, referred to as "secondary metabolites", or SMs, which do not directly support the primary functions of the plants (Verpoote, 1998).

The term "secondary metabolites" was coined by Kossel (1891). These compounds play complex roles in plant metabolism. They are broadly categorized as terpenoids, phenolic compounds, and nitrogen- and sulfur-containing compounds. SMs play a variety of roles in determining plant responses to biotic stresses like pests and pathogens (Naoumkina et al., 2008) and abiotic stressors such as UV radiation (Xu et al., 2008) and stimulating seed germination in parasitic weeds (Bouwmeester et al., 2021). There is increasing evidence that SMs secreted by roots can mediate plant-soil microbiome interaction (Pang et al., 2021). SMs are also linked to the attraction attributes of plants, thereby playing a significant role in pollination. Besides being responsible for attractive colors and scents in plants (González-Teuber and Heil, 2009), SMs are also the active and volatile organic compounds in nectar and pollen, which not only attract pollinators but also act as pathogen repellants for the plants (Sosenski and Parra-Tabla, 2019).

Plant SMs are of special interest as natural products with medicinal value. With increasing health awareness among the masses, there has been a significant rise in the use of herbal drugs as being more efficient, safe, and cost-effective as compared to synthetic drugs. Undoubtedly, the herbal medicine industry has proven to be one of the rapidly emerging industries worldwide. In this scenario, the dependence on medicinally active plants has increased tremendously, primarily for their secondary metabolites imparting therapeutic properties to the plants (Thomford et al., 2018; Nandy et al., 2022).

Plant cell and tissue culture play a very important role in the controlled production of various SMs accomplishing industrial demands (Vijaya et al., 2010). In the past few years, the development of different cell culture methods has shown a massive improvement in the production of SMs with medicinal properties as compared to sourcing these from intact plants (Chandran et al., 2020). The current chapter gives an insight into the role of various plant tissue culture techniques in SM production.

14.2 TYPES OF PLANT SECONDARY METABOLITES

The SMs are classified according to their biosynthetic routes and structure and can be broadly divided into three major groups: (1) Flavonoids, allied phenolic, and polyphenolic compounds; (2) terpenoids;

DOI: 10.1201/b23374-14

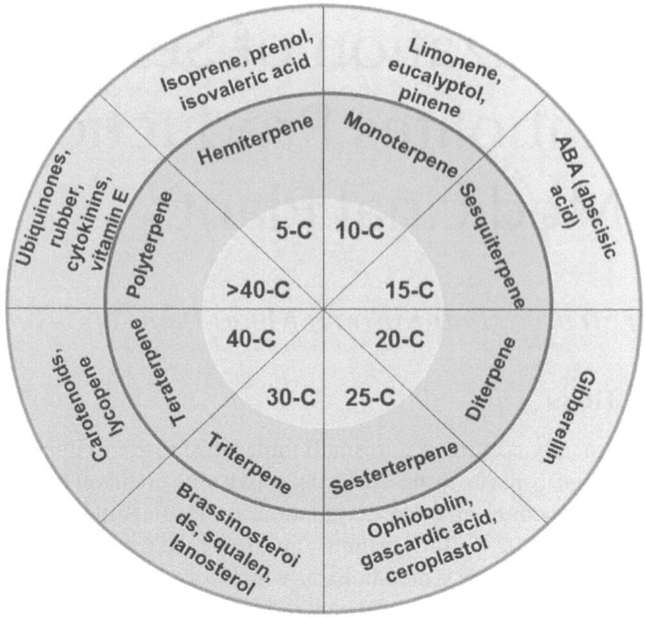

FIGURE 14.1 Classification of terpenoids.

and (3) nitrogen-containing alkaloids and sulfur-containing compounds. These compounds are linked to primary metabolites by building blocks and biosynthetic enzymes (Vuolo et al., 2019).

14.2.1 Terpenoids

Mostly consisting of rubber, essential oils, and carotenoid pigments, these are thought to be the largest group of SMs produced as an outcome of plant defense mechanisms against insects that consume them. Terpenes or isoprenoids make up the majority of substances, particularly essential oils, and have molecular structures with 2-methylbuta-1,3-diene (isoprene units) carbon backbones that can be reconfigured into cyclic structures (Hyldgaard et al., 2012). The classification of terpenes based on structural variation mostly depends on the number of isoprene units (Khayyat and Roselin, 2018). These terpenoids can be further classified into different categories shown in Figure 14.1.

14.2.2 Phenolic Compounds

Phenolic compounds synthesized by various medicinal plants exhibit properties like antioxidant, anti-inflammatory, anticarcinogenic, bactericidal, antiseptic, and antihelmintic and may also protect against oxidative stress and some diseases. These phenolic compounds (flavonoids, allied phenolic, and polyphenolic compounds) are more widely distributed in plants and are derived from pentose phosphate, shikimate, and phenylpropanoid pathways in plants (Pengelly, 2020). They can be further classified into different categories. The phenolic compounds can be further classified into different categories (Zhang et al., 2015) as shown in Figure 14.2.

14.2.3 Sulfur- and Nitrogen-Containing Compounds

These include all phytoalexins, alkaloids, defensins, glycosylate, and other substances that are generated as a defense mechanism against microbial infections. Glutathione, glucosinolaytes, phytoalexins, thionins, defensins, and allinin come under the category of sulfur-containing compounds,

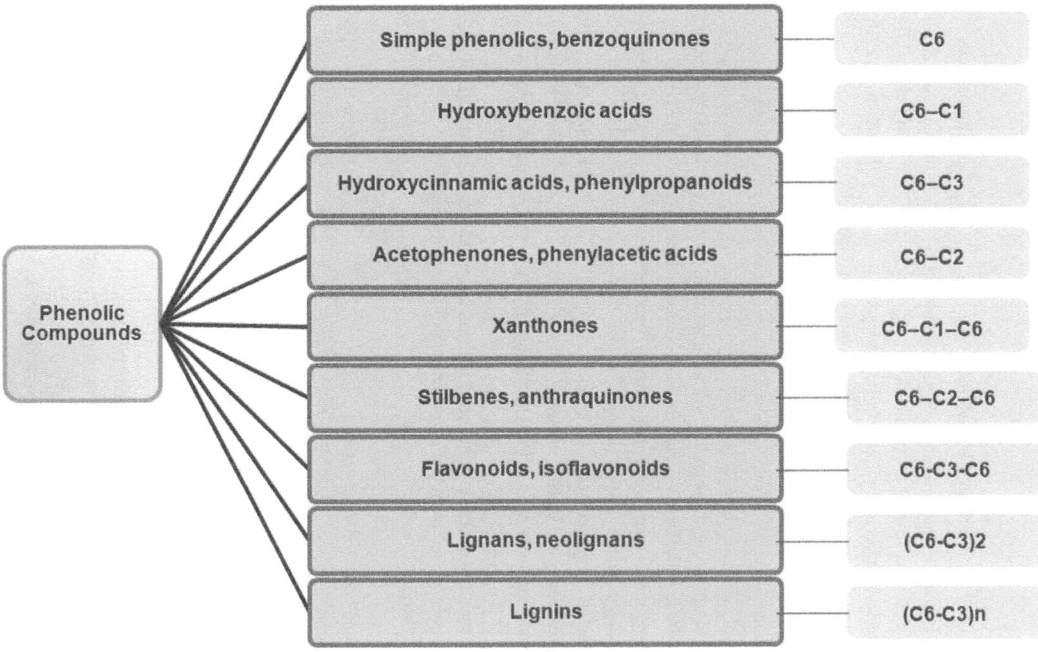

FIGURE 14.2 Classification of phenolic compounds found in plants.

while alkaloids (e.g. cocaine), cynogenic glucosides (dhurrin), and non-protein amino acids (e.g. canavanin) are the nitrogen-containing compounds. It has been shown that these defense mechanisms produced chemicals that have various qualities that might be exploited as nutritional supplements or in several pharmaceutical applications, for example, the anticancerous properties of terpenoids (Yang et al., 2018).

14.3 TISSUE CULTURE STRATEGIES TO IMPROVE SECONDARY METABOLITE PRODUCTION

Plant cell, tissue, and organ cultures are capable of producing and accumulating SMs identical to those produced by the parent plant in nature. *In vitro* plant materials have regained attention as potential factories for the production of SMs due to the exceeding demands for natural and renewable products in the current market. This has generated novel opportunities to conduct studies on the role of *in vitro* techniques in producing SMs (Karuppusamy, 2009). Discussed next are some optimization methods for secondary metabolite production in medicinal plants under *in vitro* conditions.

14.3.1 Explant Selection

Different culture methods have been tried for enhanced secondary metabolite production under *in vitro* conditions (Monisha et al., 2022). These include cell culture, shoot culture, adventitious root culture, etc. (Table 14.1).

14.3.1.1 Cell and Organ Culture

The utilization of plant cell cultures in the production of SMs is gaining much attention in the pharmaceutical, cosmetics, and nutraceutical industries (Bayraktar, 2019). The method involves the separation of *in vitro*-raised cell clumps (or calli) into individual cells on a suspension medium

TABLE 14.1
Secondary Metabolite Production from Medicinal Plants Using Different Culture Types

S. No.	Secondary Metabolites	Plant Species	Common Name	Culture Type	Medicinal Importance	References
1	Pinostrobin	*Boesenbergia rotunda*	Fingerroot	Adventitious root culture	Rheumatism, muscle pain, fever reduction, gastrointestinal disorders, flatulence, dyspepsia, peptic ulcers	Azhar et al., 2018; Yusuf et al., 2018
2	Kaempferol and myricetin	*Gynura procumbens*	Longevity spinach	Adventitious root culture	Antihypertensive, antiglycemic, antioxidant, anticancer, anti-inflammatory, antimicrobial, fertility enhancement, organ protective	Faizah et al., 2018
3	Anthraquinones, phenolic, flavonoid	*Polygonum multiforum*	Tuber fleece flower	Adventitious root culture	Treatment of backache, dizziness, liver diseases, graying of hair, constipation	Ho et al., 2018
4	Anthraquinones, phenolic	*Rubia tinctorum*	Common madder	Adventitious root culture	Treatment of kidney and bladder stones	Bicer et al., 2017
5	Alkaloids, berberine	*Argemone mexicana*	Mexican prickly poppy	Shoot culture	Treatment of tumors, warts, skin diseases, inflammation, rheumatism, jaundice, leprosy, microbial infections, malaria	Monforte-González et al., 2019
6	Ginseng	*Atropa belladonna*	Deadly nightshade	Plant cell culture	Pain reliever, muscle relaxer, anti-inflammatory, treat menstrual problems, peptic ulcers, motion sickness	Raj and Saudagar, 2019
7	Phenolic compounds	*Panax quinquefolium*	American ginseng	Cell suspension culture	Boost immunity, anticancerous	Kochan et al., 2019
8	Anthraquinone and phenols	*Rubia tinctorum*	Common madder	Adventitious root culture	Treatment of kidney and bladder stones	Demirci et al., 2021
9	Atropine (alkaloids)	*Hyoscyamus muticus*	Egyptian henbane	Cell suspension culture	Treat pain and insomnia, used as local anesthetic	Abdelazeez et al., 2022
10	Rebaudioside-A content	*Stevia rebaudiana*	Candy leaf	Shoot culture	Antihypertensive, antiobesity, antidiabetic, antioxidant, anticancer, anti-inflammatory, antimicrobial improvement of kidney function	Laha et al., 2023
11	Berberine betacyanins	*Thalictrum minus*	Lesser meadow rue	Immobilized cell culture	Anticancer, antibiotic, anti-inflammatory	Kobayashi et al., 1987
12	Echinacea polysaccharides	*Echinacea purpurea*	Purple coneflower	Cell culture	Immunostimulant, anti-inflammatory	Wagner et al., 1988
13	Podophyllotoxin	*Podophyllum* spp.	Mayapple	Cell/organ culture	Anticancerous	Giri and Narasu, 2000
14	Hypericin	*Hypericum perforatum*	St. John's wort	Cell suspension	Treatment stomach ulcers, biliary disorders, bronchial inflammation, genitourinary system, diabetes, obesity, migraines	Xu et al., 2011
15	Bacoside A	*Bacopa monnieri*	Water hyssop	Shoot	Memory improvement, insomnia, epilepsy	Naik et al., 2010; Praveen and Murthy, 2012
16	Withanolide A	*Withania somnifera*	Ashwagandha	Cell suspension/hairy root	Treatment of Alzheimer's, Parkinson's, cerebral ischemia, epilepsy, tardive dyskinesia, stress, and managing drug addiction	Praveen and Murthy, 2010
17	Xanthotoxin	*Ammi majus*	Bishop's weed	Shoot	Treatment of psoriasis and vitiligo	Purohit et al., 1995
18	Tanshinone	*Salvia castanea*	Salvia	Hairy root	Treatment of cardiovascular and cerebrovascular disease, angina pectoris, myocardial infarction, chronic hepatitis, liver fibrosis	Chodisetti et al., 2015

and utilizing them for enhanced metabolite production; it has been used for many medicinal plant species. The selection of the cell line, elicitation, metabolic engineering, and immobilization are some of the factors that need to be standardized for SM production from cells (Valdiani et al., 2019; Ozyigit et al., 2023).

It has also been reported that in many cases it becomes imperative to culture the plant organ that is specialized in the production of SM, for example, root or shoot. In such cases, *in vitro* SM production by any other method is not as competent as the original tissue culture. Root-tissue culture of *Panax ginseng* and leaf cultures of *Hypericum perforatum* have been reported to be essential for enhanced accumulation of SMs as compared to other methods (Hussain et al., 2012).

14.3.1.2 Callus Culture

Callus culture involves the culturing of dedifferentiated plant cells induced on media with relatively high auxin concentrations or a combination of auxin and cytokinin under *in vitro* conditions. Callus cultures can be embryogenic, containing differentiated embryogenically competent cells with the potential of regenerating a complete plant, while the non-embryogenic callus cultures are utilized for the production of SMs that retain homogenous clumps of dedifferentiated cells (Ptak et al., 2013; Filová, 2014).

Many active ingredients and SMs have been successfully produced utilizing the callus culture approach such as the production of medicinally vital phenolic and flavonoid compounds including caffeic acid, gallic acid, rutin, myricetin, iso-rhamnetin, quercetin, chlorogenic, vanillic acid, and luteolin from *Lycium barbarum* (Karakas, 2020); indole alkaloids, gingerol, shogaol including picrinine acetylechitamine, tubotaiwine, and echitamine from *Alstoniascholaris* (Jeet et al., 2020); phenylethanoid (salidroside, tyrosol), phenylpropanoid (rosavin and rosarin), and phenolic acids (gallic acid, coumaric acid, and cinnamic acid) from *Rhodiola imbricata* (Rattan et al., 2020); phenolic compounds (ferulic acid, isoquercitrin, rutin, quercetin, quercetin-7-O-glucoside, and luteolin) from *Hyssopus officinalis* (Babich et al., 2021); and phenylethanoids and bioactive compounds, including 1,2-benzenedicarboxylic acid (phthalic acid), 3,7,11,15- tetramethyl-2-hexadecen-1-ol, 2-hexadecen-1-ol-3,7,11,15- tetrametil, hexadecenoic acid methyl ester (methyl palmitate), n-hexadecenoic acid (palmitic acid), 9,12-octadecadienoic acid methyl ester, 9,12,15-octadecatrienoic acid methyl ester, phytol, octadecanoic acid methyl ester (methyl stearate), 9,12,15-octadecatrienoic acid (linolenic acid), and squalene, from *Mucuna pruriens* (Sweetlin and Daniel, 2020).

14.3.1.3 *In Vitro* Transformed Root and Shoot Cultures

Plant cell transformation by infection with *Agrobacterium* is a well-established method for the transformation of roots and shoots. In this process, attachment of the bacterium to the wounded plant cell leads to the insertion of the T-DNA fragment of the root-inducing (Ri) plasmid of *Agrobacterium rhizogenes* into the plant genome (Chandra, 2012). The T-DNA genes code for enzymes involved in the synthesis of opines and phytohormones auxin and cytokinin. Opines are used as carbon and nitrogen sources for the bacterium, while auxin and cytokinin trigger the differentiation of plant cells into hairy roots (Subramoni et al., 2014).

Hairy roots are more efficient producers of SMs as compared to cell and callus cultures due to their genetic stability, better and longer biosynthetic capacity, and higher biomass yield. For specific SMs, plants that naturally produce those metabolites are chosen for hair root cultures. In addition, transgenic hairy root cultures with modified biosynthesis pathways have also been used to produce valuable metabolites (Gutierrez-Valdes et al., 2020). There are numerous successful examples of SM production through hairy root culture in medicinal species, some of which include *Linum mucronatum* (Samadi et al., 2014), *Bacopa monnierri* (Bansal et al., 2015), *Tylophora indica* (Cyriac et al., 2020); *Atropa belladonna* (Biswas et al., 2023), and *Taxas baccata* (He et al., 2023).

In vitro shoot cultures are another crucial approach for the production of SMs. *Agrobacterium tumefaciens* is used to infect the aerial sections of the plants, resulting in the development of transgenic shoots (shooty teratomas). According to certain studies, shoot cultures exhibit characteristics that are almost identical to those of hairy root cultures in terms of SM production, genetic stability, and the link between growth and SM production (Massot et al., 2000).

14.3.2 Standardization of Factors for Secondary Metabolite Production

For optimal culture development and extraction of SMs from cultured explants, various factors like media optimization, culture conditions, etc., need to be standardized (Karuppusamy, 2009).

14.3.2.1 Media Optimization

Nutritional requirements for optimal growth of tissue *in vitro* may vary with the species. Tissues from different parts of a plant may have different requirements for satisfactory growth (Murashige and Skoog, 1962). Plant tissue culture media provides the inorganic nutrients and a carbohydrate (sucrose is most common) to replace the carbon that the plant normally fixes from the atmosphere by photosynthesis. To improve growth, many media also include trace amounts of certain organic compounds, notably vitamins, and plant growth regulators (Rai et al., 2022). For SM production, media optimization may involve the addition of new sources, the replacement of existing sources, the removal of some sources, and testing different combinations of one element with another.

A variety of media optimization methods have been studied to improve SM production. Applications of auxin (2,4-dichlorophenoxyacetic acid—2,4-D) and cytokinin (benzyl amino purine, BAP) have been observed to induce manipulation in the metabolome of callus culture in *Bidenspilosa* (Asteraceae), with increased production of chlorogenic acids, including caffeoyl and feruloyl derivatives of quinic acid (Ramabulana et al., 2021). Krstić-Milošević et al., (2021) for the very first time evaluated the establishment of xanthones production in differentiated shoot cultures of the endangered *Gentianella lutescens* (Gentianeceae) by modifying the concentration of sucrose, sorbitol, and abiotic elicitors salicylic acid (SA), jasmonic acid (JA), and methyl jasmonate (MeJA). A combination of the components of the Gamborg (Gamborg et al., 1968) and Murashige and Skoog (Murashige and Skoog, 1962) culture media has been modified by changing the nitrogen concentration and the NO_3:NH_4 ratio in carrot (*Daucus carota*, Apiaceae), which affected the accumulation of carotenoids in the carrot plant (Oleszkiewicz et al., 2021). The optimization of growth and production of total phenolics has been achieved by combining NO_3, as well as the NH_4:K^+ ratio and the cytokinins, BAP, and naphthalene acid (NAA) on callus culture of *Rhodiolarosea* (Crassulaceae) (Erst et al., 2022).

14.3.2.2 Optimizing Culture Conditions

It involves monitoring factors like pH, temperature, incubation time, darkness and light levels, and the impact of various compounds on the media. It is possible to consider various light and dark photoperiod circumstances and adjust incubation accordingly (Kapoor et al., 2018). The yield of SM production is greatly affected by the change in temperature, photoperiod, and pH (Rao and Ravishankar, 2002).

14.3.2.2.1 Temperature

The normal temperature range for growing cultured cells and inducing callus tissues lies between 17°C and 25°C. However, this range may vary among different plant species (Rao and Ravishankar, 2002). Comparing the cell cultures of *Cathranthus roseus* grown at 27°C, a 12-fold higher production of crude alkaloids was observed at 16°C rather than at 27°C (Courtois and Guern, 1980). Similarly, it was observed that a reduction in cultivation temperature has increased the total fatty acid content per cell in dry weight (Toivonen et al., 1992). On the contrary, compared to 24°C or 28°C, cell cultures of tobacco maintained at 32°C produced an improved yield of ubiquinone (Ikeda et al., 1977). Another study showed that while maintaining *Digitalis lanata* cell cultures, 19°C conditions favored biotransformation of digitoxin to digoxin, while a temperature of 32°C and formation of purpurea glycoside were favored (Kreis and Reinhard, 1992). In hairy root cultures of

P. ginseng, the temperature range was shown to influence the production of ginsenoside (Szakiel et al., 2011), while in the cell cultures, ginsenosides synthesis was optimized at 25 ± 1 °C (Murthy et al., 2014).

14.3.2.2.2 Illumination

Illumination shows a great impact on the yield of SM production. It has been observed in many medicinal plants such as *Prunella vulgaris* that light stress improved the production of SMs (Fazal et al., 2016). A novel area of research that is economically beneficial for the pharmaceutical and nutraceutical industry is being explored with the aid of light elicitation in improving the production of many SMs. For this, different light sources such as ultraviolet radiation (UV), light-emitting diodes (LED), and fluorescence have been reportedly utilized as efficient elicitors for producing medicinally important SMs (Ahmad et al., 2016, Dou et al., 2019). It has also been observed that the anthocyanin accumulation in cell cultures of *D. carota* and *Vitis* hybrids was affected by the strong stimulation of light (Seitz and Hinderer, 1988). The composition of sesquiterpenes has also been reported to be affected by illumination in callus cultures of *Marticaria chamomilla* (Mulder-Krieger et al., 1988). The accumulation of monoterpenes was reportedly prompted in callus cultures of *Citrus limon* by the exclusion of light (Mulder-Krieger et al., 1988). Contrastingly, the cultures of *Podophyllum hexandrum* produced better friable calli and higher podophyllotoxin when grown in the dark than in the light (Chattopadhyay et al., 2002).

14.3.2.3 Precursor Feeding

Based on the understanding of biosynthetic pathways, precursor feeding has been realized to be an important method to improve the generation of SMs in plant cell cultures (Namdeo et al., 2007). Palacio et al. (2011) found that feeding 0.5 mM of L-phenylalanine on cell cultures of the *Larrea divaricata* Cav. plant led to a 2.2-fold rise in nor-dihydroguaiaretic acid (NDGA) in comparison to the control.

14.3.2.4 Immobilization

Immobilization is defined as the localization of cells in a matrix to improve particular catalyst or process features (Rodríguez-Restrepo and Orrego, 2020). It has been suggested that better yields of SMs could arise from higher cell densities, especially when production times from frequent usage are considered and contamination risk and product inhibition are reduced (Saeed and Iqbal, 2013). Consequently, cell immobilization has been considered an effective method for SM production.

14.3.3 Contemporary/Advanced Methods

14.3.3.1 Elicitation

According to Namdeo (2007), elicitation is the most realistic possible technique for improving the output of desired SMs from plant, cell, and organ systems. Plants frequently accumulate SMs in response to various stressors, elicitors, or signal molecules. This method alters the metabolic pathways of medicinal plants for stimulating bioactive compounds while applying different *in vitro* culture procedures. This has led to the process of elicitation being widely used and economically feasible for large-scale SM production. Elicitors act as signaling molecules, and the recognition of generated signals is achieved by membrane-bound receptors on plant cells. This activation of signal transduction pathways changes the expression of regulatory players, which increases the synthesis and accumulation of phytochemicals (Miladinova-Georgieva et al., 2022). A variety of physical, chemical, and microbiological variables function as abiotic or biotic elicitors, eventually increasing the production of SMs (Ghorbanpour et al., 2014). Biotic elicitors include proteins (glycoproteins, cellulose, pectolyase, cryptogein, oligandrin, lectins, or agglutinins), carbohydrates (oligogalacturonides, chitin, chitosan), plant growth promoting-rhizobacteria (PGPR) (*Pseudomonas putida,*

Azospirillum fluorescens, *Allohizobium*, *Azorhizobium*, *Bradyrhizobium*, *Rhizobium*), fungus (*Glomus*, *Trichoderma*, *Azotobacter*, *Azospirillum*, *Phytophthora*, *Protomyces*, *Mucor*, *Fusarium*, *Claviceps*), and hormones (auxin, cytokinin, gibberellic acid, SA, MeJA, JA). While the abiotic elicitors include heavy metals like cadmium, zinc, copper, etc.; temperature (low and high both); light (intensity, irradiance, and UV radiation); salinity (osmotic, ionic, and oxidative stress); and drought (water deficit, osmotic, oxidative) (Scarpa et al., 2022). There are multiple examples of medicinal plant species such as *Fagonia indica* (Khan et al., 2017), *Polygonum multiflorum* (Ho et al., 2018), *Rubia tinctorum* (Bicer et al., 2017), and *Taxus globose* (Barrales-Cureño et al., 2022) which utilize specific types of elicitors, producing different types of plant SMs.

14.3.3.2 Nanoparticles for SMs

Nanoparticles are defined as particles with a size between 1 and 100 nm. These have been applied in plant tissue culture to improve bioactive metabolites and increase plant development and other factors (Wang et al., 2016; Ruttkay-Nedecky et al., 2017). Certain examples of metal oxide nanoparticles can be used for the improvement of SM production in plants such as titanium oxide nanoparticles in *Cicer arietinum* (Mohammed, 2015) and silver nanoparticles in *Artemisia annua* and fenugreek (Zhang et al., 2013; Jasim et al., 2017). In *P. vulgaris* L., improved production of phenolics, flavonoids, and total protein content along with induction of callus proliferation was observed using silver (AgNP) and silver-gold (AgNP-AuNP) nanoparticles in the ratios of 1:2 and 2:1, respectively (Fazal et al., 2016). Also, increased concentration of withanolide was observed in *Withania somnifera* using silver and zinc nanoparticles (AgNP + ZnNP) (Singh et al., 2019).

14.3.3.3 Metabolic Engineering

Through overexpression studies that allow modification of biosynthetic pathways, metabolic engineering gives an entirely innovative outlook to study the expression of genes involved in the manufacturing of SMs (O'Connor, 2015). This entails manipulating the genes that encode crucial and rate-limiting enzymes in the biosynthesis pathways as well as reviewing enzymatic reactions and biosynthetic processes at the gene, transcriptome, and proteomic levels (Lu et al., 2016). Shi et al. (2020) created a transgenic root line overexpressing geraniol-10-hydroxylase and secologanin synthetase genes to increase the production of the anticancer alkaloid camptothecin in the hairy roots of *Ophiorrhiza pumil*. *Ophiorrhiza* species are used in cases of snakebite, stomatitis, ulcers, wounds, gastropathy, leprosy, and amenorrhea (Krishnan et al., 2014, Prabha and Karuppusamy, 2018, Midhu et al., 2019, Martins and Nunez, 2015). Camptothecin production in the stem and roots has been reported to reach up to 2 $mg.g^{-1}$ DW under optimal growing conditions, with the modified hairy root line producing 3.5 $mg.g^{-1}$ of the substance (Lee et al., 2020). In *A. belladonna* L., production of hyoscyamine has been achieved by targeting the AbH6H gene through *A. tumefaciens* EHA105–mediated freeze-thaw (Zeng et al., 2021). In addition, improved production of phytocannabinoids has been observed while targeting the PDS gene through *Agrobacterium*-mediated transformation (Zhang et al., 2021).

14.3.4 Large-scale Production of SMs through Bioreactors

Cell, tissue, and organ cultures can now be used to yield SMs from medicinal plants, which has been made possible by the recent advancements in biotechnology research. It is crucial to generate SMs in large amounts to satisfy the rising demands for medicinal compounds in the market worldwide. It is difficult to adopt a laboratory-scale manufacturing method for profitable measures of production in this situation. Bioreactor technology has been extensively used for scaling up the production of bioactive chemicals from plants. The whole procedure takes place in a bioreactor that can be regulated to produce high-quality yields in huge quantities, making the bioreactor system far more cost-effective than growing plants in nature (Rahmat and Kang, 2019). The production of secondary metabolites follow a series of steps which can be studied through Figure 14.3.

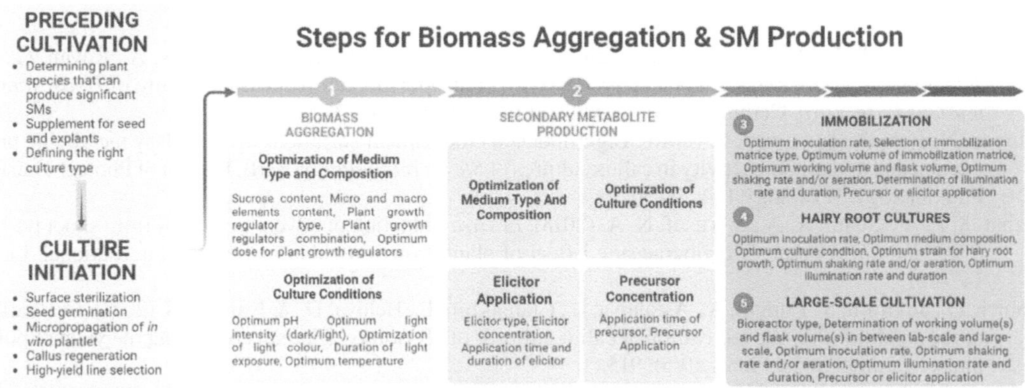

FIGURE 14.3 Stepwise description of secondary metabolite production.

To scale up the volume of SMs produced *in vitro*, there are two primary factors, optimization and engineering, that affect plant morphogenesis and biomass accumulation (Srikantan and Srivastava, 2018). There is a need for improved mass transfers of dissolved oxygen and nutrients during the scaling-up procedure, which could be maintained by controlling the oxygen supply, CO_2 exchange, and pH, as well as by optimization of minerals, carbohydrates, and growth regulators in the liquid medium (Gatti et al., 2017). By utilizing the cultivation strategy which involves the use of bioreactors for plant cell and hairy root cultures, one can achieve ease in scale-up, thorough mixing, low shear stress, and a suitable support system for organ cultures (Honda et al., 2001). Besides, the use of bioreactor systems provides rapid growth of cultures and bulk transfer of nutrients and gases leading to scaling up of the SM production and ultimately saving labor costs (Yancheva et al., 2019). Medicinal plants have been used to scale up the production of SMs utilizing different types of bioreactors such as *H. perforatum* L. (shake flask culture), *Lessertia frutescens* L. (glass ballon-type bubble bioreactors), *Aloe barbadensis* (continuous immersion bioreactor), and *Eucalyptus grandis* × *E. urophylla* (temporary immersion bioreactors) (Cui et al., 2010; Shaik et al., 2011; Cardarelli et al., 2014; Businge et al., 2017).

14.4 CONCLUSION

The SMs found in medicinal plants are a reserve of bioactive chemicals that have a variable range of pharmacological effects. A large variety of these SMs can be cultivated by utilizing plant cell and tissue culture techniques, which are highly beneficial in improving the production. Traditional methods, such as media optimization and strain improvement, as well as *in vitro* culture practices, including cell suspension culture, callus culture, hairy root culture, shoot culture, immobilization, and advanced methods, such as the use of elicitors, nanoparticles, metabolic engineering, and biotransformation have all demonstrated their ability to be dependable, affordable, and able to sustainably meet the increasing demands of industry. These alternative methods, being independent of geographic and climatic conditions, prove to be highly economical and sustainable due to the ability to conserve plants that are endangered to the point of near-extinction. Also, these methods control the diversity of SMs produced by different plants. The mass manufacturing of SMs can be accomplished using bioreactors after employing any of these methods. But the choice of which method to use will rely on the type of plant being utilized and the SMs to be produced. Over the last decade, great progress has been made in this area, although the overall production of SMs has not been improved as the scaling up faces many difficulties leading to poor commercial success. To ensure focused production with a high and steady yield of the desired secondary molecules, further research is required to understand the biosynthetic pathways and the epigenetic mechanisms regulating the biosynthesis of SMs.

REFERENCES

Abdelazeez, W. M. A., Anatolievna, K. Y., Zavdetovna, K. L., Damirovna, A. G., El-Dis, A., Rayan, G., & Arnoldovna, T. O. (2022). Enhanced productivity of atropine in cell suspension culture of *Hyoscyamus muticus* L. *In Vitro.* Cellular & Developmental Biology-Plant 1–13.

Ahmad, N., Rab, A., & Ahmad, N. (2016). Light-induced biochemical variations in secondary metabolite production and antioxidant activity in callus cultures of *Stevia rebaudiana* (Bert). Journal of Photochemistry Photobiology: B Biology 154: 51–56.

Azhar, S. Z. A., Ghani, K. A., & Yusuf, N. A. (2018). *In vitro* induction of adventitious root from shoot bud of *Boesenbergia rotunda* (Zingiberaceae): Effect of plant growth regulators. Science International 30(1): 147–151.

Babich, O., Sukhikh, S., Pungin, A., Astahova, L., Chupakhin, E., Belova, D., & Ivanova, S. (2021). Evaluation of the conditions for the cultivation of callus cultures of *Hyssopus officinalis* regarding the yield of polyphenolic compounds. Plants 10(5): 915.

Baker, S. A., & Rutter, J. (2023). Metabolites as signalling molecules. Nature Reviews Molecular Cell Biology 24(5): 355–374.

Bansal, M., Kumar, A., & Reddy, M. (2015). Production of bacoside A, a memory enhancer from hairy root cultures of *Bacopa monnieri* (L.) Wettst. Journal of Applied Research on Medicinal and Aromatic Plants 2: 92–101.

Barrales-Cureño, H. J., Ramos Valdivia, A. C., & Soto Hernández, M. (2022). Increased production of taxoids in suspension cultures of *Taxus globosa* after elicitation. Future Pharmacology 2(1): 45–54.

Bayraktar, M. (2019). Micropropagation of *Stevia rebaudiana* Bertoni using RITA bioreactor. Horticulture Science 54(4): 725–731.

Bicer, P. O., Demirci, T., Asci, O. A., & Baydar, N. G. (2017). Effects of methyl jasmonate and caffeic acid applications on secondary metabolite production in madder (*Rubia tinctorum*) root cultures. Indian Journal of Pharmaceutical Education and Research 51: 508–512.

Biswas, D., Chakraborty, A., Mukherjee, S., & Ghosh, B. (2023). Hairy root culture: A potent method for improved secondary metabolite production of solanaceous plants. Frontiers in Plant Science 14: 1197555.

Bouwmeester, H., Li, C., Thiombiano, B., Rahimi, M., & Dong, L. (2021). Adaptation of the parasitic plant lifecycle: Germination is controlled by essential host signaling molecules. Plant Physiology 185(4): 1292–1308.

Businge, E., Trifonova, A., Schneider, C., Rödel, P., & Egertsdotter, U. (2017). Evaluation of a new temporary immersion bioreactor system for micropropagation of cultivars of eucalyptus, birch and fir. Forests 8(6): 196.

Cardarelli, M., Suárez, C. M. C., & Colla, G. (2014). Influence of ozone treatments on *in vitro* propagation of *Aloe barbadensis* in continuous immersion bioreactor. Industrial Crops and Products 55: 194–201.

Chandra, S. (2012). Natural plant genetic engineer *Agrobacterium rhizogenes*: Role of t-DNA in plant secondary metabolism. Biotechnology Letters 34: 407–415.

Chandran, H., Meena, M., Barupal, T., & Sharma, K. (2020). Plant tissue culture as a perpetual source for production of industrially important bioactive compounds. Biotechnology Reports 26: 450.

Chattopadhyay, S., Srivastava, A. K., Bhojwani, S. S., & Bisaria, V. S. (2002). Production of podophyllotoxin by plant cell cultures of *Podophyllum hexandrum* in bioreactor. Journal of Bioscience and Bioengineering 93(2): 215–220.

Chodisetti, B., Rao, K., Gandi, S., & Giri, A. (2015). Gymnemic acid enhancement in the suspension cultures of *Gymnema sylvestre* by using the signaling molecules—Methyl jasmonate and salicylic acid. *In Vitro* Cellular & Developmental Biology-Plant 51: 88–92.

Courtois, D., & Guern, J. (1980). Temperature response of *Catharanthus roseus* cells cultivated in liquid medium. Plant Science Letters 17(4): 473–482.

Cui, X.-H., Chakrabarty, D., Lee, E. J., & Paek, K. Y. (2010). Production of adventitious roots and secondary metabolites by *Hypericum perforatum* L. In a bioreactor. Bioresource Technology 101: 4708–4716.

Cyriac, A., Thomas, T., & Thomas, T. D. (2020). Tylophorine: Sources, properties, applications and biotechnological production. Plant-derived Bioactives: Production, Properties and Therapeutic Applications 167–176.

Demirci, T., Aras Ascı, Ö, & Göktürk Baydar, N. (2021). Influence of salicylic acid and L-phenylalanine on the accumulation of anthraquinone and phenolic compounds in adventitious root cultures of madder (*Rubia tinctorum* L.). Plant Cell, Tissue and Organ Culture (PCTOC) 144(2): 313–324.

Dou, H., Niu, G., & Gu, M. (2019). Pre-harvest UV-B radiation and photosynthetic photon flux density interactively affect plant photosynthesis, growth, and secondary metabolites accumulation in basil (*Ocimum basilicum*) plants. Agronomy 9: 434.

Erst, A. A., Petruk, A. A., Erst, A. S., Krivenko, D. A., Filinova, N. V., Maltseva, S. Y., Kulikovsiky, M. S., & Banaev, E. V. (2022). Optimization of biomass accumulation and production of phenolic compounds in callus cultures of *Rhodiola rosea* L. Using design of experiments. Plants 11(1): 124.

Faizah, H., Tanjung, M., Purnobasuki, H., & Manuhara, Y. S. W. (2018). Biomass and flavonoid production of *Gynura procumbens* (L.). Merr adventitious root culture in baloon-type bubble-bioreactor influenced by elicitation. Asian Journal of Plant Sciences 17(2): 107–119.

Fazal, H., Abbasi, B. H., Ahmad, N., & Ali, M. (2016). Elicitation of medicinally important antioxidant secondary metabolites with silver and gold nanoparticles in callus cultures of *Prunella vulgaris* L. Applied Biochemistry and Biotechnology 180: 1076–1092.

Fazal, H., Abbasi, B. H., Ahmad, N., Ali, S. S., Akbar, F., & Kanwal, F. (2016). Correlation of different spectral lights with biomass accumulation and production of antioxidant secondary metabolites in callus cultures of medicinally important *Prunella vulgaris* L. Journal of Photochemistry Photobiology B: Biology 159: 1–7.

Filová, A. (2014). Production of secondary metabolities in plant tissue cultures. Research Journal of Agricultural Sciences 46(1): 236–245.

Gamborg, O. L., Miller, R., & Ojima, K. (1968). Nutrient requirements of suspension cultures of soybean root cells. Experimental Cell Research 50(1): 151–158.

Gatti, E., Sgarbi, E., Ozudogru, E. A., & Lambardi, M. (2017). The effect of Plantform™ bioreactor on micropropagation of *Quercus robur* in comparison to a conventional *in vitro* culture system on gelled medium, and assessment of the microenvironment influence on leaf structure. Plant Biosystems 151: 1129–1136.

Ghorbanpour, M., Khavazi, K., & Hatami, M. (2014). Chemical compositions and anti-microbial activity of *Salvia officinalis* L. Essential oil under rhizobacteria (*Pseudomonas fluorescens* and *Pseudomonas putida*) inoculation. European Journal of Soil Biology (in press).

Giri, A., & Lakshmi Narasu, M. (2000). Production of podophyllotoxin from *Podophyllum hexandrum*: A potential natural product for clinically useful anticancer drugs. Cytotechnology 34: 17–26.

González-Teuber, M., & Heil, M. (2009). Nectar chemistry is tailored for both attraction of mutualists and protection from exploiters. Plant Signaling & Behavior 4(9): 809–813.

Gutierrez-Valdes, N., Häkkinen, S. T., Lemasson, C., Guillet, M., Oksman-Caldentey, K. M., Ritala, A., & Cardon, F. (2020). Hairy root cultures—A versatile tool with multiple applications. Frontiers in Plant Science 11: 504243.

He, J., Plácido, J. P. A., Pateraki, I., Kampranis, S., Favero, B. T., & Lütken, H. (2023). Hairy root induction of *Taxus baccata* L. By natural transformation with *Rhizobium rhizogenes*. Horticulturae 9, 4. https://doi.org/10.3390/horticulturae9010004

Ho, T. T., Lee, J. D., Jeong, C. S., Paek, K. Y., & Park, S. Y. (2018). Improvement of biosynthesis and accumulation of bioactive compounds by elicitation in adventitious root cultures of *Polygonum multiflorum*. Applied Microbiology and Biotechnology 102: 199–209.

Honda, H., Liu, C., & Kobayashi, T. (2001). Large-scale plant micropropagation. Advances in Biochemical Engineering/Biotechnology 72: 157–182.

Hussain, M. S., Fareed, S., Ansari, S., Rahman, M. A., Ahmad, I. Z., & Saeed, M. (2012). Current approaches toward production of secondary plant metabolites. Journal of Pharmacy & Bioallied Sciences 4(1): 10–20.

Hyldgaard, M., Mygind, T., & Meyer, R. L. (2012). Essential oils in food preservation: Mode of action, synergies, and interactions with food matrix components. Frontiers in Microbiology 3: 12.

Ikeda, T., Matsumoto, T., & Noguchi, M. (1977). Effects of inorganic nitrogen sources and physical factors on the formation of ubiquinone by tobacco plant cells in suspension culture. Agricultural and Biological Chemistry 41(7): 1197–1201.

Jasim, B., Thomas, R., Mathew, J., & Radhakrishnan, E. K. (2017). Plant growth and diosgenin enhancement effect of silver nanoparticles in fenugreek (*Trigonella foenum-graecum* L.). Saudi Pharmaceutical Journal 25(3): 443–447.

Jeet, A., Singh, Y., Singh, P., Nimoriya, R., Bilung, C. J., Kanojiya, S., & Mishra, D. K. (2020). Strategies for indole alkaloids enrichment through callus culture from *Alstonia scholaris* (L.) R. Br. Plant Growth Regulation 90: 383–392.

Kapoor, S., Raghuvanshi, R., Bhardwaj, P., Sood, H., Saxena, S., & Chaurasia, O. P. (2018). Influence of light quality on growth, secondary metabolites production and antioxidant activity in callus culture of *Rhodiola imbricata* Edgew. Journal of Photochemistry and Photobiology B: Biology 183: 258–265.

Karakas, F. P. (2020). Efficient plant regeneration and callus induction from nodal and hypocotyl explants of goji berry (*Lycium barbarum* L.) and comparison of phenolic profiles in calli formed under different combinations of plant growth regulators. Plant Physiology Biochemistry and Biotechnology 146: 384–391.

Karuppusamy, S. (2009). A review on trends in production of secondary metabolites from higher plants by *in vitro* tissue, organ and cell cultures. Journal of Medicinal Plants Research 3: 1222–1239.

Khan, T., Abbasi, B. H., Khan, M. A., & Azeem, M. (2017). Production of biomass and useful compounds through elicitation in adventitious root cultures of *Fagonia indica*. Industrial Crops and Products 108: 451–457.

Khayyat, S. A., & Roselin, L. S. (2018). Recent progress in photochemical reaction on main components of some essential oils. Journal of Saudi Chemical Society 22(7): 855–875.

Kobayashi, Y., Fukui, H., & Tabata, M. (1987). An immobilized cell culture system for berberine production by *Thalictrum minus* cells. Plant Cell Reports 6: 185–186.

Kochan, E., Szymańska, G., Grzegorczy-Karolak, I., Szymczyk, P., & Sienkiewicz, M. (2019). Ginsenoside and phenolic compounds in hydromethanolic extracts of American ginseng cell cultures and their antioxidant properties. Acta Societatis Botanicorum Poloniae 88(4): 3638.

Kossel, A. (1891). Ueber die chemischeZusammensetzung der Zelle. Du Bois-Reymond's Archiv/Arch Anat Physiol Abt 278: 181–186.

Kreis, W., & Reinhard, E. (1992). 12β-Hydroxylation of digitoxin by suspension-cultured *Digitalis lanata* cells: Production of digoxin in 20-litre and 300-litre air-lift bioreactors. Journal of Biotechnology 26(2–3): 257–273.

Krishnan, S. A., Dileepkumar, R., Nair, A. S., & Oommen, O. V. (2014). Studies on neutralizing effect of *Ophiorrhiza mungos* root extract against *Daboia russelii* venom. Journal of Ethnopharmacology 151: 543–547.

Krstić-Milošević, D., Banjac, N., Janković, T., Vinterhalter, D., & Vinterhalter, B. (2021). *Gentianella lutescens* subsp. *carpatica* J. Holub.: Shoot propagation *in vitro* and effect of sucrose and elicitors on xanthones production. Plants 10(8): 1651.

Laha, S., Subrahmanyeswari, T., Verma, S. K., Kamble, S. N., Singh, S., Bhattacharyya, S., & Gantait, S. (2023). Biogenic synthesis, characterization and application of silver nanoparticles as biostimulator for growth and rebaudioside—A production in genetically stable stevia (*Stevia rebaudiana* Bert.) Under *in vitro* conditions. Industrial Crops and Products 197: 116520.

Lee, J. Y., Hiyama, M., Hikosaka, S., & Goto, E. (2020). Effects of concentration and temperature of nutrient solution on growth and camptothecin accumulation of *Ophiorrhiza pumila*. Plants 9(6): 793.

Lu, X., Tang, K., & Li, P. (2016). Plant metabolic engineering strategies for the production of pharmaceutical terpenoids. Frontiers in Plant Science 7: 1647.

Martins, D., & Nunez, C. V. (2015). Secondary metabolites from Rubiaceae species. Molecules 20: 13422–13495.

Massot, B., Milesi, S., Gontier, E., Bourgaud, F., & Guckert, A. (2000). Optimized culture conditions for the production of furanocoumarins by micropropagated shoots of *Ruta graveolens*. Plant Cell, Tissue and Organ Culture 62: 11–19.

Midhu, C. K., Hima, S., Binoy, J., & Satheeshkumar, K. (2019). Influence of incubation period on callus tissues for plant regeneration in *Ophiorrhiza pectinata* Arn. through somatic embryogenesis. Proceedings of National Academy of Sciences USA 89: 1439–1446.

Miladinova-Georgieva, K., Geneva, M., Stancheva, I., Petrova, M., Sichanova, M., & Kirova, E. (2022). Effects of different elicitors on micropropagation, biomass and secondary metabolite production of *Stevia rebaudiana* Bertoni—A review. Plants 12(1): 153.

Mohammed, A. E. (2015). Green synthesis, antimicrobial and cytotoxic effects of silver nanoparticles mediated by *Eucalyptus camaldulensis* leaf extract. Asian Pacific Journal of Tropical Biomedicine 5(5): 382–386.

Monforte-González, M., Serrano-Gamboa, J. G., Guízar-González, C., Miranda-Ham, M. L., & Vázquez-Flota, F. A. (2019). Alkaloid synthesis is coupled to shoot morphogenesis in *Argemone mexicana* L.(Papaveraceae) *in vitro* cultures. *In Vitro* Cellular & Developmental Biology-Plant 55: 695–701.

Monisha, K., Riddhi, M., & Parth, P. (2022). Strategies for secondary metabolite production: A review. Asian Journal of Plant Science and Research 12(5).

Mulder-Krieger, T., Verpoorte, R., Svendse, A., & Scheffer, J. (1988). Production of essential oils and flavours in plant cell and tissue cultures. A review. Plant Cell Tissue and Organ Culture 13: 85–114.

Murashige, T., & Skoog, F. (1962). A revised medium for rapid growth and bio assays with tobacco tissue cultures. Physiologia Plantarum 15(3): 473–497.

Murthy, H. N., Lee, E. J., & Paek, K. Y. (2014). Production of secondary metabolites from cell and organ cultures: Strategies and approaches for biomass improvement and metabolite accumulation. Plant Cell Tissue Organ Culture 118: 1–16.

Naik, P. M., Manohar, S. H., Praveen, N., & Murthy, H. N. (2010). Effects of sucrose and pH levels on *in vitro* shoot regeneration from leaf explants of *Bacopa monnieri* and accumulation of Bacoside A in regenerated shoots. Plant Cell, Tissue and Organ Culture (PCTOC) 100: 235–239.

Namdeo, A. G. (2007). Plant cell elicitation for production of secondary metabolites: A review. Pharmacognosy Reviews 1(1): 69–79.

Namdeo, A. G., Jadhav, T. S., Rai, P. K., Gavali, S., & Mahadik, K. R. (2007). Phcog rev.: review article precursor feeding for enhanced production of secondary metabolites: A review. Pharmacognosy Reviews 1: 227–231.

Nandy, S., Das, T., Tudu, C. K., Mishra, T., Ghorai, M., Gadekar, V. S., & Dey, A. (2022). Unravelling the multi-faceted regulatory role of polyamines in plant biotechnology, transgenics and secondary metabolomics. Applied Microbiology and Biotechnology 106(3): 905–929.

Naoumkina, M. A., He, X., & Dixon, R. A. (2008). Elicitor-induced transcription factors for metabolic reprogramming of secondary metabolism in *Medicago truncatula*. BMC Plant Biology 8(1): 1–14.

O'Connor, S. E. (2015). Engineering of secondary metabolism. Annual Review of Genetics 49: 71–94.

Oleszkiewicz, T., Kruczek, M., & Baranski, R. (2021). Repression of carotenoid accumulation by nitrogen and NH_4+ supply in carrot callus cells *in vitro*. Plants 10(9): 1813.

Ozyigit, I. I., Dogan, I., Hocaoglu-Ozyigit, A., Yalcin, B., Erdogan, A., Yalcin, I. E., Cabi, E., & Kaya, Y. (2023). Production of secondary metabolites using tissue culture-based biotechnological applications. Frontiers in Plant Science 14: 1132555.

Pagare, S., Bhatia, M., Tripathi, N., Pagare, S., & Bansal, Y. K. (2015). Secondary metabolites of plants and their role: Overview. Current Trends in Biotechnology and Pharmacy 9(3): 293–304.

Palacio, L., Cantero, J. J., Cusidó, R., & Goleniowski, M. (2011). Phenolic compound production by *Larrea divaricata* Cav. Plant cell cultures and effect of precursor feeding. Process Biochemistry 46(1): 418–422.

Pang, Z., Chen, J., Wang, T., Gao, C., Li, Z., Guo, L., & Cheng, Y. (2021). Linking plant secondary metabolites and plant microbiomes: A review. Frontiers in Plant Science 12: 621276.

Pengelly, A. (2020). The constituents of medicinal plants: An introduction to the chemistry And therapeutics of herbal medicine. Routledge.

Prabha, G., & Karuppusamy, S. (2018). Phytochemical profile and radical scavenging activity of alcoholic extract of *Ophiorrhiza radicans* Gardner (Rubiaceae)—a rare plant of Southern Western Ghats of India. Trends in Biosciences 11: 1572–1576.

Praveen, N., & Murthy, H. N. (2010). Production of withanolide-A from adventitious root cultures of *Withania somnifera*. Acta Physiologiae Plantarum 32: 1017–1022.

Praveen, N., & Murthy, H. N. (2012). Synthesis of withanolide A depends on carbon source and medium pH in hairy root cultures of *Withania somnifera*. Industrial Crops and Products 35(1): 241–243.

Ptak, A., Tahchy, A., Skrzypek, E., Wójtowicz, T., & Laurain-Mattar, D. (2013). Influence of auxins on somatic embryogenesis and alkaloid accumulation in *Leucojum aestivum* callus. Open Life Sciences 8(6): 591–599.

Purohit, M., Pande, D., Datta, A., & Srivastava, P. S. (1995). Enhanced xanthotoxin content in regenerating cultures of *Ammi majus* and micropropagation. Planta Medica 61(5): 481–482.

Rahmat, E., & Kang, Y. (2019). Adventitious root culture for secondary metabolite production in medicinal plants: A review. Journal of Plant Biotechnology 46(3): 143–157.

Rai, A. C., Kumar, A., Modi, A., & Singh, M. (Eds.). (2022). Advances in plant tissue culture: Current developments and future trends. Academic Press.

Raj, S., & Saudagar, P. (2019). Plant cell culture as alternatives to produce secondary metabolites. Natural bio-active compounds: Volume 3: Biotechnology, bioengineering, and molecular approaches. Springer 265–286.

Ramabulana, A. T., Steenkamp, P. A., Madala, N. E., & Dubery, I. A. (2021). Application of plant growth regulators modulates the profile of chlorogenic acids in cultured *Bidens pilosa* cells. Plants 10(3): 437.

Rao, S. R., & Ravishankar, G. A. (2002). Plant cell cultures: Chemical factories of secondary metabolites. Biotechnology Advances 20(2): 101–153.

Rattan, S., Sood, A., Kumar, P., Kumar, A., Kumar, D., & Warghat, A. R. (2020). Phenylethanoids, phenylpropanoids, and phenolic acids quantification vis-à-vis gene expression profiling in leaf and root derived callus lines of *Rhodiola imbricata* (Edgew.). Industrial Crops and Products 154: 112708.

Rodríguez-Restrepo, Y. A., & Orrego, C. E. (2020). Immobilization of enzymes and cells on lignocellulosic materials. Environmental Chemistry Letters 18: 787–806.

Ruttkay-Nedecky, B., Krystofova, O., Nejdl, L., & Adam, V. (2017). Nanoparticles based on essential metals and their phytotoxicity. Journal of Nanobiotechnology 15(1): 1–19.

Saeed, A., & Iqbal, M. (2013). Loofa (*Luffa cylindrica*) sponge: Review of development of the biomatrix as a tool for biotechnological applications. Biotechnology Progress 29(3): 573–600.

Samadi, A., Jafari, M., Nejhad, N. M., & Hossenian, F. (2014). Podophyllotoxin and 6-methoxy podophyllotoxin production in hairy root cultures of *Linum mucronatum* ssp. *mucronatum*. Pharmacognosy Magazine 10(38): 154–160.

Scarpa, G. M., Prota, V., Schianchi, N., & Manunta, F. (2022). *In vitro* cultures for the production of secondary metabolites. Secondary Metabolites-Trends and Reviews. IntechOpen.

Seitz, H. U., & Hinderer, W. (1988). Anthocyanins. Phytochemicals in Plant Cell Cultures 49–76.

Shaik, S., Singh, N., & Nicholas, A. (2011). Comparison of the selected secondary metabolite content present in the cancer-bush *Lessertia* (Sutherlandia) frutescens l. Extracts. African Journal of Traditional, Complementary, and Alternative Medicines 8: 429–434.

Shi, M., Gong, H., Cui, L., Wang, Q., Wang, C., Wang, Y., & Kai, G. (2020). Targeted metabolic engineering of committed steps improves anti-cancer drug camptothecin production in *Ophiorrhiza pumila* hairy roots. Industrial Crops and Products 148: 112277.

Singh, R., Singh, D. P., Gupta, P., Jain, P., Mishra, T., Kumar, A., & Shirke, P. A. (2019). Nanoparticles alter the withanolide biosynthesis and carbohydrate metabolism in *Withania somnifera* (Dunal). Industrial Crops and Products 127: 94–109.

Sosenski, P., & Parra-Tabla, V. (2019). Secondary metabolites: attracting pollinators. eLS 1–9.

Srikantan, C., & Srivastava, S. (2018). Bioreactor design and analysis for large-scale plant cell and hairy root cultivation. Hairy Roots: An Effective Tool of Plant Biotechnology 147–182.

Subramoni, S., Nathoo, N., Klimov, E., & Yuan, Z. C. (2014). Agrobacterium tumefaciens responses to plant-derived signaling molecules. Frontiers in Plant Science 5: 322.

Sweetlin, P., & Daniel, R. R. (2020). Determination of bioactive compounds in ethanolic extract of callus derived from *Mucuna pruriens* using gas chromatography and mass spectroscopic technique. Journal of Natural Remedies 21(7 (S2)): 11–16.

Szakiel, A., Paczkowski, C., & Henry, M. (2011). Influence of environmental abiotic factors on the content of saponins in plants. Phytochemistry Reviews 10: 471–491.

Thomford, N. E., Senthebane, D. A., Rowe, A., Munro, D., Seele, P., Maroyi, A., & Dzobo, K. (2018). Natural products for drug discovery in the 21st century: Innovations for novel drug discovery. International Journal of Molecular Sciences 19(6): 1578.

Toivonen, L., Laakso, S., & Rosenqvist, H. (1992). The effect of temperature on hairy root cultures of *Catharanthus roseus*: Growth, indole alkaloid accumulation and membrane lipid composition. Plant Cell Reports 11: 395–399.

Valdiani, A., Hansen, O. K., Nielsen, U. B., Johannsen, V. K., Shariat, M., Georgiev, M. I., & Abiri, R. (2019). Bioreactor-based advances in plant tissue and cell culture: Challenges and prospects. Critical Reviews in Biotechnology 39(1): 20–34.

Verpoote, R. (1998). Exploration of nature's chemodiversity: The role of secondary metabolites as leads in drug development. Drug Discovery Today 3(5): 2328.

Vijaya, S. N., Udayasri, P. V., Aswani, K. Y., Ravi, B. B., Phani, K. Y., & Vijay, V. M. (2010). Advancements in the production of secondary metabolites. Journal of Natural Products 3: 112–123.

Vuolo, M. M., Lima, V. S., & Junior, M. R. M. (2019). Phenolic compounds: Structure, classification, and antioxidant power. Bioactive Compounds 33–50.

Wagner, H., Stuppner, H., Schäfer, W., & Zenk, M. (1988). Immunologically active polysaccharides of *Echinacea purpurea* cell cultures. Phytochemistry 27(1): 119–126.

Wang, P., Lombi, E., Zhao, F. J., & Kopittke, P. M. (2016). Nanotechnology: A new opportunity in plant sciences. Trends in Plant Science 21(8): 699–712.

Xu, C., Sullivan, J. H., Garrett, W. M., & Caperna, T. J. (2008). Impact of solar ultraviolet-B on the proteome in soybean lines differing in flavonoid contents. Phytochemistry 69: 38–48.

Xu, M., Yang, B., Dong, J., Lu, D., Jin, H., Sun, L., & Xu, X. (2011). Enhancing hypericin production of hypericum perforatum cell suspension culture by ozone exposure. Biotechnology Progress 27(4): 1101–1106.

Yancheva, S., Georgieva, L., Badjakov, I., Dincheva, I., Georgieva, M., Georgiev, V., & Kondakova, V. (2019). Application of bioreactor technology in plant propagation and secondary metabolite production. Journal of Central European Agriculture 20(1): 321–340.

Yang, L., Wen, K. S., Ruan, X., Zhao, Y. X., Wei, F., & Wang, Q. (2018). Response of plant secondary metabolites to environmental factors. Molecules 23(4): 762.

Yusuf, N. A., Rahim, N. S. M., Azhar, S. Z. A., Ghani, K. A., Sommano, S., & Khalid, N. (2018). Adventitious root cultures of *Boesenbergia rotunda* as a source of pinostrobin. International Journal of Advanced Science, Engineering and Information Technology 8: 377–383.

Zeng, L., Zhang, Q., Jiang, C., Zheng, Y., Zuo, Y., Qin, J., & Deng, H. (2021). Development of *Atropa belladonna* L. Plants with high-yield hyoscyamine and without its derivatives using the CRISPR/Cas9 system. International Journal of Molecular Sciences 22(4): 1731.

Zhang, B., Cai, J., Duan, C. Q., Reeves, M. J., & He, F. (2015). A review of polyphenolics in oak woods. International Journal of Molecular Sciences 16(4): 6978–7014.

Zhang, B., Zheng, L. P., Yi Li, W., & Wen Wang, J. (2013). Stimulation of artemisinin production in *Artemisia annua* hairy roots by ag-SiO2 core-shell nanoparticles. Current Nanoscience 9(3): 363–370.

Zhang, X., Xu, G., Cheng, C., Lei, L., Sun, J., Xu, Y., & Su, J. (2021). Establishment of an agrobacterium-mediated genetic transformation and CRISPR/Cas9-mediated targeted mutagenesis in hemp (Cannabis sativa l). Plant Biotechnology Journal 19(10): 1979–1987.

15 Nanotechnological Interventions in Medicinal Plant Tissue Culture

Aarushi Gautam, Manu Pant, Abhilasha Mishra, and Gaurav Pant

15.1 INTRODUCTION

In the past decade, nanomaterials have become essential to strengthen various research fields. They have gained enormous popularity due to their plasticity, high stability, and target selectivity (Shahcheraghi et al., 2022). Nanotechnology has also become an essential part of biotechnology and is often referred to as 'nanobiotechnology' or 'bionanotechnology' where nanoparticles (NPs) are integrated into the biotechnological processes to enhance their output. In the biotechnological aspect, NPs have successful applications in drug delivery, cancer immunotherapy, cancer detection, gene modification, gene sequencing, bioremediation, the cosmetic industry, food, agriculture, etc. Plant biotechnology is the field that deals with improving or introducing new desirable traits in medicinal plants through plant tissue culture (PTC) and genetic engineering techniques. PTC is the technique of raising plants *in vitro* under aseptic conditions from a small part of the plant known as an explant. PTC has proven to be the best alternative to propagating medicinal plants that are difficult to propagate via conventional methods or are overexploited to the state of being endangered or extinct. This is further upgraded by the inclusion of NPs in the mass propagation of the medicinal plants.

NPs have immense potential and have proven beneficial for the micropropagation of medicinal plants. Many reports have been published advocating the employment of NPs in PTC to enhance *in vitro* growth and advancement of medicinal plants as shown in Figure 15.1. For instance, Zafar et al. (2016) reported that MS medium augmented with zinc oxide (ZnO) NPs in varying concentrations from 1 mg/L to 20 mg/L led to the appearance of thin white roots, and callus and roots were induced when the stem of *Brassica nigra* L., valued for its antiparkinsonian and other cognitive-enhancing properties, was inoculated on the MS medium containing ZnO NPs at 10 mg/L concentration. Another study conducted by Spinoso-Castillo et al. (2017) on a plant with high antioxidant properties, *Vanilla planifolia*, reported that MS medium containing silver (Ag) NPs at the concentrations of 25 mg/L and 50 mg/L initiated the creation of reactive oxygen species (ROS) and increased antioxidant capacity, lipid peroxidation, and total phenolic content. Many more reports have been published on applying NPs regarding various aspects of PTC of medicinal plants, as mentioned in Table 15.1. This chapter entails the most commonly used NPs in PTC and their applications in medicinal plant propagation.

15.2 COMMONLY UTILIZED NPs FOR *IN VITRO* PROPAGATION OF MEDICINAL PLANTS

15.2.1 Silver Nanoparticles (AgNPs)

Silver NPs are the most commonly sought-after NPs due to their interesting physiological properties. They are primarily known for their robust toxicity against various microbes (Elechiguerra

DOI: 10.1201/b23374-15

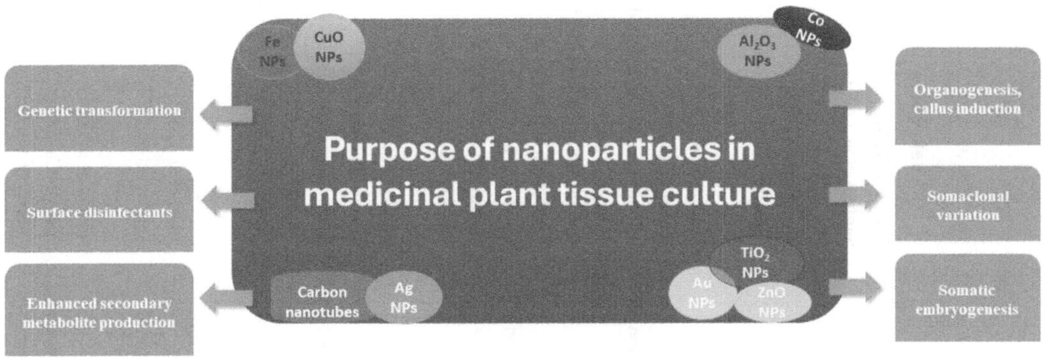

FIGURE 15.1 Applications of NPs in medicinal plant tissue culture.

et al., 2005). AgNPs are also popular for their nutritional and antimicrobial properties in the tissue culture of medicinal plants. The application of AgNPs in the PTC was first reported by Abdi et al. (2008) in *Valeriana officinalis*, a medicinally important herb. They designed experiments to evaluate the antimicrobial activity of nanosilver (NS) on *V. officinalis* where the stem explants were treated with NS solution before and after surface sterilization at three different concentrations (25, 50, 100 mg/L) for different durations (30, 60, and 180 minutes). The best results were observed when the explants were inoculated on MS medium supplemented with 5 mg/L Kn and 0.1 mg/l NAA and the highest rate of disinfected explants (89%) when treated with NS solution after surface sterilization at 100 mg/L for 180 minutes. It was reported that biologically synthesized AgNPs improved seed germination and growth of *Boswellia ovalifoliolata* trees, which have been used in Indian medicine (Ayurveda) to treat osteoarthritis, asthma, bronchitis, diabetes, etc. (Savithramma et al., 2012). AgNPs have also shown positive effects in somatic embryogenesis, callus induction, genetic transformation, and organogenesis (Mahendran et al., 2019). An experiment conducted by Tung et al. (2022) noted that the shoots regenerated and elongated from protocorm-like bodies (PLBs) of *Anthurium andraeanum* 'Tropical' when transferred on ¼ MS medium containing 0.5 mg/L NAA, 1 g/L activated charcoal, and 7 g/L AgNPs resulted in efficiently promoting the plantlet growth.

15.2.2 Gold Nanoparticles (AuNPs)

Gold NPs (Au) are commonly used NPs due to their low toxicity, inert nature, and biocompatibility (Hammami and Alabdallah, 2021). In PTC, AuNPs have shown great efficiency in enhanced somatic embryogenesis, DNA delivery, seed germination, and seedling growth and increased protein content and antioxidant enzyme activity. When *Linum usitatissimum* L. (common name: flax) calli and regenerated cells were subjected to the direct delivery of Ag and AuNPs at 1 mg/L concentration, it was observed that AuNPs enhanced somatic embryogenesis, regeneration zones, and callus growth when compared to AgNPs (Kokina et al., 2013). Vijayakumar et al. (2010) reported that sharp carbon-supported AuNPs efficiently delivered the GUS gene at 200–2000 ng concentration in *Leucaena leucocephala*, a well-known medicinal tree popularly known for its anticancer, anti-inflammatory, antihistaminic, antiandrogenic, and antidiabetic properties (Meena Devi et al., 2013; Zayed and Samling, 2016; Zayed et al., 2019). A study governed by Fazal et al. (2016) reported that Au, Ag, and AgAuNPs in combination with or without NAA positively affected *Prunella vulgaris* calli. AgAu (1:2) and AgAu (2:1) NPs along with NAA (2 mg/L) in MS medium improved the callus proliferation rate to 100% when collated to the control (95%). Furthermore, supplementing AuNPs (30 μg/L) in MS medium with or without NAA (2 mg/L) boosted the biomass aggregation twofold in the callus during the log phase of the cells.

TABLE 15.1
Function of Nanoparticles in the *in Vitro* Propagation of Medicinal Plants

Plant Species	Nanoparticle	Optimal Concentration	Effect on Plant	References
Aloe vera	Ag	0.63 mg/L	Improved aloin in cell suspension cultures	Raei et al. (2014)
Araucaria excelsa R. Br. Var. glauca	Ag	200–400 mg/L	Decreased bacterial contamination	Sarmast et al. (2011)
Artemisia annua	Ag-SiO_2	900 mg/L	Increased artemisinin content in hairy root cultures	Zhang et al. (2013)
Bacopa monnieri	Ag	16×10^{-3} 16×10^{-10}	Decreased the contamination	Kalsaitkar et al. (2014)
Brassica juncea	Ag	25–400 mg/L	Improved seedling development	Sharma et al. (2012)
Brassica napus	Au (magnetic)	7.5–30 µg/L	Transported *GUS* gene to protoplasts	Hao et al. (2013)
Calendula officinalis	$AgNO_3$	0.3–1.2 mg/L	Enhanced fresh and dry wt. of the callus and secondary metabolite production	Al-Oubaidi and Mohammed-Ameen, 2014
Cicer arietinum	TiO_2	0.5–6 mg/L	Enhanced flavonoid and phenolic production	Al-Oubaidi and Kasid, 2015
Cynodon dactylon	Ag	100–200 mg/L	Ag+ thymol in combination reduced bacterial contamination	Taghizadeh and Solgi, 2014
Hordeum vulgare	TiO_2	10–60 µg/mL	Improved callus induction and callus growth, reduced bacterial contamination	Mandeh et al. (2012)
Hyoscyamus reticulatus	Fe_3O_4	450–3600 mg/L	Increased hyoscyamine and scopolamine content in hairy root cultures	Moharrami et al. (2017)
Leucaena leucocephela	Sharp carbon-supported Au	200–2000 ng	Transported *GUS* gene	Vijayakumar et al. (2010)
Linum usitatissimmum L.	C	1–3 g/L	Improved no. of tetraploid cells and methylation	Kokina et al. (2012)
	Au/Ag	500 mg/L	Improved somaclonal variation, callus induction, and development	Kokina et al. (2013)
Mentha longifolia	Cu Co	0.5 mg/L 0.8 mg/L	Enhanced shoot number, number of internodes, height of shoots, and essential oil production	Talankova-Sereda et al. (2016)
Prunella vulgaris	Ag Au AgAu	30 µg/L	Increased callus multiplication, protein production, flavonoids and phenolics content, antioxidant activity	Fazal et al. (2016)
Stevia rebaudiana	ZnO	0.1–1000 mg/L	Increased shoot induction, steviol glycoside content	Javed et al. (2017)
Tecomella undulata	Ag	5–80 mg/L	Enhanced rate of shoot formation, no. of shoots per explant, and survival rate of plantlets	Aghdaei et al. (2012)
Valeriana officinalis	Ag	25–100 mg/L	Decreased bacterial contamination	Abdi et al. (2008)
Vanilla planifolia	Ag	25–200 mg/L	Decreased bacterial contamination, enhanced shoot induction, antioxidant activity	Spinoso-Castillo et al. (2017)
Verbena bipinnatifida Nutt.	$CuSO_4$	5–15 µg/L	Enhanced shoot and root growth, fresh weight, and total phenolic production	Genady et al. (2016)

15.2.3 Silicon Nanoparticles (SiNPs)

Silicon (Si) is the second most bountiful element after oxygen found in the lithosphere. Although silicon is regarded as a non-essential element for plants, reports suggest that silicon fortifies the physical barrier of plants by accumulating on plant cell walls and boosts the capability of disease resistance and tolerance (Ma and Yamaji, 2008). Reports also suggest that Si and SiNPs lower oxidative stress by preparing defense reactions under biotic and abiotic stress (Zhu and Gong, 2014; Debona et al., 2017; Sousa et al., 2019; El-Shetehy et al., 2020). Hussain et al. (2013) also reported that SiNPs have a minute particle size and straightforwardly infiltrate plant cell walls and organelles. Cokul Raj et al. (2023) studied the impact of SiNPs on the *in vitro* flowering and pollen viability of the phytochemically rich aromatic herb *Vitex negundo* L. *In vitro* shoots were supplied with shoot proliferation medium containing MS + 0.5 mg/L BAP + 0.25mg/L IAA and SiNPs (50 mg/l–300 mg/L). Results showed that in morphometric evaluation, SiNPs at 100 mg/L concentration stimulated flower bud induction within 2 weeks of culture. Incorporation of SiNPs in the *in vitro* culture of *V. negundo* ensured viable pollens, perpetual flowering, ameliorated *in vitro* rooting, and acclimatization. The effect of SiNPs on *Hemidesmus indicus*, an essential and threatened medicinal plant prized for its antipyretic, diaphoretic, and blood purifying properties, was observed by Manokari et al. (2023). In the experiment, *in vitro* shoots from nodal explants of *H. indicus* were cultured on MS medium augmented with 0.5 mg/L Kn + 0.1 mg/L IAA 1 mg/L BAP and 1 mg/L SiNPs. The results revealed that upon addition of SiNPs in the MS medium, the shoots possessed a hypodermis, functional stomata, the prominent density of cells in cortex and vascular elements in stem and leaves and multilayered mesophyll. SiNPs (1 mg/L) also boosted rooting up to 99% when added to ¼ MS rooting medium.

15.2.4 Titanium Nanoparticles (TiNPs)

Titanium NPs are usually used to stimulate chlorophyll formation and carbohydrate production, improving the plant's photosynthetic rate and overall yield (Shi et al., 2013; Irshad et al., 2021). TiO_2 NPs have a high specific area, absorption capacity, photocatalytic activity, and easy operation. They are also known for their antimicrobial activity and promote growth (Vatankhah et al., 2023). The effect of different concentrations of TiO_2 NPs was observed on the plants. It was concluded that at lower concentrations (50, 100 mg/L), TiO_2 NPs stimulated the plant growth, while at higher concentrations (2500 mg/L), the NPs restricted the plant growth (Chahardoli et al., 2022). *Bacopa monnieri*, a popular neurostimulator, was subjected to the treatment of TiO_2 NPs at varying concentrations. Full leaf (FL), distal portion half-leaf (DPHL), and proximal portion half-leaf (PPHL) explants were cultured on the MS medium augmented with BA (1 mg/L) and various concentrations of TiO_2 NPs (2–10 mg/L). The most preferred results for shoot regeneration were procured in TiO_2 NPs at 8 mg/L for DPHL; FL medium containing 2 mg/L produced more shoots when compared to control treatments (Aasim et al., 2023).

15.2.5 Carbon Nanoparticles (Carbon NPs)

Carbon nanomaterials like single-walled carbon nanotubes (SWCNTs), carbon NPs, and multi-walled carbon nanotubes (MWCNTs) have been employed, and their effects have been studied on the plants (García- Sánchez et al., 2015; Mukherjee et al., 2016). Carbon nanotubes (CNTs) are considered to be essential for biomedical engineering (Cha et al., 2013), as they retain exceptional structural and physicochemical properties like having hollow cavities and mechanical strength along with high electrical and thermal conductivity (Hone, 2001). When applied to seedlings, SWCNTs boosted the seedling growth in *Ficus carica* (fig), widely used to treat gastrointestinal, cardiovascular, and respiratory disorders (Flores et al., 2013). The impact of MWCNTs on *in vitro* propagation of two medicinal plants, *Nepeta cataria* and *Salvia sclarea* L., were observed for callus induction and shoot regeneration by Likhani Mehrjardi et al. (2022). Seeds of both species were inoculated on MS medium containing 0.1 mg/L Kn and 1 mg/L NAA and MS medium containing 1 mg/L BA and 0.1 mg/L IAA supplemented with varied concentrations of MWCNTs (20, 60, 80, and 200 μg/L) to instigate callus regeneration. Maximum callus was procured in the medium containing 20 μg/L MWCNT in *N. cartaria*. In contrast, in *S. sclarea* maximum callus was obtained in the medium augmented with 80 μg/L MWCNT, and maximal shoot proliferation was observed in the medium supplemented with 80 μg/L MWCNT.

Apart from these NPs, some other metal and non-metal NPs are employed in plant tissue culture, and their positive and negative effects are being studied on different plant species. For instance, Cu and CoNPs showed impressive results in *Mentha longifolia*, an effective plant for treating gastrointestinal, urinary, and respiratory disorders, by increasing the shoot number, shoot height, number of internodes, and essential oil output (Talankova-Sereda et al., 2016). This indicates the possibility of exploring many other NPs for *in vitro* propagation of medicinal plants.

15.3 APPLICATIONS OF NPs IN MICROPROPAGATION OF MEDICINAL PLANTS

15.3.1 Surface Disinfectants

Microbial contamination is a massive problem in establishing *in vitro* cultures. Very often, microbial contaminants retard the growth of explants in the culture medium, leading to the death of explants or hindering the initiation process. Before the culture initiation stage, explants are sterilized with different agents to eliminate any contamination. Various surface disinfectants, such as sodium hypochlorite (NaOCl), mercuric chloride ($HgCl_2$), bleach, hydrogen peroxide (H_2O_2), ethanol (EtOH), bromine water (BW), silver nitrate ($AgNO_3$), antibiotics, calcium hypochlorite (CaOCl), and fungicides, are used to treat the explants to ensure proper sterilization of the explants (Leifert and Cassells, 2001). Sometimes exposure to these harsh chemicals is not nearly enough to eliminate microbial contamination, as the concentration of the sterilizing agents and exposure time of the treatment needs to be standardized. In some cases, exposure to antibiotics for a long duration of time has led to phytotoxicity (Leifert et al., 1992; Leifert et al., 1994; da Silva et al., 2003; Qin et al., 2011; Tambarussi et al., 2015) in some plant species. The introduction of metal and non-metal NPs as sterilizing agents in combination with other sterilizing agents, alone or in the culture medium, has proven beneficial in eliminating microbial contamination, as shown in Figure 15.2. The inclusion of silica-silver NPs, Al_2O_3, Ag, nickel, TiO_2, ZnO, and iron oxide (Fe_3O_4) has been proclaimed to possess antibacterial activity against a variety of plant microbes (Beyth et al., 2015; Kim et al., 2017). A study noted that explants immersed in AgNPs and then media supplemented with AgNPs were able to decrease the contamination in *in vitro* culture of *Araucaria excelsa* (syn.: *A. heterophylla*), a medicinal tree known for its antipyretic, gastroprotection, anti-inflammation, and antitumor properties (Abdel-Sattar et al., 2009; Elkady and Ayoub, 2018; Elshamy et al., 2020). It was observed that the primary plagiotropic stem (PPS) explants, after surface sterilization, when immersed in AgNPs (200 mg/L) for 1 hour 30 minutes reduced the contamination rate from 61.5% to 11.3%. Further, supplementing the culture medium with AgNPs (400 mg/L) decreased the contamination from 81.25% to 18.75% (Sarmast et al., 2011). The endogenous bacterial contamination of callus cultures of *B. monnieri* declined after supplementing the MS medium with 160 mg/L Ag NPs (Kalsaitkar et al., 2014). Taghizadeh and Solgi (2014) reported that treating nodal explants of *Cyondon dactylon*, known for its antioxidant and diuretic properties, with a combination of Ag NPs (200 mg/L) and thymol (100 mg/L) for 60 minutes reduced the bacterial contamination in the explants.

15.3.2 *In Vitro* Organogenesis and Embryogenesis

The inclusion of nanoparticles in *in vitro* propagation for shoot regeneration and growth and callus induction has given promising results. Many reports have been published indicating the positive

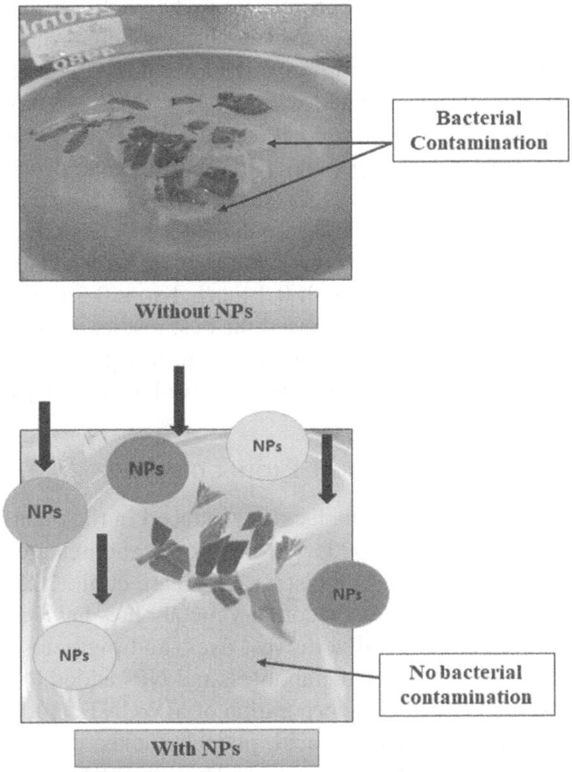

FIGURE 15.2 Elimination of bacterial contamination by NPs.

impact of NPs on the *in vitro* growth of plants. Aghdaei et al. (2012) reported the positive effects of AgNPs in *Tecomella undulata*, commonly known for its anticancer and hepatoprotective properties. AgNPs successfully enhanced the shoot number, callus induction, and shoot induction percentage when added to the MS medium in 10 mg/L along with 2.5 mg/L BAP and 0.1 mg/L IAA. The growth characteristics of *Brassica juncea* seeds were boosted by the incorporation of 50 mg/L AgNPs in the MS basal medium due to the decline in the amount of proline, hydrogen peroxide, and malondialdehyde (MAD). (Sharma et al., 2012). Genady et al. (2016) reported that augmenting copper sulfate ($CuSO_4$) NPs (5 µg/L) in the MS medium exponentially improved the fresh weight (39% as compared to control), shoot length (52% as compared to control), and root length (21% as compared to control) in *Verbena bipinnatifida* seeds, a plant popular for its healing properties. According to Talankova-Sereda et al. (2016), supplementing 0.5 mg/L CuNPs and 0.8 mg/L CoNPs in the modified MS medium improved the shoot number, length of shoots, and rooting in *M. longifolia*. Javed et al. (2017) noted the best shoot induction frequency (89.6%) in the MS medium when supplemented with 1 mg/L ZnO NPs in the nodal explants of *Stevia rebaudiana* (a natural sweetener). The effect of ZnO NPs and ZnO submicron particles (ZnO SMPs) was observed on the seed germination and growth of *Allium cepa* (onion). NPs and SMPs were added into the MS medium as a water suspension in the range of 50, 100, 200, 400, 800, 1600, and 3200 mg/L. Outstanding seed germination results were recorded when the seeds were germinated on MS medium containing 800 mg/L ZnO NPs and ZnO SMPs (56% and 52%, respectively) (Tymoszuk and Wojnarowicz, 2020).

Integration of MWCNTs in the medium containing 1 mg/L 2,4-D boosted the callus growth up to 64% in *Nicotiana tabacum* explants by stimulating the genes associated with water transport (*NtPIP1*), cell division (*CycB*), and cell wall extension (*NtLRX1*), as proclaimed by Khodakovskaya et al. (2012). It was observed by Ghorbanpour and Hadian (2015) that the inclusion of MWCNTs in the leaf culture of *Satureja khuzestanica*, popularly known for its analgesic and anti-inflammatory properties, had a positive effect on callus multiplication when cultured on B5 medium at the concentration of 50 μg/mL. *Thymus daenensis* is a valuable medicinal plant extensively utilized in traditional systems of medicine as an antimicrobial bioresource. The explants of *T. daenensis*, when cultured on the MS medium augmented with 2,4-D + BAP + MWCNTs, resulted in maximal development when the medium constituted 100 μg/L MWCNTs (Samadi et al., 2023). *Panax vietnamensis* (Vietnamese ginseng) is a popular perennial medicinal herb and has been overexploited to the threat of extinction. *P. vietnamensis* was raised *in vitro* by incorporating the AgNPs in the culture medium and studying its impact on somatic embryogenesis and *in vitro*-raised plantlets with the rhizome. The highest number of somatic embryos (140) were formed from *in vitro* leaf-derived callus on MS medium augmented with 0.2 mg/L Kn + 0.5 mg/L NAA + 1 mg/L 2,4-D and 1.6 mg/L AgNPs. Plants with rhizomes showed the best results in the SH medium (Schenk and Hildebrandt, 1972) augmented with 1 mg/L NAA and 1.2 mg/L Ag NPs; the size of the rhizome was also found to be superior as compared to the medium deprived of AgNPs (Manh Cuong et al., 2021).

15.3.3 Enhancing Secondary Metabolite Production

Plant secondary metabolites are the small molecules produced in the plants through the secondary metabolic pathway (Selvakesavan et al., 2023). Secondary metabolites are an integral part of plant defense mechanisms against abiotic and biotic stresses (Khare et al., 2020; Mahajan et al., 2020). Apart from defense against stresses, various secondary metabolites have medicinal values, are used in cosmetic industries, and added as food additives, dyes, and pigments. PTC has attested to be beneficial for the fabrication of secondary metabolites, and NPs added to the nutrient medium are considered nutrient sources and elicitors. Incorporation of MWCNTs (100 μg/L or 250 μg/L) in B5 medium enhanced the content of rosmarinic acid, phenolic acids, caffeic acid, and flavonoids in the leaf culture of *S. khuzestanica* (Ghorbanpour and Hadian, 2015). The amount of essential oil content in the calli of *Calendula officinalis*, famous for treating dermatitis and skin inflammations, exponentially improved when cultured in MS medium containing 0.3 mg/L AgNPs (Al-Oubaidi and Mohammed-Ameen, 2014). Biotic elicitors (*Bacillus cereus* and *Staphylococcus aureus*) and abiotic elicitors ($AgNO_3$ and Ag NPs) were tested for atropine production in hairy root cultures of *Datura metel*. The cultures were subjected to both elicitors for 12, 24, and 48 hours, and the results revealed that the most efficient elicitor for atropine production was AgNPs (Shakeran et al., 2015). Similarly, callus cultures of *P. vulgaris*, known for its anti-inflammatory properties, when treated with Au-Ag NPs in a 1:3 ratio amassed the maximal amount of phenolic compounds and flavonoids (Fazal et al., 2016). The hairy root cultures of *Hyoscyamus reticulatus*, valued for its tropane alkaloids, when exposed to the treatment of FeNPs (450–3600 mg/L) to observe the effect of an increase in the amount of scopolamine and hyoscyamine, a significant increase in the amount of both secondary metabolites was observed when cultures were exposed to 900 mg/L Fe NPs for 24 hours and 450 mg/L Fe NPs for 48 hours (Moharrami et al., 2017). Chung et al. (2018) examined the effects of Ag^+ and AgNPs to increase biomass aggregation and phenolic yield in the hairy root cultures of *Cucumis anguria*, commonly used in folk medicines to treat stomach infections. AgNPs were successful in enhancing the production of flavonols, hydroxybenzoic acid, and hydroxycinnamic acid; moreover, anticancer, antimicrobial, and antioxidant activities were enhanced upon integration of AgNPs; meanwhile, Ag^+ was unsuccessful in eliciting any such response. Chung et al. (2019) noted the effects of incorporating the CuO NPs on the fabrication of gymnemic acid II (GAII) and phenolic content in the cell suspension culture (CSC) of the antidiabetic medicinal plant *Gymnema sylvestre*. Seeds of *G. sylvestre* were cultured on MS medium consolidated with 2,4-D (0.5 mg/L to 3 mg/L) and Kn (0.1 mg/L) for callus induction. Friable callus was transferred to liquid MS medium fortified with 2,4-D (2.0 mg/L) and Kn (0.1 mg/L) for CSC initiation and then treated with CuONPs at three different concentrations (1, 3, or 5 mg/L). The highest yield of GAII and phenolic content was obtained when the shoots were treated with 3 mg/L CuONPs. *Sequoia sempervirens* L., commonly known as redwood, is an ornamental and medicinal tree possessing antitumor effects against breast, colon, and lung tumors (Arafat, 2018). Nano-oxides of Al, Fe, Ti, and Zn at three different concentrations (2.5, 5, and 10 mg/L) were supplemented to the MS medium with 75% of the original composition containing 0.2 mg/L BA individually, and their effects were studied in terms of the morphological characters, total chlorophyll content, and total phenolic and flavonoid content. It was reported that *in vitro* shoots exposed to 5 mg/L TiNPs recorded the maximum amount of chlorophyll content (220.35 mg/100 g fresh weight), total phenolic content (3.41 ± 0.25 mg gallic acid equivalent/g dry weight), and total flavonoid content (1.67 ± 0.12 mg catechin equivalent/g dry weight) (El-Sayed et al., 2021). Joshi et al. (2022) subcultured *in vitro* shoots of *Nardostachys jatamansi* (a multipurpose medicinal plant) on ½ strength MS medium fortified with 1 mg/L IAA + citrate Au and CTAB AuNPs. The results revealed a substantial increase in the total chlorophyll and chlorophyll a, b, and carotenoid content along with an increase in total flavonoids, phenolics, DPPH and SOD enzyme activity, and internal concentrations of IAA, GA_3, and ABA. The effects of various metals (Au, Ag, Cu, Pd) and metal oxides (ZnO, CuO, TiO_2, CeO_2) nanoparticles on secondary metabolite production were observed on the cell suspension cultures of *Hypericum perforatum*, widely used in traditional and pharmaceutical preparations. AgNPs were successful in eliciting cellular agglomeration of bisxanthone (540.3-fold), gancaonin O (214.2-fold), and fusarokyrin (98.6-fold). Other NPs that successfully instigated other compounds are hyperxanthone C (Ag), gallic acid (ZnO NPs), and apigenin (Cu NPs) (Kruszka et al., 2022). These reports prove that NPs play a major role in eliciting secondary metabolites in various medicinal plants, as mentioned in Table 15.2.

15.3.4 Inducing Somaclonal Variations

The variations observed in the somatic cells in *in vitro* conditions are termed somaclonal variations and may prove to be significant in developing new and improved plant varieties. NP exposure to plant cells greatly impacts the DNA integrity and mitotic index and alters the DNA and protein expression in plants (Atha et al., 2012; Landa et al., 2015; Tripathi et al., 2017). Kokina et al. (2012) investigated the effects of CNPs. MS medium supplemented with CNPs and calli of *L. usitatissimum* (flax), when cultured on the medium, showed that the ploidy of cells was impacted, tetraploid cells increased, and DNA methylation in the cells was also found to be higher. Variations were found in the callus morphology and anatomy due to the alterations in protein and DNA profiling when the callus of *Solanum nigrum* was cultured on a nutrient medium augmented with Ag NPs (Ewais et al., 2015). The rate of somaclonal variations was found to be high in both callus and regenerated shoots of *L. usitatissimum* cultured on the nutrient medium containing Au and Ag NPs (Kokina et al., 2017).

15.3.5 Genetic Transformation

Delivering foreign genes in a host cell is called transformation, which can be achieved either by direct (vectorless) or indirect (vector-mediated) methods. Vectorless methods include particle bombardment, electroporation, gene gun, polyfection, etc., while the vector-mediated method includes the use of *Agrobacterium* to transfer the foreign gene into the plant cell. However, these methods fall short due to the vector being host-specific and some direct methods require protoplasts; in addition, the transformation and regeneration rate of protoplasts is low. Nanotechnology has shown the potential to overcome these barriers, as they can easily pass through plant cell walls to ensure proper delivery of DNA or RNA, leading to stable transformation (Lv et al., 2020). Mesoporous silica NPs have been reported to transfer DNA into protoplasts through endocytosis; in addition, the gold-capped mesoporous silica NPs, DNA, and chemicals were delivered in the callus and biolistic gun (Torney et al., 2007). Poly(amidoamine) dendrimer NPs were used by Pasupathy et al. (2008) to create a unique gene delivery technique in plants. The turf-grass cells were modified by introducing plasmid DNA, which encodes green fluorescent protein (GFP). By modifying the molar ratio

TABLE 15.2
Impact of NPs on the Production of Secondary Metabolites in Medicinal Plants

Nanoparticle Type	Nanoparticle Concentration	Plant Species	Explant/Culture Medium	Impact on Secondary Metabolite	Reference
			Metals		
Ag	8 mg/L 10 mg/L	*Glycyrrhiza glabra*	Seed/MS media	Increased glycyrrhizin content	Tahoori et al. (2019)
Co	5 mg/L	*Artemisia annua*	Cell suspension culture/ MS media	Twofold enhancement in artemisinin content	Ghasemi et al. (2015)
Ag	2 mg/L	*Isatis constricta*	Node/MS media	Improved tryptanthrin and indigo yield	Karakaş (2020)
Ag	45 mg/L	*Stevia rebaudiana* Bertoni	Callus/MS media	67% enhanced stevioside production	Golkar et al. (2019)
Ag	90 µg/L	*Caralluma tuberculata*	Callus/MS media	Enhanced production of flavonoids and phenols	Ali et al. (2019)
Ag	5 mg/L	*Corylus avellana* L.	Cell suspension culture/ MS media	Enhanced Taxol production	Jamshidi et al. (2016)
Cu	5 µM	*Verbena bipinnatifida*	Node/MS media	Enhanced total phenolic content by twofold	Genady et al. (2016)
			Bimetallic		
Ag + Au	1:3 ratio	*Prunella vulgaris* L.	Leaf/MS media	Improved total flavonoid (4%) and total phenolic (23%) production	Fazal et al. (2016)
Cu + Au	–	*Stevia rebaudiana* Bertoni	Adventitious root/MS media	20% enhancement in flavonoid and 54% enhancement in phenolic content	Ghazal et al. (2018)
Au + Ag	10–50 mg/dm^3	*Lavandula angustifolia*	Cell suspension/MS media	Enhanced essential oil yield by 2.3-fold	Wesołowska et al. (2019)

			Metal oxides		
CuO	3 mg/L	*Gymnema sylvestre*	Cell suspension culture/ MS media	Improved gymnemic acid, total flavonoid, and total phenolic content production	Chung and Thiruvengadam, 2019
ZnO	75 mg/L	*Echinacea purpurea*	Leaf/MS media	Increased total flavonoid content	Karimi et al. (2018)
ZnO	100 mg/L	*Hyoscyamus reticulatus*	Cotyledon/MS media	Total phenolic compounds (3.2-fold) and tropane alkaloids (1.2-fold) yield enhanced	Asl et al. (2019)
Mn_2O_3	25 mg/L	*Atropa belladonna*	Shoot tip/MS media	Improved yield of alkaloid (23%), total flavonoid (32%), and total phenolic (12%)	Tian et al. (2018)
Fe_3O_4	450, 900 mg/L	*Hyoscyamus reticulatus*	Cotyledon/MS media	Improved yield of hyoscyamine and scopolamine by fivefold	Moharrami et al. (2017)
Fe_3O_4 ZnO	100 ppb	*Hypericum perforatum*	Cell suspension culture/ MS media	Both NPs resulted in increased production of 3-fold hypericin and 13-fold hyperforin	Sharafi et al. (2013)
ZnO	150 mg/L	*Thymus kotschyanus* *Thymus daensis*	Callus/MS media	Increased thymol and carvacrol production	Mosavat et al. (2019)
CuO	5 mg/L	*Trigonella foenum-gaecum*	Seed/MS media	Improved total flavonoid and phenolic production	Ain et al. (2018)
TiO_2	120 mg/L	*Aloe vera*	Cell suspension culture/ MS media	Enhanced aloin production	Raei et al. (2014)
CuO	10 mg/L	*Stevia rebaudiana* Bertoni	Node/MS media	Enhanced yield of total flavonoid, steviol glycoside, and total phenolic content	Javed et al. (2017)
			Carbon-based		
SWCNTs	0.002 g/L	*Simmondsia chinensis*	Node/MS media	Enhanced the amount of phenolics (23.17 mg GAE/g dry weight), flavonoids (20.66 mg QE/g dry weight), tannins (6.35 mg TE/g dry weight)	Gaafar et al. (2018)
MWCNTs	100 µg/mL	*Satureja khuzestanica*	Callus/B5 basal media	3% and 12% enhanced the production of flavonoids and phenolic compounds, respectively	Ghorbanpour and Hadian, 2015

of the dendrimer to plasmid DNA and the pH of the culture medium, the transfection efficiency was further raised. The *GUS* gene, when transferred in *B. juncea* cells, revealed that the most efficient transformation was done in the presence of NP-mediated gene transfer (80.7%), followed by *Agrobacterium*-mediated transfer (54.4%) and then by naked DNA (8%). pCambia 1301, carrying the *GUS* gene, was successfully transferred into *B. juncea* with the aid of calcium phosphate (CaP) NPs (Naqvi et al., 2012). In a study governed by Hao et al. (2013) magnetic Au NPs were shown to deliver plasmid DNA in protoplasts and cells of carrot and canola plants. Ardekani et al. (2014) reported that he GFP gene-carrying pBI121 was efficiently delivered by CaP NPs in tobacco cells. Antibiotics such as cefotaxime, carbenicillin, rifampicin, etc., are used for the elimination of the bacteria in *Agrobacterium*-mediated gene transformation after co-cultivation. It has been shown that antibiotics are phytotoxic to the explants and compromise the regeneration capacity and genetic stability of the plants. However, supplementing the culture medium (LB) with Ag NPs (10 μg/mL) retarded the growth of *Agrobacterium tumifaciens* and *Agrobacterium rhizogenes*, and the bacteria were completely obliterated after the co-cultivation during *Agrobacterium*-mediated gene transfer in *T. undulata* (Sarmast and Salehi, 2016).

15.4 PHYTOTOXICITY OF NANOPARTICLES

The benefits of nanomaterials in the *in vitro* propagation of medicinal plants have encouraged researchers to utilize nanotechnology for the enhancement of various attributes in PTC. However, the technique is not without disadvantages. Reports have suggested that often NPs have proven to be toxic to plants. It is speculated that the chemical makeup, dosage, size, stability, and type of nanomaterials, as well as the culture medium's composition, application approach, explant type, and plant species, have an impact on NP phytotoxicity (Kim et al., 2017). Nanotoxicology is an emerging branch of nanotechnology that deals with the toxicity of nanomaterials in various aspects. In plants, NPs have been reported to have toxic effects on germination, growth, and advancement (Wang et al., 2019; Balážová et al., 2020). Research has revealed that NPs can stimulate reactive oxygen species (ROS) and cause hormonal imbalances in plants, which have a negative impact on plant growth and development (Balážová et al., 2020). Introduction of NPs in cell suspension cultures led to a decrease in cell viability due to the changes in nucleic acid expression, DNA damage, upregulation of ROS species, and triggering electrolyte leakage (Lin et al., 2009; Tan et al., 2009; Jamshidi et al., 2016). The content of stevioside in the shoot cultures of *S. rebaudiana* declined upon exposure to the Zn NPs (50–1000 mg/L) (Desai et al., 2015). Balážová et al. (2020) revealed that NPs may trigger autophagy and programmed cell death in tobacco cells, which leads to a decline in cell viability and growth of cells. Similarly, in *B. nigra*, supplementing MS medium with ZnO NPs (500–1500 mg/L) resulted in the inhibition of seed germination, while stem explants of *B. nigra* were able to form roots in the presence of ZnO (1–20 mg/L) when added into the MS medium (Zafar et al., 2016).

These reports suggest that an ideal dose of NPs needs to be standardized for the plants, as a high concentration dose has a major negative impact on plant growth and development. The ideal safe NP doses that remain within the favorable effects on plant development with less detrimental effects on the environment and plants themselves require more thorough dose-dependent research. For plant nanotechnology, a set of safety analysis and toxicological risk identification guidelines, comprising exposure routes and safe exposure dosages, should be developed and carefully followed (Ruttkay-Nedecky et al., 2017). It is also essential to study the life cycle of NPs, their manufacturing, risk association, disposal, and toxicity to the plants and environment for every individual NP. The potential advantages and disadvantages of each NP must be known to an individual prior to its application in PTC.

15.5 CONCLUSION AND FUTURE PROSPECTS

Nanotechnology has without a doubt created a positive impact on plant biotechnology by enhancing the organogenesis process, callus induction, increasing the secondary metabolite production and somaclonal variations, effective genetic transformation, and elimination of *in vitro* contamination in medicinal plants. Plant nanobiotechnology has the potential to effectively change the fate of plants in a positive direction. An important aspect of the inclusion of NPs in the tissue culture is to standardize the concentration and exposure time of NPs to the plants. Many novel new and effective NPs are being manufactured to enhance the growth of valuable medicinal plants and their secondary metabolites and to conserve them before their extinction. The effect of NPs on callus proliferation and multiplication still needs to be studied in more depth; their role as elicitors and mechanisms of action on the improved production of secondary metabolites are still being studied. Furthermore, many more metal and metal oxide NPs have yet to be explored for their role in the tissue culture of medicinal plants. However, nanotoxicity is a major issue that cannot be ignored, calling for regular monitoring of the effect of NPs and honest submissions of NPs' physical and biological impact on the plants and environment.

REFERENCES

Aasim, M., Korkmaz, E., Culu, A., Kahveci, B., & Sonmezoglu, O. A. (2023). TiO2 nanoparticle synthesis, characterization and application to shoot regeneration of water hyssop (*Bacopa monnieri* L. Pennel) *in vitro*. *Biotechnic & Histochemistry*, *98*(1), 29–37.

Abdel-Sattar, E., Monem, A. R. A., Ezzat, S. M., El-Halawany, A. M., & Mouneir, S. M. (2009). Chemical and biological investigation of *Araucaria heterophylla* Salisb. Resin. *Zeitschrift Für Naturforschung C*, *64*(11–12), 819–823.

Abdi, G., Salehi, H., & Khosh-Khui, M. (2008). Nanosilver: A novel nanomaterial for removal of bacterial contaminants in valerian (*Valeriana officinalis* L.) Tissue culture. *Acta. Physiologiae Plantarum*, *30*, 709–14.

Aghdaei, M., Sarmast, M. K., & Salehi, H. (2012). Effects of silver nanoparticles on *Tecomella undulata. (Roxb.) Seem. Micropropagation*, *26*(1), 21–24.

Ain, N. U., Haq, I. U., Abbasi, B. H., Javed, R., & Zia, M. (2018). Influence of PVP/PEG impregnated CuO NPs on physiological and biochemical characteristics of *Trigonella foenum-graecum* L. *IET Nanobiotechnology*, *12*(3), 349–356.

Ali, A., Mohammad, S., Khan, M. A., Raja, N. I., Arif, M., Kamil, A., & Mashwani, Z. U. R. (2019). Silver nanoparticles elicited *in vitro* callus cultures for accumulation of biomass and secondary metabolites in *Caralluma tuberculata*. *Artificial Cells, Nanomedicine, and Biotechnology*, *47*(1), 715–724.

Al-Oubaidi, H. K. M., & Kasid, N. M. (2015). Increasing phenolic and flavonoids compounds of *Cicer arietinum* L. from embryo explant using titanium dioxide nanoparticle *in vitro*. *World Journal of Pharmaceutical Research*, *4*(11), 1791–1799.

Al-Oubaidi, H. K. M., & Mohammed-Ameen, A. S. (2014). The effect of (AgNO3) NPs on increasing of secondary metabolites of *Calendula officinalis* L. *in vitro*. *International Journal of Pharmacy Practice*, *5*, 267–272.

Arafat, M. A. M. (2018). *Comparative pharmacognostical study and tissue culture of Sequoia sempervirens (D. Don Endl) and Taxodium distichum (L. Rich) cultivated in Egypt (Doctoral dissertation, MSc dissertation, Cairo University, Cairo, Egypt)*.

Ardekani, M. R. S., Abdin, M. Z., Nasrullah, N. A. Z. I. M. A., & Samim, M. (2014). Calcium phosphate nanoparticles a novel non-viral gene delivery system for genetic transformation of tobacco. *International Journal of Pharmacy and Pharmaceutical Sciences*, *6*(6), 605–609.

Asl, K. R., Hosseini, B., Sharafi, A., & Palazon, J. (2019). Influence of nano-zinc oxide on tropane alkaloid production, h6h gene transcription and antioxidant enzyme activity in *Hyoscyamus reticulatus* L. hairy roots. *Engineering in Life Sciences*, *19*(1), 73–89.

Atha, D. H., Wang, H., Petersen, E. J., Cleveland, D., Holbrook, R. D., Jaruga, P., & Nelson, B. C. (2012). Copper oxide nanoparticle mediated DNA damage in terrestrial plant models. *Environmental Science & Technology*, *46*(3), 1819–1827.

Balážová, Ľ, Baláž, M., & Babula, P. (2020). Zinc oxide nanoparticles damage tobacco BY-2 cells by oxidative stress followed by processes of autophagy and programmed cell death. *Nanomaterials*, *10*(6), 1066.

Beyth, N., Houri-Haddad, Y., Domb, A., Khan, W., & Hazan, R. (2015). Alternative antimicrobial approach: Nano-antimicrobial materials. *Evidence-based Complementary and Alternative Medicine*, *2015*(9), 1–16.

Cha, C., Shin, S. R., Annabi, N., Dokmeci, M. R., & Khademhosseini, A. (2013). Carbon-based nanomaterials: Multifunctional materials for biomedical engineering. *ACS Nano*, *7*(4), 2891–2897.

Chahardoli, A., Sharifan, H., Karimi, N., & Kakavand, S. N. (2022). Uptake, translocation, phytotoxicity, and hormetic effects of titanium dioxide nanoparticles (TiO2NPs) in *Nigella arvensis* L. *Science of The Total Environment*, *806*, 151222.

Chung, I. M., Rajakumar, G., & Thiruvengadam, M. (2018). Effect of silver nanoparticles on phenolic compounds production and biological activities in hairy root cultures of *Cucumis anguria*. *Acta Biologica Hungarica*, *69*, 97–109.

Chung, I. M., & Thiruvengadam, M. (2019). Impact of copper oxide nanoparticles on enhancement of gymnemic acid and phenolic compounds using cell suspension culture of *Gymnema sylvestre* (Retz.) R. *British Journal of Applied Physics*, *9*, 213–223.

Cokul Raj, M., Manokari, M., Arumugam, N., Dey, A., Faisal, M., Alatar, A. A., & Shekhawat, M. S. (2023). Silicon nanoparticles mediated *in vitro* flowering and study of pollen viability in *Vitex negundo* L. *Silicon*, *15*(11), 1–11.

da Silva, J. A. T., Nhut, D. T., Tanaka, M., & Fukai, S. (2003). The effect of antibiotics on the *in vitro* growth response of chrysanthemum and tobacco stem transverse thin cell layers (tTCLs). *Scientia Horticulturae*, *97*(3-4), 397–410.

Debona, D., Rodrigues, F. A., & Datnoff, L. E. (2017). Silicon's role in abiotic and biotic plant stresses. *Annual Review of Phytopathology*, *55*, 85–107.

Desai, C. V., Desai, H. B., Suthar, K. P., Singh, D., Patel, R. M., & Taslim, A. (2015). Phytotoxicity of zinc-nanoparticles and its influence on stevioside production in *Stevia rebaudiana* Bertoni. *Applied Biological Research*, *17*(1), 1–7.

Elechiguerra, J. L., Burt, J. L., Morones, J. R., Camacho-Bragado, A., Gao, X., Lara, H. H., & Yacaman, M. J. (2005). Interaction of silver nanoparticles with HIV-1. *Journal of Nanobiotechnology*, *3*, 1–10.

Elkady, W. M., & Ayoub, I. M. (2018). Chemical profiling and antiproliferative effect of essential oils of two *Araucaria* species cultivated in Egypt. *Industrial Crops and Products*, *118*, 188–195.

El-Sayed, I. M., Salama, W. H., Salim, R. G., & Taha, L. S. (2021). Relevance of nanoparticles on micropropagation, antioxidant activity and molecular characterization of *Sequoia sempervirens* L. plant. *Jordan Journal of Biological Sciences*, *14*(2), 373–382.

Elshamy, A. I., Ammar, N. M., Hassan, H. A., Al-Rowaily, S. L., Ragab, T. I., El Gendy, A. E. N. G., & Abd-ElGawad, A. M. (2020). Essential oil and its nanoemulsion of *Araucaria heterophylla* resin: Chemical characterization, anti-inflammatory, and antipyretic activities. *Industrial Crops and Products*, *148*, 112272.

El-Shetehy, M., Moradi, A., Maceroni, M., Maceroni, M., Reinhardt, D., Petri-Fink, A., Rothen-Rutishauser, B., Mauch, F., & Schwab, F. (2020). Silica nanoparticles enhance disease resistance in *Arabidopsis* plants. *Nature Nanotechnology*, *16*, 344–353.

Ewais, E. A., Desouky, S. A., & Elshazly, E. H. (2015). Evaluation of callus responses of *Solanum nigrum* L. exposed to biologically synthesized silver nanoparticles. *Nanoscience and Nanotechnology*, *5*(3), 45–56.

Fazal, H., Abbasi, B. H., Ahmad, N., & Ali, M. (2016). Elicitation of medicinally important antioxidant secondary metabolites with silver and gold nanoparticles in callus cultures of *Prunella vulgaris* L. *Applied Biochemistry and Biotechnology*, *180*, 1076–1092.

Flores, D., Chaves, J. S., Chacón, R., & Schmidt, A. (2013, May). A novel technique using SWCNTs to enhanced development and root growth of fig plants (*Ficus carica*). In *Technical Proceedings of the NSTI nanotechnology conference and expo (NSTI-Nanotech'13)* (Vol. 3, pp. 167–170).

Gaafar, A. A., Taha, R. A., Abou-Baker, N. H., Shaaban, E. A., & Salama, Z. A. (2018). Evaluation of regeneration, active ingredients and antioxidant activities in jojoba tissue cultures as affected by carbon nanotubes. *Bioscience Research*, *15*(3), 2383–2392.

García-Sánchez, S., Bernales, I., & Cristobal, S. (2015). Early response to nanoparticles in the *Arabidopsis* transcriptome compromises plant defence and root-hair development through salicylic acid signalling. *BMC Genomics*, *16*(1), 1–17.

Genady, E. A., Qaid, E. A., & Fahmy, A. H. (2016). Copper sulfate nanoparticles *in vitro* applications on *Verbena bipinnatifida* nutt. Stimulating growth and total phenolic content increments. *International Journal of Pharmaceutical Research and Allied Sciences*, *5*, 196–202.

Ghasemi, B., Hosseini, R., & Nayeri, F. D. (2015). Effects of cobalt nanoparticles on artemisinin production and gene expression in *Artemisia annua*. *Turkish Journal of Botany*, *39*(5), 769–777.

Ghazal, B., Saif, S., Farid, K., Khan, A., Rehman, S., Reshma, A., & Ahmad, N. (2018). Stimulation of secondary metabolites by copper and gold nanoparticles in submerge adventitious root cultures of *Stevia rebaudiana* (Bert.). *IET Nanobiotechnology*, *12*(5), 569–573.

Ghorbanpour, M., & Hadian, J. (2015). Multi-walled carbon nanotubes stimulate callus induction, secondary metabolites biosynthesis and antioxidant capacity in medicinal plant *Satureja khuzestanica* grown *in vitro*. *Carbon*, *94*, 749–759.

Golkar, P., Moradi, M., & Garousi, G. A. (2019). Elicitation of stevia glycosides using salicylic acid and silver nanoparticles under callus culture. *Sugar Tech*, *21*, 569–577.

Hammami, I., & Alabdallah, N. M. (2021). Gold nanoparticles: Synthesis properties and applications. *Journal of King Saud University Science*, *33*(7), 101560.

Hao, Y., Yang, X., Shi, Y., Song, S., Xing, J., Marowitch, J., & Chen, J. (2013). Magnetic gold nanoparticles as a vehicle for fluorescein isothiocyanate and DNA delivery into plant cells. *Botany*, *91*(7), 457–466.

Hone, J. (2001). Phonons and Thermal Properties of Carbon Nanotubes. In *Carbon Nanotubes: Synthesis, Structure, Properties, and Applications* (pp. 273–286). Berlin, Heidelberg: Springer Berlin Heidelberg.

Hussain, H. I., Yi, Z., Rookes, J. E., Kong, L. X., & Cahill, D. M. (2013). Mesoporous silica nanoparticles as a biomolecule delivery vehicle in plants. *Journal of Nanoparticle Research*, *15*, 1676.

Irshad, M. A., Nawaz, R., Ur Rehman, M. Z., Adrees, M., Rizwan, M., Ali, S., & Tasleem, S. (2021). Synthesis, characterization and advanced sustainable applications of titanium dioxide nanoparticles: A review. *Ecotoxicology and Environmental Safety*, *212*, 111978.

Jamshidi, M., Ghanati, F., Rezaei, A., & Bemani, E. (2016). Change of antioxidant enzymes activity of hazel (*Corylus avellana* L.) cells by AgNPs. *Cytotechnology*, *68*, 525–530.

Javed, R., Usman, M., Yücesan, B., Zia, M., & Gürel, E. (2017). Effect of zinc oxide (ZnO) nanoparticles on physiology and steviol glycosides production in micropropagated shoots of *Stevia rebaudiana* Bertoni. *Plant Physiology and Biochemistry*, *110*, 94–99.

Joshi, S., Dar, A. I., Acharya, A., & Joshi, R. (2022). Charged gold nanoparticles promote *in vitro* proliferation in *Nardostachys jatamansi* by differentially regulating chlorophyll content, hormone concentration, and antioxidant activity. *Antioxidants*, *11*(10), 1962.

Kalsaitkar, P., Tanna, J., Kumbhare, A., Akre, S., Warade, C., & Gandhare, N. (2014). Silver nanoparticles induced effect on in-vitro callus production in *Bacopa monnieri*. *Asian Journal of Biological and Life Sciences*, *3*(3), 167–172.

Karakaş, Ö (2020). Effect of silver nanoparticles on production of indole alkaloids in *Isatis constricta*. *Iranian Journal of Science and Technology, Transactions A: Science*, *44*(3), 621–627.

Karimi, N., Behbahani, M., Dini, G., & Razmjou, A. (2018). Enhancing the secondary metabolite and anticancer activity of *Echinacea purpurea* callus extracts by treatment with biosynthesized ZnO nanoparticles. *Advances in Natural Sciences: Nanoscience and Nanotechnology*, *9*(4), 045009.

Khare, S., Singh, N. B., Singh, A., Hussain, I., Niharika, K. M., Yadav, V., & Amist, N. (2020). Plant secondary metabolites synthesis and their regulations under biotic and abiotic constraints. *Journal of Plant Biology*, *63*, 203–216.

Khodakovskaya, M. V., De Silva, K., Biris, A. S., Dervishi, E., & Villagarcia, H. (2012). Carbon nanotubes induce growth enhancement of tobacco cells. *ACS Nano*, *6*(3), 2128–2135.

Kim, D. H., Gopal, J., & Sivanesan, I. (2017). Nanomaterials in plant tissue culture: The disclosed and undisclosed. *RSC Advances*, *7*(58), 36492–36505.

Kokina, I., Gerbreders, V., Sledevskis, E., & Bulanovs, A. (2013). Penetration of nanoparticles in flax (*Linum usitatissimum* L.) calli and regenerants. *Journal of Biotechnology*, *165*(2), 127–132.

Kokina, I., Jahundoviča, I., Mickeviča, I., Jermaļonoka, M., Strautiņš, J., Popovs, S., & Gerbreders, V. (2017). Target transportation of auxin on mesoporous Au/SiO_2 nanoparticles as a method for somaclonal variation increasing in flax (*L. usitatissimum* L.). *Journal of Nanomaterials*, *2017*(1), 7143269.

Kokina, I., Sļedevskis, Ē, Gerbreders, V., Grauda, D., Jermaļonoka, M., Valaine, K., & Rashal, I. (2012). Reaction of flax (*Linum usitatissimum* L.) calli culture to supplement of medium by carbon nanoparticles. *Proceedings of the Latvian Academy of Sciences. Section B. Natural, Exact, and Applied Sciences.*, *66*(4–5), 200–209.

Kruszka, D., Selvakesavan, R. K., Kachlicki, P., & Franklin, G. (2022). Untargeted metabolomics analysis reveals the elicitation of important secondary metabolites upon treatment with various metal and metal oxide nanoparticles in *Hypericum perforatum* L. cell suspension cultures. *Industrial Crops and Products*, *178*, 114561.

Landa, P., Prerostova, S., Petrova, S., Knirsch, V., Vankova, R., & Vanek, T. (2015). The transcriptomic response of *Arabidopsis thaliana* to zinc oxide: A comparison of the impact of nanoparticle, bulk, and ionic zinc. *Environmental Science & Technology*, *49*(24), 14537–14545.

Leifert, C., Camotta, H., & Waites, W. M. (1992). Effect of combinations of antibiotics on micropropagated *Clematis, Delphinium, Hosta, Iris and Photinia*. *Plant Cell, Tissue and Organ Culture*, *29*, 153–160.

Leifert, C., & Cassells, A. C. (2001). Microbial hazards in plant tissue and cell cultures. *In Vitro Cellular & Developmental Biology-Plant*, *37*, 133–138.

Leifert, C., Waites, B., Keetley, J. W., Wright, S. M., Nicholas, J. R., & Waites, W. M. (1994). Effect of medium acidification on filamentous fungi, yeasts and bacterial contaminants in *delphinium* tissue cultures. *Plant Cell, Tissue and Organ Culture*, *36*, 149–155.

Likhani Mehrjardi, H. A., Jonoubi, P., Majd, A., & Haji Hosseini, R. (2022). Effects of multi-wall carbon nano tube (MWCNT) on *Nepeta cataria* and *Salvia sclarea in vitro* culture. *Iranian Journal of Plant Physiology*, *12*(4), 4339–4346.

Lin, C., Fugetsu, B., Su, Y., & Watari, F. (2009). Studies on toxicity of multi-walled carbon nanotubes on *Arabidopsis* T87 suspension cells. *Journal of Hazardous Materials*, *170*(2–3), 578–583.

Lv, Z., Jiang, R., Chen, J., & Chen, W. (2020). Nanoparticle-mediated gene transformation strategies for plant genetic engineering. *The Plant Journal*, *104*(4), 880–891.

Ma, J. F., & Yamaji, N. (2008). Functions and transport of silicon in plants. *Cellular and Molecular Life Sciences*, *65*, 3049–3057.

Mahajan, M., Kuiry, R., & Pal, P. K. (2020). Understanding the consequence of environmental stress for accumulation of secondary metabolites in medicinal and aromatic plants. *Journal of Applied Research on Medicinal and Aromatic Plants*, *18*, 100255.

Mahendran, D., Geetha, N., & Venkatachalam, P. (2019). Role of silver nitrate and silver nanoparticles on tissue culture medium and enhanced the plant growth and development. *In vitro Plant Breeding towards Novel Agronomic Traits: Biotic and Abiotic Stress Tolerance*, 59–74.

Mandeh, M., Omidi, M., & Rahaie, M. (2012). *In vitro* influences of TiO_2 nanoparticles on barley (*Hordeum vulgare* L.) tissue culture. *Biological Trace Element Research*, *150*, 376–380.

Manh Cuong, D., Cong Du, P., Tung, H. T., Ngan, H. T. M., Luan, V. Q., Phong, T. H., & Nhut, D. T. (2021). Silver nanoparticles as an effective stimulant in micropropagation of *Panax vietnamensis*—a valuable medicinal plant. *Plant Cell, Tissue and Organ Culture (PCTOC)*, *146*(3), 577–588.

Manokari, M., Faisal, M., Alatar, A. A., Barboni, D., & Shekhawat, M. S. (2023). Silicon nanoparticles on the enhanced micropropagation of *Hemidesmus indicus* (L.) R. Br. ex Schult. *Silicon*, *16*(2), 775–782.

Meena Devi, V. N., Ariharan, V. N., & Nagendra Prasad, P. (2013). Nutritive value and potential uses of *Leucaena leucocephala* as biofuel—a mini review. *Research Journal of Pharmaceutical, Biological and Chemical Sciences*, *4*(1), 515–521.

Moharrami, F., Hosseini, B., Sharafi, A., & Farjaminezhad, M. (2017). Enhanced production of hyoscyamine and scopolamine from genetically transformed root culture of *Hyoscyamus reticulatus* L. Elicited by iron oxide nanoparticles. *In Vitro Cellular & Developmental Biology-Plant*, *53*, 104–111.

Mosavat, N., Golkar, P., Yousefifard, M., & Javed, R. (2019). Modulation of callus growth and secondary metabolites in different *Thymus* species and *Zataria multiflora* micropropagated under ZnO nanoparticles stress. *Biotechnology and Applied Biochemistry*, *66*(3), 316–322.

Mukherjee, A., Majumdar, S., Servin, A. D., Pagano, L., Dhankher, O. P., & White, J. C. (2016). Carbon nanomaterials in agriculture: A critical review. *Frontiers in Plant Science*, *7*, 172.

Naqvi, S., Maitra, A. N., Abdin, M. Z., Akmal, M. D., Arora, I., & Samim, M. D. (2012). Calcium phosphate nanoparticle mediated genetic transformation in plants. *Journal of Materials Chemistry*, *22*(8), 3500–3507.

Pasupathy, K., Lin, S., Hu, Q., Luo, H., & Ke, P. C. (2008). Direct plant gene delivery with a poly (amidoamine) dendrimer. *Biotechnology Journal: Healthcare Nutrition Technology*, *3*(8), 1078–1082.

Qin, Y. H., Teixeira da Silva, J. A., Bi, J. H., Zhang, S. L., & Hu, G. B. (2011). Response of *in vitro* strawberry to antibiotics. *Plant Growth Regulation*, *65*, 183–193.

Raei, M., Angaji, S. A., Omidi, M., & Khodayari, M. (2014). Effect of abiotic elicitors on tissue culture of *Aloe vera*. *International Journal of Biosciences*, *5*(1), 74–81.

Ruttkay-Nedecky, B., Krystofova, O., Nejdl, L., & Adam, V. (2017). Nanoparticles based on essential metals and their phytotoxicity. *Journal of Nanobiotechnology*, *15*(1), 1–19.

Samadi, S., Saharkhiz, M. J., Azizi, M., Samiei, L., & Ghorbanpour, M. (2023). Exploring potential of multi-walled carbon nanotubes to establish efficient callogenesis, elicitation of phenolic compounds and anti-oxidative activities in thyme plants (*Thymus daenensis*): An *in vitro* assay. *South African Journal of Botany*, *157*, 602–613.

Sarmast, M. K., & Salehi, H. (2016). Silver nanoparticles: An influential element in plant nanobiotechnology. *Molecular Biotechnology*, *58*, 441–449.

Sarmast, M., Salehi, H., & Khosh-Khui, M. (2011). Nanosilver treatment is effective in reducing bacterial contaminations of *Araucaria excelsa* r. Br. var. glauca explants. *Acta Biologica Hungarica*, *62*(4), 477–484.

Savithramma, N., Ankanna, S., & Bhumi, G. (2012). Effect of nanoparticles on seed germination and seedling growth of *Boswellia ovalifoliolata* an endemic and endangered medicinal tree taxon. *Nano Vision*, *2*(1), 2.

Schenk, R. U., & Hildebrandt, A. C. (1972). Medium and techniques for induction and growth of monocotyledonous and dicotyledonous plant cell cultures. *Canadian Journal of Botany*, *50*(1), 199–204.

Selvakesavan, R. K., Kruszka, D., Shakya, P., Mondal, D., & Franklin, G. (2023). Impact of Nanomaterials on Plant Secondary Metabolism. *Nanomaterial Interactions with Plant Cellular Mechanisms and Macromolecules and Agricultural Implications*, 133–170.

Shahcheraghi, N., Golchin, H., Sadri, Z., Tabari, Y., Borhanifar, F., & Makani, S. (2022). Nano-biotechnology, an applicable approach for sustainable future. *Biotech*, *12*(3), 65.

Shakeran, Z., Keyhanfar, M., Asghari, G., & Ghanadian, M. (2015). Improvement of atropine production by different biotic and abiotic elicitors in hairy root cultures of *Datura metel*. *Turkish Journal of Biology*, *39*(1), 111–118.

Sharafi, E., Khayam Nekoei, S. M., Fotokian, M. H., Davoodi, D., Mirzaei, H. H., & Hasanloo, T. (2013). Improvement of hypericin and hyperforin production using zinc and iron nano-oxides as elicitors in cell suspension culture of St John's wort (*Hypericum perforatum* L.). *Journal of Medicinal Plants and By-Product*, *2*(2), 177–184.

Sharma, P., Bhatt, D., Zaidi, M. G. H., Saradhi, P. P., Khanna, P. K., & Arora, S. (2012). Silver nanoparticle-mediated enhancement in growth and antioxidant status of *Brassica juncea*. *Applied Biochemistry and Biotechnology*, *167*, 2225–2233.

Shi, H., Magaye, R., Castranova, V., & Zhao, J. (2013). Titanium dioxide nanoparticles: A review of current toxicological data. *Particle and Fibre Toxicology*, *10*, 1–33.

Sousa, A., Saleh, A. M., Habeeb, T. H., Hassan, Y. M., Zrieq, R., Wadaan, M. A. M., Hozzein, W. N., Selim, S., Matos, M., & Abd Elgawad, H. (2019). Silicon dioxide nanoparticles ameliorate the phytotoxic hazards of aluminum in maize grown on acidic soil. *Science of The Total Environment*, *693*(13), 3636.

Spinoso-Castillo, J. L., Chavez-Santoscoy, R. A., Bogdanchikova, N., Pérez-Sato, J. A., Morales-Ramos, V., & Bello-Bello, J. J. (2017). Antimicrobial and hormetic effects of silver nanoparticles on *in vitro* regeneration of vanilla (*Vanilla planifolia* Jacks. ex Andrews) using a temporary immersion system. *Plant Cell, Tissue and Organ Culture (PCTOC)*, *129*, 195–207.

Taghizadeh, M., & Solgi, M. (2014). The application of essential oils and silver nanoparticles for sterilization of bermudagrass explants in *in vitro* culture. *International Journal of Horticultural Science and Technology*, *1*(2), 131–140.

Tahoori, F., Ahmad, M. A. J. D., Nejadsattari, T., Ofoghi, H., & Iranbakhsh, A. (2019). Qualitative and quantitative study of quercetin and glycyrrhizin in *in vitro* culture of liquorice (*Glycyrrhiza glabra* L.) and elicitation with $AgNO_3$. *Notulae Botanicae Horti Agrobotanici Cluj-Napoca*, *47*(1), 143–151.

Talankova-Sereda, T. E., Liapina, K. V., Shkopinskij, E. A., Ustinov, A. I., Kovalyova, A. V., Dulnev, P. G., & Kucenko, N. I. (2016). The Influence of Cu и Co Nanoparticles on Growth Characteristics and Biochemical Structure of *Mentha longifolia in Vitro*. In *Nanophysics, Nanophotonics, Surface Studies, and Applications: Selected Proceedings of the 3rd International Conference Nanotechnology and Nanomaterials (NANO2015), August 26–30, 2015* (pp. 427–436). Lviv, Ukraine: Springer International Publishing.

Tambarussi, E. V., Rogalski, M., Nogueira, F. T. S., Brondani, G. E., De Martin, V. D. F., & Carrer, H. (2015). Influence of antibiotics on indirect organogenesis of teak. *Annals of Forest Research*, *58*(1), 177–183.

Tan, X. M., Lin, C., & Fugetsu, B. (2009). Studies on toxicity of multi-walled carbon nanotubes on suspension rice cells. *Carbon*, *47*(15), 3479–3487.

Tian, H., Ghorbanpour, M., & Kariman, K. (2018). Manganese oxide nanoparticle-induced changes in growth, redox reactions and elicitation of antioxidant metabolites in deadly nightshade (*Atropa belladonna* L.). *Industrial Crops and Products*, *126*, 403–414.

Torney, F., Trewyn, B. G., Lin, V. S. Y., & Wang, K. (2007). Mesoporous silica nanoparticles deliver DNA and chemicals into plants. *Nature Nanotechnology*, *2*(5), 295–300.

Tripathi, D. K., Singh, S., Singh, S., Pandey, R., Singh, V. P., Sharma, N. C., & Chauhan, D. K. (2017). An overview on manufactured nanoparticles in plants: Uptake, translocation, accumulation and phytotoxicity. *Plant Physiology and Biochemistry*, *110*, 2–12.

Tung, H. T., Suong, P. T., Khai, H. D., Luan, V. Q., Cuong, D. M., Hien, V. T., & Nhut, D. T. (2022). Protocorm-like body formation, stem elongation, and enhanced growth *of Anthurium andraeanum* 'Tropical' plantlet on medium containing silver nanoparticles. *In Vitro Cellular & Developmental Biology-Plant*, *58*, 70–79.

Tymoszuk, A., & Wojnarowicz, J. (2020). Zinc oxide and zinc oxide nanoparticles impact on *in vitro* germination and seedling growth in *Allium cepa* L. *Materials*, *13*(12), 2784.

Vatankhah, A., Aliniaeifard, S., Moosavi-Nezhad, M., Abdi, S., Mokhtarpour, Z., Reezi, S., & Fanourakis, D. (2023). Plants exposed to titanium dioxide nanoparticles acquired contrasting photosynthetic and morphological strategies depending on the growing light intensity: A case study in radish. *Scientific Reports*, *13*(1), 5873.

Vijayakumar, P. S., Abhilash, O. U., Khan, B. M., & Prasad, B. L. (2010). Nanogold-loaded sharp-edged carbon bullets as plant-gene carriers. *Advanced Functional Materials*, *20*(15), 2416–2423.

Wang, Q., Li, C., Wang, Y., & Que, X. (2019). Phytotoxicity of graphene family nanomaterials and its mechanisms: A review. *Frontiers in Chemistry*, *7*, 292.

Wesołowska, A., Jadczak, P., Kulpa, D., & Przewodowski, W. (2019). Gas chromatography-mass spectrometry (GC-MS) analysis of essential oils from AgNPs and AuNPs elicited *Lavandula angustifolia in vitro* cultures. *Molecules*, *24*(3), 606.

Zafar, H., Ali, A., Ali, J. S., Haq, I. U., & Zia, M. (2016). Effect of ZnO nanoparticles on *Brassica nigra* seedlings and stem explants: Growth dynamics and antioxidative response. *Frontiers in Plant Science*, *7*, 535.

Zayed, M. Z., & Samling, B. (2016). Phytochemical constituents of the leaves of *Leucaena leucocephala* from Malaysia. *International Journal of Pharmacy and Pharmaceutical Sciences*, *8*(12), 174–179.

Zayed, M. Z., Wu, A., & Sallam, S. M. (2019). Comparative phytochemical constituents of *Leucaena leucocephala* (lam.) Leaves, fruits, stem barks, and wood branches grown in Egypt using GC-MS method coupled with multivariate statistical approaches. *BioResources*, *14*(1), 996–1013.

Zhang, B., Zheng, L. P., Yi Li, W., & Wen Wang, J. (2013). Stimulation of artemisinin production in *Artemisia annua* hairy roots by ag-SiO2 core-shell nanoparticles. *Current Nanoscience*, *9*(3), 363–370.

Zhu, Y., & Gong, H. (2014). Beneficial effects of silicon on salt and drought tolerance in plants. *Agronomy for Sustainable Development*, *34*, 455–472.

16 Bioinformatics and Its Applications in Plant Tissue Culture

Akansha Chauhan, Mohamed Sabri Bensaad, Manu Pant, Mohamed Amine Kahoul, Debasis Mitra, Pradeep Kumar Das Mohapatra, and Periyasamy Panneerselvam

16.1 INTRODUCTION

Plant tissue culture plays a crucial role in the establishment and propagation of medicinal plants. It is a powerful technique that involves the growth of plant cells, tissues, or organs in aseptic conditions, facilitating the complete regrowth of plants from small plant parts, i.e., explants. Since ancient times, about 80% of the global population has relied on traditional formulations of medicinal plants for essential healthcare and managing various illnesses (Pant, 2014). The growing need for standardized medicinal plant-derived products underscores the requirement for large-scale replication through the implementation of plant tissue culture techniques. It allows the selective clonal propagation of disease-free medicinal plants and ensures that the new plants possess the same desirable medicinal properties and chemical constituents as the parent plant. Many medicinal plants produce valuable secondary metabolites, such as flavonoids, alkaloids, and essential oils, that have medicinal properties (Wink, 2012; Jamwal et al., 2018). Refining tissue culture methods can be fine-tuned to amplify the synthesis of these secondary metabolites, leading to increased yields of distinct medicinal compounds (Vashisht et al., 2015).

The integration of bioinformatics with plant tissue culture of medicinal plants offers a powerful approach for enhanced production of bioactive compounds, understanding their regulatory mechanisms, and accelerating drug discovery and development processes. The computational tools and data analysis methods enhance our understanding and optimization of tissue culture techniques in plants. By analyzing genomic, transcriptomic, and metabolomic data, bioinformatics helps to identify key genes, pathways, and regulatory networks responsible for plant growth, development, and the synthesis of bioactive compounds. These insights aid in selecting suitable plant materials, improving culture conditions, and accelerating the production of elite medicinal plants with desired traits for therapeutic applications. Bioinformatics serves as a powerful tool to guide precision breeding and targeted genetic modifications, ultimately optimizing plant tissue culture and advancing the field of medicinal plant biotechnology. Genome sequencing and annotation, identification of medicinally relevant genes, gene expression profiling, microRNA identification, gene editing, comparative genomics, data integration, and visualization are the successful applications of bioinformatic tools that allow us to access the in-depth information to stimulate the synthesis of desirable compounds (Klie and Nikoloski, 2012; Weber and Kim, 2016; Yang et al., 2019; Liu et al., 2020; Zhang et al., 2021). Various databases provided by the National Center for Biotechnology Information (NCBI), like GenBank, RefSeq, and Sequence Read Archive (SRA), are invaluable resources for accessing genomic and transcriptomic data relevant to rare medicinal plants and herbs (Gauthier et al., 2019). The present chapter highlights basic information on bioinformatics and its implications for other sectors.

DOI: 10.1201/b23374-16

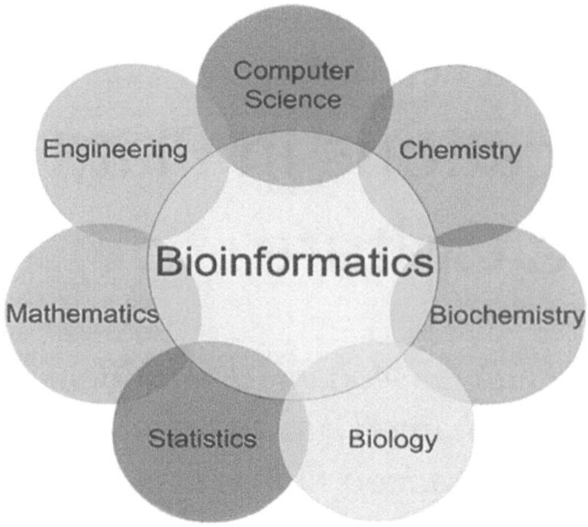

FIGURE 16.1 Fundamental relationship between bioinformatics and other fields of study.

Nowadays, scientific breakthroughs depend on advances in computer science and in particular on a recent discipline called bioinformatics, which emerged from the fusion of biology, statistics, and informatics (Mitra et al., 2022). This discipline served as a bridge between scientists to share crucial biological information and to gain time, thereby facilitating collaboration at the international level.

Bioinformatics employs computational tools and analytical techniques to gather and interpret diverse biological data, encompassing genomics, transcriptomics, proteomics, and genetic advancements (Figure 16.1). These contributions were instrumental in the successful realization of the Human Genome Project (Klimontov et al., 2023), a well-known project that helped the scientific community in the identification and characterization process of about 100,000 genes (Sanderson et al., 2022), offering a reliable database for future experimental research. The most recent achievement of bioinformatics was revealed by a study made by Gendoo (2023), in which the capacity of this discipline to simplify the comprehension of molecular changes across larger systems was demonstrated. Bioinformatics also played a crucial role in vaccine design and development processes, especially during the pre-selection of peptide antivirals to lower incidences of allergenic reactions (Nandy and Basak, 2019), and the effectiveness of bioinformatics tools was principally proven during the coronavirus pandemic (COVID-19) situation.

Bioinformatics is not only limited to medical fields but is also playing a significant role in other sectors. For example, in agriculture, the combination of bioinformatics and biotechnology allowed scientists to track in an accurate way the origin of microbial contamination but also to develop an accurate multicriteria scorecard for microbial bioaerosol data (Miaow et al., 2021). This discipline also allowed biotechnologists to make an important and safe step in genetically engineered food development, aiming to enhance the quality of our everyday food, predict the possible allergens, and search for metabolic capacities including virulence and fermentation processes (Doxey et al., 2019; Lopez-Pedrouso et al., 2023). Table 16.1 summarizes the impact of bioinformatics on educational, healthcare, agricultural, energy, and environmental sectors.

16.2 BIOINFORMATICS APPLICATION IN PLANTS RESEARCH

Plants are vital to maintaining the balance of ecosystems, as they are an essential part of the biosphere. They form the base and first link in the food chain, thus ensuring the survival of all living

TABLE 16.1
Impact of Bioinformatics on Educational, Healthcare, Agricultural, Energy, and Environmental Sectors

Sectors	Bioinformatics Impact	Sources
Educational	Simplify the transmission and comprehension of information as well as the creation of a modern thinking way.	Mulder et al., 2018
	Access to technology at early age and the creation of a multiskilled generation.	Magana et al., 2014
	Modernize and personalize learning processes and educational programs of countries by providing free access to data and content in real time.	Mulder et al., 2018
	Develop creativity and spirit of research by allowing the visualization of numerical data.	Marangoni et al., 2023
	Analyze and interpret a large scale of complex datasets.	Quagliaro et al., 2023
	Scientific hypothesis materialization and emergence of interdisciplinary collaboration.	Marangoni et al., 2023
Healthcare	Simulation of the toxicity profile, dose response, and rational index of drug desirability.	Xia, 2017
	Screening and refinement of novel drug targets.	Xia, 2017
	Gene expression profiling and search for the open reading frame (ORF) of a specific gene.	Li et al., 2023
	Deduce the intron/exon structure of a gene as well as its respective expressed sequence tagging (EST) profile.	Girardini et al., 2023
	Identification of the possible active sites, structural, and functional domains of proteins.	Kumar et al., 2023
	Prediction of the secondary and tertiary structure of hypothetical proteins as a therapeutic approach against severe infection.	Rahman et al., 2023
	Prediction of post-translational modifications and secondary structures related to proteins.	Kumar et al., 2023
	Simulation of sequence homology, physical properties of amino acids, and prediction of mutations or gene variants.	Li et al., 2023
	Phylogenetic tree construction and genome-wide association studies (GWAS).	Kasap et al., 2023
Agricultural	Enhancing plant resilience to both biotic and abiotic stressors.	Wang et al., 2023
	Prevent microbial contamination and enhancement of the nutritional quality in depleted soils.	Miaow et al., 2021
	Control and improve the genetic elements inserted into conventional genetically modified organisms (GMOs).	López-Pedrouso et al., 2023
	Facilitate the simulation and incorporation of micronutrients into plants for enhanced human and animal health.	Naravane and Tagkopoulos, 2023
	Increase the production performance of agro-based industries.	Ferrocino et al., 2023
	Improvement of phyto-remediation capabilities and insect-resistant crops.	Yadav et al., 2023
Energy and environment	Model the evolution of an animal population in a given environment.	Mendoza et al., 2022
	Algorithmic calculation and estimation of carbon footprint.	Grealey et al., 2022
	Elaboration of common environmental resource data center (ERDC) interface and protection of endangered species.	Ju and Zhang, 2015
	Elaboration of next-generation biofuel that is more respectful to the ecosystem.	Zhou et al., 2011

organisms. In pharmacological terms, medicinal plants are very useful in offering a wide range of medicinal and therapeutic properties (Nugraha et al., 2022). Plant biology has also significantly contributed to major advances in science, such as the identification of the basis of cell morphogenesis and the relationship between the cell and the organism, the concept of *de novo* organogenesis, the heritability of traits, and epigenetic inactivation (Jonsson et al., 2022; Neves et al., 2023). Moreover, several active compounds in plants provide the foundation for the creation of novel drugs, hence the name "phytopharmaceuticals" (Dincheva et al., 2023).

However, the structural complexity of certain phytocompounds requires the intervention of modern approaches to reveal their full potential, and in this context, bioinformatics represents the key. Several computational approaches allowed scientists to refine extensive libraries of biocompounds into smaller sets of anticipated active compounds suitable for experimental testing (Sliwoski et al., 2013). Bioinformatics also helped in the characterization of the mechanism of many diseases and validate new and cost-effective drugs made of plant compounds (Ma et al., 2020; Wang et al., 2023). Several omics approaches contributed to discovering genes and pathways potentially linked to significant bioactive compounds but have also been involved in the quality control and identification process of herbal medicines (Wang et al., 2022). The physicochemical and biological aspects of several phytocompounds have been revealed due to the contribution of the algorithmic bioinformatics systems, which help in the crucial determination of a drug's pharmacokinetic and pharmacodynamics profile (Bensaad et al., 2022).

Another good example to illustrate the implication of bioinformatics in plant research is a Japanese online platform called KEGG (http://www.kegg.jp/) that allows the prediction of the molecular interaction and reaction networks of phytocompounds and to simulate plant genomics and metabolomics processes (Kanehisa, 2016), as the webserver contains a huge database, principally retrieved from the Kyoto Encyclopedia of Genes and Genomes. It is important to underline that the actual plant molecular marker databases also revealed additional information related to the genetic, genomic, or phenotypic aspects of many plant species, which could be helpful to ameliorate plant breeding (Lai et al., 2015). The obtained data also helped in the development process of harmless procedures resistance to insect damage. Botanists have recognized that for effective plant protection, it is essential to employ both computational and in silico genomics technologies to associate transcription factors and promoter regions with gene-enzyme resistance to diseases (Mu et al., 2022).

Recently, bioinformatics allowed the creation of next-generation biofuel, which is more respectful to the ecosystem. Kumar and his collaborators (2023) suggested that the application of technology can considerably help in the management process of microalgal biomass and their incorporation in biofuel production due to their high lipid content (Ebhodaghe et al., 2022). The principal applications of bioinformatics in plant research are summarized in Table 16.2 and Figure 16.2.

The NCBI databases, such as PubMed and PubMed Central, comprise an extensive assortment of scientific articles and research papers related to medicinal plants and plant tissue culture. NCBI's GenBank provides access to DNA and RNA sequences of various medicinal plants. By exploring the genetic makeup of these plants, one can gain insights into their molecular characteristics and identify genes or markers responsible for stress tolerance, secondary metabolites, and other phytochemicals. Tools like Basic Local Alignment Search Tool (BLAST) can help in comparing sequences, while Primer-BLAST can design primers for experiments (Mohimani et al., 2014). These tools facilitate the design of experiments and the analysis of results related to medicinally important plants. Some bioinformatic tools, including Phytozome, MISA, GATK, and Geneious, grant access to the genetic composition of these plants. They offer a comprehensive insight into the potential future production of plant compounds with significant medicinal value. This potential of targeted plants can be enhanced by genetic engineering, precursor feeding, hormone, and media optimization according to the requirement by applying plant tissue culture techniques (Figure 16.3).

TABLE 16.2
Principal Bioinformatics Tools Used in Plant Tissue Culture Studies

Bioinformatics Tools/ Platforms	Links	Roles	Sources
ITIS®	https://www.gbif.org/dataset/9ca92552-f23a-41a8-a140-01abaa31c931	Identification of plant transposon insertion sites and prediction of plant sensitivity to stressful tissue culture environment.	Jiang et al., 2015; Fabian et al., 2020
CultureTrax®	https://www.culturetrax.com/	Simulation of plant's cell culture and stem cell culture.	Canfield et al., 2018
Mega 7®	http://www.megasoftware.net/	3D characterization of SoPCS gene and prediction of *Saccharum officinarum* tissue tolerance to cadmium. Development of plant tissue culture more resistant to toxic heavy metals.	Kolahi et al., 2018
MAFFT®HMMER®	https://mafft.cbrc.jp/alignment/software/ http://hmmer.org/download.html	Searching for sequence homologs and alignments. Identification of the molecular mechanism of the WOX11 gene and thus regeneration from injured areas of separated leaves or stems, as well as *in vitro* plant tissue culture.	Yang et al., 2022
ANNs/SVMs	https://www.kaggle.com/code/jchen2186/machine-learning-with-iris-dataset/input	Optimization of length and number of microshoots or roots, biomass in plant cell cultures or hairy root culture.	Hesami et al., 2020
MLPNSGA-II	https://sourceforge.net/directory/windows/?q=multi-layer+perceptron	Optimization of the chrysanthemum's somatic embryogenesis process and plant tissue culture modeling accuracy.	Hesami et al., 2020
ANFIS	https://pypi.org/project/anfis/	Prediction of cell culture–responsive Taxol biosynthesis in plant and modeling of plant culture filtrate cell with methyl-β-cyclodextrin.	Ji et al., 2023
SketchUp®	https://help.sketchup.com/fr/downloading-sketchup	Development of novel plant tissue culture systems by ensuring a coherent evolution from 3D printing to prototype.	Shukla et al., 2017
CFD®	https://sim-flow.com/download/cfd-simulation-software/	Determination of oxygen mass transfer in the growth domain of a complex system of hairy root culture of plants.	Liu et al., 2011
DoEANNs	https://www.statease.com/software/design-expert/https://www.kaggle.com/code/jchen2186/machine-learning-with-iris-dataset/input	Improvement of micropropagation and conservation processes of plants. Maximization of data quality related to various tissue culture process.	Nezami and Gallego, 2023
GEP®	https://www.gene-expression-programming.com/	Prediction of plant tissue culture–related factors including control of *in vitro* proliferation rate, shoot length, shoot tip necrosis, and vitrification index.	Ji et al., 2023

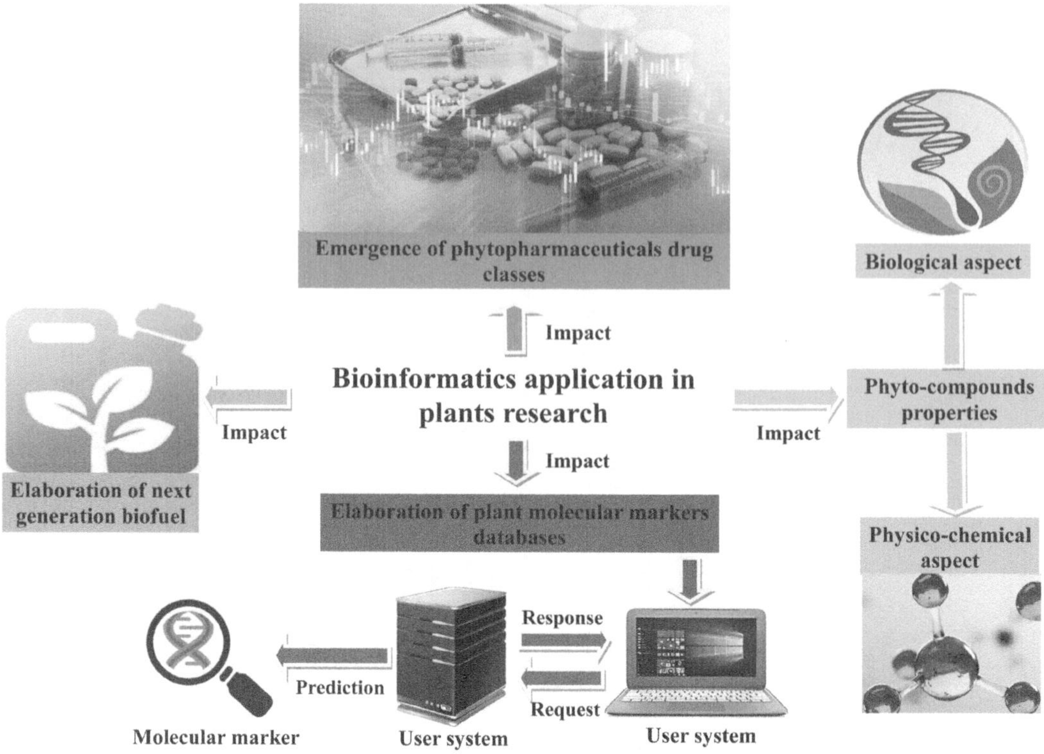

FIGURE 16.2 Major applications of bioinformatics in plant research.

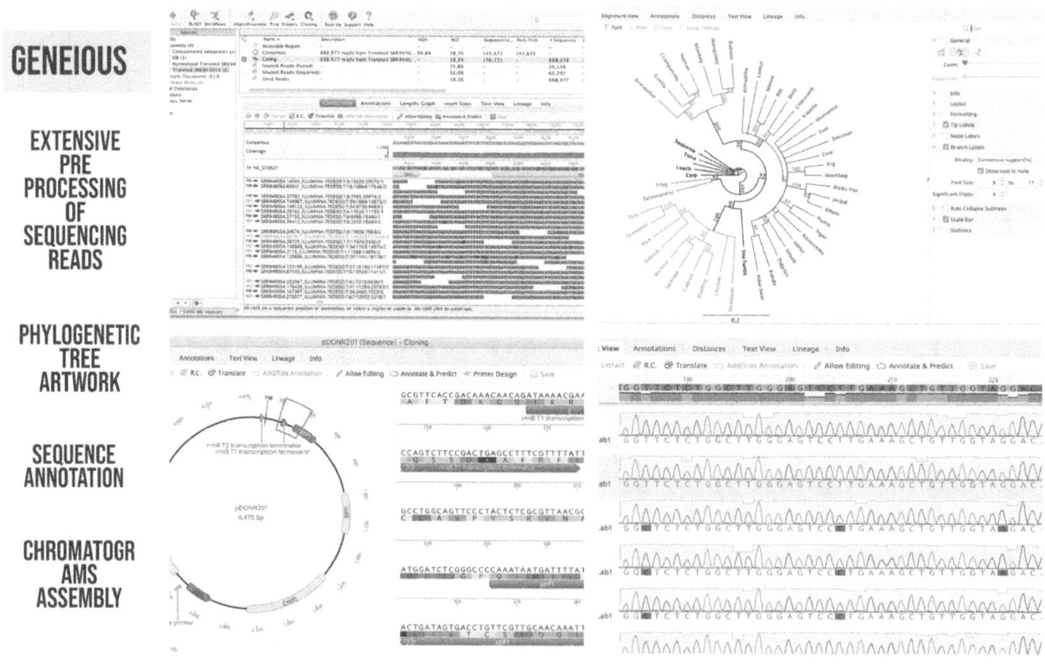

FIGURE 16.3 An overview of the Geneious software used for plant-based research.

16.3 BIOINFORMATICS DATA-BASED APPLICATION IN PLANT CULTURE RESEARCH

Many biotechnological approaches involving metabolic engineering systems have emerged to design novel transgenic cell lines but also to optimize key factors, including cell line selection, and culture conditions to ameliorate the metabolite production in plant cell cultures (Cusido et al., 2014). This also allowed the development of a more accurate enzymatic tool, facilitating the biochemical conversion of phenolic natural products through plant cell cultures to ensure congruent and stable phenolic production (Mohamed et al., 2022). Moreover, the plant cell suspension culture (PCSC) platform allowed the emergence of next-generation metabolic engineering techniques and thus the development of more accurate PCSC techniques (Arya et al., 2020). Table 16.3 shows some bioinformatics tools that are employed in terms of enhanced secondary metabolite production in plant tissue culture.

Several bioinformatics platforms and databases have emerged and simplified the comprehension of the molecular aspect of plants. As an example, the DRAGO2 database (http://prgdb.org/prgdb/drago2) focused on annotate resistance genes of plants, while Bio-Analytic Resource (http://bar.utoronto.ca/), PlantGDB (http://www.plantgdb.org/), and iLoci (https://github.com/BrendelGroup/AEGeAn), respectively, allowed the comprehension of molecular markers, mapping, genomics profiles, and gene expression of plants (Duvick et al., 2008; Standage et al., Tan et al., 2022). Currently, genomic databases called Ensembl Plants (https://plants.ensembl.org/index.html) allow the scientific community to produce functional annotations for genes in over 33 plant species (Tan et al., 2022), which is a non-negligible achievement.

Regarding the utilization of bioinformatics tools in plant tissue culture research, numerous studies have already illustrated the influence of this domain. A bioinformatics tool called ITIS® (https://www.itis.gov/) has been successfully integrated to study various bodies such as bacteria, insects, cell culture, mammals, yeast, and plants and is currently used to predict plants' phenotypic changes via the identification of transposon insertion sites (Jiang et al., 2015). The information holds significant importance since the application of transposon insertion sequencing technology can positively influence long-term tissue-cultured cell lines. A study made by Fabian et al. (2020) revealed that a transposon named TCUP is capable of reactivation and could be considered a key indicator of plant sensitivity to stressful tissue culture environments, which is an important criterion during clonal propagation of plants.

Another bioinformatics software called Mega 7® (http://www.megasoftware.net/) has been used to predict the 3D structure of the SoPCS gene for determining the response of the plant species *Saccharum officinarum* to cadmium (Kolahi et al., 2018), which could be applied in the near future to generate plants resistant to toxic effects of cadmium (Cd) by plant tissue culture and thus ensure a coherent plant growth process. Furthermore, the integration of artificial intelligence technology and optimization algorithms, principally referred as AI-OA, has the potential to enhance various phases of plant tissue culture progressively. Research conducted by Hesami et al. (2020) unveiled that artificial neural networks (ANNs) and support vector machines (SVMs), which can be downloaded via the link (https://www.kaggle.com/code/jchen2186/machine-learning-with-iris-dataset/input), can be valuable for optimizing factors such as the length and number of micro shoots or roots, hairy root culture, biomass in plant cell cultures, and environmental conditions. This optimization aims to achieve maximum productivity and efficiency in plant cultivation. Furthermore, CultureTrax®, a cloud-based software application (https://www.culturetrax.com/), has demonstrated its ability to standardize and transfer a complex cell culture protocol, and thus guarantee a coherent simulation of a plant's cell culture and stem cell culture (Canfield et al., 2018).

It is also interesting that other approaches combining the algorithm systems of two machine learning methods, Multilayer Perceptron (MLP) (https://sourceforge.net/directory/windows/?q=multi-layer±perceptron) and Non-dominated Sorting Genetic Algorithm-II (NSGA-II) (https://pymoo.org/algorithms/moo/nsga2.html), have the potential to enhance the somatic embryogenesis

TABLE 16.3
High-Throughput Metabolic Modelling Tools and Software to Facilitate Medicinal Plants Engineering for Secondary Metabolite Production

S. No.	Software/Tool	Links	Application	References
1	CLC Genomics Workbench	https://digitalinsights.qiagen.com/products-overview/discovery-insights-portfolio/analysis-and-visualization/qiagen-clc-genomics-workbench/	Help to analyze and visualize genomic data, including genome assembly, annotation, and gene expression analysis	Schilbert et al., 2020
2	Artemis	https://www.sanger.ac.uk/tool/artemis/	Genome viewer and annotation tool that aids in the visualization and annotation of genomes	Carver et al., 2012
3	Phytozome	https://phytozome-next.jgi.doe.gov/	Helps to access annotated plant genomes, gene expression data, and comparative genomics information	Goodstein et al., 2012; Zhang et al., 2019
4	MapMan	https://mapman.gabipd.org/mapman	Helps in visualizing and analyzing metabolic pathways and regulatory networks in plants	Schwacke et al., 2019
5	Geneious	https://www.geneious.com/	Helps researchers with sequence alignment, primer design, and genetic analysis related to plant tissue culture experiments	Tahir et al., 2018; Zhang and Zhang, 2021
6	MISA (MIcroSAtellite identification tool)	https://bio.tools/misa	Specialized software for identifying and characterizing microsatellites (simple sequence repeats) in genomic sequences	Jiang et al., 2023
7	SOAPdenovo	https://www.animalgenome.org/bioinfo/resources/manuals/SOAP.html	Useful for assembling genomes of non-model plant species, which may be essential for tissue culture research on diverse plant species	Luo et al 2012; Xie et al., 2014;
8	PhyML	http://www.atgc-montpellier.fr/phyml/	Provides phylogenetic information, which can be instrumental in making informed decisions about the selection of plant materials, designing breeding strategies, and understanding the genetic basis of desirable traits	Abdelsalam et al., 2022;
9	PAML	http://abacus.gene.ucl.ac.uk/software/	Software for phylogenetic analysis of DNA or protein sequences	Xu et al., 2022
10	GATK	https://gatk.broadinstitute.org/hc/en-us	Proper data preprocessing, alignment, and variant calling pipelines can be established using GATK to analyze genomic data from plant tissue culture experiments effectively	Hoops et al., 2018

of chrysanthemums accurately, offering a relevant computational methodology for future investigations in plant tissue culture (Hesami et al., 2020). Two-dimensional axisymmetric computational fluid dynamics (CFD®), a bioinformatics software, which can be downloaded via this link (https://sim-flow.com/download/cfd-simulation-software/), has theatrically and experimentally resolved some issues related to the mass transfer of oxygen within the growth domain of a complex system of hairy root culture (Liu et al., 2011).

16.4 RECENT ADVANCES AND FUTURE PERSPECTIVES IN BIOINFORMATICS FOR PLANT TISSUE CULTURE

Plant tissue culture techniques can provide a controlled, sustainable, and cost-effective means of producing vegetal products, shortening time to market, and accelerating commercial applications (Loyola-Vargas and Ochoa-Alejo, 2018). Traditionally, natural products have served as the primary reservoir of lead compounds for antimicrobial drugs. Additionally, they find applications in various other fields, including insecticides, anthelmintics, flavorings, anticancer drugs, pain killers, cosmeceuticals, and crop protection. Recently, many studies have been able to elucidate the tissue culture process of several plant species via the contribution of biotechnology and bioinformatics, which allowed the coherent simulation and reproduction of culture conditions including light, temperature, humidity, and the molecular mechanism of plant hormones involved in its growth regulation (Ji et al., 2023; Wijerathna-Yapa and Hiti-Bandaralage, 2023).

The new generation of computer-based tools such as design of experiments (DoE) (https://www.statease.com/software/design-expert/) and ANNs (https://www.kaggle.com/code/jchen2186/machine-learning-with-iris-dataset/input) has considerably improved micropropagation and conservation processes of plants (Nezami and Gallego, 2023) but has also maximized data quality, which allowed scientists to develop specific culture media for pistachio and to identify the smallest set of factors that exert a notable impact on a particular process (Nezami and Gallego, 2023).

Interestingly, both multi-omics domain monitoring technologies and nanomaterials are currently integrated to improve plant tissue culture but also to enhance the accessibility and yield of natural metabolites. A recent study on *in silico* genome-wide identification process underlined that a variety of computational tools, including multiple sequence alignment program (MAFFT®) (https://mafft.cbrc.jp/alignment/software/) and HMMER® (http://hmmer.org/download.html), which are principally utilized to search sequence databases for homologous sequences and to perform sequence alignments, allowed the identification of 34 MsWOX genes, including WOX11 (Yang et al., 2022). These two bioinformatics software programs also allowed scientists to elucidate the molecular mechanism of this gene regeneration from the injured areas of separated leaves or stems and to *in vitro* tissue culture of plants and callus formation (Wan et al., 2023). This gene is also crucial to clonally propagate crop plants using both direct and indirect *de novo* root regeneration (DNRR) arising from pluripotent cells within the pericycle tissue (Jing et al., 2020).

Simultaneously, another recent bioinformatics approach by Ji and his team (2023) on the plant species *Corylus avellana* using the Adaptive Neuro-Fuzzy Inference System–genetic algorithm (ANFIS) (https://pypi.org/project/anfis/) allowed a coherent prediction of cell culture–responsive Taxol biosynthesis and a proper modeling of culture filtrate cell wall alone or in combination with methyl-β-cyclodextrin. The same team reported also the effective role of a hybrid modeling approach called gene expression programming (GEP) (https://www.gene-expression-programming.com/) in plant tissue culture, especially to control *in vitro* proliferation rate, shoot length, shoot tip necrosis, and vitrification index (Ji et al., 2023). The SketchUp software, which integrates a 3D printing system, is currently used as a prototype to develop novel plant tissue culture systems since this technology allows one to control of variable fluence rate/spectra LED lighting to provide more consistent and appropriately randomized experimental plant biology conditions, thereby enhancing energy efficiency (Shukla et al., 2017).

Other data-mining bioinformatics approaches are also under development to predict plant tissue–specific expression patterns and pathways of H2S metabolism in different cell compartments by analyzing the expression profile, transcriptional, and post-transcriptional regulation of genes related to the previously cited process (Mondal et al., 2022). However, it has become mandatory to develop bioinformatics approaches targeting the quantitative traits of plants and to simulate a more accurate means of embryogenic and regenerable tissue culture response.

16.5 CONCLUSION

Humans have historically turned to plants to address a variety of issues, but the solutions adopted often resulted in unintended consequences, including overharvesting and environmental degradation. The adverse outcomes serve as a lesson, emphasizing the importance of pursuing sustainable solutions to our problems. The amalgamation of bioinformatics and plant tissue culture provides researchers with a means to gain a comprehensive and systematic understanding of medicinally important plants. Through the application of computational techniques, genomic data analysis, and advanced molecular modeling, researchers can unravel the complex genetic and biochemical pathways responsible for the synthesis of bioactive compounds in medicinal plants. This will be a boon to open new avenues for sustainable agriculture, drug discovery, and healthcare, which will ultimately benefit both human beings and the environment. As technology and research methodology continue to evolve, the field of bioinformatics will undoubtedly play a pivotal role in shaping the future of medicinal plant utilization and production.

REFERENCES

Abdelsalam, N. R., Hasan, M. E., Javed, T., Rabie, S. M., El-Wakeel, H. E. D. M., Zaitoun, A. F., & Shah, A. N. (2022). Endorsement and phylogenetic analysis of some Fabaceae plants based on DNA barcoding. Molecular Biology Reports, 49(6), 5645–5657. https://doi.org/10.1007/s11033-022-07574-z

Arya, S. S., Rookes, J. E., Cahill, D. M., & Lenka, S. K. (2020). Next-generation metabolic engineering approaches towards development of plant cell suspension cultures as specialized metabolite producing biofactories. Biotechnology Advances, 45, 107635. https://doi.org/10.1016/j.biotechadv.2020.107635

Bensaad, M. S., Dassamiour, S., Hambaba, L., Melakhessou, M. A., Sami, R., Al-Mushhin, A. M., Benajiba, N., & Al Masoudi, L. M. (2022). In silico investigation and BSA denaturation inhibitory activity of ethyl acetate and n-butanol extracts of *Centaurea tougourensis* Boiss. and Reut. International Journal of Pharmacology. 18 (6), 1296–1308. https://doi.org/10.3923/ijp.2022.1296.1308

Canfield, S. G., Jin, G., Palecek, S. P., & Sampsell, T. (2018). Software to improve transfer and reproducibility of cell culture methods. Biotechniques, 65(5), 289–292. https://doi.org/10.2144/btn-2018-0062

Carver, T., Harris, S. R., Berriman, M., Parkhill, J., & McQuillan, J. A. (2012). Artemis: an integrated platform for visualization and analysis of high-throughput sequence-based experimental data. Bioinformatics, 28(4), 464–469. https://doi.org/10.1093/bioinformatics/btr703

Cusido, R. M., Onrubia, M., Sabater-Jara, A. B., Moyano, E., Bonfill, M., Goossens, A., Angeles Pedreño, M., & Palazon, J. (2014). A rational approach to improving the biotechnological production of taxanes in plant cell cultures of Taxus spp. Biotechnology Advances, 32(6), 1157–1167. https://doi.org/10.1016/j.biotechadv.2014.03.002

Dincheva, I., Badjakov, I., & Galunska, B. (2023). New insights into the research of bioactive compounds from plant origins with nutraceutical and pharmaceutical potential. Plants (Basel, Switzerland), 12(2), 258. https://doi.org/10.3390/plants12020258

Doxey, A. C., Mansfield, M. J., & Lobb, B. (2019). Exploring the evolution of virulence factors through bioinformatic data mining. mSystems, 4(3), e00162–19. https://doi.org/10.1128/mSystems.00162-19

Duvick, J., Fu, A., Muppirala, U., Sabharwal, M., Wilkerson, M. D., Lawrence, C. J., Lushbough, C., & Brendel, V. (2008). PlantGDB: A resource for comparative plant genomics. Nucleic Acids Research, 36(Database issue), D959–D965. https://doi.org/10.1093/nar/gkm1041

Ebhodaghe, S. O., Imanah, O. E., & Ndibe, H. 2022. Biofuels from microalgae biomass: A review of conversion processes and procedures. Arabian Journal of Chemistry. 15(2), 103591. https://doi.org/10.1016/j.arabjc.2021.103591

Fabian, B. K., Tetu, S. G., & Paulsen, I. T. (2020). Application of transposon insertion sequencing to agricultural science. Frontiers in Plant Science, 11, 291. https://doi.org/10.3389/fpls.2020.00291

Ferrocino, I., Rantsiou, K., McClure, R., Kostic, T., de Souza, R. S. C., Lange, L., FitzGerald, J., Kriaa, A., Cotter, P., Maguin, E., Schelkle, B., Schloter, M., Berg, G., Sessitsch, A., Cocolin, L., & Microbiome Support Consortium. (2023). The need for an integrated multi-OMICs approach in microbiome science in the food system. Comprehensive Reviews in Food Science and Food Safety, 22(2), 1082–1103. https://doi.org/10.1111/1541-4337.13103

Gauthier, J., Vincent, A. T., Charette, S. J., & Derome, N. (2019). A brief history of bioinformatics. Briefings in Bioinformatics, 20(6), 1981–1996. https://doi.org/10.1093/bib/bby063

Gendoo, D. M. A. (2023). Overview of bioinformatics software and databases for metabolic engineering. Methods in molecular biology (Humana Press, New York, NY), 2553, 265–274. https://doi.org/10.1007/978-1-0716-2617-7_13

Girardini, K. N., Olthof, A. M., & Kanadia, R. N. (2023). Introns: The "dark matter" of the eukaryotic genome. Frontiers in Genetics, 14, 1150212. https://doi.org/10.3389/fgene.2023.1150212

Goodstein, D. M., Shu, S., Howson, R., Neupane, R., Hayes, R. D., Fazo, J., & Rokhsar, D. S. (2012). Phytozome: A comparative platform for green plant genomics. Nucleic Acids Research, 40(D1), D1178–D1186. doi: 10.1093/nar/gkr944

Grealey, J., Lannelongue, L., Saw, W. Y., Marten, J., Méric, G., Ruiz-Carmona, S., & Inouye, M. (2022). The carbon footprint of bioinformatics. Molecular Biology and Evolution, 39(3), msac034. https://doi.org/10.1093/molbev/msac034

Hagen, J. B. (2000). The origins of bioinformatics. Nature Reviews Genetics, 1(3), 231–236. https://doi.org/10.1038/35042090

Hesami, M., Naderi, R., & Tohidfar, M. (2020). Introducing a hybrid artificial intelligence method for high-throughput modeling and optimizing plant tissue culture processes: The establishment of a new embryogenesis medium for chrysanthemum, as a case study. Applied Microbiology and Biotechnology, 104(23), 10249–10263. https://doi.org/10.1007/s00253-020-10978-1

Hesami, M., Naderi, R., Tohidfar, M., & Yoosefzadeh-Najafabadi, M. (2020). Development of support vector machine-based model and comparative analysis with artificial neural network for modeling the plant tissue culture procedures: Effect of plant growth regulators on somatic embryogenesis of chrysanthemum, as a case study. Plant Methods, 16, 112. https://doi.org/10.1186/s13007-020-00655-9

Hoopes, G. M., Hamilton, J. P., Kim, J., Zhao, D., Wiegert-Rininger, K., Crisovan, E., & Buell, C. R. (2018). Genome assembly and annotation of the medicinal plant *Calotropis gigantea*, a producer of anticancer and antimalarial cardenolides. G3: Genes, Genomes, Genetics, 8(2), 385–391. https://doi.org/10.1534/g3.117.300331

Jamwal, K., Bhattacharya, S., & Puri, S. (2018). Plant growth regulator mediated consequences of secondary metabolites in medicinal plants. Journal of Applied Research on Medicinal and Aromatic Plants, 9, 26–38. https://doi.org/10.1016/j.jarmap.2017.12.003

Ji, B., Xuan, L., Zhang, Y., Mu, W., Paek, K. Y., Park, S. Y., Wang, J., & Gao, W. (2023). Application of data modeling, instrument engineering and nanomaterials in selected Medid the scientific recinal plant tissue culture. Plants (Basel, Switzerland), 12(7), 1505. https://doi.org/10.3390/plants12071505

Jiang, C., Chen, C., Huang, Z., Liu, R., & Verdier, J. (2015). ITIS, a bioinformatics tool for accurate identification of transposon insertion sites using next-generation sequencing data. BMC Bioinformatics, 16(1), 72. https://doi.org/10.1186/s12859-015-0507-2

Jiang, M., Yan, S., Ren, W., Xing, N., Li, H., Zhang, M., & Ma, W. (2023). Genetic diversity of the Chinese medicinal plant astragali radix based on transcriptome-derived SSR markers. Electronic Journal of Biotechnology, 62, 13–20. https://doi.org/10.1016/j.ejbt.2022.12.001

Jing, T., Ardiansyah, R., Xu, Q., Xing, Q., & Müller-Xing, R. (2020). Reprogramming of cell fate during root regeneration by transcriptional and epigenetic networks. Frontiers in Plant Science, 11, 317. https://doi.org/10.3389/fpls.2020.00317

Jonsson, K., Hamant, O., & Bhalerao, R. P. (2022). Plant cell walls as mechanical signaling hubs for morphogenesis. Current Biology: CB, 32(7), R334–R340. https://doi.org/10.1016/j.cub.2022.02.036

Ju, F., & Zhang, T. (2015). Experimental design and bioinformatics analysis for the application of metagenomics in environmental sciences and biotechnology. Environmental Science & Technology, 49(21), 12628–12640. https://doi.org/10.1021/acs.est.5b03719

Kanehisa, M. (2016). KEGG bioinformatics resource for plant genomics and metabolomics. Methods in molecular biology (Humana Press, New York, NY), 1374, 55–70. https://doi.org/10.1007/978-1-4939-3167-5_3

Kasap, E. Y., Parfenova, OK., Kurkin, R. V., & Grishin, D. V. (2023). Bioinformatic analysis of the coding region of the melatonin receptor 1b gene as a reliable DNA marker to resolve interspecific mammal phylogenetic relationships. Mathematical Biosciences and Engineering, 20(3), 5430–5447. https://doi.org/10.3934/mbe.2023251

Klie, S., & Nikoloski, Z. (2012). The choice between MapMan and gene ontology for automated gene function prediction in plant science. Frontiers in Genetics, 3, 115. https://doi.org/10.3389/fgene.2012.00115

Klimontov, V. V., Koshechkin, K. A., Orlova, N. G., Sekacheva, M. I., & Orlov, Y. L. (2023). Medical genetics, genomics and bioinformatics-2022. International Journal of Molecular Sciences, 24(10), 8968. https://doi.org/10.3390/ijms24108968

Kolahi, M., Yazdi, M., Goldson-Barnaby, A., & Tabandeh, M. R. (2018). In silico prediction, phylogenetic and bioinformatic analysis of SoPCS gene, survey of its protein characterization and gene expression in response to cadmium in saccharum officinarum. Ecotoxicology and Environmental Safety, 163, 7–18. https://doi.org/10.1016/j.ecoenv.2018.07.032

Kumar, A., Kaynak, B. T., Dorman, K. S., Doruker, P., & Jernigan, R. L. (2023). Predicting allosteric pockets in protein biological assemblages. Bioinformatics (Oxford, England), 39(5), btad275. https://doi.org/10.1093/bioinformatics/btad275

Kumar, N., Banerjee, C., Negi, S., & Shukla, P. (2023). Microalgae harvesting techniques: Updates and recent technological interventions. Critical Reviews in Biotechnology, 43(3), 342–368. https://doi.org/10.1080/07388551.2022.2031089

Lai, K., Lorenc, M. T., & Edwards, D. (2015). Molecular marker databases. Methods in molecular biology (Humana Press, New York, NY), 1245, 49–62. https://doi.org/10.1007/978-1-4939-1966-6_4

Li, Y., Chen, M., Chen, Q., Yuan, M., Zeng, X., Zeng, Y., He, M., Wang, B., & Han, B. (2023). Bioinformatics identification of therapeutic gene targets for gastric cancer. Advances in Therapy, 40(4), 1456–1473. https://doi.org/10.1007/s12325-023-02428-x

Liu, C., Hao, J., Qiu, M., Pan, J., & He, Y. (2020). Genome-wide identification and expression analysis of the MYB transcription factor in Japanese plum (*Prunus salicina*). Genomics, 112(6), 4875–4886. https://doi.org/10.1016/j.ygeno.2020.08.018

Liu, R., Sun, W., & Liu, C. Z. (2011). Computational fluid dynamics modeling of mass transfer behavior in a bioreactor for hairy root culture. I. Model development and experimental validation. Biotechnology Progress, 27(6), 1661–1671. https://doi.org/10.1002/btpr.682

López-Pedrouso, M., Lorenzo, J. M., Alché, J. D., Moreira, R., & Franco, D. (2023). Advanced proteomic and bioinformatic tools for predictive analysis of allergens in novel foods. Biology, 12(5), 714. https://doi.org/10.3390/biology12050714

Loyola-Vargas, V. M., & Ochoa-Alejo, N. (2018). An introduction to plant tissue culture: Advances and perspectives. Methods in molecular biology (Humana Press, New York, NY), 1815, 3–13. https://doi.org/10.1007/978-1-4939-8594-4_1

Luo, R., Liu, B., Xie, Y., Li, Z., Huang, W., Yuan, J., & Wang, J. (2012). SOAPdenovo2: An empirically improved memory-efficient short-read de novo assembler. Gigascience, 1(1), 2047–217X. doi: 10.1186/2047-217X-1-18

Ma, X., Meng, Y., Wang, P., Tang, Z., Wang, H., & Xie, T. (2020). Bioinformatics-assisted, integrated omics studies on medicinal plants. Briefings in Bioinformatics, 21(6), 1857–1874. https://doi.org/10.1093/bib/bbz132

Magana, A. J., Taleyarkhan, M., Alvarado, D. R., Kane, M., Springer, J., & Clase, K. (2014). A survey of scholarly literature describing the field of bioinformatics education and bioinformatics educational research. CBE Life Sciences Education, 13(4), 607–623. https://doi.org/10.1187/cbe.13-10-0193

Marangoni, R., Bevilacqua, V., Cannataro, M., Mele, B. H., Mauri, G., & Marabotti, A., & BITS Training & Teaching Group. (2023). An overview of bioinformatics courses delivered at the academic level in Italy: Reflections and recommendations from BITS. PLoS Computational Biology, 19(2), e1010846. https://doi.org/10.1371/journal.pcbi.1010846

Mendoza, R. M., Kim, S. H., Vasquez, R., Hwang, I. C., Park, Y. S., Paik, H. D., Moon, G. S., & Kang, D. K. (2022). Bioinformatics and its role in the study of the evolution and probiotic potential of lactic acid bacteria. Food Science and Biotechnology, 32(4), 389–412. https://doi.org/10.1007/s10068-022-01142-8

Miaow, K., Lacap-Bugler, D., & Buckley, H. L. (2021). Identifying optimal bioinformatics protocols for aerosol microbial community data. PeerJ, 9, e12065. https://doi.org/10.7717/peerj.12065

Mitra, D., Mitra, D., Bensaad, M. S., Sinha, S., Pant, K., Pant, M., Priyadarshini, A., Singh, P., Dassamiour, S., Hambaba, L., Panneerselvam, P., & Mohapatra, P. K. D., 2022. Evolution of bioinformatics and its impact on modern bio-science in the twenty-first century: Special attention to pharmacology, plant science and drug discovery. Computational Toxicology, 24, 100248–100263. https://doi.org/10.1016/j.comtox.2022.100248

Mohamed, T. A., Ali, S. K., Elshamy, A. I., Saleh, I. A., Ibrahim, M. A. A., Atia, M. A. M., Alshammari, S. O., Mohamed, A. E. H., Hussien, T. A., Hamed, A. R., Saedi, H. R. E., Abdel-Azim, N. S., Shams, K. A., Efferth, T., Saker, M., Paré, P. W., & Hegazy, M. F. (2022). Plant cell cultures: An enzymatic tool for polyphenolic and flavonoid transformations. Phytomedicine: International Journal of Phytotherapy and Phytopharmacology, 100, 154019. https://doi.org/10.1016/j.phymed.2022.154019

Mohimani, H., Kersten, R. D., Liu, W. T., Wang, M., Purvine, S. O., Wu, S., & Dorrestein, P. C. (2014). Automated genome mining of ribosomal peptide natural products. ACS Chemical Biology, 9(7), 1545–1551. https://doi.org/10.1021/cb500199h

Mondal, R., Madhurya, K., Saha, P., Chattopadhyay, S. K., Antony, S., Kumar, A., Roy, S., & Roy, D. (2022). Expression profile, transcriptional and post-transcriptional regulation of genes involved in hydrogen sulphide metabolism connecting the balance between development and stress adaptation in plants: A data-mining bioinformatics approach. Plant Biology (Stuttgart, Germany), 24(4), 602–617. https://doi.org/10.1111/plb.13378

Mu, H., Wang, B., & Yuan, F. (2022). Bioinformatics in plant breeding and research on disease resistance. Plants (Basel, Switzerland), 11(22), 3118. https://doi.org/10.3390/plants11223118

Mulder, N., Schwartz, R., Brazas, M. D., Brooksbank, C., Gaeta, B., Morgan, S. L., Pauley, M. A., Rosenwald, A., Rustici, G., Sierk, M., Warnow, T., & Welch, L. (2018). The development and application of bioinformatics core competencies to improve bioinformatics training and education. PLoS Computational Biology, 14(2), e1005772. https://doi.org/10.1371/journal.pcbi.1005772

Nandy, A., & Basak, S. C. (2019). Bioinformatics in design of antiviral vaccines. Encyclopedia of Biomedical Engineering, 280–290. https://doi.org/10.1016/B978-0-12-801238-3.10878-5

Naravane, T., & Tagkopoulos, I. (2023). Machine learning models to predict micronutrient profile in food after processing. Current Research in Food Science, 6, 100500. https://doi.org/10.1016/j.crfs.2023.100500

Neves, M., Correia, S., & Canhoto, J. (2023). Ethylene inhibition reduces de novo shoot organogenesis and subsequent plant development from leaf explants of *Solanum betaceum* Cav. Plants (Basel, Switzerland), 12(9), 1854. https://doi.org/10.3390/plants12091854

Nezami, E., & Gallego, P. P. (2023). History, phylogeny, biodiversity, and new computer-based tools for efficient micropropagation and conservation of pistachio (Pistacia spp.) Germplasm. Plants (Basel, Switzerland), 12(2), 323. https://doi.org/10.3390/plants12020323

Nugraha, A. S., Agustina, R. P., Mirza, S., Rani, D. M., Winarto, N. B., Triatmoko, B., Pratama, A. N. W., Keller, P. A., & Wangchuk, P. (2022). Phytochemistry and pharmacology of medicinal plants used by the Tenggerese Society in Java Island of Indonesia. Molecules (Basel, Switzerland), 27(21), 7532. https://doi.org/10.3390/molecules27217532

Pant, B. (2014). Application of plant cell and tissue culture for the production of phytochemicals in medicinal plants. In Infectious diseases and nanomedicine II: First international conference (ICIDN–2012), dec. 15-18, 2012, Kathmandu, Nepal (pp. 25–39). India: Springer.

Quagliaro, P., Dziri, S., Magdoud El Alaoui, F., Saint Louis, P., de Pontual, L., Marin, J., Carbonnelle, E., & Billard-Pomares, T. (2023). Performances of bioinformatics tools for the analysis of sequencing data of mycobacterium tuberculosis complex strains. Tuberculosis (Edinburgh, Scotland), 139, 102324. https://doi.org/10.1016/j.tube.2023.102324

Rahman, M. F., Hasan, R., Biswas, M. S., Shathi, J. H., Hossain, M. F., Yeasmin, A., Abedin, M. Z., & Hossain, M. T. (2023). A bioinformatics approach to characterize a hypothetical protein Q6S8D9_SARS of SARS-CoV. Genomics & Informatics, 21(1), e3. https://doi.org/10.5808/gi.22021

Sanderson, S. C., Lewis, C., Hill, M., Peter, M., McEntagart, M., Gale, D., Morris, H., Moosajee, M., Searle, B., Hunter, A., Patch, C., & Chitty, L. S. (2022). Decision-making, attitudes, and understanding among patients and relatives invited to undergo genome sequencing in the 100,000 genomes project: A multisite survey study. Genetics in Medicine: Official Journal of the American College of Medical Genetics, 24(1), 61–74. https://doi.org/10.1016/j.gim.2021.08.010

Schilbert, H. M., Rempel, A., & Pucker, B. (2020). Comparison of read mapping and variant calling tools for the analysis of plant NGS data. Plants, 9(4), 439. https://doi.org/10.3390/plants9040439

Schwacke, R., Ponce-Soto, G. Y., Krause, K., Bolger, A. M., Arsova, B., Hallab, A., & Usadel, B. (2019). MapMan4: A refined protein classification and annotation framework applicable to multi-omics data analysis. Molecular Plant, 12(6), 879–892. https://doi.org/10.1016/j.molp.2019.01.003

Shukla, M. R., Singh, A. S., & Piunno, K. et al. Application of 3D printing to prototype and develop novel plant tissue culture systems. Plant Methods 13, 6 (2017). https://doi.org/10.1186/s13007-017-0156-8

Sliwoski, G., Kothiwale, S., Meiler, J., & Lowe, E. W. Jr (2013). Computational methods in drug discovery. Pharmacological Reviews, 66(1), 334–395. https://doi.org/10.1124/pr.112.007336

Standage, D. S., Lai, T., & Brendel, V. P. (2022). iLoci: Robust evaluation of genome content and organization for provisional and mature genome assemblies. NAR Genomics and Bioinformatics, 4(1), lqac013. https://doi.org/10.1093/nargab/lqac013

Tahir, A., Hussain, F., Ahmed, N., Ghorbani, A., & Jamil, A. (2018). Assessing universality of DNA barcoding in geographically isolated selected desert medicinal species of Fabaceae and Poaceae. PeerJ, 6, e4499. https://doi.org/10.7717/peerj.4499

Tan, Y. C., Kumar, A. U., Wong, Y. P., & Ling, A. P. K. (2022). Bioinformatics approaches and applications in plant biotechnology. Journal, Genetic Engineering & Biotechnology, 20(1), 106.

Vashisht, I., Mishra, P., Pal, T., Chanumolu, S., Singh, T. R., & Chauhan, R. S. (2015). Mining NGS transcriptomes for miRNAs and dissecting their role in regulating growth, development, and secondary metabolites production in different organs of a medicinal herb, *Picrorhiza kurroa*. Planta, 241, 1255–1268. https://doi.org/10.1007/s00425-015-2255-y

Wan, Q., Zhai, N., Xie, D., Liu, W., & Xu, L. (2023). WOX11: The founder of plant organ regeneration. Cell Regeneration (London, England), 12(1), 1. https://doi.org/10.1186/s13619-022-00140-9

Wang, C., Chen, W., Aili, M., Zhu, L., & Chen, Y. (2023). tRNA-derived small RNAs in plant response to biotic and abiotic stresses. Frontiers in Plant Science, 14, 1131977. https://doi.org/10.3389/fpls.2023.1131977

Wang, J., Li, X., Chang, H., & Si, N. (2023). Network pharmacology and bioinformatics study on the treatment of renal fibrosis with Persicae Semen-Carthami Flos drug pair. Medicine, 102(8), e32946. https://doi.org/10.1097/MD.0000000000032946

Wang, P., Schumacher, A. M., & Shiu, S. H. (2022). Computational prediction of plant metabolic pathways. Current Opinion in Plant Biology, 66, 102171. https://doi.org/10.1016/j.pbi.2021.102171

Weber, T., & Kim, H. U. (2016). The secondary metabolite bioinformatics portal: Computational tools to facilitate synthetic biology of secondary metabolite production. Synthetic and Systems Biotechnology, 1(2), 69–79. https://doi.org/10.1016/j.synbio.2015.12.002

Wijerathna-Yapa, A., & Hiti-Bandaralage, J. (2023). Tissue culture-a sustainable approach to explore plant stresses. Life (Basel, Switzerland), 13(3), 780. https://doi.org/10.3390/life13030780

Wink, M. (2012). Medicinal plants: A source of anti-parasitic secondary metabolites. Molecules, 17(11), 12771–12791. https://doi.org/10.3390/molecules171112771

Xia, X. (2017). Bioinformatics and drug discovery. Current Topics in Medicinal Chemistry, 17(15), 1709–1726. https://doi.org/10.2174/1568026617666161116143440

Xie, Y., Wu, G., Tang, J., Luo, R., Patterson, J., Liu, S., & Wang, J. (2014). SOAPdenovo-trans: De novo transcriptome assembly with short RNA-seq reads. Bioinformatics, 30(12), 1660–1666. https://doi.org/10.1093/bioinformatics/btu077

Xu, S., Li, L., Luo, X., Chen, M., Tang, W., Zhan, L., & Yu, G. (2022). Ggtree: A serialized data object for visualization of A phylogenetic tree and annotation data. IMeta, 1(4), e56. https://doi.org/10.1002/imt2.56

Yadav, R., Singh, G., Santal, A. R., & Singh, N. P. (2023). Omics approaches in effective selection and generation of potential plants for phytoremediation of heavy metal from contaminated resources. Journal of Environmental Management, 336, 117730. https://doi.org/10.1016/j.jenvman.2023.117730

Yang, K., Li, Y., Wang, S., Xu, X., Sun, H., Zhao, H., & Gao, Z. (2019). Genome-wide identification and expression analysis of the MYB transcription factor in Moso bamboo (*Phyllostachys edulis*). PeerJ, 6, e6242. https://doi.org/10.7717/peerj.6242

Yang, T., Gao, T., Wang, C., Wang, X., Chen, C., Tian, M., & Yang, W. (2022). In silico genome wide identification and expression analysis of the WUSCHEL-related homeobox gene family in *Medicago sativa*. Genomics & Informatics, 20(2), e19. https://doi.org/10.5808/gi.22013

Zhang, Q., & Zhang, D. (2021). Complete chloroplast genome sequence of medicinal plant potentilla *Lineata treviranus* (Rosaceae) from Yunnan, China. Mitochondrial DNA Part B, 6(8), 2422–2424. doi: 10.1080/23802359.2021.1951137

Zhang, Y., Shen, Q., Leng, L., Zhang, D., Chen, S., Shi, Y., & Chen, S. (2021). Incipient diploidization of the medicinal plant Perilla within 10,000 years. Nature Communications, 12(1), 5508. https://doi.org/10.1038/s41467-021-25681-6

Zhang, Y., Zheng, L., Zheng, Y., Zhou, C., Huang, P., Xiao, X., & Li, C. (2019). Assembly and annotation of a draft genome of the medicinal plant *Polygonum cuspidatum*. Frontiers in Plant Science, 10, 1274. https://doi.org/10.3389/fpls.2019.01274

Zhou, M., Zheng, X., Wang, J., & Dong, S. (2011). 'Non-destructive' biocomputing security system based on gas-controlled biofuel cell and potentially used for intelligent medical diagnostics. Bioinformatics (Oxford, England), 27(3), 399–404. https://doi.org/10.1093/bioinformatics/btq678

Index

Note: Page numbers in **bold** and *italics* refer to tables and figures, respectively.

D

E

I

J

K

L

N

O

P

Q

R

S

W

X

Y

Z